MAPS OF MEANING

MAPS OF MEANING

The Architecture of Belief

JORDAN B. PETERSON

ROUTLEDGE
New York and London

Published in 1999 by
Routledge
29 West 35th St.
New York, NY 10001

Published in Great Britain by
Routledge
11 New Fetter Lane
London EC4P 4EE

Copyright © 1999 by Routledge

Library of Congress Cataloging-in-Publication Data.

Peterson, Jordan B.
        Maps of meaning : the architecture of belief / Jordan B. Peterson.
            p. cm.
        Includes bibliographical references and index.
        ISBN 0-415-92221-6 (hardcover).—ISBN 0-415-92222-4 (pbk.)
        1. Archetype (Psychology)   2. Meaning (Psychology)   I. Title.
    BF175.5.A72P48   1999
    150'.1—dc21

                                                                                98-37486
                                                                                   CIP

CONTENTS

FIGURES

I will utter things which have been kept
secret from the foundation of the world.

(Matthew 13:35)

Preface

DESCENSUS AD INFEROS

Something we cannot see protects us from something we do not understand. The thing we cannot see is culture, in its intrapsychic or internal manifestation. The thing we do not understand is the chaos that gave rise to culture. If the structure of culture is disrupted, unwittingly, chaos returns. We will do anything—anything—to defend ourselves against that return.

> "The very fact that a general problem has gripped and assimilated the whole of a person is a guarantee that the speaker has really experienced it, and perhaps gained something from his sufferings. He will then reflect the problem for us in his personal life and thereby show us a truth." [1]

I was raised under the protective auspices, so to speak, of the Christian church. This does not mean that my family was explicitly religious. I attended conservative Protestant services during childhood with my mother, but she was not a dogmatic or authoritarian believer, and we never discussed religious issues at home. My father appeared essentially agnostic, at least in the traditional sense. He refused to even set foot in a church, except during weddings and funerals. Nonetheless, the historical remnants of Christian morality permeated our household, conditioning our expectations and interpersonal responses, in the most intimate of manners. When I grew up, after all, most people still attended church; furthermore, all the rules and expectations that made up middle-class society were Judeo-Christian in nature. Even the increasing number of those who could not tolerate formal ritual and belief still implicitly accepted—still acted out—the rules that made up the Christian game.

When I was twelve or so my mother enrolled me in confirmation classes, which served as introduction to adult membership in the church. I did not like attending. I did not like the attitude of my overtly religious classmates (who were few in number) and did not desire their lack of social standing. I did not like the school-like atmosphere of the confirmation classes. More importantly, however, I could not swallow what I was being taught. I asked the minister, at one point, how he reconciled the story of Genesis with the creation theories of modern science. He had not undertaken such a reconciliation; furthermore, he seemed more convinced, in his heart, of the evolutionary viewpoint. I was looking for an excuse to leave, anyway, and that was the last straw. Religion was for the ignorant, weak and superstitious. I stopped attending church and joined the modern world.

Although I had grown up in a Christian environment—and had a successful and happy childhood, in at least partial consequence—I was more than willing to throw aside the structure that had fostered me. No one really opposed my rebellious efforts, either, in church or at home—in part because those who were deeply religious (or who might have wanted to be) had no intellectually acceptable counter-arguments at their disposal. After all, many of the basic tenets of Christian belief were incomprehensible, if not clearly absurd. The virgin birth was an impossibility; likewise, the notion that someone could rise from the dead.

Did my act of rebellion precipitate a familial or a social crisis? No. My actions were so predictable, in a sense, that they upset no one, with the exception of my mother (and even she was soon resigned to the inevitable). The other members of the church—my "community"—had become absolutely habituated to the increasingly more frequent act of defection, and did not even notice.

Did my act of rebellion upset me, personally? Only in a manner I was not able to perceive, until many years later. I developed a premature concern with large-scale political and social issues, at about the same time I quit attending church. Why were some countries, some people, rich, happy and successful, while others were doomed to misery? Why were the forces of NATO and the Soviet Union continually at each other's throats? How was it possible for people to act the way the Nazis had during World War II? Underlying these specific considerations was a broader, but at the time ill-conceptualized question: how did evil—particularly group-fostered evil—come to play its role in the world?

I abandoned the traditions that supported me, at about the same time I left childhood. This meant that I had no broader socially constructed "philosophy" at hand to aid my understanding as I became aware of the existential problems that accompany maturity. The final consequences of that lack took years to become fully manifest. In the meantime, however, my nascent concern with questions of moral justice found immediate resolution. I started working as a volunteer for a mildly socialist political party, and adopted the party line.

Economic injustice was at the root of all evil, as far as I was concerned. Such injustice could be rectified, as a consequence of the rearrangement of social organizations. I could play a part in that admirable revolution, carrying out my ideological beliefs. Doubt vanished; my role was clear. Looking back, I am amazed at how stereotypical my actions—reactions— really were. I could not rationally accept the premises of religion as I understood them. I turned, in consequence, to dreams of political utopia, and personal power. The same ideological trap caught millions of others, in recent centuries.

When I was seventeen I left the town I grew up in. I moved nearby and attended a small college, which offered the first two years of undergraduate education. I involved myself there in university politics—which were more or less left wing at that time—and was elected to the college board of governors. The board was composed of politically and ideologically conservative people: lawyers, doctors, and businessmen. They were all well (or at least practically) educated, pragmatic, confident, outspoken; they had all accomplished something worthwhile and difficult. I could not help but admire them, even though I did not share their political stance. I found the fact of my admiration unsettling.

I had attended several left-wing party congresses, as a student politician and active party worker. I hoped to emulate the socialist leaders. The left had a long and honorable history in Canada, and attracted some truly competent and caring people. However, I could not generate much respect for the numerous low-level party activists I encountered at these meetings. They seemed to live to complain. They had no career, frequently, and no family, no completed education—nothing but ideology. They were peevish, irritable, and little, in every sense of the word. I was faced, in consequence, with the mirror image of the problem I encountered on the college board: I did *not* admire many of the individuals who believed the same things I did. This additional complication furthered my existential confusion.

My college roommate, an insightful cynic, expressed skepticism regarding my ideological beliefs. He told me that the world could not be completely encapsulated within the boundaries of socialist philosophy. I had more or less come to this conclusion on my own, but had not admitted so much in words. Soon afterward, however, I read George Orwell's *Road to Wigan Pier*. This book finally undermined me—not only my socialist ideology, but my faith in ideological stances themselves. In the famous essay concluding that book (written for—and much to the dismay of—the British Left Book Club) Orwell described the great flaw of socialism, and the reason for its frequent failure to attract and maintain democratic power (at least in Britain). Orwell said, essentially, that socialists did not really like the poor. They merely hated the rich.[2] His idea struck home instantly. Socialist ideology served to mask resentment and hatred, bred by failure. Many of the party activists I had encountered were using the ideals of social justice to rationalize their pursuit of personal revenge.

Whose fault was it that I was poor or uneducated and unadmired? Obviously, the fault of the rich, well-schooled and respected. How convenient, then, that the demands of revenge and abstract justice dovetailed! It was only right to obtain recompense from those more fortunate than me.

Of course, my socialist colleagues and I weren't out to hurt anyone. Quite the reverse. We were out to improve things—but we were going to start with other people. I came to see the temptation in this logic, the obvious flaw, the danger—but could also see that it did not exclusively characterize socialism. Anyone who was out to change the world by changing others was to be regarded with suspicion. The temptations of such a position were too great to be resisted.

It was not *socialist* ideology that posed the problem, then, but ideology as such. Ideology divided the world up simplistically into those who thought and acted properly, and those who did not. Ideology enabled the believer to hide from his own unpleasant and inadmissible fantasies and wishes. Such realizations upset my beliefs (even my faith in beliefs), and the plans I had formulated as a consequence of these beliefs. I could no longer tell who was good and who was bad, so to speak—so I no longer knew whom to support, or whom to fight. This state of affairs proved very troublesome, pragmatically as well as philosophically. I wanted to become a corporate lawyer—had written the Law School Admissions Test, had taken two years of appropriate preliminary courses. I wanted to learn the ways of my enemies, and embark on a political career. This plan disintegrated. The world obviously did not need another lawyer, and I no longer believed that I knew enough to masquerade as a leader.

I became simultaneously disenchanted with the study of political science, my erstwhile major. I had adopted that discipline so I could learn more about the structure of human beliefs (and for the practical, career-oriented reasons described previously). It remained very interesting to me when I was at junior college, where I was introduced to the history of political philosophy. When I moved to the main campus at the University of Alberta, however, my interest disappeared. I was taught that people were motivated by rational forces; that human beliefs and actions were determined by economic pressures. This did not seem sufficient explanation. I could not believe (and still do not) that commodities—"natural resources," for example—had intrinsic and self-evident value. In the absence of such value, the worth of things had to be socially or culturally (or even individually) determined. This act of determination appeared to me *moral*—appeared to me to be a consequence of the moral philosophy adopted by the society, culture or person in question. What people valued, economically, merely reflected what they believed to be important. This meant that real motivation had to lie in the domain of value, of morality. The political scientists I studied with did not see this, or did not think it was relevant.

My religious convictions, ill-formed to begin with, disappeared when I was very young. My confidence in socialism (that is, in political utopia) vanished when I realized that the world was not merely a place of economics. My faith in ideology departed, when I began to see that ideological identification itself posed a profound and mysterious problem. I could not accept the theoretical explanations my chosen field of study had to offer, and no longer had any practical reasons to continue in my original direction. I finished my three-year bachelor's degree, and left university. All my beliefs—which had lent order to the chaos of my existence, at least temporarily—had proved illusory; I could no longer see the sense in things. I was cast adrift; I did not know what to do or what to think.

But what of others? Was there evidence anywhere that the problems I now faced had been solved, by anyone, in any acceptable manner? The customary behavior and attitudes of my friends and family members offered no solution. The people I knew well were no more resolutely goal-directed or satisfied than I was. Their beliefs and modes of being seemed merely to disguise frequent doubt and profound disquietude. More disturbingly, on the more general plane, something truly insane was taking place. The great societies of the world were feverishly constructing a nuclear machine, with unimaginably destructive capabilities. Someone or something was making terrible plans. Why? Theoretically normal and well-adapted people were going about their business prosaically, as if nothing were the matter. Why weren't they disturbed? Weren't they paying attention? Wasn't I?

My concern with the general social and political insanity and evil of the world—sublimated by temporary infatuation with utopian socialism and political machination—returned with a vengeance. The mysterious fact of the Cold War increasingly occupied the forefront of my consciousness. How could things have come to such a point?

> History is just a madhouse
> it's turned over all the stones
> and its very careful reading
> leaves you little that's unknown

I couldn't understand the nuclear race: what could possibly be worth risking annihilation—not merely of the present, but of the past and the future? *What could possibly justify the threat of total destruction?*

Bereft of solutions, I had at least been granted the gift of a problem.

I returned to university and began to study psychology. I visited a maximum security prison on the outskirts of Edmonton, under the supervision of an eccentric adjunct professor at the University of Alberta. His primary job was the psychological care of convicts. The prison was full of murderers, rapists and armed robbers. I ended up in the gym, near the weight room, on my first reconnaissance. I was wearing a long wool cape, *circa* 1890, which I had bought in Portugal, and a pair of tall leather boots. The psychologist who was accompanying me disappeared, unexpectedly, and left me alone. Soon I was surrounded by unfamiliar men, some of whom were extremely large and tough-looking. One in particular stands out in my memory. He was exceptionally muscular, and tattooed over his bare chest. He had a vicious scar running from his collarbone to his midsection. Maybe he had survived open-heart surgery. Or maybe it was an ax wound. The injury would have killed a lesser man, anyway—someone like me.

Some of the prisoners, who weren't dressed particularly well, offered to trade their clothes for mine. This did not strike me as a great bargain, but I wasn't sure how to refuse. Fate rescued me, in the form of a short, skinny, bearded man. He said that the psychologist had sent him, and he asked me to accompany him. He was only one person, and many others (much larger) currently surrounded me and my cape. So I took him at his word. He led me outside the gym doors, and into the prison yard, talking quietly but reasonably about something innocuous (I don't recall what) all the while. I kept glancing back hopefully at the open doors behind us as we got further and further away. Finally my supervisor appeared, and motioned me back. We left the bearded prisoner, and went to a private office. The psychologist told me that the harmless-appearing little man who had escorted me out of the gym had murdered two policemen after he had forced them to dig their own graves. One of the policemen had little children and had begged for his life on their behalf while he was digging—at least according to the murderer's own testimony.

This really shocked me.

I had read about this sort of event, of course—but it had never been made *real* for me. I had never met someone even tangentially affected by something like this, and had certainly not encountered anyone who had actually done something so terrible. How could the man I had talked to—who was so apparently normal (and so seemingly inconsequential)—have done such an awful thing?

Some of the courses I was attending at this time were taught in large lecture theaters, where the students were seated in descending rows, row after row. In one of these courses— *Introduction to Clinical Psychology*, appropriately enough—I experienced a recurrent compulsion. I would take my seat behind some unwitting individual and listen to the professor speak. At some point during the lecture, I would unfailingly feel the urge to stab the point of my pen into the neck of the person in front of me. This impulse was not overwhelming— luckily—but it was powerful enough to disturb me. What sort of terrible person would have an impulse like that? Not me. I had never been aggressive. I had been smaller and younger than my classmates for most of my life.

I went back to the prison, a month or so after my first visit. During my absence, two prisoners had attacked a third, a suspected informer. They held or tied him down and pulverized one of his legs with a lead pipe. I was taken aback, once again, but this time I tried something different. I tried to imagine, *really imagine*, what I would have to be like to do such a thing. I concentrated on this task for days and days—and experienced a frightening revelation. The truly appalling aspect of such atrocity did not lie in its impossibility or remoteness, as I had naively assumed, but in its *ease*. I was not much different from the violent prisoners—not *qualitatively* different. I could do what they could do (although I hadn't).

This discovery truly upset me. I was not who I thought I was. Surprisingly, however, the desire to stab someone with my pen disappeared. In retrospect, I would say that the behavioral urge had manifested itself in explicit knowledge—had been translated from emotion and image to concrete realization—and had no further "reason" to exist. The "impulse" had only occurred, because of the question I was attempting to answer: "How can men do terrible things to one another?" I meant *other* men, of course—*bad men*—but I had still asked the question. There was no reason for me to assume that I would receive a predictable or personally meaningless answer.

At the same time, something odd was happening to my ability to converse. I had always enjoyed engaging in arguments, regardless of topic. I regarded them as a sort of game (not that this is in any way unique). Suddenly, however, I couldn't talk—more accurately, I couldn't *stand listening to myself talk*. I started to hear a "voice" inside my head, commenting on my opinions. Every time I said something, it said something—something critical. The voice employed a standard refrain, delivered in a somewhat bored and matter-of-fact tone:

You don't believe that.

That isn't true.

You don't believe that.

That isn't true.

The "voice" applied such comments to almost every phrase I spoke.

I couldn't understand what to make of this. I knew the source of the commentary was part of me, but this knowledge only increased my confusion. *Which* part, precisely, *was me—the talking part* or *the criticizing part?* If it was the talking part, then what was the criticizing part? If it was the criticizing part—well, then: how could virtually everything I said be untrue? In my ignorance and confusion, I decided to experiment. I tried only to say things that my internal reviewer would pass unchallenged. This meant that I really had to listen to what I was saying, that I spoke much less often, and that I would frequently stop, midway through a sentence, feel embarrassed, and reformulate my thoughts. I soon noticed that I felt much less agitated and more confident when I only said things that the "voice" did not object to. This came as a definite relief. My experiment had been a success; I was the criticizing part. Nonetheless, it took me a long time to reconcile myself to the idea that almost all my thoughts weren't real, weren't true—or, at least, weren't mine.

All the things I "believed" were things I thought sounded good, admirable, respectable, courageous. They weren't my things, however—I had stolen them. Most of them I had taken from books. Having "understood" them, abstractly, I presumed I had a right to them—presumed that I could adopt them, as if they were mine: presumed that they were *me*. My head was stuffed full of the ideas of others; stuffed full of arguments I could not logically refute. I did not know then that an irrefutable argument is not necessarily true, nor that the right to identify with certain ideas had to be earned.

I read something by Carl Jung, at about this time, that helped me understand what I was experiencing. It was Jung who formulated the concept of *persona*: the mask that "feigned individuality."[3] Adoption of such a mask, according to Jung, allowed each of us—and those around us—to believe that we were authentic. Jung said:

When we analyse the persona we strip off the mask, and discover that what seemed to be individual is at bottom collective; in other words, that the persona was only a mask of the collective psyche. Fundamentally the persona is nothing real: it is a compromise between individual and society as to what a man should appear to be. He takes a name, earns a title, exercises a function, he is this or that. In a certain sense all this is real, yet in relation to the essential individuality of the person concerned it is only a secondary reality, a compromise formation, in making which others often have a greater share than he. The persona is a semblance, a two-dimensional reality, to give it a nickname.[4]

Despite my verbal facility, I was not real. I found this painful to admit.

I began to dream absolutely unbearable dreams. My dream life, up to this point, had been relatively uneventful, as far as I can remember; furthermore, I have never had a particularly good visual imagination. Nonetheless, my dreams became so horrible and so emotionally grip-ping that I was often afraid to go to sleep. I dreamt dreams vivid as reality. I could not escape from them or ignore them. They circulated, in general, around a single theme: that of nuclear war, and total devastation—around the worst evils that I, or something in me, could imagine:

My parents lived in a standard ranch-style house, in a middle-class neighborhood, in a small town in northern Alberta. I was sitting in the darkened basement of this house, in the family room, watching TV, with my cousin Diane, who was in truth—in waking life—the most beautiful woman I had ever seen. A newscaster suddenly interrupted the program. The television picture and sound distorted, and static filled the screen. My cousin stood up and went behind the TV to check the electrical cord. She touched it, and started convulsing and froth-ing at the mouth, frozen upright by intense current.

A brilliant flash of light from a small window flooded the basement. I rushed upstairs. There was nothing left of the ground floor of the house. It had been completely and cleanly sheared away, leaving only the floor, which now served the basement as a roof. Red and orange flames filled the sky, from horizon to horizon. Nothing was left as far as I could see, except skeletal black ruins sticking up here and there: no houses, no trees, no signs of other human beings or of any life whatsoever. The entire town and everything that surrounded it on the flat prairie had been completely obliterated.

It started to rain mud, heavily. The mud blotted out everything, and left the earth brown, wet, flat and dull, and the sky leaden, even gray. A few distraught and shell-shocked people started to gather together. They were carrying unlabeled and dented cans of food, which contained nothing but mush and vegetables. They stood in the mud looking exhausted and disheveled. Some dogs emerged, out from under the basement stairs, where they had inexplicably taken residence. They were standing upright, on their hind legs. They were thin, like greyhounds, and had pointed noses. They looked like creatures of ritual—like Anubis, from the Egyptian tombs. They were carrying plates in front of them, which contained pieces of seared meat. They wanted to trade the meat for the cans. I took a plate. In the center of it was a circular slab of flesh four inches in diameter and one inch thick, foully cooked, oily, with a marrow bone in the center of it. Where did it come from?

I had a terrible thought. I rushed downstairs to my cousin. The dogs had butchered her, and were offering the meat to the survivors of the disaster.

I dreamed apocalyptic dreams of this intensity two or three times a week for a year or more, while I attended university classes and worked—as if nothing out of the ordinary was going on in my mind. Something I had no familiarity with was happening, however. I was being affected, simultaneously, by events on two "planes." On the first plane were the nor-mal, predictable, everyday occurrences that I shared with everybody else. On the second plane, however (unique to me, or so I thought) existed dreadful images and unbearably intense emotional states. This idiosyncratic, subjective world—which everyone normally treated as illusory—seemed to me at that time to lie somehow *behind* the world everyone knew and regarded as real. But what did *real* mean? The closer I looked, the less comprehen-sible things became. Where *was* the real? What was at the bottom of it all? I did not feel I could live without knowing.

My interest in the Cold War transformed itself into a true obsession. I thought about the suicidal and murderous preparation of that war every minute of every day, from the moment I woke up until the second I went to bed. How could such a state of affairs come about? *Who was responsible?*

I dreamed that I was running through a mall parking lot, trying to escape from something. I was running through the parked cars, opening one door, crawling across the front seat, opening the other, moving to the next. The doors on one car suddenly slammed shut. I was in the passenger seat. The car started to move by itself. A voice said harshly, "there is no way out of here." I was on a journey, going somewhere I did not want to go. I was not the driver.

I became very depressed and anxious. I had vaguely suicidal thoughts, but mostly wished that everything would just go away. I wanted to lie down on my couch, and sink into it, literally, until only my nose was showing—like the snorkel of a diver above the surface of the water. I found my awareness of things unbearable.

I came home late one night from a college drinking party, self-disgusted and angry. I took a canvas board and some paints. I sketched a harsh, crude picture of a crucified Christ—glaring and demonic—with a cobra wrapped around his naked waist, like a belt. The picture disturbed me—struck me, despite my agnosticism, as sacrilegious. I did not know what it meant, however, or why I had painted it. Where in the world had it come from?[5] I hadn't paid any attention to religious ideas for years. I hid the painting under some old clothes in my closet and sat cross-legged on the floor. I put my head down. It became obvious to me at that moment that I had not developed any real understanding of myself or of others. Everything I had once believed about the nature of society and myself had proved false, the world had apparently gone insane, and something strange and frightening was happening in my head. James Joyce said, "History is a nightmare from which I am trying to awake."[6] For me, history literally *was* a nightmare. I wanted above all else at that moment to wake up and make my terrible dreams go away.

I have been trying ever since then to make sense of the human capacity, *my capacity*, for evil—particularly for those evils associated with belief. I started by trying to make sense of my dreams. I couldn't ignore them, after all. Perhaps they were trying to tell me something? I had nothing to lose by admitting the possibility. I read Freud's *Interpretation of Dreams* and found it useful. Freud at least took the topic seriously—but I could not regard my nightmares as wish-fulfillments. Furthermore, they seemed more *religious* than sexual in nature. I knew, vaguely, that Jung had developed specialized knowledge of myth and religion, so I started through his writings. His thinking was granted little credence by the academics I knew, but they weren't particularly concerned with dreams. I couldn't help being concerned by mine. They were so intense I thought they might derange me. (What was the alternative? To believe that the terrors and pains they caused me were not *real?*)

Much of the time I could not understand what Jung was getting at. He was making a point I could not grasp, speaking a language I did not comprehend. Now and then, however, his statements struck home. He offered this observation, for example:

It must be admitted that the archetypal contents of the collective unconscious can often assume gro-
tesque and horrible forms in dreams and fantasies, so that even the most hard-boiled rationalist is not
immune from shattering nightmares and haunting fears.[7]

The second part of that statement certainly seemed applicable to me, although the first ("the
archetypal contents of the collective unconscious") remained mysterious and obscure. Still,
this was promising. Jung at least recognized that the things that were happening to me *could
happen.* Furthermore, he offered some hints as to their cause. So I kept reading. I soon came
across the following hypothesis. Here was a potential solution to the problems I was facing—or at least the description of a place to look for such a solution:

The psychological elucidation of ... [dream and fantasy] images, which cannot be passed over in
silence or blindly ignored, leads logically into the depths of religious phenomenology. The history of
religion in its widest sense (including therefore mythology, folklore, and primitive psychology) is a
treasure-house of archetypal forms from which the doctor can draw helpful parallels and enlightening
comparisons for the purpose of calming and clarifying a consciousness that is all at sea. It is absolutely
necessary to supply these fantastic images that rise up so strange and threatening before the mind's eye
with some kind of context so as to make them more intelligible. Experience has shown that the best
way to do this is by means of comparative mythological material.[8]

The study of "comparative mythological material" in fact made my horrible dreams disappear. The cure wrought by this study, however, was purchased at the price of complete and
often painful transformation: what I believe about the world, now—and how I act, in consequence—is so much at variance with what I believed when I was younger that I might as
well be a completely different person.

I discovered that beliefs make the world, in a very real way—that beliefs *are* the world, in
a more than metaphysical sense. This discovery has not turned me into a moral relativist,
however: quite the contrary. I have become convinced that the world-that-is-belief is orderly; that there are universal moral absolutes (although these are structured such that a diverse
range of human opinion remains both possible and beneficial). I believe that individuals and
societies who flout these absolutes—in ignorance or in willful opposition—are doomed to
misery and eventual dissolution.

I learned that the meanings of the most profound substrata of belief systems can be rendered explicitly comprehensible, even to the skeptical rational thinker—and that, so rendered, can be experienced as fascinating, profound and necessary. I learned why people wage
war—why the desire to maintain, protect and expand the domain of belief motivates even
the most incomprehensible acts of group-fostered oppression and cruelty—and what might
be done to ameliorate this tendency, despite its universality. I learned, finally, that the terrible
aspect of life might actually be a necessary precondition for the existence of life—and that it
is possible to regard that precondition, in consequence, as comprehensible and acceptable. I
hope that I can bring those who read this book to the same conclusions, without demanding

any unreasonable "suspension of critical judgment"—excepting that necessary to initially encounter and consider the arguments I present. These can be summarized as follows:

The world can be validly construed as a forum for action, as well as a place of things. We describe the world as a place of things, using the formal methods of science. The techniques of narrative, however—myth, literature and drama—portray the world as a forum for action. The two forms of representation have been unnecessarily set at odds, because we have not yet formed a clear picture of their respective domains. The domain of the former is the objective world—what is, from the perspective of intersubjective perception. The domain of the latter is the world of value—what is and what should be, from the perspective of emotion and action.

The world as forum for action is composed, essentially, of three constituent elements, which tend to manifest themselves in typical patterns of metaphoric representation. First is unexplored territory—the Great Mother, nature, creative and destructive, source and final resting place of all determinate things. Second is explored territory—the Great Father, culture, protective and tyrannical, cumulative ancestral wisdom. Third is the process that mediates between unexplored and explored territory—the Divine Son, the archetypal individual, creative exploratory Word and vengeful adversary. We are adapted to this world of divine characters, much as to the objective world. The fact of this adaptation implies that the environment is in "reality" a forum for action, as well as a place of things.

Unprotected exposure to unexplored territory produces fear. The individual is protected from such fear as a consequence of ritual imitation of the Great Father—as a consequence of the adoption of group identity, which restricts the meaning of things, and confers predictability on social interactions. When identification with the group is made absolute, however—when everything has to be controlled, when the unknown is no longer allowed to exist—the creative exploratory process that updates the group can no longer manifest itself. This restriction of adaptive capacity dramatically increases the probability of social aggression.

Rejection of the unknown is tantamount to "identification with the devil," the mythological counterpart and eternal adversary of the world-creating exploratory hero. Such rejection and identification is a consequence of Luciferian pride, which states: *all that I know is all that is necessary to know.* This pride is totalitarian assumption of omniscience—is adoption of God's place by "reason"— is something that inevitably generates a state of personal and social being indistinguishable from hell. This hell develops because creative exploration—impossible, without (humble) acknowledgment of the unknown—constitutes the process that constructs and maintains the protective adaptive structure that gives life much of its acceptable meaning.

"Identification with the devil" amplifies the dangers inherent in group identification, which tends of its own accord towards pathological stultification. Loyalty to personal interest—subjective meaning—can serve as an antidote to the overwhelming temptation constantly posed by the possibility of denying anomaly. Personal interest—subjective meaning—reveals itself at the juncture of explored and unexplored territory, and is indicative of participation in the process that ensures continued healthy individual and societal adaptation.

Loyalty to personal interest is equivalent to identification with the archetypal hero—the "savior"—who upholds his association with the creative Word in the face of death, and despite group pressure to conform. Identification with the hero serves to decrease the unbearable motivational valence of the unknown; furthermore, provides the individual with a standpoint that simultaneously transcends and maintains the group.

Similar summaries precede each chapter (and subchapter). Read as a unit, they comprise a complete but compressed picture of the book. These should be read first, after this preface. In this manner, the whole of the argument I am offering might come quickly to aid comprehension of the parts.

1

MAPS OF EXPERIENCE

Object and Meaning

The world can be validly construed as forum for action, or as place of things.

The former manner of interpretation—more primordial, and less clearly understood—finds its expression in the arts or humanities, in ritual, drama, literature and mythology. The world as forum for action is a place of value, a place where all things have meaning. This meaning, which is shaped as a consequence of social interaction, is implication for action, or—at a higher level of analysis—implication for the configuration of the interpretive schema that produces or guides action.

The latter manner of interpretation—the world as place of things—finds its formal expression in the methods and theories of science. Science allows for increasingly precise determination of the consensually validatable properties of things, and for efficient utilization of precisely determined things as tools (once the direction such use is to take has been determined, through application of more fundamental narrative processes).

No complete world-picture can be generated without use of both modes of construal. The fact that one mode is generally set at odds with the other means only that the nature of their respective domains remains insufficiently discriminated. Adherents of the mythological worldview tend to regard the statements of their creeds as indistinguishable from empirical "fact," even though such statements were generally formulated long before the notion of objective reality emerged. Those who, by contrast, accept the scientific perspective—who assume that it is, or might become, complete—forget that an impassable gulf currently divides what is from what should be.

> We need to know four things:
> what there is,
> what to do about what there is,

1

that there is a difference between knowing *what there is*, and knowing *what to do about what there is*
and what that difference is.

To explore something, to "discover what it is"—that means most importantly to discover its
significance for motor output, within a particular social context, and only more particularly
to determine its precise objective sensory or material nature. This is knowledge in the most
basic of senses—and often constitutes sufficient knowledge.

Imagine that a baby girl, toddling around in the course of her initial tentative investigations, reaches up onto a countertop to touch a fragile and expensive glass sculpture. She
observes its color, sees its shine, feels that it is smooth and cold and heavy to the touch.
Suddenly her mother interferes, grasps her hand, tells her *not* to *ever touch* that object. The
child has just learned a number of specifically consequential things about the sculpture—has
identified its sensory properties, certainly. More importantly, however, she has determined
that approached in the wrong manner, the sculpture is dangerous (at least in the presence of
mother); has discovered as well that the sculpture is regarded more highly, in its present unaltered configuration, than the exploratory tendency—at least (once again) by mother. The
baby girl has simultaneously encountered an object, from the empirical perspective, *and its
socioculturally determined status*. The empirical object might be regarded as those sensory properties "intrinsic" to the object. The *status of the object*, by contrast, consists of its meaning—
consists of its implication for behavior. Everything a child encounters has this dual nature,
experienced by the child as part of a unified totality. Everything *is* something, and *means*
something—and the distinction between essence and significance is not necessarily drawn.

The significance of something—specified in actuality as a consequence of exploratory
activity undertaken in its vicinity—tends "naturally" to become assimilated to the object
itself. The object, after all, is the proximal cause or the stimulus that "gives rise" to action
conducted in its presence. For people operating naturally, like the child, what something signifies is more or less inextricably *part* of the thing, part of its magic. The magic is of course
due to apprehension of the specific cultural and intrapsychic significance of the thing, and
not to its objectively determinable sensory qualities. Everyone understands the child who
says, for example, "I saw a scary man"; the child's description is immediate and concrete, even
though he or she has attributed to the object of perception a quality that is in fact context-
dependent and subjective. It is difficult, after all, to realize the subjective nature of fear, and
not to feel threat as part of the "real" world.

The automatic attribution of meaning to things—or the failure to distinguish between
them initially—is a characteristic of narrative, of myth, not of scientific thought. Narrative
accurately captures the nature of raw experience. Things *are* scary, people *are* irritating,
events *are* promising, food *is* satisfying—at least in terms of our basic experience. The modern mind, which regards itself as having transcended the domain of the magical, is nonetheless still endlessly capable of "irrational" (read motivated) reactions. We fall under the spell of

experience whenever we attribute our frustration, aggression, devotion or lust to the person or situation that exists as the proximal "cause" of such agitation. We are not yet "objective," even in our most clear-headed moments (and thank God for that). We become immediately immersed in a motion picture or a novel, and willingly suspend disbelief. We become impressed or terrified, despite ourselves, in the presence of a sufficiently powerful cultural figurehead (an intellectual idol, a sports superstar, a movie actor, a political leader, the pope, a famous beauty, even our superior at work)—in the presence, that is, of anyone who sufficiently embodies the oft-implicit values and ideals that protect us from disorder and lead us on. Like the medieval individual, we do not even need the person to generate such affect. The icon will suffice. We will pay vast sums of money for articles of clothing worn or personal items used or created by the famous and infamous of our time.[9]

The "natural," pre-experimental, or mythical mind is in fact *primarily* concerned with meaning—which is essentially implication for action—and not with "objective" nature. The formal object, as conceptualized by modern scientifically oriented consciousness, might appear to those still possessed by the mythic imagination—if they could "see" it at all—as an irrelevant shell, as all that was left after everything intrinsically intriguing had been stripped away. For the pre-experimentalist, the thing is most truly the significance of its sensory properties, as they are experienced in subjective experience—in affect, or emotion. And, in truth—in real life—to know what something *is* still means to know two things about it: its *motivational relevance*, and the specific nature of its sensory qualities. The two forms of knowing are not identical; furthermore, experience and registration of the former necessarily precedes development of the latter. Something must have emotional impact before it will attract enough attention to be explored and mapped in accordance with its sensory properties. Those sensory properties—of prime import to the experimentalist or empiricist—are meaningful only insofar as they serve as cues for determining specific affective relevance or behavioral significance. We need to know what things *are* not to know what they are but to *keep track of what they mean*—to understand what they signify for our behavior.

It has taken centuries of firm discipline and intellectual training, religious, proto-scientific and scientific, to produce a mind capable of concentrating on phenomena that are not yet or are no longer immediately intrinsically gripping—to produce a mind that regards *real* as something separable from *relevant*. Alternatively, it might be suggested that all the myth has not yet vanished from science, devoted as it is to human progress, and that it is this nontrivial remainder that enables the scientist to retain undimmed enthusiasm while endlessly studying his fruitflies.

How, precisely, did people think, not so very long ago, before they were experimentalists? What were things before they were objective things? These are very difficult questions. The "things" that existed prior to the development of experimental science do not appear valid either as things *or* as the meaning of things to the modern mind. The question of the nature of the substance of *sol*—the sun—(to take a single example) occupied the minds of those who practiced the pre-experimental "science" of alchemy for many hundreds of years. We would no longer presume even that the sun has a uniform substance, unique to it, and would certainly take exception to the properties attributed to this hypothetical element by the medieval alchemist, if we allowed its existence. Carl Jung, who spent much of the latter part of his life studying medieval thought patterns, characterized *sol*:

The sun signifies first of all gold, whose [alchemical] sign it shares. But just as the "philosophical" gold is not the "common" gold, so the sun is neither just the metallic gold nor the heavenly orb. Sometimes the sun is an active substance hidden in the gold and is extracted [alchemically] as the *tinctura rubea* (red tincture). Sometimes, as the heavenly body, it is the possessor of magically effective and transformative rays. As gold and a heavenly body it contains an active sulphur of a red colour, hot and dry. Because of this red sulphur the alchemical sun, like the corresponding gold, is red. As every alchemist knew, gold owes its red color to the admixture of Cu (copper), which he interpreted as Kypris (the Cyprian, Venus), mentioned in Greek alchemy as the transformative substance. Redness, heat, and dryness are the classical qualities of the Egyptian Set (Greek Typhon), the evil principle, which, like the alchemical sulphur, is closely connected with the devil. And just as Typhon has his kingdom in the forbidden sea, so the sun, as *sol centralis*, has its sea, its "crude perceptible water," and as *sol coelestis* its "subtle imperceptible water." This sea water (*aqua pontica*) is extracted from sun and moon....

The active sun-substance also has favourable effects. As the so-called "balsam" it drips from the sun and produces lemons, oranges, wine, and, in the mineral kingdom, gold.[10]

We can barely understand such a description, contaminated as it is by imaginative and mythological associations peculiar to the medieval mind. It is precisely this fantastical contamination, however, that renders the alchemical description worth examining—not from the perspective of the history of science, concerned with the examination of outdated objective ideas, but from the perspective of psychology, focused on the interpretation of subjective frames of reference.

"In it [the "Indian Ocean," in this example] are images of heaven and earth, of summer, autumn, winter, and spring, male and female. If thou callest this spiritual, what thou doest is probable; if corporeal, thou sayest the truth; if heavenly, thou liest not; if earthly, thou hast well spoken."[11] The alchemist could not separate his subjective ideas about the nature of things—that is, his *hypotheses*—from the things themselves. His hypotheses, in turn—products of his imagination—were derived from the unquestioned and unrecognized "explanatory" presuppositions that made up his culture. The medieval man lived, for example, in a universe that was *moral*—where everything, even ores and metals, strived above all for perfection.[12] Things, for the alchemical mind, were therefore characterized in large part by their *moral* nature—by their impact on what we would describe as affect, emotion or motivation; were therefore characterized by their *relevance* or *value* (which is impact on affect). Description of this relevance took narrative form, mythic form—as in the example drawn from Jung, where the sulphuric aspect of the sun's substance is attributed negative, demonic characteristics. It was the great feat of science to strip *affect* from *perception*, so to speak, and to allow for the description of experiences purely in terms of their consensually apprehensible features. However, it is the case that the affects generated by experiences are *real*, as well.

The alchemists, whose conceptualizations intermingled affect with sense, dealt with affect as a matter of course (although they did not "know" it—not *explicitly*). We have removed the affect from the thing, and can therefore brilliantly manipulate the thing. We are still victims, however, of the uncomprehended emotions generated by—we would say, in the presence of—the thing. We have lost the mythic universe of the pre-experimental mind, or have at least ceased to further its development. That loss has left our increased technological power ever more dangerously at the mercy of our still unconscious systems of valuation.

Prior to the time of Descartes, Bacon and Newton, man lived in an animated, spiritual world, saturated with meaning, imbued with moral purpose. The nature of this purpose was revealed in the stories people told each other—stories about the structure of the cosmos and the place of man. But now we think empirically (at least we think we think empirically), and the spirits that once inhabited the universe have vanished. The forces released by the advent of the experiment have wreaked havoc within the mythic world. Jung states:

How totally different did the world appear to medieval man! For him the earth was eternally fixed and at rest in the center of the universe, encircled by the course of a sun that solicitously bestowed its warmth. Men were all children of God under the loving care of the Most High, who prepared them for eternal blessedness; and all knew exactly what they should do and how they should conduct themselves in order to rise from a corruptible world to an incorruptible and joyous existence. Such a life no longer seems real to us, even in our dreams. Natural science has long ago torn this lovely veil to shreds.[13]

Even if the medieval individual was not in all cases tenderly and completely enraptured by his religious beliefs (he was a great believer in hell, for example), he was certainly not plagued by the plethora of rational doubts and moral uncertainties that beset his modern counterpart. Religion for the pre-experimental mind was not so much a matter of faith as a matter of fact—which means that the prevailing religious viewpoint was not merely one compelling theory among many.

The capacity to maintain explicit belief in religious "fact," however, has been severely undermined in the last few centuries—first in the West, and then everywhere else. A succession of great scientists and iconoclasts has demonstrated that the universe does not revolve around man, that our notion of separate status from and "superiority" to the animal has no empirical basis, and that there is no God in heaven (nor even a heaven, as far as the eye can see). In consequence, we no longer believe our own stories—no longer even believe that those stories served us well in the past. The objects of revolutionary scientific discovery— Galileo's mountains on the lunar orb; Kepler's elliptical planetary orbits—manifested themselves in apparent violation of mythic order, predicated as it was on the presumption of heavenly perfection. The new phenomena produced by the procedures of experimentalists could not *be*, could not exist, from the perspective defined by tradition. Furthermore—and more importantly—the new theories that arose to make sense of empirical reality posed a severe threat to the integrity of traditional models of reality, which had provided the world with determinate meaning. The mythological cosmos had man at its midpoint; the objective universe was heliocentric at first, and less than that later. Man no longer occupies center stage. The world is, in consequence, a completely different place.

The mythological perspective has been overthrown by the empirical; or so it appears. This should mean that the morality predicated upon such myth should have disappeared, as well, as belief in comfortable illusion vanished. Friedrich Nietzsche made this point clearly, more than a hundred years ago:

When one gives up Christian belief [for example] one thereby deprives oneself of the *right* to Christian morality.... Christianity is a system, a consistently thought out and *complete* view of things. If one breaks out of it a fundamental idea, the belief in God, one thereby breaks the whole thing to pieces: one has nothing of any consequence left in one's hands. Christianity presupposes that man does not know, *cannot* know what is good for him and what evil: he believes in God, who alone knows. Christian morality is a command: its origin is transcendental; it is beyond all criticism, all right to criticize; it possesses truth only if God is truth—it stands or falls with the belief in God. If [modern Westerners] really do believe they know, of their own accord, "intuitively," what is good and evil; if they consequently think they no longer have need of Christianity as a guarantee of morality; that is merely the *consequence* of the ascendancy of Christian evaluation and an expression of the *strength* and *depth* of this ascendancy: so that the origin of [modern] morality has been forgotten, so that the highly conditional nature of its right to exist is no longer felt.[14]

If the presuppositions of a theory have been invalidated, argues Nietzsche, then the theory has been invalidated. But in this case the "theory" survives. The fundamental tenets of the Judeo-Christian moral tradition continue to govern every aspect of the actual individual behavior and basic values of the typical Westerner—even if he is atheistic and well-educated, even if his abstract notions and utterances appear iconoclastic. He neither kills nor steals (or if he does, he hides his actions, even from his own awareness), and he tends, in theory, to treat his neighbor as himself. The principles that govern his society (and, increasingly, all others[15]) remain predicated on mythic notions of individual value—intrinsic right and responsibility—despite scientific evidence of causality and determinism in human motivation. Finally, in his mind—even when sporadically criminal—the victim of a crime still cries out to heaven for "justice," and the conscious lawbreaker still *deserves* punishment for his or her actions.

Our systems of post-experimental thought and our systems of motivation and action therefore co-exist in paradoxical union. One is "up-to-date"; the other, archaic. One is scientific; the other, traditional, even superstitious. We have become atheistic in our description, but remain evidently religious—that is, *moral*—in our disposition. What we accept as true and how we act are no longer commensurate. We carry on as if our experience has meaning—as if our activities have transcendent value—but we are unable to justify this belief intellectually. We have become trapped by our own capacity for abstraction: it provides us with accurate descriptive information but also undermines our belief in the utility and meaning of existence. This problem has frequently been regarded as tragic (it seems to me, at least, ridiculous)—and has been thoroughly explored in existential philosophy and literature. Nietzsche described this modern condition as the (inevitable and necessary) consequence of the "death of God":

Have you not heard of that madman who lit a lantern in the bright morning hours, ran to the market place, and cried incessantly, "I seek God! I seek God!" As many of those who do not believe in God were standing around just then, he provoked much laughter.

Why, did he get lost? said one. Did he lose his way like a child? said another. Or is he hiding? Is he afraid of us? Has he gone on a voyage? or emigrated? Thus they yelled and laughed.

The madman jumped into their midst and pierced them with his glances. "Whither is God," he cried. "I shall tell you. *We have killed him*—you and I. All of us are his murderers. But how have we done this? How were we able to drink up the sea? Who gave us the sponge to wipe away the entire horizon? What did we do when we unchained this earth from its sun? Whither is it moving now? Whither are we moving now? Away from all suns? Are we not plunging continuously? Backward, side-ward, forward, in all directions? Is there any up or down left? Are we not straying as through an infinite nothing? Do we not feel the breath of empty space? Has it not become colder? Is not night and more night coming on all the while? Must not lanterns be lit in the morning? Do we not hear anything yet of the noise of the grave-diggers who are burying God? Do we not smell anything yet of God's decomposition? Gods too decompose.

"God is dead. God remains dead. And we have killed him. How shall we, the murderers of all murderers, comfort ourselves? What was holiest and most powerful of all that the world has yet owned has bled to death under our knives. Who will wipe this blood off us? What water is there for us to clean ourselves? What festivals of atonement, what sacred games shall we have to invent? Is not the greatness of this deed too great for us? Must not we ourselves become gods simply to seem worthy of it?"[16]

We find ourselves in an absurd and unfortunate situation—when our thoughts turn, involun-tarily, to consideration of our situation. It seems impossible to believe that life is intrinsically, religiously meaningful. We continue to act and think "as if"—as if nothing fundamental has really changed. This does not change the fact that our integrity has vanished.

The great forces of empiricism and rationality and the great technique of the experiment have killed myth, and it cannot be resurrected—or so it seems. We still *act out* the precepts of our forebears, nonetheless, although we can no longer justify our actions. Our behavior is shaped (at least in the ideal) by the same mythic rules—*thou shalt not kill, thou shalt not covet*—that guided our ancestors for the thousands of years they lived without benefit of formal empirical thought. This means that those rules are so powerful—so necessary, at least—that they maintain their existence (and expand their domain) even in the presence of explicit theories that undermine their validity. That is a mystery. And here is another:

How is it that complex and admirable ancient civilizations could have developed and flour-ished, initially, if they were predicated upon nonsense? (If a culture survives, and grows, does that not indicate in some profound way that the ideas it is based upon are valid? If myths are mere superstitious proto-theories, why did they work? Why were they remembered? Our great rationalist ideologies, after all—fascist, say, or communist—demonstrated their essential uselessness within the space of mere generations, despite their intellectually compelling nature. Traditional societies, predicated on religious notions, have survived—essentially unchanged, in some cases, for tens of thousands of years. How can this longevity be understood?) Is it actually sensible to argue that persistently successful traditions are based on ideas that are simply wrong, regardless of their utility?

Is it not more likely that we just do not know how it could be that traditional notions are *right*, given their *appearance* of extreme irrationality?

Is it not likely that this indicates modern philosophical ignorance, rather than ancestral philosophical error?

We have made the great mistake of assuming that the "world of spirit" described by those who preceded us was the modern "world of matter," primitively conceptualized. This is not true—at least not in the simple manner we generally believe. The cosmos described by mythology was *not* the same place known to the practitioners of modern science—but that does not mean it was not *real*. We have not yet found God above, nor the devil below, because we do not yet understand where "above" and "below" might be found.

We do not know what our ancestors were talking about. This is not surprising, because they did not "know," either (and it didn't really matter to them that they did not know). Consider this archaic creation myth[17] from Sumer—the "birthplace of history":

So far, no cosmogonic text properly speaking has been discovered, but some allusions permit us to reconstruct the decisive moments of creation, as the Sumerians conceived it. The goddess Nammu (whose name is written with the pictograph representing the primordial sea) is presented as "the mother who gave birth to the Sky and the Earth" and the "ancestress who brought forth all the gods." The theme of the primordial waters, imagined as a totality at once cosmic and divine, is quite frequent in archaic cosmogonies. In this case too, the watery mass is identified with the original Mother, who, by parthenogenesis, gave birth to the first couple, the Sky (An) and the Earth (Ki), incarnating the male and female principles. This first couple was united, to the point of merging, in the *hieros gamos* [mystical marriage]. From their union was born En-lil, the god of the atmosphere. Another fragment informs us that the latter separated his parents.... The cosmogonic theme of the separation of sky and earth is also widely disseminated.[18]

This myth is typical of archaic descriptions of reality. What does it mean to say that the Sumerians believed that the world emerged from a "primordial sea," which was the mother of all, and that the sky and the earth were separated by the act of a deity? We do not know. Our abysmal ignorance in this regard has not been matched, however, by a suitable caution. We appear to have made the presumption that stories such as these—myths—were equivalent in function and intent (but were inferior methodologically) to empirical or post-experimental description. It is this fundamentally absurd insistence that, above all, has destabilized the effect of religious tradition upon the organization of modern human moral reasoning and behavior. The "world" of the Sumerians was not objective reality, as we presently construe it. It was simultaneously more and less—more, in that this "primitive" world contained phenomena that we do not consider part of "reality," such as affect and meaning; less, in that the Sumerians could not describe (or conceive of) many of those things the processes of science have revealed to us.

Myth is *not* primitive proto-science. It is a qualitatively different phenomenon. Science might be considered "description of the world with regards to those aspects that are consensually apprehensible" or "specification of the most effective mode of reaching an end (given a defined end)." Myth can be more accurately regarded as "description of the world as it *signifies* (for *action*)." The mythic universe is *a place to act*, not *a place to perceive*. Myth describes things in terms of their unique or shared affective valence, their value, their motivational significance. The Sky (An) and the Earth (Ki) of the Sumerians are not the sky and earth of modern man, therefore; they are the Great Father and Mother of all things (including the thing—En-lil, who is actually a process—that in some sense gave rise to them).

We do not understand pre-experimental thinking, so we try to explain it in terms that we do understand—which means that we explain it away, define it as nonsense. After all, we think scientifically—so we believe—and we think we know what that means (since scientific thinking can in principle be defined). We are familiar with scientific thinking and value it highly—so we tend to presume that it is all there is to thinking (presume that all other "forms of thought" are approximations, at best, to the ideal of scientific thought). But this is not accurate. Thinking also and more fundamentally is *specification of value*, specification of implication for behavior. This means that *categorization*, with regards to value—determination (or even perception) of what constitutes a single thing, or class of things—is the act of *grouping together according to implication for behavior*.

The Sumerian category of Sky (An), for example, is a domain of phenomena with similar implications for behavioral output, or for affect; the same can be said for the category of Earth (Ki), *and all other mythic categories*. The fact that the "domain of the Sky" has implications for action—has motivational significance—makes it a *deity* (which is something that controls behavior, or at least that must be served). Comprehension of the fact that such a classification system actually has meaning necessitates learning to think differently (necessitates, as well, learning to think about thinking differently).

The Sumerians were concerned, above all, with how to act (were concerned with the value of things). Their descriptions of reality (to which we attribute the qualities of proto-science) in fact comprised their summary of the world *as phenomenon—as place to act*. They did not "know" this—not *explicitly*—any more than we do. But it was still true.

The empirical endeavor is devoted to objective description of *what is*—to determination of what it is about a given phenomena that can be consensually validated and described. The objects of this process may be those of the past, the present, or the future, and may be static or dynamic in nature: a good scientific theory allows for prediction and control of becoming (of "transformation") as well as being. However, the "affect" that an encounter with an "object" generates is not a part of what that object *is*, from this perspective, and therefore must be eliminated from further consideration (along with anything else subjective)—must be at least eliminated from definition as a *real aspect of the object*.

The painstaking empirical process of identification, communication and comparison has proved to be a strikingly effective means for specifying the nature of the relatively invariant features of the collectively apprehensible world. Unfortunately, this useful methodology cannot be applied to determination of *value*—to consideration of *what should be*, to specification of the direction that things *should* take (which means, to description of the future we should construct, as a consequence of our actions). Such acts of valuation necessarily constitute moral decisions. We can use information generated in consequence of the application of science to guide those decisions, but not to tell us if they are correct. We lack a process of verification, in the moral domain, that is as powerful or as universally acceptable as the experimental (empirical) method in the realm of description. This absence does not allow us to sidestep the problem. No functioning society or individual can avoid rendering moral judgment, regardless of what might be said or imagined about the necessity of such judgment. Action *presupposes* valuation, or its implicit or "unconscious" equivalent. To act is literally to manifest preference about one set of possibilities, contrasted with an infinite set of alternatives. If we wish to live, we must act. Acting, we value. Lacking omniscience, painfully, we must make decisions, in the absence of sufficient information. It is, traditionally speaking, our knowledge of good and evil, our moral sensibility, that allows us this ability. It is our mythological conventions, operating implicitly or explicitly, that guide our choices. But what are these conventions? How are we to understand *the fact of their existence?* How are we to understand *them?*

It was Nietzsche, once again, who put his finger on the modern problem, central to issues of valence or meaning: not, as before "how to act, from within the confines of a particular culture," but "whether to believe that the question of how to act could even be reasonably asked, let alone answered":

Just because our moral philosophers knew the facts of morality only very approximately in arbitrary extracts or in accidental epitomes—for example, as the morality of their environment, their class, their church, the spirit of their time, their climate and part of the world—just because they were poorly informed and not even very curious about different peoples, times, and past ages—they never laid eyes on the real problems of morality; for these emerge only when we compare many moralities. In all "science of morals" so far one thing was *lacking*, strange as it may sound: the problem of morality itself; what was lacking was any suspicion that there was something problematic here.[19]

This "problem of morality"—*is there anything moral, in any realistic general sense, and if so, how might it be comprehended?*—is a question that has now attained paramount importance. We have the technological power to do anything we want (certainly, anything destructive; potentially, anything creative); commingled with that power, however, is an equally profound existential uncertainty, shallowness and confusion. Our constant cross-cultural interchanges and our capacity for critical reasoning have undermined our faith in the traditions of our fore-bears, perhaps for good reason. However, the individual cannot live without belief—without action and valuation—and science cannot provide that belief. We must nonetheless put our faith into something. Are the myths we have turned to since the rise of science more sophisticated, less dangerous, and more complete than those we rejected? The ideological structures that dominated social relations in the twentieth century appear no less absurd, on the face of it, than the older belief systems they supplanted; they lacked, in addition, any of the incomprehensible mystery that necessarily remains part of genuinely artistic and creative production. The fundamental propositions of fascism and communism were rational, logical, statable, comprehensible—and terribly wrong. No great ideological struggle presently tears at the soul of the world, but it is difficult to believe that we have outgrown our gullibility. The rise of the New Age movement in the West, for example—as compensation for the decline of traditional spirituality—provides sufficient evidence for our continued ability to swallow a camel, while straining at a gnat.

Could we do better? Is it possible to understand what might reasonably, even admirably, be believed, after understanding that we must believe? Our vast power makes self-control (and, perhaps, self-comprehension) a necessity—so we have the motivation, at least in principle. Furthermore, the time is auspicious. The third Christian millennium is dawning—at the end of an era when we have demonstrated, to the apparent satisfaction of everyone, that certain forms of social regulation just do not work (even when judged by their own criteria for success). We live in the aftermath of the great statist experiments of the twentieth century, after all, conducted as Nietzsche prophesied:

In the doctrine of socialism there is hidden, rather badly, a "will to negate life"; the human beings or races that think up such a doctrine must be bungled. Indeed, I should wish that a few great experiments might prove that in a socialist society life negates itself, cuts off its own roots. The earth is large enough and man still sufficiently unexhausted; hence such a practical instruction and *demonstratio ad absurdum* would not strike me as undesirable, even if it were gained and paid for with a tremendous expenditure of human lives.[20]

There appears to exist some "natural" or even—dare it be said?—some "absolute" constraints on the manner in which human beings may act as individuals and in society. Some moral presuppositions and theories are *wrong*; human nature is not infinitely malleable.

It has become more or less evident, for example, that pure, abstract rationality, ungrounded in tradition—the rationality that defined Soviet-style communism from inception to dissolution—appears absolutely unable to determine and make explicit just what it is that should guide individual and social behavior. Some systems do not work, even though they make abstract sense (even more sense than alternative, currently operative, incomprehensible, haphazardly evolved systems). Some patterns of interpersonal interaction—which constitute the state, insofar as it exists as a model for social behavior—do not produce the ends they are supposed to produce, cannot sustain themselves over time, and may even produce contrary ends, devouring those who profess their value and enact them. Perhaps this is because planned, logical and intelligible systems fail to make allowance for the irrational, transcendent, incomprehensible and often ridiculous aspect of human character, as described by Dostoevsky:

Now I ask you: what can be expected of man since he is a being endowed with such strange qualities? Shower upon him every earthly blessing, drown him in a sea of happiness, so that nothing but bubbles of bliss can be seen on the surface; give him economic prosperity, such that he should have nothing else to do but sleep, eat cakes and busy himself with the continuation of his species, and even then out of sheer ingratitude, sheer spite, man would play you some nasty trick. He would even risk his cakes and would deliberately desire the most fatal rubbish, the most uneconomical absurdity, simply to introduce into all this positive good sense his fatal fantastic element. It is just his fantastic dreams, his vulgar folly that he will desire to retain, simply in order to prove to himself—as though that were so necessary—that men still are men and not the keys of a piano, which the laws of nature threaten to control so completely that soon one will be able to desire nothing but by the calendar.

And that is not all: even if man really were nothing but a piano-key, even if this were proved to him by natural science and mathematics, even then he would not become reasonable, but would purposely do something perverse out of simple ingratitude, simply to gain his point. And if he does not find means he will contrive destruction and chaos, will contrive sufferings of all sorts, only to gain his point! He will launch a curse upon the world, and as only man can curse (it is his privilege, the primary distinction between him and other animals), maybe by his curse alone he will attain his object—that is, convince himself that he is a man and not a piano-key! If you say that all this, too, can be calculated and tabulated, chaos and darkness and curses, so that the mere possibility of calculating it all before-hand would stop it all, and reason would reassert itself, then man would purposely go mad in order to be rid of reason and gain his point! I believe in it, I answer for it, for the whole work of man really seems to consist in nothing but proving to himself every minute that he is a man and not a piano-key! It may be at the cost of his skin, it may be by cannibalism! And this being so, can one help being tempted to rejoice that it has not yet come off, and that desire still depends on something we don't know?[21]

We also presently possess in accessible and complete form the traditional wisdom of a large part of the human race—possess accurate description of the myths and rituals that contain and condition the implicit and explicit values of almost everyone who has ever lived. These myths are centrally and properly concerned with the nature of successful human existence. Careful comparative analysis of this great body of religious philosophy might allow us to provisionally determine the nature of essential human motivation and morality—if we were willing to admit our ignorance and take the risk. Accurate specification of underlying mythological commonalities might comprise the first developmental stage in the conscious evolution of a truly universal system of morality. The establishment of such a system, acceptable to empirical and religious minds alike, could prove of incalculable aid in the reduction of intrapsychic, interindividual and intergroup conflict. The grounding of such a comparative analysis within a psychology (or even a neuropsychology) informed by strict empirical research might offer us the possibility of a form of convergent validation, and help us overcome the age-old problem of deriving the *ought* from the *is*; help us see how *what we must do* might be inextricably associated with *what it is that we are.*

Proper analysis of mythology, of the type proposed here, is not mere discussion of "historical" events enacted upon the world stage (as the traditionally religious might have it), and it is not mere investigation of primitive belief (as the traditionally scientific might presume). It is, instead, the examination, analysis and subsequent incorporation of an edifice of meaning, which contains within it hierarchical organization of experiential valence. The mythic imagination is concerned with the world in the manner of the phenomenologist, who seeks to discover the nature of subjective reality, instead of concerning himself with description of the objective world. Myth, and the drama that is part of myth, provide answers in image to the following question: "how can the current state of experience be conceptualized in abstraction, with regards to its *meaning?*" [which means its (subjective, biologically predicated, socially constructed) emotional relevance or motivational significance]. Meaning means implication for behavioral output; logically, therefore, myth presents information relevant to the most fundamental of moral problems: *"what should be? (what should be done?)"* The desirable future (the object of *what should be*) can be conceptualized only in relationship to the present, which serves at least as a necessary point of contrast and comparison. To get somewhere in the future presupposes being somewhere in the present; furthermore, the desirability of *the place traveled to* depends on the valence of the place vacated. The question of *"what should be?"* (what line should be traveled?) therefore has contained within it, so to speak, three subqueries, which might be formulated as follows:

1) *What is?* What is the nature (meaning, the *significance*) of the current state of experience?
2) *What should be?* To what (desirable, valuable) end should that state be moving?
3) *How should we therefore act?* What is the nature of the specific processes by which the present state might be transformed into that which is desired?

Active apprehension of the goal of behavior, conceptualized in relationship to the interpreted present, serves to constrain or provide determinate framework for the evaluation of ongoing events, which emerge as a consequence of current behavior. The goal is an imaginary state, consisting of "a place" of desirable motivation or affect—a state that only exists in fantasy, as something (potentially) preferable to the present. (Construction of the goal therefore means establishment of a theory about the ideal relative status of motivational states—about the *good.*) This imagined future constitutes a *vision of perfection*, so to speak, generated in the light of all current knowledge (at least under optimal conditions), to which specific and general aspects of ongoing experience are continually compared. This vision of perfection is the promised land, mythologically speaking—conceptualized as a spiritual domain (a psychological state), a political utopia (a state, literally speaking), or both, simultaneously.

We answer the question *"what should be?"* by formulating an image of the desired future.

We cannot conceive of that future, except in relationship to the (interpreted) present—and it is our interpretation of the emotional acceptability of the present that comprises our answer to the question "*what is?*" ["what is the nature (meaning, the *significance*) of the current state of experience?"].

We answer the question "*how then should we act?*" by determining the most efficient and self-consistent strategy, all things considered, for bringing the preferred future into being.

Our answers to these three fundamental questions—modified and constructed in the course of our social interactions—constitutes our knowledge, insofar as it has any behavioral relevance; constitutes our knowledge, from the mythological perspective. The structure of the mythic *known*—what is, what should be, and how to get from one to the other—is presented in *Figure 1: The Domain and Constituent Elements of the Known*.

The known is explored territory, a place of stability and familiarity; it is the "city of God," as profanely realized. It finds metaphorical embodiment in myths and narratives describing the community, the kingdom or the state. Such myths and narratives guide our ability to understand the particular, bounded motivational significance of the present, experienced in relation to some identifiable desired future, and allow us to construct and interpret appropriate patterns of action, from within the confines of that schema. We all produce determinate models of what is, and what should be, and how to transform one into the other. We produce these models by balancing our own desires, as they find expression in fantasy and action, with those of the others—individuals, families and communities—that we habitually encounter. "How to act," constitutes the most essential aspect of the social contract; the domain of the *known* is, therefore, the "territory" we inhabit with all those who share our implicit and explicit traditions and beliefs. Myths describe the existence of this "shared and determinate territory" as a fixed aspect of existence—which it is, as the fact of culture is an unchanging aspect of the human environment.

"Narratives of the known"—patriotic rituals, stories of ancestral heroes, myths and symbols of cultural or racial identity—describe established territory, weaving for us a web of meaning that, shared with others, eliminates the necessity of dispute over meaning. All those who know the rules, and accept them, can play the game—without fighting over the rules of the game. This makes for peace, stability, and potential prosperity—a good game. The good, however, is the enemy of the better; a more compelling game might always exist. Myth portrays what is known, and performs a function that if limited to that, might be regarded as paramount in importance. But myth also presents information that is far more profound—almost unutterably so, once (I would argue) properly understood. We all produce models of what is and what should be, and how to transform one into the other. We change our behavior, when the consequences of that behavior are not what we would like. But sometimes mere alteration in behavior is insufficient. We must change not only what we do, but what we think is important. This means reconsideration of the nature of the motivational significance of the present, and reconsideration of the ideal nature of the future. This is a radical, even revolutionary transformation, and it is a very complex process in its realization—but mythic thinking has represented the nature of such change in great and remarkable detail.

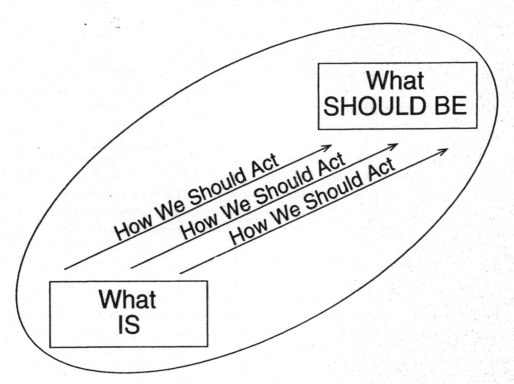

Figure 1: The Domain and Constituent Elements of the Known

The basic grammatical structure of transformational mythology, so to speak, appears most clearly revealed in the form of the "way" (as in the "American Way of Life"). The great literary critic Northrop Frye comments upon the idea of the way, as it manifests itself in literature and religious writing:

Following a narrative is closely connected with the central literary metaphor of the journey, where we have a person making the journey and the road, path, or direction taken, the simplest word for this being 'way.' Journey is a word connected with *jour* and *journee*, and metaphorical journeys, deriving as they mostly do from slower methods of getting around, usually have at their core the conception of the day's journey, the amount of space we can cover under the cycle of the sun. By a very easy extension of metaphor we get the day's cycle as a symbol for the whole of life. Thus in Housman's poem "Reveille" ("Up, lad: when the journey's over/ There'll be time enough to sleep") the awakening in the morning is a metaphor of continuing the journey of life, a journey ending in death. The prototype for the image is the Book of Ecclesiastes, which urges us to work while it is day, before the night comes when no man can work....

The word "way" is a good example of the extent to which language is built up on a series of metaphorical analogies. The most common meaning of "way" in English is a method or manner of procedure, but method and manner imply some sequential repetition, and the repetition brings us to the metaphorical kernel of a road or path.... In the Bible "way" normally translates the Hebrew *derek* and the Greek *hodos*, and throughout the Bible there is a strong emphasis on the contrast between a straight way that takes us to our destination and a divergent way that misleads or confuses. This metaphorical contrast haunts the whole of Christian literature: we start reading Dante's *Commedia*, and the third line speaks of a lost or erased way: "Che la diritta *via* era smarita." Other religions have the same metaphor: Buddhism speaks of what is usually called in English an eightfold path. In Chinese Taoism the Tao is usually also rendered "way" by Arthur Waley and others, though I understand that the character representing the word is formed of radicals meaning something like "head-going." The sacred book of Taoism, the *Tao te Ching*, begins by saying that the Tao that can be talked about is not the real Tao: in other words we are being warned to beware of the traps in metaphorical language, or, in a common Oriental phrase, of confusing the moon with the finger pointing at it. But as we read on we find that the Tao can, after all, be to some extent characterized: the way is specifically the "way of the valley," the direction taken by humility, self-effacement, and the kind of relaxation, or non-action, that makes all action effective.[22]

The "way" is the path of life and its purpose.[23] More accurately, the content of the way is the specific path of life. The form of the way, its most fundamental aspect, is the apparently intrinsic or heritable possibility of positing or of being guided by a central idea. This apparently intrinsic form finds its expression in the tendency of each individual, generation after generation, to first ask and subsequently seek an answer to the question "what is the meaning of life?"

The central notion of the way underlies manifestation of four more specific myths, or classes of myths, and provides a more complete answer, in dramatic form, to the three questions posed previously [*what is the nature* (meaning, the significance) *of current being?, to what* (desirable) *end should that state be moving?* and, finally, *what are the processes by which the present state might be transformed into that which is desired?*] The four classes include:

(1) myths describing a current or pre-existent stable state (sometimes a paradise, sometimes a tyranny);
(2) myths describing the emergence of something anomalous, unexpected, threatening and promising into this initial state;
(3) myths describing the dissolution of the pre-existent stable state into chaos, as a consequence of the anomalous or unexpected occurrence;
(4) myths describing the regeneration of stability [paradise regained (or, tyranny regenerated)], from the chaotic mixture of dissolute previous experience and anomalous information.

The metamythology of the way, so to speak, describes the manner in which specific ideas (myths) about the present, the future, and the mode of transforming one into the other are initially constructed, and then reconstructed, in their entirety, when that becomes necessary.

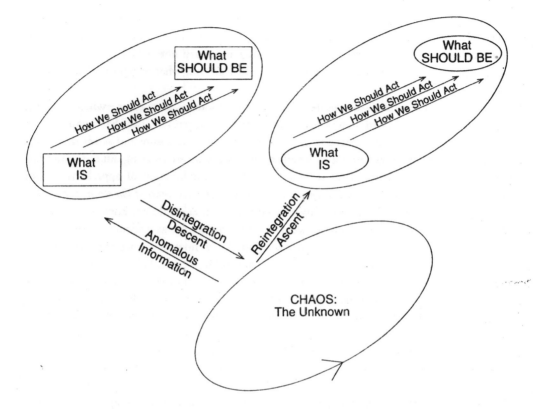

Figure 2: The Metamythological Cycle of the Way

The traditional Christian (and not just Christian) notion that man has fallen from an original "state of grace" into his current morally degenerate and emotionally unbearable condition—accompanied by a desire for the "return to Paradise"—constitutes a single example of this "metamyth." Christian morality can therefore be reasonably regarded as the "plan of action" whose aim is re-establishment, or establishment, or attainment (sometimes in the "hereafter") of the "kingdom of God," the ideal future. The idea that man needs redemption—and that re-establishment of a long-lost Paradise might constitute such redemption—appear as common themes of mythology, among members of exceedingly diverse and long-separated human cultures.[24] This commonality appears because man, eternally self-conscious, suffers eternally from his existence, and constantly longs for respite.

Figure 2: The Metamythological Cycle of the Way schematically portrays the "circle" of the way, which "begins" and "ends" at the same point—with establishment of conditional, but determinate moral knowledge (belief). Belief is *disruptible*, because finite—which is to say that the infinite mystery surrounding human understanding may break into our provisional models of how to act at any time and point, and disrupt their structure. The manner in which we act as children, for example, may be perfectly appropriate for the conditions of childhood; the processes of maturation change the conditions of existence, introducing anomaly where only certainty once stood, making necessary not only a change of plans, but reconceptualization of where those plans might lead, and what or who they refer to, in the present.

The known, our current story, protects us from the unknown, from *chaos*—which is to say, provides our experience with determinate and predictable structure. Chaos has a nature all of its own. That nature is experienced as *affective valence*, at first exposure, not as *objective property*. If something unknown or unpredictable occurs, while we are carrying out our motivated plans, we are first *surprised*. That surprise—which is a combination of apprehension and curiosity—comprises our *instinctive emotional response to the occurrence of something we did not desire*. The appearance of something unexpected is proof that we do not know how to act— by definition, as it is the production of what we want that we use as evidence for the integrity of our knowledge. If we are somewhere we don't know how to act, we are (probably) in trouble—we might learn something new, but we are still in trouble. When we are in trouble, we get scared. When we are in the domain of the known, so to speak, there is no reason for fear. Outside that domain, panic reigns. It is for this reason that we dislike having our plans disrupted, and cling to what we understand. This conservative strategy does not always work, however, because what we understand about the present is not necessarily sufficient to deal with the future. This means that we have to be able to modify what we understand, even though to do so is to risk our own undoing. The trick, of course, is to modify and yet to remain secure. This is not so simple. Too much modification brings chaos. Too little modification brings stagnation (and then, when the future we are unprepared for appears—chaos).

Involuntary exposure to chaos means accidental encounter with the forces that undermine the known world. The affective consequences of such encounter can be literally overwhelming. It is for this reason that individuals are highly motivated to avoid sudden manifestations of the unknown. And this is why individuals will go to almost any length to ensure that their protective cultural "stories" remain intact.

2

MAPS OF MEANING

Three Levels of Analysis

———————————————○———————————————

uman beings are prepared, biologically, to respond to anomalous information—to novelty. This instinctive response includes redirection of attention, generation of emotion (fear first, generally speaking, then curiosity), and behavioral compulsion (cessation of ongoing activity first, generally speaking, then active approach and exploration). This pattern of instinctive response drives learning—particularly, but not exclusively, the learning of appropriate behavior. All such learning takes place—or took place originally—as a consequence of contact with novelty, or anomaly.

What is novel is of course dependent on what is known—is necessarily defined in opposition to what is known. Furthermore, what is known is always known conditionally, since human knowledge is necessarily limited. Our conditional knowledge, insofar as that knowledge is relevant for the regulation of emotion, consists of our models of the emotional significance of the present, defined in opposition to an idealized, hypothetical or fantasied future state. We evaluate the "unbearable present" in relationship to the "ideal future." We act to transform "where we are" into "where we would like to be."

When our attempts to transform the present work as planned, we remain firmly positioned in the *domain of the known* (metaphorically speaking). When our behaviors produce results that we did not want, however—that is, when we err—we move into the *domain of the unknown*, where more primordial emotional forces rule. "Small-scale" errors force us to reconstruct our plans, but allow us to retain our goals and our conceptualizations of present conditions. Catastrophic errors, by contrast, force us not only to re-evaluate our means, but our starting points and our ends. Such revaluation necessarily involves extreme emotional dysregulation.

The "domain of the known" and the "domain of the unknown" can reasonably be regarded as permanent constituent elements of human experience—even of the human environment. Regardless of culture, place and time, human individuals are forced to adapt to the fact of culture (the domain of the known, roughly speaking) and the fact of its ultimate insufficiency (as the domain of the unknown necessarily remains extant, regardless of extent of previous "adaptation"). The human brain—and the higher animal brain—appears therefore to have adapted itself to the eternal presence of these two "places"; the brain has one mode of operation when in explored territory, and another when in unexplored territory. In the unexplored world, caution—expressed in fear and behavioral immobility—initially predominates, but may be superseded by curiosity—expressed in hope, excitement and, above all, in creative exploratory behavior. Creative exploration of the unknown, and consequent generation of knowledge, is construction or update of patterns of behavior and representation, such that the unknown is transformed from something terrifying and compelling into something beneficial (or, at least, something irrelevant). The presence of capacity for such creative exploration and knowledge generation may be regarded as the third, and final, permanent constituent element of human experience (in addition to the domain of the "known" and "unknown").

Mythological representations of the world—which are representations of reality as a forum for action—portray the dynamic interrelationship between all three constituent elements of human experience. The eternal unknown—nature, metaphorically speaking, creative and destructive, source and destination of all determinant things—is generally ascribed an affectively ambivalent feminine character (as the "mother" and eventual "devourer" of everyone and everything). The eternal known, in contrast—culture, defined territory, tyrannical and protective, predictable, disciplined and restrictive, cumulative consequence of heroic or exploratory behavior—is typically considered masculine (in contradistinction to "mother" nature). The eternal knower, finally—the process that mediates between the known and the unknown—is the knight who slays the dragon of chaos, the hero who replaces disorder and confusion with clarity and certainty, the sun god who eternally slays the forces of darkness, and the "word" that engenders cosmic creation.

Normal and Revolutionary Life: Two Prosaic Stories

We tell ourselves stories about who we are, where we would like to be, and how we are going to get there. These stories regulate our emotions, by determining the significance of all the things we encounter and all the events we experience. We regard things that get us on our way as positive, things that impede our progress as negative, and things that do neither as irrelevant. Most things are irrelevant—and that is a good thing, as we have limited attentional resources.

Inconveniences interfere with our plans. We do not like inconveniences, and will avoid dealing with them. Nonetheless, they occur commonly—so commonly, in fact, that they might be regarded as an integral, predictable, and constant feature of the human environment. We have adapted to this feature—have the intrinsic resources to cope with inconveniences. We benefit, become stronger, in doing so.

Ignored inconveniences accumulate, rather than disappear. When they accumulate in sufficient numbers, they produce a catastrophe—a self-induced catastrophe, to be sure, but one that may be indistinguishable from an "act of God." Inconveniences interfere with the integrity of our plans—so we tend to pretend that they are not there. Catastrophes, by contrast, interfere with the integrity of our whole stories, and massively dysregulate our emotions. By their nature, they are harder to ignore—although that does not stop us from trying to do so.

Inconveniences are common; unfortunately, so are catastrophes—self-induced and otherwise. We are adapted to catastrophes, like inconveniences, as constant environmental features. We can resolve catastrophe, just as we can cope with inconvenience—although at higher cost. As a consequence of this adaptation, this capacity for resolution, catastrophe can rejuvenate. It can also destroy.

The more ignored inconveniences in a given catastrophe, the more likely it will destroy.

Enough has been learned in the last half-century of inquiry into intellectual and emotional function to enable the development of a provisional general theory of emotional regulation. Description of the role that reaction to novelty or anomaly plays in human information processing is clearly central to such a theory. A compelling body of evidence suggests that our affective, cognitive and behavioral responses to the unknown or unpredictable are "hardwired"; suggests that these responses constitute inborn structural elements of the processes of consciousness itself. We attend, involuntarily, to those things that occur contrary to our predictions—that occur despite our desires, as expressed in expectation. That involuntary attention comprises a large part of what we refer to when we say "consciousness." Our initial attention constitutes the first step in the process by which we come to adjust our behavior and our interpretive schemas to the world of experience—assuming that we do so; constitutes as well the first step we take when we modify the world to make it what we desire, instead of what it is currently.

Modern investigation into the role of novelty in emotion and thought began with the Russians—E.N. Sokolov, O. Vinogradova, A.R. Luria (and, more recently, E. Goldberg)—who adopted an approach to human function that is in many ways unique. Their tradition apparently stems from Pavlov, who viewed the reflex arc as a phenomenon of central importance, and from the Marxist intellectual legacy, which regarded work—creative action—as the defining feature of man. Whatever the specific historical precedents, it is most definitely the case that the Russians have regarded motor output and its abstract equivalents as the critically relevant aspect of human existence. This intellectual position distinguished them, historically, from their Western counterparts, who tend(ed) to view the brain as an information-processing machine, akin to the computer. Psychologists in the West have concentrated their energies on determining how the brain determines what is out there, so to speak, from the objective viewpoint. The Russians, by contrast, have devoted themselves to the role of the brain in governing behavior, and in generating the affects or emotions associated with that behavior. Modern animal experimentalists—most notably Jeffrey Gray[25]—have adopted the Russian line, with striking success. We now know, at least in broad outline, how we respond to those (annoying, irritating, frightening, promising) things that we do not expect.

The pioneering Russian psychophysiologist E.N. Sokolov began work on the "reflex basis" of attention in the 1950s. By the early '60s, this work had advanced to the point where he could formulate the following key propositions. First:

One possible approach to analyzing the process of reflection is to consider the nervous system as a mechanism which models the external world by specific changes that occur in its internal structure. In this sense a distinct set of changes in the nervous system is isomorphic with the external agent that it reflects and resembles. As an internal model that develops in the nervous system in response to the effect of agents in the environment, the image performs the vital function of modifying the nature of behavior, allowing the organism to predict events and actively adjust to its environment.[26]

And second:

My first encounter with phenomena which indicated that the higher divisions of the central nervous system form models of external agents involved the study of reactions to "novel" [stimulus features. I characterized these reactions as] *orienting reflexes*. The peculiar feature of the orienting reflex is that after several applications of the same stimulus (generally five to fifteen) the response disappears (or, as the general expression goes, "is extinguished"). However, the slightest possible change in the stimulus is sufficient to awaken the response. . . . Research on the orienting reflex indicates that it does not occur as a direct result of incoming excitation; rather, it is produced by signals of discrepancy which develop when afferent [that is, *incoming*] signals are compared with the trace formed in the nervous system by an earlier signal.[27]

Sokolov was concerned primarily with the modeling of events in the objective external world—assuming, essentially, that when we model, we model *facts*. Most of the scholars who have followed his lead have adopted this central assumption, at least implicitly. This position requires some modification. We do model facts, but we *concern* ourselves with valence, or value. It is therefore the case that our maps of the world contain what might be regarded as two distinct types of information: sensory and affective. It is not enough to know that something *is*. It is equally necessary to know what it *signifies*. It might even be argued that animals—and human beings—are *primarily* concerned with the affective or emotional significance of the environment.

Along with our animal cousins, we devote ourselves to fundamentals: will this (new) thing eat me? Can I eat it? Will it chase me? Should I chase it? Can I mate with it? We *may* construct models of "objective reality," and it is no doubt useful to do so. We *must* model meanings, however, in order to survive. Our most fundamental maps of meaning—maps which have a narrative structure—portray *the motivational value of our current state*, conceived of in *contrast to a hypothetical ideal*, accompanied *by plans of action*, which are our pragmatic notions about how to get what we want.

Description of these three elements—current state, ideal future state, and means of active mediation—constitute the necessary and sufficient preconditions for the weaving of the most simple narrative, which is a means for describing the valence of a given environment, in reference to a temporally and spatially bounded set of action patterns. Getting to point "b" presupposes that you are at point "a"—you can't plan movement in the absence of an initial position. The fact that point "b" constitutes the end goal means that it is valenced more highly than point "a"—that it is a place more desirable, when considered against the necessary contrast of the current position. It is the perceived improvement of point "b" that makes the whole map meaningful or affect-laden; it is the capacity to construct hypothetical or abstract end points, such as "b"—and to contrast them against "the present"—that makes human beings capable of using their cognitive systems to modulate their affective reactions.[28]

The domain mapped by a functional narrative (one that, when enacted, produces the results desired) might reasonably be regarded as "explored territory," as events that occur "there" are predictable. Any place where enacted plans produce unexpected, threatening or punishing consequences, by contrast, might be regarded as "unexplored territory." What happens "there" does not conform to our wishes. This means that a familiar place, where unpredictable things start happening, is no longer familiar (even though it might be the same place with regards to its strict spatial location, from the "objective" perspective). We know how to act in some places, and not in others. The plans we put into action sometimes work, and sometimes do not work. The experiential domains we inhabit—our "environments," so to speak—are therefore permanently characterized by the fact of the predictable and controllable, in juxtaposition with the unpredictable and uncontrollable. The universe is composed of "order" and "chaos"—at least from the metaphorical perspective. Oddly enough, however, it is to this "metaphorical" universe that our nervous system appears to have adapted.

What Sokolov discovered, to put it bluntly, is that human beings (and other animals far down the phylogenetic chain) are characterized by an innate response to what they cannot predict, do not want, and cannot understand. Sokolov identified the central characteristics of how we respond to the unknown—to the strange category of *all events that have not yet been categorized.* The notion that we respond in an "instinctively patterned" manner to the appearance of the unknown has profound implications. These can best be first encountered in narrative form.

Normal Life

> "If problems are accepted,
> and dealt with before they arise,
> they might even be prevented before confusion begins.
> In this way peace may be maintained."[30]

You work in an office; you are climbing the corporate ladder. Your daily activity reflects this superordinate goal. You are constantly immersed in one activity or another designed to produce an elevation in your status from the perspective of the corporate hierarchy. Today, you have to attend a meeting that may prove vitally important to your future. You have an image in your head, so to speak, about the nature of that meeting and the interactions that will characterize it. You imagine what you would like to accomplish. Your image of this potential future is a *fantasy*, but it is based, insofar as you are honest, on all the relevant information derived from past experience that you have at your disposal. You have attended many meetings. You know what is likely to happen, during any given meeting, within reasonable bounds; you know how you will behave, and what effect your behavior will have on others. Your model of the desired future is clearly predicated on what you currently know.

You also have a model of the present, constantly operative. You understand *your* (somewhat subordinate) *position* within the corporation, which is your importance relative to others above and below you in the hierarchy. You understand the significance of those experiences that occur regularly while you are during your job: you know who you can give orders to, who you have to listen to, who is doing a good job, who can safely be ignored, and so on. You are always comparing this present (unsatisfactory) condition to that of your ideal, which is you, increasingly respected, powerful, rich and happy, free of anxiety and suffering, climbing toward your ultimate success. You are unceasingly involved in attempts to transform the present, as you currently understand it, into the future, as you hope it will be. Your actions are designed to produce your ideal—designed to transform the present into something ever more closely resembling what you want. Your are confident in your model of reality, in your story; when you put it into action, you get *results*.

You prepare yourself mentally for your meeting. You envision yourself playing a centrally important role—resolutely determining the direction the meeting will take, producing a powerful impact on your co-workers. You are in your office, preparing to leave. The meeting is taking place in another building, several blocks away. You formulate provisional plans of behavior designed to get you there on time. You estimate travel time at fifteen minutes.

You leave your office on the twenty-seventh floor, and you wait by the elevator. The minutes tick by—more and more of them. The elevator fails to appear. You had not taken this possibility into account. The longer you wait, the more nervous you get. Your heart rate starts increasing, as you prepare for action (action unspecified, as of yet). Your palms sweat. You flush. You berate yourself for failing to consider the potential impact of such a delay. Maybe you are not as smart as you think you are. You begin to revise your model of yourself. No time for that now: you put such ideas out of your head and concentrate on the task at hand.

The unexpected has just become manifest—in the form of the missing elevator. You planned to take it to get where you were going; it did not appear. Your original plan of action is not producing the effects desired. It was, *by your own definition*, a bad plan. You need another one—and quickly. Luckily you have an alternate strategy at your disposal. The stairs! You dash to the rear of the building. You try the door to the stairwell. It is locked. You curse the maintenance staff. You are frustrated and anxious. The unknown has emerged once again. You try another exit. Success! The door opens. Hope springs forth from your breast. You still might make it on time. You rush down the stairs—all twenty-seven floors—and onto the street.

You are, by now, desperately late. As you hurry along, you monitor your surroundings: is progress toward your goal continuing? Anyone who gets in your way inconveniences you—elderly women, playful, happy children, lovers out for a stroll. You are a good person, under most circumstances—at least in your own estimation. Why, then, do these innocent people aggravate you so thoroughly? You near a busy intersection. The crosswalk light is off. You fume and mutter away stupidly on the sidewalk. Your blood pressure rises. The light finally changes. You smile and dash forward. Up a slight rise you run. You are not in great physical shape. Where did all this energy come from? You are approaching the target building. You glance at your watch. Five minutes left: no problem. A feeling of relief and satisfaction floods you. You are *there*; in consequence, you are not an idiot. If you believed in God, you would thank Him.

Had you been early—had you planned appropriately—the other pedestrians and assorted obstacles would not have affected you at all. You might have even appreciated them—at least the good-looking ones—or may at least not have classified them as obstacles. Maybe you would have even used the time to enjoy your surroundings (unlikely) or to think about other issues of real importance—like tomorrow's meeting.

You continue on your path. Suddenly, you hear a series of loud noises behind you—noises reminiscent of a large motorized vehicle hurtling over a small concrete barrier (much like a curb). You are safe on the sidewalk—or so you presumed a second ago. Your meeting fantasies vanish. The fact that you are late no longer seems relevant. You stop hurrying along, instantly, arrested in your path by the emergence of this new phenomenon. Your auditory system localizes the sounds in three dimensions. You involuntarily orient your trunk, neck, head and eyes toward the place in space from which the sounds apparently emanate.[31] Your pupils dilate, and your eyes widen.[32] Your heart rate speeds up, as your body prepares to take adaptive action—once the proper path of that action has been specified.[33]

You actively explore the unexpected occurrence, once you have oriented yourself toward it, with all the sensory and cognitive resources you can muster. You are generating hypotheses about the potential cause of the noise even before you turn. Has a van jumped the curb? The image flashes through your mind. Has something heavy fallen from a building? Has the wind overturned a billboard or street sign? Your eyes actively scan the relevant area. You see a truck loaded with bridge parts heading down the street, just past a pothole in the road. The mystery is solved. You have determined the specific motivational significance of what just seconds ago was the dangerous and threatening unknown, and it is zero. A loaded truck hit a bump. Big deal! Your heart slows down. Thoughts of the impending meeting re-enter the theater of your mind. Your original journey continues as if nothing has happened.

What is going on? Why are you frightened and frustrated by the absence of the expected elevator, the presence of the old woman with the cane, the carefree lovers, the loud machinery? Why are you so emotionally and behaviorally variable?

Detailed description of the processes governing these common affective occurrences provides the basis for proper understanding of human motivation. What Sokolov and his colleagues essentially discovered was that the unknown, experienced in relationship to your currently extant model of present and future, has *a priori* motivational significance—or, to put it somewhat differently, that the unknown could serve *as an unconditioned stimulus*.

What is the *a priori* motivational significance of the unknown? Can such a question even be asked? After all, the unknown by definition has not yet been explored. Nothing can be said, by the dictates of standard logic, about something that has not yet been encountered. We are not concerned with sensory information, however—nor with particular material attributes—but with *valence*. Valence, in and of itself, might be most simply considered as bipolar: negative or positive (or, of course, as neither). We are familiar enough with the ultimate potential range of valence, negative and positive, to place provisional borders around possibility. The *worst* the unknown could be, in general, is death (or, perhaps, lengthy suffering followed by death); the fact of our vulnerable mortality provides the limiting case. The *best* the unknown could be is more difficult to specify, but some generalizations might prove acceptable. We would like to be wealthy (or at least free from want), possessed of good health, wise and well-loved. The greatest good the unknown might confer, then, might be regarded as that which would allow us to transcend our innate limitations (poverty, ignorance, vulnerability), rather than to remain miserably subject to them. The emotional "area" covered by the unknown is therefore very large, ranging from that which we fear most to that which we desire most intently.

The unknown is, of course, defined in contradistinction to the known. Everything *not* understood or *not* explored is unknown. The relationship between the oft- (and unfairly) separated domains of "cognition" and "emotion" can be more clearly comprehended in light of this rather obvious fact. It is the absence of an *expected* satisfaction, for example, that is punishing, hurtful[34]—the emotion is generated as a default response to sudden and unpredictable alteration in the theoretically comprehended structure of the world. It is the man *expecting* a raise because of his outstanding work—the man configuring a desired future on the basis of his understanding of the present—who is hurt when someone "less deserving" is promoted before him ("one is best punished," after all, "for one's virtues"[35]). The man whose expectations have been dashed—who has been threatened and hurt—is likely to work less hard in the future, with more resentment and anger. Conversely, the child who has not completed her homework is thrilled when the bell signaling class end rings, before she is called upon. The bell signals *the absence of an expected punishment*, and therefore induces positive affect, relief, happiness.[36]

It appears, therefore, that *the image of a goal* (a fantasy about the nature of the desired future, conceived of in relationship to a model of the significance of the present) provides much of the framework determining the motivational significance of ongoing current events. The individual uses his or her knowledge to construct a hypothetical state of affairs, where the motivational balance of ongoing events is optimized: where there is sufficient satisfaction, minimal punishment, tolerable threat and abundant hope, all balanced together *properly* over the short and longer terms. This optimal state of affairs might be conceptualized as a pattern of career advancement, with a long-term state in mind, signifying perfection, as it might be attained profanely (richest drug dealer, happily married matron, chief executive officer of a large corporation, tenured Harvard professor). Alternatively, perfection might be regarded as the absence of all unnecessary things, and the pleasures of an ascetic life. The point is that some desirable future state of affairs is conceptualized in fantasy and used as a target point for operation in the present. Such operations may be conceived of as links in a chain (with the end of the chain anchored to the desirable future state).

A meeting (like the one referred to previously) might be viewed by those participating in it as one link in the chain which hypothetically leads to the paradisal state of corporate chief executive officer (or to something less desirable but still good). The (well-brought-off) meeting, as *subgoal*, would therefore have the same motivational significance as the goal, although at lesser intensity (as it is only one small part of a large and more important whole). The *exemplary* meeting will be conceptualized in the ideal—like all target states—as a dynamic situation where, *all things considered*, motivational state is optimized. The meeting is imagined, a representation of the desired outcome is formulated, and a plan of behavior designed to bring about that outcome is elaborated and played out. The "imagined meeting" is fantasy, but fantasy based on past knowledge (assuming that knowledge has in fact been generated, and that the planner is able and willing to use it).

The affective systems that govern response to punishment, satisfaction, threat and promise all have a stake in attaining the ideal outcome. Anything that interferes with such attainment (little old ladies with canes) will be experienced as threatening and/or punishing; anything that signifies increased likelihood of success (open stretches of sidewalk) will be experienced as promising[37] or satisfying. It is for this reason that the Buddhists believe that everything is *Maya*, or illusion:[38] *the motivational significance of ongoing events is clearly determined by the nature of the goal toward which behavior is devoted*. That goal is conceptualized in episodic imagery—in fantasy. We constantly compare the world at present to the world idealized in fantasy, render affective judgment, and act in consequence. Trivial promises and satisfactions indicate that we are doing well, are progressing toward our goals. An unexpected opening in the flow of pedestrians appears before us, when we are in a hurry; we rush forward, pleased at the occurrence. We get somewhere a little faster than we had planned and feel satisfied with our intelligent planning. Profound promises or satisfactions, by contrast, validate our global conceptualizations—indicate that our emotions are likely to stay regulated on the path we have chosen. Trivial threats or punishments indicate flaws in our *means* of attaining desired ends. We modify our behavior accordingly and eliminate the threat. When the elevator does not appear at the desired time, we take the stairs. When a stoplight slows us down, we run a bit faster, once it shuts off, than we might have otherwise. Profound threats and punishments (read: trauma) have a qualitatively different nature. Profound threats or punishments undermine our ability to believe *that our conceptualizations of the present are valid* and *that our goals are appropriate*. Such occurrences disturb our belief in our *ends* (and, not infrequently, in our starting points).

We construct our idealized world, in fantasy, according to all the information we have at our disposal. We use what we know to build an image of what we could have and, therefore,

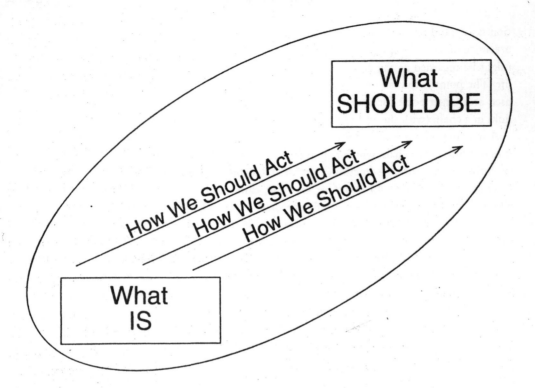

Figure 3: Normal Life

of what we should do. But we compare our interpretation of the world as it unfolds in the present to *the desired world, in imagination*, not to mere expectation; we compare what we have (in interpretation) to what we *want*, rather than to what we merely think *will be*. Our goal setting, and consequent striving, is motivated: we chase what we *desire*, in our constant attempts to optimize our affective states. (Of course, we use our behavior to ensure that our dreams come true; that is healthy "adaptation." But we still compare what is happening to what we want—to what we desire to be—not merely to what we cold-bloodedly expect.)

The maps that configure our motivated behavior have a certain comprehensible structure. They contain two fundamental and mutually interdependent poles, one *present*, the other *future*. The present is sensory experience as it is currently manifested to us—as we currently understand it—granted motivational significance according to our current knowledge and desires. The future is an image or partial image of perfection, to which we compare the present, insofar as we understand its significance. *Wherever there exists a mismatch between the two, the unexpected or novel occurs* (by definition), grips our attention, and activates the intrapsychic systems that govern fear and hope.[39] We strive to bring novel occurrences back into the realm of predictability or to exploit them for previously unconsidered potential by altering our behavior or our patterns of representation. We conceive of a path connecting present to future. This path is "composed" of the behaviors required to produce the transformations we desire—required to turn the (eternally) insufficient present into the (ever-receding) paradisal future. This path is *normally* conceived of as linear, so to speak, as something analogous to Thomas Kuhn's notion of *normal science*, wherein known patterns of behavior operating upon an understood present will produce a future whose desirability is an unquestioned given.[40]

Anything that interferes with our potential means to a specified end is punishing or threatening, in the rather trivial sense described previously. Encounter with punishments or threats of this category merely oblige us to choose an alternative mean from among the number we generally have present. A similar situation obtains for promises and satisfactions. When a means produces the end desired (or furthers progress along that path) we experience satisfaction (and hope—as an interim end *accomplished* also signifies increased likelihood of success, farther out in the future). Such satisfaction brings our particular behaviors to an end; we switch goals and continue into the future. Modification of our means, as a consequence of the motivational significance of the outcomes of those means, might be considered *normal* adaptation. The structure of normal adaptation is schematically portrayed in *Figure 3: Normal Life*. We posit a goal, in image and word, and we compare present conditions to that goal. We evaluate the significance of ongoing events in light of their perceived relationship to the goal. We modify our behavioral outputs—our means—when necessary, to make the attainment of our goal ever more likely. We modify our actions within the game but accept the rules without question. We move in a linear direction from present to future.

Revolutionary Life

The *revolutionary* model of adaptation—again, considered akin to Kuhn's *revolutionary science*[41]—is more complex. Let us presume that you return from your meeting. You made it on time and, as far as you could tell, everything proceeded according to plan. You noticed that your colleagues appeared a little irritated and confused by your behavior as you attempted to control the situation, but you put this down to jealousy on their part—to their inability to comprehend the majesty of your conceptualizations. You are satisfied, in consequence—satisfied temporarily—so you start thinking about tomorrow, as you walk back to work. You return to your office. There is a message on your answering machine. The boss wants to see you. You did not expect this. Your heart rate speeds up a little: good or bad, this news demands *preparation for action*.[42] What does she want? Fantasies of potential future spring up. Maybe she heard about your behavior at the meeting and wants to congratulate you on your excellent work. You walk to her office, apprehensive but hopeful.

You knock and stroll in jauntily. The boss looks at you and glances away somewhat unhappily. Your sense of apprehension increases. She motions for you to sit, so you do. What is going on? She says, "I have some bad news for you." This is not good. This is not what you wanted. Your heart rate is rising unpleasantly. You focus all of your attention on your boss. "Look," she says, "I have received a number of very unfavorable reports regarding your behavior at meetings. All of your colleagues seem to regard you as a rigid and overbearing negotiator. Furthermore, it has become increasingly evident that you are unable to respond positively to feedback about your shortcomings. Finally, you do not appear to properly understand the purpose of your job or the function of this corporation."

You are shocked beyond belief, paralyzed into immobility. Your vision of the future with this company vanishes, replaced by apprehensions of unemployment, social disgrace and fail-ure. You find it difficult to breathe. You flush and perspire profusely; your face is a mask of barely suppressed horror. You cannot believe that your boss is such a bitch. "You have been with us for five years," she continues, "and it is obvious that your performance is not likely to improve. You are definitely not suited for this sort of career, and you are interfering with the progress of the many competent others around you. In consequence, we have decided to ter-minate your contract with us, effective immediately. If I were you, I would take a good look at myself."

You have just received unexpected information, but of a different order of magnitude than the petty anomalies, irritations, threats and frustrations that disturbed your equilibrium in the morning. You have just been presented with incontrovertible evidence that your charac-terizations of the present and of the ideal future are seriously, perhaps irreparably, flawed. Your presumptions about the nature of the world are in error. The world you know has just crumbled around you. Nothing is what it seemed; everything is unexpected and new again. You leave the office in shock. In the hallway, other employees avert their gaze from you, in embarrassment. Why did you not see this coming? How could you have been so mistaken in your judgment?

Maybe everyone is out to get you.

Better not think that.

You stumble home, in a daze, and collapse on the couch. You can't move. You are hurt and terrified. You feel like you might go insane. Now what? How will you face people? The com-fortable, predictable, rewarding present has vanished. The future has opened up in front of you like a pit, and you have fallen in. For the next month, you find yourself unable to act. Your spirit has been extinguished. You sleep and wake at odd hours; your appetite is dis-turbed. You are anxious, hopeless and aggressive, at unpredictable intervals. You snap at your family and torture yourself. Suicidal thoughts enter the theater of your imagination. You do not know what to think or what to do: you are the victim of an internal war of emotion.

Your encounter with the terrible unknown has shaken the foundations of your worldview. You have been exposed, involuntarily, to the *unexpected and revolutionary*. Chaos has eaten your soul. This means that your long-term goals have to be reconstructed, and the motiva-tional significance of events in your current environment re-evaluated—literally *revalued*. This capacity for complete revaluation, in the light of new information, is even more partic-ularly human than the aforementioned capability for exploration of the unknown and gener-ation of new information. Sometimes, in the course of our actions, we elicit phenomena

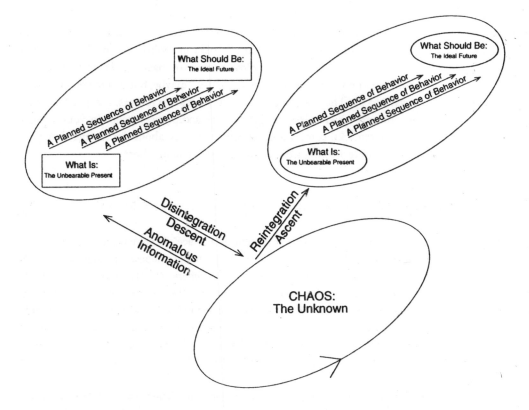

Figure 4: Revolutionary Adaptation

whose very existence is impossible, according to our standard methods of construal (which are at base a mode of attributing motivational significance to events). Exploration of these new phenomena, and integration of our findings into our knowledge, occasionally means reconceptualization of that knowledge[43] (and consequent re-exposure to the unknown, no longer inhibited by our mode of classification).[44] This means that simple movement from present to future is occasionally interrupted by a complete breakdown and reformulation, a reconstitution of what the present *is* and what the future *should be*. The ascent of the individual, so to speak, is punctuated by periods of dissolution and rebirth.[45] The more general model of human adaptation—conceptualized most simply as *steady state, breach, crisis, redress*[46]—therefore ends up looking like *Figure 4: Revolutionary Adaptation*. The processes of revolutionary adaptation, enacted and represented, underlie diverse cultural phenomena ranging from the rites of "primitive" initiation[47] to the conceptions of sophisticated religious systems.[48] Indeed, our very cultures are erected upon the foundation of a single great story: *paradise, encounter with chaos, fall and redemption.*

A month after you were fired, a new idea finds its way into your head. Although you never let yourself admit it, you didn't really like your job. You only took it because you felt that it was expected of you. You never put your full effort into it, because you really wanted to do something else—something other people thought was risky or foolish. You made a bad decision, a long time ago. Maybe you needed this blow, to put you back on the path. You start imagining a new future—one where you are not so "secure," maybe, but where you are doing what you actually want to do. The possibility of undisturbed sleep returns, and you start eating properly again. You are quieter, less arrogant, more accepting—except in your weaker moments. Others make remarks, some admiring, some envious, about the change they perceive in you. You are a man recovering from a long illness—a man reborn.

NEUROPSYCHOLOGICAL FUNCTION: THE NATURE OF THE MIND

It is reasonable to regard the world, as forum for action, as a "place"—a place made up of the familiar, and the unfamiliar, in eternal juxtaposition. The brain is actually composed, in large part, of two subsystems, adapted for action in that place. The right hemisphere, broadly speaking, responds to novelty with caution, and rapid, global hypothesis formation. The left hemisphere, by contrast, tends to remain in charge when things—that is, explicitly categorized things—are unfolding according to plan. The right hemisphere draws rapid, global, valence-based, metaphorical pictures of novel things; the left, with its greater capacity for detail, makes such pictures explicit and verbal. Thus the exploratory capacity of the brain "builds" the world of the familiar (the known), from the world of the unfamiliar (the unknown).

When the world remains known and familiar—that is, when our beliefs maintain their validity—our emotions remain under control. When the world suddenly transforms itself into something new, however, our emotions are dysregulated, in keeping with the relative novelty of that transformation, and we are forced to retreat or to explore once again.

The Valence of Things

> "Anyone who considers the basic drives of man ... will find that all of them have done philosophy at some time—and that every one of them would like only too well to represent just *itself* as the ultimate purpose of existence and the legitimate *master* of all the other drives. For every drive wants to be master—and it attempts to philosophize in *that spirit*."[49]

> "It is true that man was created in order to serve the gods, who, first of all, needed to be fed and clothed."[50]

We can make lists of *general* goods and bads, which might appear reasonable to others, because we tend to make judgments of meaning in relatively standard and predictable ways.

Food, to take a simple example, is *good*, assuming it is palatably prepared, while a blow on the head is *bad* in direct proportion to its force. The list of general goods and bads can be extended with little effort. Water, shelter, warmth and sexual contact are good; diseases, droughts, famines and fights are bad. The essential similarities of our judgments of meaning can easily lead us to conclude that the goodness or badness of things or situations is something more or less fixed. However, the fact of subjective interpretation—and its effects on evaluation and behavior—complicate this simple picture. We will work, expend energy and overcome obstacles to gain a good (or to avoid something bad). But we won't work for food, at least not very hard, if we have enough food; we won't work for sex, if we are satisfied with our present levels of sexual activity; and we might be very pleased to go hungry, if that means our enemy will starve. Our predictions, expectations and desires condition our evaluations to a finally unspecifiable degree. Things have no absolutely *fixed* significance, despite our ability to generalize about their value. It is our personal preferences, therefore, that determine the import of the world (but these preferences have constraints!).

The meaning we attribute to objects or situations is not stable. What is important to one man is not necessarily important to another; likewise, the needs and desires of the child differ from those of the adult. The meaning of things depends to a profound and ultimately undeterminable degree upon the relationship of those things to the goal we currently have in mind. Meaning shifts when goals change. Such change necessarily transforms the contingent expectations and desires that accompany those goals. We experience "things" personally and idiosyncratically, despite broad interpersonal agreement about the value of things. The goals we pursue singly—the outcomes we expect and desire as individuals—determine the meaning of our experience. The existential psychotherapist Viktor Frankl relates a story from his experiences as a Nazi death camp inmate that makes this point most strikingly:

Take as an example something that happened on our journey from Auschwitz to the camp affiliated with Dachau. We became more and more tense as we approached a certain bridge over the Danube which the train would have to cross to reach Mauthausen, according to the statement of experienced traveling companions. Those who have never seen anything similar cannot possibly imagine the dance of joy performed in the carriage by the prisoners when they saw that our transport was not crossing the bridge and was instead heading "only" for Dachau.

And again, what happened on our arrival in that camp, after a journey lasting two days and three nights? There had not been enough room for everybody to crouch on the floor of the carriage at the same time. The majority of us had to stand all the way, while a few took turns at squatting on the scanty straw which was soaked with human urine. When we arrived the first important news that we heard from older prisoners was that this comparatively small camp (its population was 2,500) had no "oven," no crematorium, no gas! That meant that a person who had become a "Moslem" [no longer fit for work] could not be taken straight to the gas chamber, but would have to wait until a so-called "sick convoy" had been arranged to return to Auschwitz. This joyful surprise put us all in a good mood. The wish of the senior warden of our hut in Auschwitz had come true: we had come, as quickly as possible, to a camp which did not have a "chimney"—unlike Auschwitz. We laughed and cracked jokes in spite of, and during, all we had to go through in the next few hours.

When we new arrivals were counted, one of us was missing. So we had to wait outside in the rain and cold wind until the missing man was found. He was at last discovered in a hut, where he had fallen asleep from exhaustion. Then the roll call was turned into a punishment parade. All through the night and late into the next morning, we had to stand outside, frozen and soaked to the skin after the strain of our long journey. And yet we were all very pleased! There was no chimney in this camp and Auschwitz was a long way off.[51]

Nothing produces terror and fear like a concentration camp—unless the camp encountered is better than the camp expected. Our hopes, desires and wishes—which are always conditional—define the context within which the things and situations we encounter take on determinate significance; define even the context within which we understand "thing" or "situation." We presume that things have a more or less fixed meaning, because we share a more or less fixed "condition" with others—at least with those others who are familiar to us, who share our presumptions and worldviews. Those (culturally determined) things we take for granted—and which are, therefore, invisible—determine our affective responses to "environmental stimuli." We assume that such things are permanent attributes of the world; but they are not. Our situations—and, therefore, our "contexts of interpretation"—can change dramatically, at any moment. We are indeed fortunate (and, generally, oblivious of that fortune) when they do not.

It is not possible to finally determine how or whether something is meaningful by observing the *objective features* of that thing. Value is not invariant, in contrast to objective reality; furthermore, it is not possible to derive an *ought* from an *is* (this is the "naturalistic fallacy" of David Hume). It is possible, however, to determine the *conditional meaning* of something, by observing how behavior (one's own behavior, or someone else's) is conducted in the presence of that thing (or in its absence). "Things" (objects, processes) emerge—into subjective experience, at least—as a consequence of behaviors. Let us say, for the sake of example, that behavior "a" produces phenomenon "b" (always remembering that we are talking about behavior in a particular context). Behavior "a" consequently increases in frequency. It can be deduced, then, that phenomenon "b" is regarded as positive, by the agent under observation, in the particular "context" constituting the observed situation. If behavior "a" decreases in frequency, the opposite conclusion can be reasonably reached. The observed agent regards "b" as negative.

The behavioral psychologist B.F. Skinner originally defined a reinforcer as a stimulus which produced a change in the frequency of a given behavior.[52] He was loathe to become concerned with the internal or intrapsychic whys and wherefores of reinforcement, preferring instead to work by definition. If a stimulus increased the rate at which a given behavior was manifested, it was positive. If it decreased the rate of that behavior, it was negative. Of course, Skinner recognized that the valence of a given stimulus was context-dependent. An animal had to be "food-deprived" (in normal parlance, *hungry*) before food could serve as a positive reinforcer. And as the animal being fed became less food-deprived, the valence and potency of the reinforcer *food* decreased.

Skinner believed that discussions of an animal's (or a human's) internal state were unnecessary. If you knew an animal's reinforcement history, you could determine what "stimuli" were likely to have positive or negative valence. The fundamental problem with this argument is one of parsimony. It is impossible to know an animal's "reinforcement history"—particularly if that animal is as complex and long-lived as a human being. This is tantamount to saying, "you must know everything that has ever happened to that animal"; analogous to the old determinist claim that "if you knew the present position and momentum of every particle in the universe, you could determine all future positions and momenta." You can't know all present positions and momenta: the measurement problems are insurmountable, and the uncertainty principle makes it impossible anyway. Likewise, you don't have access to the "reinforcement history," and even if you did, measuring it would alter it. (I am not making an formal "uncertainty" claim for psychology; just drawing what I hope is a useful analogy).

Skinner addressed this problem by limiting his concern to experimental situations *so simple that only immediate reinforcement history played a context-determining role.* This "implicit" limit enabled him to sidestep the fundamental issue, and to make inappropriate generalizations. It didn't matter how a rat related to his mother six months earlier if you could make him "food-deprived" enough. The (short-term) fact of the food deprivation, for example, overrode individual rat differences—at least in the experimental condition under question—and could therefore usefully be ignored. Similarly, if you starve human beings, you can be reasonably sure that they will become concerned with food. However, even in this extreme case, you cannot predict how this concern will manifest itself, or what (ethical) considerations might play an intermediate, or even determining, role. Alexander Solzhenitsyn examined this very problem during the time he spent in the Soviet "Gulag Archipelago" (the Soviet prison camp system):

At the Samarka Camp in 1946 a group of intellectuals had reached the very brink of death: They were worn down by hunger, cold, and work beyond their powers. And they were even deprived of sleep. They had nowhere to lie down. Dugout barracks had not yet been built. Did they go and steal? Or squeal? Or whimper about their ruined lives? No! Foreseeing the approach of death in days rather than weeks, here is how they spent their last sleepless leisure, sitting up against the wall: Timofeyev-Ressovsky gathered them into a "seminar," and they hastened to share with one another what one of them knew and the others did not—they delivered their last lectures to each other. Father Savely—spoke of "unshameful death," a priest academician—about patristics, one of the Uniate fathers—about something in the area of dogmatics and canonical writings, an electrical engineer—on the principles of the energetics of the future, and a Leningrad economist—on how the effort to create principles of Soviet economics had failed for lack of new ideas. From one session to the next, participants were missing—they were already in the morgue.

That is the sort of person who can be interested in all this while already growing numb with approaching death—now that is an intellectual![53]

Past experience—learning—does not merely *condition*; rather, such experience determines the precise nature of the framework of reference or context that will be brought to bear on the analysis of a given situation. This cognitive frame of reference acts as the *intermediary* between past learning, present experience and future desire. This intermediary is a valid object of scientific exploration—a phenomenon as real as anything abstracted is real—and is far more parsimonious and accessible, as such a phenomenon, than the simple noninterpreted (and nonmeasurable, in any case) sum total of reinforcement history. Frameworks of reference, influenced in their structure by learning, specify the valence of ongoing experience; determine what might be regarded, in a given time and place, as good, bad or indifferent. Furthermore, inferences about the nature of the framework of reference governing the behavior of others (that is, looking at the world through the eyes of another) may produce results that are more useful, more broadly generalizable (as "insights" into the "personality" of another), and less demanding of cognitive resources than attempts to understand the details of a given reinforcement history.

Valence can be positive or negative, as the early behaviorists noted. Positive and negative are not opposite ends of a continuum, however—not in any straightforward way.[54] The two "states" appear orthogonal, although (perhaps) mutually inhibitory. Furthermore, positive and negative are not simple: each can be subdivided, in a more or less satisfactory manner, at least once. Positively valued things, for example, can be *satisfying* or *promising* (can serve as consummatory or incentive rewards, respectively[55]). Many satisfying things are consumable, in the literal sense, as outlined previously. Food, for example, is a consummatory reward to the hungry—which means that it is valued under such circumstances as a satisfaction. Likewise, water satisfies the individual deprived of liquid. Sexual contact is rewarding to the lustful, and warmth is desirable to those without shelter. Sometimes more complex stimuli are satisfying or rewarding as well. It all depends on what is presently desired, and how that desire plays itself out. A mild verbal reprimand might well foster feelings of relief in the individual who expects a severe physical beating—which is to say, technically, that the *absence of an expected punishment* can serve quite effectively as a reward (it is in fact the form of reward that the tyrant prefers). Regardless of their form, attained satisfactions produce satiation, calm and somnolent pleasure, and (temporary) cessation of the behaviors directed to that particular end—although behaviors that culminate in a satisfactory conclusion are more likely to be manifested, in the future, when "instinctive" or "voluntary" desire re-emerges.

Promises, which are also positive, might be regarded as more *abstractly* meaningful than satisfactions, as they indicate potential rather than actuality. Promises—cues of consummatory rewards or satisfactions—indicate the imminent attainment of something desired or potentially desirable. Their more abstract quality does not make them secondary or *necessarily learned*, however, as was once thought; our response to potential satisfaction is often as basic or primary as our response to satisfaction itself. Promises (cues of satisfaction) have been regarded, technically, as incentive rewards, because they induce *forward locomotion*—which is merely movement toward the place that the cue indicates satisfaction will occur.[56] Curiosity,[57] hope[58] and excited pleasure tend to accompany exposure to cues of reward (and are associated with subsequent forward locomotion).[59] Behaviors that produce promises—like those that result in satisfactions—also increase in frequency, over time.[60]

Negatively valued things—which have a structure that mirrors that of their positive counterparts—can either be *punishing* or *threatening*.[61] Punishments—a diverse group of stimuli or contexts, as defined immediately below—all appear to share one feature (at least from the perspective of the theory outlined in this manuscript): they indicate the temporary or final impossibility of implementing one or more means or attaining one or more ends. Some stimuli are almost universally experienced as punishing, because their appearance indicates reduced likelihood of carrying through virtually any imaginable plan—of obtaining almost every satisfaction, or potential desirable future. Most things or situations that produce bodily injury fall into this category. More generally, punishments might be conceived of as involuntary states of deprivation (of food, water, optimal temperature,[62] or social contact[63]); as disappointments[64] or frustrations[65] (which are *absences of expected rewards*[66]), and as stimuli sufficiently intense to produce damage to the systems encountering them. Punishments stop action, or induce retreat or escape (backward locomotion),[67] and engender the emotional state commonly known as *pain* or *hurt*. Behaviors, which culminate in punishment and subsequent hurt, tend to *extinguish*—to decrease in frequency, over time.[68]

Threats, which are also negative, indicate potential, like promises—but potential for punishment, for hurt, for pain. Threats—cues of punishment—are stimuli that indicate enhanced likelihood of punishment and hurt.[59] Threats are abstract, like promises; however, like promises, *they are not necessarily secondary or learned.*[70] Unexpected phenomena, for example—which constitute innately recognizable threats—stop us in our tracks, and make us feel *anxiety*.[71] So, arguably, do certain innate fear stimuli—like snakes.[72] Behaviors that culminate in the production of cues of punishment—that create situations characterized by anxiety—tend to decrease in frequency over time (much like those that produce immediate punishment).[73]

Satisfactions and their cues are *good*, simply put; punishments and threats are *bad*. We tend to move forward[74] (to feel hope, curiosity, joy) and then to consume (to make love, to eat, to drink) in the presence of good things; and to pause (and feel anxious), then withdraw, move backwards (and feel pain, disappointment, frustration, loneliness) when faced by things we do not like. In the most basic of situations—when we know what we are doing, when we are engaged with the familiar—these fundamental tendencies suffice. Our actual situations, however, are almost always more complex. If things or situations were straightforwardly or simply positive or negative, good or bad, we would not have to make judgments regarding them, would not have to think about our behavior, and how and when it should be modified—indeed, would not have to *think* at all. We are faced, however, with the constant problem of ambivalence in meaning, which is to say that a thing or situation might be bad and good simultaneously (or good in two conflicting manners; or bad, in two conflicting manners).[75] A cheesecake, for example, is *good* when considered from the perspective of food deprivation or hunger, but *bad* when considered from the perspective of social desirability and the svelte figure that such desirability demands. The newly toilet-trained little boy who

has just wet his bed might well simultaneously feel satisfaction at the attainment of a biologically vital goal and apprehension as to the likely socially constructed interpersonal consequence of that satisfaction. Nothing comes without a cost, and the cost has to be factored in,
when the meaning of something is evaluated. Meaning depends on context; contexts—stories, in a word—constitute goals, desires, wishes. It is unfortunate, from the perspective of
conflict-free adaptation, that we have many goals—many stories, many visions of the ideal
future—and that the pursuit of one often interferes with our chances (or someone else's
chances) of obtaining another.

We solve the problem of contradictory meanings by interpreting the value of things from
within the confines of our stories—which are adjustable maps of experience and potential,
whose specific contents are influenced by the demands of our physical being. Our central
nervous systems are made up of many "hard-wired" or automatized subsystems, responsible
for biological regulation—for maintaining homeostasis of temperature, ensuring proper
caloric intake, and monitoring levels of plasma carbon dioxide (for example). Each of these
subsystems has a job to do. If that job is not done within a certain variable span of time, the
whole game comes to a halt, perhaps permanently. Nothing gets accomplished then. We *must*
therefore perform certain actions if we are to survive. This does not mean, however, that our
behaviors are *determined*—at least not in any simplistic manner. The subsystems that make
up our shared structure—responsible, when operative, for our instincts (thirst, hunger, joy,
lust, anger, etc.)—do not appear to directly grip control of our behavior, do not transform us
into driven automatons. Rather, they appear to influence our fantasies, our plans, and alter
and modify the content and comparative importance of our goals, our ideal futures (conceived of in comparison to our "unbearable" presents, as they are currently construed).

Each basic subsystem has its own particular, singular image of what constitutes the
ideal—the most valid goal at any given moment. If someone has not eaten in several days,
his vision of the (immediately) desirable future will likely include the image of eating.
Likewise, if someone has been deprived of water, she is likely to make drinking her goal. We
share fundamental biological structure, as human beings, so we tend to agree, broadly, about
what should be regarded as valuable (at least in a specified context). What this means, essentially, is that we can make *probabilistic* estimates about those things that a given individual
(and a given culture) might regard as desirable, at any moment. Furthermore, we can
increase the accuracy of our estimates by programmed deprivation (because such deprivation
specifies interpretive context). Nonetheless, we can never be *sure*, in the complex normal
course of events, just what it is that someone will want.

Judgment regarding the significance of things or situations becomes increasingly complicated when the fulfillment of one biologically predicated goal interferes with the pursuit or
fulfillment of another.[76] To what end should we devote our actions, for example, when we
are simultaneously lustful and guilty, or cold, thirsty and frightened? What if the only way to
obtain food is to steal it, say, from someone equally hungry, weak and dependent? How is our
behavior guided when our desires compete—which is to say, when wanting one thing makes
us likely to lose another or several others? There is no reason to presume, after all, that each

of our particularly specialized subsystems will agree, at any one time, about what constitutes the most immediately desirable "good." This lack of easy agreement makes us *intrinsically prone to intrapsychic conflict* and associated affective (emotional) dysregulation. We manipulate our environments and our beliefs to address this conflict—we change ourselves, or the things around us, to increase our hope and satisfaction, and to decrease our fear and pain.

It is up to the "higher" cortical systems—the phylogenetically newer, more "advanced" executive[77] portions of the brain—to render judgment about the relative value of desired states (and, similarly, to determine the proper order for the manifestation of means[78]). These advanced systems must take all states of desire into account, optimally, and determine the appropriate path for the expression of that desire. We make *decisions* about what is to be regarded as valuable, at any given time, but the neurological subsystems that keep us alive, which are singularly responsible for our maintenance, in different aspects, all have a voice in those decisions—a vote. Every part of us, kingdom that we are, depends on the healthy operation of every other part. To ignore one good, therefore, is to risk all. To ignore the demands of one necessary subsystem is merely to ensure that it will speak later with the voice of the unjustly oppressed; is to ensure that it will grip our fantasy, unexpectedly, and make of the future something unpredictable. Our "optimal paths" therefore, must be properly *inclusive*, from the perspective of our internal community, our basic physiology. The valuations and actions of others, additionally, influence our personal states of emotion and motivation as we pursue our individual goals, inevitably, in a social context. The goal, *writ large*, toward which our higher systems work must therefore be construction of a state where all our needs and the needs of others are simultaneously met. This higher goal, to which we all theoretically aspire, is a complex (and oft-implicit) fantasy—a vision or map of the promised land. This map, this story—this *framework of reference* or *context of interpretation*—is the (ideal) future, contrasted necessarily with the (unbearable) present, and includes concrete plans, designed to turn the latter into the former. The mutable meanings that make up our lives depend for their nature on the explicit structure of this interpretive context.

We select what we *should* value from among those things we *must* value. Our selections are therefore predictable, in the broad sense. This must be, as we must perform certain actions in order to live. But the predictability is limited. The world is complex enough not only so that a given problem may have many valid solutions, but so that even the definition of "solution" may vary. The particular most appropriate or likely choices of people, including ourselves, cannot (under normal circumstances, at least) be accurately determined beforehand. Nonetheless, despite our final and ineradicable ignorance, we act—judging from moment to moment what is to be deemed worthy of pursuit, determining what can be ignored, at least temporarily, during that pursuit. We are capable of acting and of producing the results we desire because we render *judgment of value*, using every bit of information at our disposal. We determine that something is worth having, at a given time and place, and make the possession of that thing our goal. And as soon as something has become our goal—no matter what that something is—*it appears to adopt the significance of satisfaction (of consummatory reward)*. It appears sufficient for something to be truly *regarded* as valuable for it to adopt the emotional aspect of value. It is in this manner that our higher-order verbal-cognitive systems serve to regulate our emotions. It is for this reason that we can play or work toward "merely symbolic" ends, for this reason that drama and literature[79] (and even sporting events) can have such profound vicarious effects on us. The mere fact that something is desired, however, does not necessarily mean that its attainment will sustain life (as a "true" satisfaction might)—or that pure regard will make something into what it is not. It is therefore necessary (if you wish to exist, that is) to construct goals—models of the desired future—that are *reasonable* from the perspective of previous experience, grounded in biological necessity. Such goals take into account the necessity of coping with our intrinsic limitations; of satisfying our inherited biological subsystems; of appeasing those transpersonal "gods," who eternally demand to be clothed and fed.

The fact that goals *should* be reasonable does not mean that they have to be or will be (at least in the short term)—or that what constitutes "reasonable" can be easily or finally determined. One man's meat is another man's poison; the contents of the ideal future (and the interpreted present) may and do vary dramatically between individuals. An anorexic, for example, makes her goal an emaciation of figure that may well be incompatible with life. In consequence, she regards food as something to be avoided—as something punishing or threatening. This belief will not protect her from starving, although it will powerfully affect her short-term determination of the valence of chocolate. The man obsessed with power may sacrifice everything—including his family—to the attainment of his narrow ambition. The empathic consideration of others, a time-consuming business, merely impedes his progress with regard to those things he deems of ultimate value. His faith in the value of his progress therefore makes threat and frustration even of love. Our beliefs, in short, can change our reactions to everything—even to those things as primary or fundamental as food and family. We remain indeterminately constrained, however, by the fact of our biological limits.

It is particularly difficult to specify the value of an occurrence when it has one meaning, from one frame of reference (with regard to one particular goal), and a different or even opposite meaning, from another equally or more important and relevant frame. Stimuli that exist in this manner constitute *unsolved problems of adaptation*. They present us with a mystery, which is what to do in their presence (whether to pause, consume, stop, or move backwards or forwards, at the most basic of levels; whether to feel anxious, satisfied, hurt or hopeful). Some things or situations may be evidently satisfying or punishing, at least from the currently extant "framework of reference," and can therefore be regarded (valued, acted toward) in an uncomplicated manner. Other things and situations, however, remain rife with contradictory or indeterminate meanings. (Many things, for example, are satisfying or promising in the short term but punishing in the medium to long term.) Such circumstances provide evidence that our systems of valuation are not yet sophisticated enough to foster complete adaptation—demonstrate to us incontrovertibly that our processes of evaluation are still incomplete:

A brain in a vat is at the wheel of a runaway trolley, approaching a fork in the track. The brain is hooked up to the trolley in such a way that the brain can determine which course the trolley will take. There are only two options: the right side of the fork, or the left side. There is no way to derail or stop the trolley, and the brain is aware of this. On the right side of the track there is a single railroad worker, Jones, who will definitely be killed if the brain steers the trolley to the right. If Jones lives he will go on to kill five men for the sake of thirty orphans (one of the five men he will kill is planning to destroy a bridge that the orphans' bus will be crossing later that night). One of the orphans who will be killed would have grown up to become a tyrant who made good, utilitarian men do bad things, another would have become John Sununu, a third would have invented the pop-top can.

If the brain in the vat chooses the left side of the track, the trolley will definitely hit and kill another railman, Leftie, and will hit and destroy ten beating hearts on the track that would have been transplanted into ten patients at the local hospital who will die without donor hearts. These are the only hearts available, and the brain is aware of this. If the railman on the left side of the track lives, he, too, will kill five men—in fact, the same five that the railman on the right would kill. However, Leftie will kill the five as an unintended consequence of saving ten men: he will inadvertently kill the five men as he rushes the ten hearts to the local hospital for transplantation. A further result of Leftie's act is that the busload of orphans will be spared. Among the five men killed by Leftie is the man responsible for putting the brain at the controls of the trolley. If the ten hearts and Leftie are killed by the trolley, the ten prospective heart-transplant patients will die and their kidneys will be used to save the lives of twenty kidney transplant patients, one of whom will grow up to cure cancer and one of whom will grow up to be Hitler. There are other kidneys and dialysis machines available, but the brain does not know this.

Assume that the brain's choice, whatever it turns out to be, will serve as an example to other brains in vats, and thus the effects of its decision will be amplified. Also assume that if the brain chooses the right side of the fork, an unjust war free of war crimes will ensue, whereas if the brain chooses the left fork, a just war fraught with war crimes will result. Furthermore, there is an intermittently active Cartesian demon deceiving the brain in such a way that the brain is never sure that it is being deceived.

Question: Ethically speaking, what should the brain do?[80]

We cannot act in two ways at one time—cannot move forwards and backwards, cannot stop and go, simultaneously. When faced with stimuli whose meaning is indeterminate we are therefore placed in conflict. Such conflict must be resolved, before adaptive action may take place. We can actually only do one thing, at one time, although we may be motivated by confusing, threatening, dangerous or unpredictable circumstances to attempt many incommensurate things simultaneously.

Unexplored Territory: Phenomenology and Neuropsychology

The dilemma of contradictory simultaneous meanings can be *solved* in only two related ways (although it can be avoided in many others). We can *alter our behaviors*, in the difficult situation, so that those behaviors no longer produce consequences we do not desire or cannot interpret. Alternatively, we can *reframe our contexts of evaluation* (our goals and our interpretations of the present), so that they no longer produce paradoxical implications, with regard to the significance of a given situation. These processes of behavioral modification and reframing constitute *acts of effortful revaluation*, which means thorough, exploratory reconsideration of what has been judged previously to be appropriate or important.

Things or situations with indeterminate meanings therefore challenge our adaptive competence; force us to revaluate our present circumstances and alter our ongoing behaviors. Such circumstances arise when something we have under control, from one perspective, is troublesome or otherwise out of control from another. Out of control means, most basically, unpredictable: something is beyond us when our interactions with it produce phenomena whose properties could not be determined, beforehand. Unexpected or novel occurrences, which emerge when our plans do not turn out the way we hope they would, therefore constitute an important—perhaps the most important—subset of the broader class of stimuli of indeterminate meaning. Something unexpected, or novel, necessarily occurs in relationship to what is known; is always identified and evaluated with respect to our currently operative plan [which is to say that a familiar thing in an unexpected place (or at an unexpected time) is actually something *unfamiliar*]. The wife of an adulterous husband, for example, is well-known to him, perhaps, when she is at home. The fact of her, and her behavior, constitutes *explored territory*. She is an entirely different sort of phenomenon, however, from the perspective of affect (and implication for behavioral output) if she makes an unexpected appearance at his favorite motel room, in the midst of a tryst. What will the husband do, in his wife's presence, when she surprises him? First, he will be taken aback, in all likelihood—then he will concoct a story that makes sense of his behavior (if he can manage it, on such short notice). He has to think up something new, *do something he has never done before.* He has to manage his wife, who he thinks he has fooled—his wife, whose mere unexpected presence at the motel is proof of her endless residual mystery. Our habitual patterns of action only suffice for things and situations of determinate significance—by definition: we only know how to act in the presence of the familiar. The appearance of the unexpected pops us out of unconscious, axiomatic complacency and forces us (painfully) to *think*.

The implications of novel or unpredictable occurrences are unknown, by definition. This observation carries within it the seeds of a difficult and useful question: what is the significance of the unknown? It might seem logical to assume that the answer is none—something unexplored cannot have meaning, because none has yet been attributed to it. The truth, however, is precisely opposite. Those things we do not understand nonetheless signify. If you can't tell what something means, because you don't know what it is, what then does it mean? It is not nothing—we are in fact frequently and predictably upset by the unexpected. Rather, it could be anything, and that is precisely the crux of the problem. Unpredictable things are not irrelevant, prior to the determination of their specific meaning. Things we have not yet explored have significance, prior to our adaptation to them, prior to our classification of their relevance, prior to our determination of their implication for behavior. Things not predicted, not desired, that occur while we are carrying out our carefully designed plans—such things come loaded, *a priori*, with meaning, both positive and negative. The appearance of unexpected things or situations indicates, at least, that our plans are in error, at some stage of their design—in some trivial way, if we are lucky; in some manner that might be devastating to our hopes and wishes, to our self-regard, if we are not.

Unexpected or unpredictable things—novel things, more exactly (the class of novel things, most particularly)—have a potentially infinite, unbounded range of significance. What does something that might be anything mean? In the extremes, it means, *the worst that could be* (or, at least, the worst you can imagine) and, conversely, *the best that could be* (or the best you can conceive of). Something new might present the possibility for unbearable suffering, followed by meaningless death—a threat virtually unbounded in significance. That new and apparently minor but nonetheless strange and worrisome ache you noticed this morning, for example, while you were exercising, might just signify the onset of the cancer that will slowly and painfully kill you. Alternatively, something unexpected might signify inconceivable opportunity for expansion of general competence and well-being. Your old, boring but secure job unexpectedly disappears. A year later, you are doing what you really want to do, and your life is incomparably better.

An unexpected thing or situation appearing in the course of goal-directed behavior constitutes a stimulus that is intrinsically problematic: novel occurrences are, simultaneously, cues for punishment (threats) and cues for satisfaction (promises).[81] This paradoxical *a priori* status is represented schematically in *Figure 5: The Ambivalent Nature of Novelty*. Unpredictable things, which have a paradoxical character, accordingly activate two antithetical emotional systems, whose mutually inhibitory activities provide basic motivation for abstract cognition, whose cooperative endeavor is critical to the establishment of permanent memory, and whose physical substrates constitute universal elements of the human nervous system. The most rapidly activated[82] of these two systems governs inhibition of ongoing behavior, cessation of currently goal-directed activity;[83] the second, equally powerful but somewhat more conservative,[84] underlies exploration, general behavioral activation[85] and forward locomotion.[86] Operation of the former appears associated with anxiety, with fear and apprehension, with negative affect—universal subjective reactions to the threatening and unexpected.[87] Operation of the latter, by contrast, appears associated with hope, with curiosity and interest, with positive affect—subjective responses to the promising and unexpected.[88] The process of exploring the emergent unknown is therefore guided by the interplay between the emotions of curiosity/hope/excitement on the one hand and anxiety on the other—or, to describe the phenomena from another viewpoint, between the different motor systems responsible for approach (forward locomotion) and inhibition of ongoing behavior.

The "ambivalent unknown" comes in two "forms," so to speak (as alluded to earlier). "Normal" novelty emerges within the "territory" circumscribed by the choice of a particular end-point or goal (which is to say, after getting to specific point "b" has been deemed the most important possible activity *at this time and in this place*). Something "normally" novel constitutes an occurrence which leaves the current departure point and goal intact, but

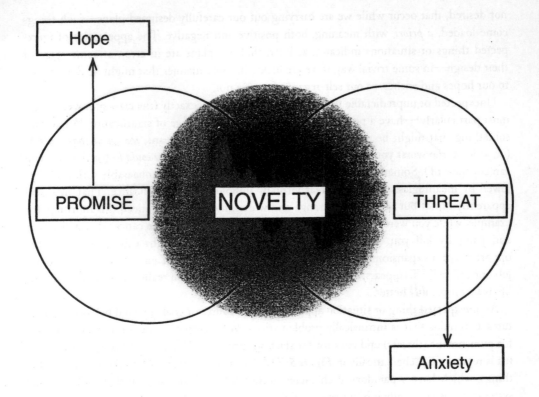

Figure 5: The Ambivalent Nature of Novelty

indicates that the *means* of achieving that goal have to be modified. Let us say, for example, that you are in your office. You are accustomed to walking down an unobstructed hallway to get to the elevator. You are so used to performing this activity that you can do it "automatically"—so you often read while walking. One day, while reading, you stumble over a chair someone left in the middle of the hallway. This is normal novelty. You don't have to alter your current goal, except in a temporary and trivial manner; you are not likely to get too upset by the unexpected obstacle. Getting to the elevator is still a real possibility, even within the desired time frame; all you have to do is walk around the chair (or move it somewhere else, if you are feeling particularly altruistic). *Figure 6: Emergence of "Normal Novelty" in the Course of Goal-Directed Behavior* provides an abstracted representation of this process of trivial adaptation.

Revolutionary novelty is something altogether different. Sometimes the sudden appearance of the unexpected means taking path "b" to grandma's house, instead of path "a." Sometimes that appearance means emergent doubt about the very existence of grandma (think "wolf" and "Red Riding Hood"). Here is an example: I am sitting alone in my office,

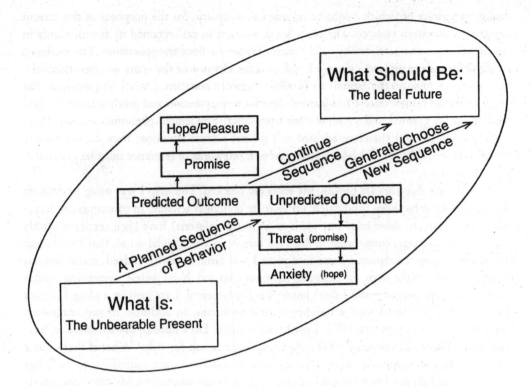

Figure 6: Emergence of "Normal Novelty" in the Course of Goal-Directed Behavior

in a high-rise building, alone at night. I suddenly fantasize: "I am going to take the elevator down three floors and get something to eat" (more accurately, hunger suddenly grips my imagination, and uses it for its own purposes). This fantasy constitutes a spatially and temporally bounded image of the ideal future—an "actual" possible future, carved out as a discriminable (and thus usable) object, from the infinite domain of potential possible futures. I use this definite image to evaluate the events and processes that constitute the interpreted present, as it unfolds around me, as I walk toward the elevator (on my way to the cafeteria). I *want* to make reality match my fantasy—to subdue my motivation (to please the gods, so to speak). If the unexpected occurs—say, the elevator is not operating—the mismatch temporarily stops me. I replace my current plan with an alternative behavioral strategy, designed to obtain the same end. This means that I do not reconfigure the temporally and spatially bounded map that I am using to evaluate my circumstances—that I am using to regulate my emotions. All I have to do is change *strategy*.

I decide to take the stairs to the cafeteria. If the stairs are blocked by construction, I am in more serious trouble. My original fantasy—"go down to the cafeteria and eat"—was predicated on an implicit presumption: *I can get downstairs.* This presumption, which I wasn't really even aware of (which might be regarded as *axiomatic*, for the purposes of the current operation), has been violated. The story "go downstairs to eat" retained its function only in an environment characterized by valid means of between-floor transportation. The existence of these means constituted a given—I had used the elevator or the stairs so often that their very presence took on the aspect of a justifiably ignored constant. Once I had mastered the stairs or the elevator—once I had learned their location, position and mechanisms—I could take them for granted and presume their irrelevance. Predictable phenomena (read "thoroughly explored, and therefore adapted to") do not attract attention; they do not require "consciousness." No new behavioral strategies or frameworks of reference must be generated, in their presence.

Anyway: the elevators are broken; the stairs are blocked. The map I was using to evaluate my environment has been invalidated: my *ends* are no longer tenable. In consequence, necessarily, the means to those ends (my plans to go to the cafeteria) have been rendered utterly irrelevant. I no longer know what to do. This means, in a nontrivial sense, that I no longer know *where I am*. I presumed I was in a place I was familiar with—indeed, many familiar things (the fact of the floor, for example) have not changed. Nonetheless, something fundamental has been altered—and I don't know *how* fundamental. I am now in a place I cannot easily leave. I am faced with a number of new problems, in addition to my unresolved hunger—at least in potential (Will I get home tonight? Do I have to get someone to "rescue" me? Who *could* rescue me? Who do I telephone to ask for help? What if there was a fire?). My old plan, my old "story" ("I am going downstairs to get something to eat") has vanished, and I do not know how to evaluate my current circumstances. My emotions, previously constrained by the existence of a temporarily valid plan, re-emerge in a confused jumble. I am anxious ("what will I do? What if there *was* a fire?"), frustrated ("I'm certainly not going to get any more work done tonight, under these conditions!") angry ("who could have been stupid enough to block all the exits?"), and curious ("just what the hell is going on around here, anyway?"). Something unknown has occurred and blown all my plans. An emissary of chaos, to speak metaphorically, has disrupted my emotional stability. *Figure 7: Emergence of "Revolutionary Novelty" in the Course of Goal-Directed Behavior* graphically presents this state of affairs.

The plans we formulate are mechanisms designed to bring the envisioned perfect future into being. Once formulated, plans govern our behavior—*until we make a mistake*. A mistake, which is the appearance of a thing or situation not envisioned, provides evidence for the incomplete nature of our plans—indicates that those plans and the presumptions upon which they are erected are in error and must be updated (or, heaven forbid, abandoned). As long as everything is proceeding according to plan, we remain on familiar ground—but when we err, *we enter unexplored territory*.

What is known and what unknown is always relative because what is unexpected depends entirely upon what we expect (desire)—on what we had previously *planned and presumed*. The unexpected constantly occurs because it is impossible, in the absence of omniscience, to formulate an entirely accurate model of what actually is happening or of what should hap-

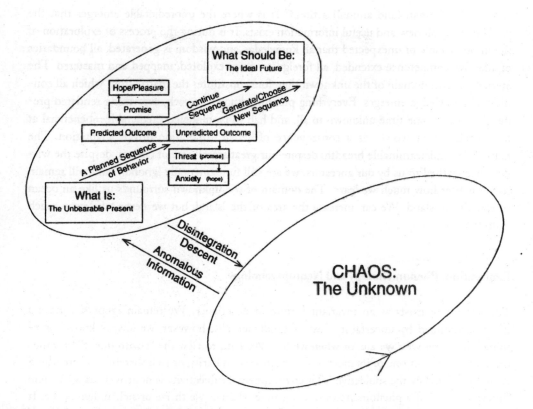

Figure 7: Emergence of "Revolutionary Novelty" in the Course of Goal-Directed Behavior

pen; it is impossible to determine what results ongoing behavior will finally produce. Errors in representation of the unbearable present and the ideal, desired future are inevitable, in consequence, as are errors in implementation *and* representation of the means by which the former can be transformed into the latter. The infinite human capacity for error means that encounter with the unknown is inevitable, in the course of human experience; means that the likelihood of such encounter is as certain, regardless of place and time of individual existence, as death and taxation. The (variable) existence of the unknown, paradoxically enough, *can therefore be regarded as an environmental constant*. Adaptation to the "existence" of this domain must occur, therefore, in every culture, and in every historical period—regardless of the particulars of any given social or biological circumstance.

Deviations from desired outcome constitute (relatively) novel events, indicative of errors in presumption, either at the level of analysis of current state, process or ideal future. Such mismatches—unpredictable, nonredundant or novel occurrences—constantly comprise the most intrinsically meaningful, interesting elements of the human experiential field. This interest and meaning signifies the presence of new information and constitutes a prepotent stimulus for human (and animal) action.[89] It is where the unpredictable emerges that the possibility for all new and useful information exists. It is during the process of exploration of the unpredictable or unexpected that all knowledge and wisdom is generated, all boundaries of adaptive competence extended, all foreign territory explored, mapped and mastered. The eternally extant domain of the unknown therefore constitutes the matrix from which all conditional knowledge emerges. Everything presently known to each, everything rendered predictable, was at one time unknown to all, and had to be rendered predictable—beneficial at best, irrelevant at worst—as a consequence of active exploration-driven adaptation. The matrix is of indeterminable breadth: despite our great storehouse of culture, despite the wisdom bequeathed to us by our ancestors, we are still fundamentally ignorant, and will remain so, no matter how much we learn. The domain of the unknown surrounds us like an ocean surrounds an island. We can increase the area of the island, but we never take away much from the sea.

Exploration: Phenomenology and Neuropsychology

The unfamiliar exists as an invariant feature of experience. We remain ignorant, and act while surrounded by uncertainty. Just as fundamentally, however, we always know something, no matter who we are, or when we live. We tend to view the "environment" as something objective, but one of its most basic features—familiarity, or lack thereof—is something virtually defined by the subjective. This environmental subjectivity is nontrivial, as well: mere "interpretation" of a phenomenon can determine whether we thrive or sicken, live or die. It appears, indeed, that the categorization or characterization of the environment as unknown/known (nature/culture, foreign/familiar) might be regarded as more "fundamental" than any objective characterization—if we make the presumption that what we have adapted to is, by definition, reality. For it is the case that the human brain—and the brain of higher animals—has specialized for operation in the "domain of order" and the "domain of chaos." And it is impossible to understand the fact of this specialization, unless those domains are regarded as more than mere metaphor.

We normally use our conceptions of cognitive processes to illuminate the working of the brain (we use our models of thought to determine "what must be the case" physiologically). However, neuropsychological investigation has advanced to the point where the reverse procedure is equally useful. What is known about brain function can illuminate our conceptions of cognition (indeed, of "reality" itself) and can provide those conceptions with suitable "objective constraints." Enlightenment thought strove to separate "reason" and "emotion"; empirical investigations into the structure and function of the brain—given great initial impetus by the consequences of that separation—have demonstrated instead that the two realms are mutually interdependent, and essentially integral.[90] We live in a universe characterized by the constant interplay of *yang* and *yin*, chaos and order: emotion provides us with an initial guide when we don't know what we are doing, when reason alone will not suffice.[91]

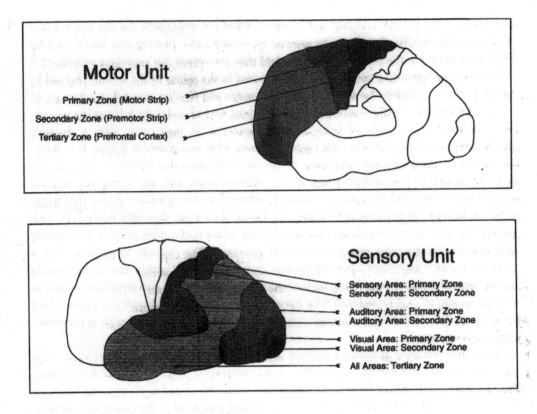

Figure 8: The Motor and Sensory Units of the Brain

"Cognition," by contrast, allows us to construct and maintain our ordered environments, and keep chaos—and affect—in check.

The brain may be usefully regarded as composed of three primary units—motor, sensory and affective—or as constituting a matched pair of hemispheres, right and left. Each manner of conceptual subdivision has its theoretical advantages; furthermore, the two are not mutually exclusive. We will attend to the description of the units, portrayed schematically in *Figure 8: The Motor and Sensory Units of the Brain*, first.

Most neocortical (and many subcortical) structures have attained their largest and most complex level of development in *homo sapiens*. This is true, in particular, of the motor unit,[92] which comprises the anterior or forward half of the comparatively newer neocortex (and which is composed of the motor, premotor and prefrontal lobes). This level of heightened development accounts in part for increased human intelligence, behavioral versatility and breadth of experience, both actual and potential, and underlies our capacity to formulate plans and intentions, organize them into programs of action and regulate their execution.[93]

The sensory unit,[94] which comprises the posterior half of the neocortex (and which is composed of the parietal, occipital and temporal lobes), is responsible for the construction of the separate worlds of our sensory systems (primarily sight, hearing and touch) and for their integration into the unified perceptual field that constitutes our conscious experience.[95] The sensory unit processes the information generated in the course of the actions planned by the motor unit, and builds the world of the recognizable and familiar out of that information.

The "limbic unit," finally—phylogenetically ancient, tucked under the folds of the neocortex—compares[96] the nature of behavioral consequences, as they occur, with a dynamic model, extant in fantasy, of what was *supposed to occur*, what was *desired to happen*. It is therefore signaling of *motivational significance*, or *affective importance*, that constitutes what is perhaps the major responsibility of the limbic system—that, and the (integrally related) inculcation and renewal of memory ("integrally related," as it is *significant* events that transform knowledge—that are stored in memory [more accurately, that *alter* memory]). This process of signaling necessarily involves comparison of the undesirable present, as currently understood, with the ideal future, as currently imagined. The capacity to generate such a contrast appears dependent upon operations undertaken deep within the comparatively ancient central portion of the brain, particularly in the tightly integrated structures known as the hippocampus[97] and amygdala.[98] The nature of this comparative process can perhaps best be understood, in introduction, through consideration of a phenomenon known as the *event-related cortical potential*.

The brain constantly produces a shifting pattern of electrical activity in the course of its operations. The electroencephalogram (the EEG) provides a rough picture of that pattern. The individual undergoing EEG examination has electrodes placed in an array on his scalp. These electrodes allow the patterns of electrical activity, generated in the course of neurological activity, to be detected, monitored and, to some degree, localized. (The brain produces enough electrical activity to be detected through the skull and tissue surrounding it, although the interference produced by that surrounding tissue makes evaluation of the EEG difficult). The rather limited capacities of EEG technology have been greatly extended by the analytic capacities of the computer. The *cortical event-related potential* is a measure of brain activity derived by computer from EEG recordings averaged at different delays after the subject being evaluated has been presented with some sort of stimulus. The nature of this stimulus may vary. In the simplest case, it is merely something sensory, like a tone presented repeatedly through stereo headphones. In more complex cases the event-related potential is monitored following presentation of a stimulus with affective valence—which means following something that must be "discriminated, recognized, or otherwise evaluated."[99] Perhaps the simplest way to produce an event of this sort is to randomly and rarely insert a tone that differs in frequency into a repetitious sequence of otherwise predictable tones (although the stimulus might just as easily be visual or tactile). These oddball events are characterized by (relative) novelty (novelty is always relative) and evoke a pattern of cortical electrical activity that differs from that produced by the predictable tones. Any event that has specific or known implications for alteration in ongoing behavior will also produce a potential like the oddball.

The average cortical event-related potential produced by infrequent or otherwise mean-ingful events is a waveform with a characteristic time course and shape. Most attention has been paid to elements of this waveform that occur within the first half-second (500 millisec-onds) after stimulus occurrence. As the first half-second passes, the polarity of the waveform shifts. Peaks and valleys occur at different, more or less standard times (and in essentially predictable "locations") and have therefore been identified and named. Event-related poten-tials (ERPs) are negative (N) or positive (P) depending on polarity and are numbered according to their occurrence in time. The earliest aspects of the ERP (<200 msec) vary with change in the purely sensory quality of an event. The waveforms named N200 (negative 200 msec) and P300 (positive 300 msec), by contrast, vary with the *affective significance and mag-nitude* of the stimulus, and can even be evoked by the absence of an event that was expected, but that did not appear. The psychophysiologist Eric Halgren states:

One may summarize the cognitive conditions that evoke the N2/P3 as being the presentation of stim-uli that are novel or that are signals for behavioral tasks, and thus need to be attended to, and processed. These evoking conditions and functional consequences are identical to those that have been found for the orienting reflex.[100]

Halgren considers the N2/P3 and the autonomic orienting reflex "different parts of an over-all organismic reaction complex evoked by stimuli that merit further evaluation."[101] He terms this overall response pattern the orienting *complex*. A substantial body of evidence sug-gests that the amygdalic and hippocampal systems are critically involved in production of the N2/P3 waveforms, although other brain systems also participate. (It is also of great interest to note that an additional waveform, the N4, is produced when human experimental subjects are exposed to abstracted symbols with integral significance, such as written, spoken or signed words and faces, in a meaningful context.[102] In such a context, the N4 occurs after the N2 but before the P3, and increases in magnitude as a function of the difficulty of integrat-ing the word with the context in which it appears. The amygdala and hippocampus are also directly responsible for the production of this waveform—and, therefore, for contextual syn-thesis, which is a vital aspect of the derivation of meaning, which is significance for behavior, given the desire to attain a particular goal.)

The processes that reveal themselves behaviorally in the orienting complex and electro-physiologically in the N2/N4/P3 waveform appear to play a central part in the manifold processes we experience (and understand) as *consciousness*. Another psychophysiologist, Arne Ohman,[103] has posited that orienting initiates a sequence of "controlled processing," which is difficult, slow, accompanied by awareness, sequential and generative (and which is referred to as *exploratory behavior* in this document), contrasted with "automatic processing," which is habitual, "unconscious" and immediate (and which occurs in "explored territory"). The orienting complex is apparently manifested only when a given experimental subject becomes aware of some relationship between sensory input and motor action. Likewise, the N2/P3 waveform appears only when the experimental stimulus utilized "has captured the subject's attention and reached his or her awareness."[104] Consciousness (affiliated tightly with orienting, for the purposes of the present argument) therefore appears as a phenomenon critically involved in and vital to the evaluation of novelty—appears vital to placement of the unpredictable into a defined and determinate context as a consequence of behavioral modification undertaken in the territory of the unknown. This means that consciousness plays a centrally important role in the *generation of the predictable and comprehended world from the domain of the unexpected*. Such response, placement and generation remains forever mediated by the twin forces of hope/curiosity and anxiety—forces produced, noncoincidentally, by the same structures that govern "reflexive" orientation and exploratory motor output.

The constant and universal presence of the incomprehensible in the world has elicited adaptive response from us and from all other creatures with highly developed nervous systems. We have evolved to operate successfully in a world eternally composed of the predictable, in paradoxical juxtaposition with the unpredictable. The combination of *what we have explored and what we have still to evaluate* actually comprises our environment, insofar as its nature can be broadly specified—and it is to that environment that our physiological structure has become matched. One set of the systems that comprise our brain and mind governs activity, when we are guided by our plans—when we are in the domain of the *known*. Another appears to operate when we face something unexpected—when we have entered the realm of the *unknown*.[105]

The "limbic unit" generates the orienting reflex, among its other tasks. It is the orienting reflex, which manifests itself in emotion, thought and behavior, that is at the core of the fundamental human response to the novel or unknown. This reflex takes a biologically determined course, ancient in nature, primordial as hunger or thirst, basic as sexuality, extant similarly in the animal kingdom, far down the chain of organic being. The orienting reflex is the general instinctual reaction to the category of all occurrences which have not yet been categorized—is response to the unexpected, novel or unknown *per se*, and not to any discriminated aspect of experience, any specifically definable situation or thing. The orienting reflex is at the core of the process that generates (conditional) knowledge of sensory phenomena *and* motivational relevance or valence. Such knowledge is most fundamentally how to behave, and what to expect as a consequence, in a particular situation, defined by culturally modified external environmental circumstance and equally modified internal motivational state. It is also information about what *is*, from the objective perspective—is the record of that sensory experience occurring in the course of ongoing behavior.

The orienting reflex substitutes for particular learned responses when the incomprehensible suddenly makes its appearance. The occurrence of the unpredictable, the unknown, the source of fear and hope, creates a seizure of ongoing specifically goal-directed behavior. Emergence of the unexpected constitutes evidence for the incomplete nature of the story currently guiding such behavior; comprises evidence for error at the level of working description of current state, representation of desired future state or conception of the means to transform the former into the latter. Appearance of the unknown motivates curious, hopeful exploratory behavior, regulated by fear, as means to update the memory-predicated working model of reality (to update the *known*, so to speak, which is defined or familiar territory). The simultaneous production of two antithetical emotional states, such as hope and fear, means conflict, and the unexpected produces intrapsychic conflict like nothing else. The magnitude and potential intensity of this conflict cannot be appreciated under normal circumstances, because under normal circumstances—in defined territory—things are going according to plan. It is only when our goals have been destroyed that the true significance of the decontextualized object or experience is revealed—and such revelation makes itself known first in the form of fear.[106] We are protected from such conflict—from subjugation to instinctive terror—by the historical compilation of adaptive information generated in the course of previous novelty-driven exploration. We are protected from unpredictability by our culturally determined beliefs, by the stories we share. These stories tell us how to presume and how to act to maintain the determinate, shared and restricted values that compose our familiar worlds.

The orienting reflex—the involuntary gravitation of attention to novelty—lays the groundwork for the emergence of (voluntarily controlled) exploratory behavior.[107] Exploratory behavior allows for classification of the general and (*a priori*) motivationally significant unexpected into specified and determinate domains of motivational relevance. In the case of something with actual (post-investigation) significance, relevance means context-specific punishment or satisfaction, or their putatively "second-order" equivalents: threat or promise (as something threatening implies punishment, as something promising implies satisfaction). This is categorization, it should be noted, in accordance with implication for motor output, or behavior, rather than with regard to sensory (or, formalized, objective) property.[108] We have generally presumed that the purpose of exploration is production of a picture of the objective qualities of the territory explored. This is evidently—but only partially—true. However, the reasons we produce such pictures (are motivated to produce such pictures) are not usually given sufficient consideration. Every explorable subterritory, so to speak, has its sensory aspect, but it is the emotional or motivational relevance of the new domain that is truly important. We need to know only that something is hard and glowing red as a means of keeping track of the fact that it is hot, and therefore dangerous—that it is punishing, if contacted. We need to know the feel and look of objects so that we can keep track of what can be eaten and what might eat us.

When we explore a new domain, we are mapping the motivational or affective significance of the things or situations that are characteristic of our goal-directed interactions within that domain, and we use the sensory information we encounter to identify what is important. It is the determination of *specific* meaning, or emotional significance, in previously unexplored territory—not identification of the objective features—that allows us to inhibit the novelty-induced terror and curiosity emergence of that territory otherwise automatically elicits. We feel comfortable somewhere new, once we have discovered that nothing exists there that will threaten or hurt us (more particularly, when we have adjusted our behavior and schemas of representation so that nothing there is likely to or able to threaten or hurt us). The consequence of exploration that allows for emotional regulation (that generates security, essentially) is not objective description, as the scientist might have it, but categorization of the implications of an unexpected occurrence for specification of means and ends. Such categorization is what an object "is," from the perspective of archaic affect and subjective experience. The orienting reflex, and the exploratory behavior following its manifestation, also allows for the differentiation of the unknown into the familiar categories of objective reality. However, this ability is a late development, emerging only four hundred years ago,[109] and cannot be considered basic to "thinking." Specification of the collectively apprehensible sensory qualities of something—generally considered, in the modern world, as the essential aspect of the description of reality—merely serves as an aid to the more fundamental process of *evaluation*, determining the precise nature of relevant or potentially relevant phenomena.

When things are going according to plan—that is, when our actions fulfill our desires—we feel secure, even happy. When nothing is going wrong, the cortical systems expressly responsible for the organization and implementation of goal-directed behavior remain firmly in control. When cortically generated plans and fantasies go up in smoke, however, this control vanishes. The comparatively ancient "limbic" hippocampal and amygdalic systems leap into action, modifying affect, interpretation and behavior. The hippocampus appears particularly specialized for comparing the (interpreted) reality of the present, as it manifests itself in the subjective sphere, with the fantasies of the ideal future constructed by the motor unit (acting in turn as the higher-order mediator—the king, so to speak—of all the specialized subsystems that compose the more fundamental or primary components of the brain). These desire-driven fantasies might be regarded as motivated hypotheses about the relative likelihood of events produced in the course of ongoing goal-directed activity. What you *expect* to happen—really, what you *want* to happen, at least in most situations—is a model you generate, using what you already know, in combination with what you are learning while you act. The hippocampal comparator[110] constantly and "unconsciously" checks what is "actually" happening against what is supposed to happen. This means that the comparator contrasts the "unbearable present," *insofar as it is comprehended* (because it is a model, too), against the ideal future, as it is imagined; means that it compares the interpreted outcome of active behavior with an image of the intended consequences of that behavior. Past experience—skill and representation of the outcome of skill (or memory, as it is applied)—governs behavior, until error is committed. When something occurs that is not intended—when the actual outcome, as interpreted, does not match the desired outcome, as posited—the hippocampus shifts mode and prepares to update cortical memory storage. Behavioral control shifts from the cortex to the limbic system—apparently, to the amygdala, which governs the provisional determination of the affective significance of unpredictable events, and has powerful output to centers of motor control.[111] This shift of control allows the activation of structures governing orienting, heightened intensity of sensory processing and exploration.

The "higher" cortex controls behavior until the unknown emerges—until it makes a mistake in judgment, until memory no longer serves—until the activity it governs produces a mismatch between what is desired and what actually occurs. When such a mismatch occurs, appropriate affect (fear and curiosity) emerges. But how can situation-relevant emotion attach itself to what has by definition not yet been encountered? Traditionally, significance is attached to previously irrelevant things or situations as a consequence of learning, which is to say that things mean *nothing* until their meaning is learned. No learning has taken place, however, in the face of the unknown—yet emotion reveals itself, in the presence of error. It appears, therefore, that the kind of emotion that the unpredictable arouses is not learned—which is to say that the novel or unexpected comes preloaded with affect. Things are not irrelevant, as a matter of course. They are *rendered* irrelevant, as a consequence of (successful) exploratory behavior. When they are first encountered, however, they are meaningful. It is the amygdala, at bottom, that appears responsible for the (disinhibited) generation of this *a priori* meaning—terror and curiosity.

The amygdala appears to automatically respond to all things or situations, *unless told not to*. It is told not to—is functionally inhibited—when ongoing goal-directed behaviors produce the desired (intended) results.[112] When an error occurs, however—indicating that current memory-guided motivated plans and goals are insufficient—the amygdala is released from inhibition and labels the unpredictable occurrence with meaning. Anything unknown is dangerous and promising, simultaneously: evokes anxiety, curiosity, excitement and hope *automatically* and *prior to what we would normally regard as exploration or as (more context-specific) classification*. The operations of the amygdala are responsible for ensuring that the unknown is regarded with respect, as the default decision. The amygdala says, in effect, "if you don't know what it signifies, you'd better pay attention to it." Attention constitutes the initial stage of exploratory behavior, motivated by amygdalic operation—composed of the interplay between anxiety,[113] which impels caution in the face of novelty-threat, and hope, which compels approach to novelty-promise.[114] Caution-regulated approach allows for the update of memory in the form of skill and representation. Exploration-updated memory inhibits the production of *a priori* affect. On familiar ground—in explored territory—we feel no fear (and comparatively little curiosity).

The desired output of behavior (what should be) is initially posited; if the current strategy fails, the approach and exploration system is activated,[115] although it remains under the governance of anxiety. The approach system (and its equivalent, in abstraction) generates (1) alternative sequences of behavior, whose goal is the production of a solution to the present dilemma; (2) alternative conceptualizations of the desired goal; or (3) re-evaluation of the motivational significance of the current state. This means (1) that a new strategy for attaining the desired goal might be invented, or (2) that a replacement goal, serving the same function, might be chosen; or (3) that the behavioral strategy might be abandoned, due to the cost of its implementation. In the latter case, the whole notion of what constitutes "reality," at least with regard to the story or frame of reference currently in use, might have to be

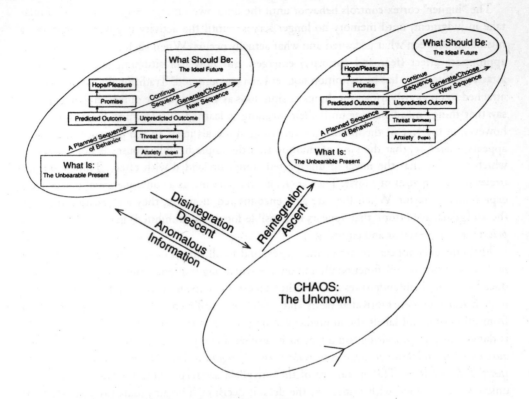

Figure 9: The Regeneration of Stability from the Domain of Chaos

reconstructed. This most troublesome state of affairs is schematically presented, in its successful form, in *Figure 9: The Regeneration of Stability from the Domain of Chaos*.[116]

Exploratory activity culminates normally in restriction, expansion, or transformation of the behavioral repertoire. In exceptional, non-normal circumstances—that is, when a major error has been committed—such activity culminates in *revolution*, in modification of the entire story guiding affective evaluation and behavioral programming. Such revolutionary modification means update of modeled reality, past, present and future, through incorporation of information generated during exploratory behavior. Successful exploration transforms the unknown into the expected, desired and predictable; establishes appropriate behavioral measures (and expectations of those measures) for next contact. Unsuccessful exploration, by contrast—avoidance or escape—leaves the novel object firmly entrenched in its initial, "natural," anxiety-provoking category. This observation sets the stage for a fundamental realization: human beings do not learn to fear new objects or situations, or even really "learn" to fear something that previously appeared safe, when it manifests a dangerous property. Fear is the *a priori* position, the natural response to everything for which no structure of behavioral adaptation has been designed and inculcated. Fear is the *innate* reaction to everything that has not been rendered predictable, as a consequence of successful, creative exploratory behavior undertaken in its presence, at some time in the past. LeDoux states:

It is well established that emotionally neutral stimuli can acquire the capacity to evoke striking emotional reaction following temporal pairing with an aversive event. Conditioning does not create new emotional responses but instead simply allows new stimuli to serve as triggers capable of activating existing, often hard-wired, species-specific emotional reactions. In the rat, for example, a pure tone previously paired with footshock evokes a conditioned fear reaction consisting of freezing behavior accompanied by a host of autonomic adjustments, including increases in arterial pressure and heart rate.[117] Similar responses are expressed when laboratory rats are exposed to a cat for the first time, but following amygdala lesions such responses are no longer present,[118] suggesting that the responses are genetically specified (since they appear when the rat sees a cat, a natural predator, for the first time) and involve the amygdala. The fact that electrical stimulation of the amygdala is capable of eliciting the similar response patterns[119] further supports the notion that the responses are hard-wired.[120]

Fear is not conditioned; security is unlearned, in the presence of particular things or contexts, as a consequence of violation of explicit or implicit presupposition. Classical behavioral psychology is wrong in the same manner our folk presumptions are wrong: fear is not secondary, not learned; security is secondary, learned. Everything not explored is tainted, *a priori*, with apprehension. Any thing or situation that undermines the foundations of the familiar and secure is therefore to be feared.[121]

It is difficult for us to formulate a clear picture of the subjective effects of the systems that dominate our initial response to the truly unpredictable, because we strive with all our might to ensure that everything around us remains normal. Under "normal" conditions, therefore, these primordial systems never operate with their full force. It might be said, with a certain amount of justification, that we devote our entire lives to making sure that we never have to face anything unknown, in the revolutionary sense—*at least not accidentally*. Our success in doing so deludes us about the true nature, power and intensity of our potential emotional responses. As civilized people, we are secure. We can predict the behaviors of others (that is, if they share our stories); furthermore, we can control our environments well enough to ensure that our subjection to threat and punishment remains at a minimum. It is the cumulative consequences of our adaptive struggle—our cultures—which enable this prediction and control. The existence of our cultures, however, blinds us to the nature of our true (emotional) natures—at least to the range of that nature, and to the consequences of its emergence.

Experimental examinations of the orienting reflex have not shed much light on our true potential for emotional response, in the past, because they generally took place under exceptionally controlled circumstances. Subjects evaluated for their responses to "novelty" are generally presented with stimuli that are novel only in the most "normal" of manners. A tone, for example, which differs unpredictably from another tone (or which appears at a relatively unpredictable time) is still a tone, something experienced a thousand times before and something experienced in a lab, in a hospital or university, under the jurisdiction of trustworthy personnel devoted to minimizing the anxiety-provoking nature of the experimental procedure. The controlled circumstances of the experiment (which are, in fact, the implicit and therefore invisible theoretical presumptions of the experiment) have led us to minimize the importance of the orienting reflex, and to misunderstand the nature of its disappearance.

Orienting signifies "attention," not terror, in the standard lab situation, and its gradual elimination with repeated stimulus presentation is regarded as "habituation"—as something boring, akin to automatic acclimation, adjustment or desensitization. Habituation is not a passive process, however, at least at higher cortical levels of processing. It just looks passive *when observed under relatively trivial circumstances*. It is in reality always the consequence of active exploration and subsequent modification of behavior, or interpretive schema. The (relatively) novel target laboratory tone, for example, is investigated for its underlying structure by the cortical systems involved in audition. These systems actively analyze the component elements of every sound.[122] The subject is led to "expect" or predict one sort of sound and gets another. The unexpected other has indeterminate significance, in that particular context, and is therefore regarded as (comparatively) meaningful—threatening and promising. The unexpected tone is presented repeatedly. The exploratory subject notes that the repetitions signify nothing, in the context that defines the experimental situation (nothing punishing, satisfying, threatening or promising), and ceases to react. He has not merely "habituated" to the stimuli. He has mapped its context-dependent significance, which is zero. This process appears trivial *because the experimental situation makes it so*. In real life, it is anything but boring.

Classical work conducted on animal "emotion" and motivation has taken place under circumstances reminiscent of the artificially constrained situations that define most work on human orienting. Animals, usually rats, are trained to be afraid—or to inhibit their behavior—in the presence of a neutral stimulus paired repeatedly with an "unconditioned" punishment [a stimulus whose motivational valence is negative, in the supposed absence of learning (or, at least, in the absence of interpretation)]. The rat is placed in the experimental environment and is allowed to familiarize himself with his surroundings. The neutral stimulus might be a light; the unconditioned stimulus, an electric shock. The light goes on; the floor of the rat's cage is briefly electrified. This sequence occurs repeatedly. Soon the rat "freezes" as soon as the light appears. He has developed a "conditioned response," manifesting behavioral inhibition (and fear, theoretically) to something that was previously neutral. Procedures of this sort effectively *produce* fear. The implicit contextual constraints or axioms of these procedures, however, lead researchers to draw odd conclusions about the nature of the "acquisition" of fear.

Such experiments first imply that fear in a given situation is necessarily something learned. Second, they imply that fear exists as a consequence of exposure to punishment, and only because of that exposure. The problem with this interpretation is that the rat was inevitably afraid as soon as he was placed in the new experimental environment, even though nothing terrible had yet happened there. After he is allowed to explore, he calms down. It is only then that he is regarded as normal. The experimenter then jars the rat out of his

acquired normalcy by presenting him with something unexpected and painful—the unconditioned stimulus, in conjunction with the neutral stimulus. He then "learns" to be afraid.
Really what has happened is that the unexpected occurrence forces the rat to reattain the
state he was in (or that same state, in an exaggerated manner) when he first entered the cage.
The fact of the electric shock, in conjunction with the light, indicates to the rat (reminds the
rat) that he is, once again, in unexplored territory. His fear, in unexplored territory, is just as
normal as his complacency in environments that he has mapped and that hold no danger.
We regard the calm rat as the real rat because we project our misinterpretations of our own
habitual nature onto our experimental animals. It is as D.O. Hebb states:

[The urbanity characterizing ourselves,] ... the civilized, amiable, and admirable part of mankind, well
brought up and not constantly in a state of fear ... depends as much on our successfully avoiding dis-
turbing stimulation as on a lowered sensitivity [to fear-producing stimuli].... [T]he capacity for emotional breakdown may [well] be self-concealing, leading [animals and human beings] to find or create
an environment in which the stimuli to excessive emotional response are at a minimum. So effective is
our society in this regard that its members—especially the well-to-do and educated ones—may not
even guess at some of their own potentialities. One usually thinks of education, in the broad sense, as
producing a resourceful, emotionally stable adult, without respect to the environment in which these
traits are to appear. To some extent this may be true. But education can be seen as being also the means
of establishing a protective social environment in which emotional stability is possible. Perhaps it
strengthens the individual against unreasonable fears and rages, but it certainly produces a uniformity of
appearance and behavior which reduces the frequency with which the individual member of the society
encounters the causes of such emotion. On this view, the susceptibility to emotional disturbance may
not be decreased. It may in fact be increased. The protective cocoon of uniformity, in personal appearance, manners, and social activity generally, will make small deviations from custom appear increasingly
strange and thus (if the general thesis is sound) increasingly intolerable. The inevitable small deviations
from custom will bulk increasingly large, and the members of the society, finding themselves tolerating
trivial deviations well, will continue to think of themselves as socially adaptable.[123]

Our emotional regulation depends as much (or more) on the stability and predictability of
the social environment (on the maintenance of our cultures) as on "interior" processes, classi-
cally related to the strength of the ego or the personality. Social order is a necessary precondition for psychological stability: it is primarily our companions and their actions (or
inactions) that stabilize or destabilize our emotions.

A rat (a person) is a complacent creature in explored territory. When in unexplored territory, however, it is anything but calm. A rat moved from its home cage to a new and
unknown environment—a new cage, for example—will first freeze (even though it has never
been punished, in the new situation). If nothing terrible happens to it (nothing punishing,
threatening or additionally unpredictable) it will begin to sniff, to look around, to move its
head, to gather new information about the intrinsically frightening place it now inhabits.
Gradually, it starts to move about. It will explore the whole cage with increasing confidence.

It is mapping the new environment for affective valence. It wants to find out: is there anything here that will kill me? Anything here I can eat? Anyone else here—someone hostile or friendly? A potential mate? The rat is interested in determining whether the new place contains anything of determinate interest to a rat, and it explores, to the best of its capacity, to make that judgment. It is not primarily interested in the "objective" nature of the new circumstances—a rat cannot actually determine what is objective and what is merely "personal opinion." Nor does it care. It just wants to know what it should do.

What happens if an animal encounters something truly unexpected—something that should just not be, according to its current frame of reference or system of belief? The answer to this question sheds substantial light on the nature of the orienting reflex, in its full manifestation. Modern experimental psychologists have begun to examine the response of animals to natural sources of mystery and threat. They allow the animals to set up their own environments, realistic environments, and then expose them to the kinds of surprising circumstances they might encounter in real life. The appearance of a predator in previously safe space (space previously explored, that is, and mapped as useful or irrelevant) constitutes one type of realistic surprise. Blanchard and colleagues describe the naturalistic behavior of rats, under such conditions:

When a cat is presented to established mixed-sex groups of laboratory rats living in a visible burrow system, the behaviors of the subjects change dramatically, in many cases for 24 hours or more.[124] The initial active defensive behavior, flight to the tunnel/chamber system, is followed by a period of immobility during which the rats make 22 kHz ultrasonic vocalizations, which apparently serve as alarm cries, at a high rate.[125] As freezing breaks up, proxemic avoidance of the open area gradually gives way to a pattern of "risk assessment" of the area where the cat was encountered. Subjects poke their heads out of the tunnel openings to scan the open area where the cat was presented, for minutes or hours before emerging, and when they do emerge, their locomotory patterns are characterized by [behaviors that theoretically reduce their visibility and vulnerability to predators and by] very short "corner runs" into and out of the open area. These risk assessment activities appear to involve active gathering of information about the possible danger source,[126] providing a basis for a gradual return to nondefensive behaviors.[127] Active risk assessment is not seen during early post-cat exposure, when freezing and avoidance of the open area are the dominant behaviors, but rises to a peak about 7–10 hours later, and then gradually declines. Nondefensive behaviors[128] such as eating, drinking and sexual and aggressive activity tend to be reduced over the same period.[129]

The unexpected appearance of a predator where nothing but defined territory previously existed terrifies the rats—badly enough that they "scream" about it, persistently, for a long period of time. Once this initial terror abates—which occurs only if nothing else horrible or punishing happens—curiosity is disinhibited, and the rats return to the scene of the crime. The space "renovelized" by the fact of the cat has to be transformed once again into explored territory *as a consequence of active modification of behavior (and representational schema), not by passive desensitization to the unexpected.* The rats run across the territory "contaminated" by the presence of the cat, to find out if anything dangerous (to running rats) still lurks there. If the answer is "no," then the space is defined, once again, as home territory (which is that place where commonplace behaviors produce desired ends). The rats transform the dangerous unknown into familiar territory as a consequence of voluntary exploration. In the absence of such exploration, terror reigns unchecked.

It is just as illuminating to consider the responses of rats to their kin, who constitute "explored territory," in contrast to their attitude toward "strangers," whose behavior is not predictable. Rats are highly social animals, perfectly capable of living with their familiar compatriots in peace. They do not like members of other kin groups, however; they will hunt them down and kill them. Accidental or purposeful intruders are dealt with in the same manner. Rats identify one another by smell. If an experimenter removes a well-loved rat from its familial surroundings, scrubs it down, provides it with a new odor, and returns it to its peers, it will be promptly dispatched by those who once loved it. The "new" rat constitutes "unexplored territory"; his presence is regarded as a threat (not unreasonably) to everything currently secure.[130] Chimpanzees, perfectly capable of killing "foreign devils" (even those who were once familiar), act in much the same manner.[131]

Explored Territory: Phenomenology and Neuropsychology

When we explore, we transform the indeterminate status and meaning of the unknown thing that we are exploring into something determinate—in the worst case, rendering it nonthreatening, nonpunishing; in the best, manipulating and/or categorizing it so that it is useful. Animals perform this transformation in the course of actual action, which is to say that they construct their worlds by shifting their positions and changing their actions in the face of the unknown, and by mapping the consequences of those shifts and changes in terms of their affective or motivational valence. When an animal encounters an unexpected situation, such as a new object placed in its cage, it first freezes, watching the object. If nothing terrible happens while it is immobile, it moves, slowly and at a distance, monitoring the thing for its reactions to these cautious exploratory activities. Perhaps the animal sniffs at the thing, or scratches at it—trying to determine what it might be good (or bad) for. It maps the utility and valence of the object, conceived in relationship to its ongoing activity (and, perhaps, to possible patterns of activity in the future). The animal builds its world of significances from the information generated in the course of—as a consequence of—ongoing exploratory behavior. The application of experimental search programs, drawn primarily from the reservoir of learned (imitated) and instinctual behavior, or manifested as trial and error, involves behavioral alteration (exploration, play) and subsequent transformation of sensory and affective input. When an animal actively explores something new, it changes the sensory quality and motivational significance of that aspect of its experience as a consequence of its exploratory strategy. This means that the animal exhibits a variety of behaviors in a given mysterious situation and monitors the results. It is the organized interpretation of these results and the behaviors that produce them *that constitute the world*, past, present and future, of the animal (in conjunction with the unknown, of course—which constantly supersedes the capacity for representation).

It is not too much to say that the animal elicits the properties of the object, sensory and affective, (or even brings them into being) through its capacity for creative investigation.[132] Animals that are relatively simple—compared, say, to higher-order primates, including man—are limited in the behaviors they manifest by the structure of their physiology. A rat cannot pick anything up, for example, to examine it in detail—and does not in addition have the visual capacity to focus intensely on the kinds of tiny features we can perceive. Higher-order nonhuman primates have a more developed grip, however, which enables more detailed exploration, and, in addition, have a relatively sophisticated prefrontal cortex. This means that such primates can evoke more features from the world, directly, and that they are increasingly capable of modeling and acting. The prefrontal cortex is the newest part of the motor unit, and "grew" out of the direct motor control centers, in the course of cortical evolution.[133] More sophistication in development of the prefrontal centers means, in part, heightened capability for *abstract* exploration, which means investigation in the absence of actual movement, which means the capacity to learn from the observation of others and through consideration of potential actions before they emerge in behavior. This means increasing capability for *thought*, considered as abstracted action and representation.[134] Action and thought produce phenomena. Novel acts and thoughts necessarily produce new phenomena. Creative exploration, concrete and abstract, is therefore linked in a direct sense to being. Increased capacity for exploration means existence in a qualitatively different— even new—world. This entire argument implies, of course, that more complex and behaviorally flexible animals inhabit ("construct," if you will[135]) a more complex universe.

Humans possess cortical development—prefrontal and otherwise—that is unique in its great mass and, more importantly, in its structure. Various indices of development have been used to signify the nature of the relationship between the brain and intelligence. Sheer mass is one measure, degree of surface convolution another. The former measure is contaminated by size of animal. Larger animals tend to have more absolutely massive brains. This does not necessarily make them smarter. Brain mass corrected for body size constitutes the encephalization quotient, a common rough measure of animal intelligence.[136] Degree of surface convolution constitutes an additionally useful measure. The gray matter of the brain, which theoretically does much of the work associated with intelligence, occupies the brain's surface, which has been dramatically increased in area by folding. Some representatives of the cetacean family (dolphins and whales) have encephalization quotients similar to and brain surfaces more convoluted than man's,[137] although the thickness of the cetacean neocortex is about half that of the human.[138] Consideration of this high level of nervous development has led to speculation about the potential superhuman range of cetacean ability.[139] However, it is structure and organization of cortex, not simply mass, or even relative mass or surface area, that most clearly defines the nature and reach of a species' experience and competence. More particularly, it is embodiment of the brain that matters. Brain structure necessarily reflects

Figure 10: The Motor Homunculus

embodiment, despite the archaic presumption of the independence of spirit and matter (or soul and body, or mind and body), because the body is, in a primary sense, the environment to which the brain has adapted.

The body is specifically represented in the neocortex. This representation is often given schematic form as the *homunculus*, or "little man." The homunculus was "discovered" by Wilder Penfield,[140] who mapped the surface of the cortices of his neurosurgical patients by stimulating them electrically, painstakingly, point after point. He did this to find out what different sections of the brain were doing, so that he could do the least damage possible when attempting to surgically treat epilepsy or cancer or other forms of brain abnormality. He would probe the surface of the brain of one of his (awake) patients with an electrode (patients undergoing neurosurgery are frequently awake, as the brain feels no pain) and monitor the results, either directly or by asking the patient what he or she experienced. Sometimes such stimulation would produce visions, sometimes elicit memories; other times, it produced movements or sensations. Penfield determined, in this manner, how the body was mapped onto the central nervous system—how it was incarnated, so to speak, in intrapsychic representation. He established, for example, that homunculi come in two forms, motor and sensory—the former associated with the primary zone of the motor unit, the latter associated with the primary zone of the sensory area of the sensory unit. The motor form—represented schematically in *Figure 10: The Motor Homunculus*—is of most interest to us, because our discussion centers on motor output. The motor homunculus is a very odd little "creature." Its face (particularly mouth and tongue) and hands (particularly thumbs) are grossly disproportionate to the rest of its "body." This is because comparatively large areas of the motor cortex are given over to control of the face and hands, which are capable of an immense number of complex and sophisticated operations. The motor homunculus is an interesting figure. It might be regarded as the body, insofar as the body has anything to do with the brain. It is useful to consider the structure of the homunculus, because it is in some profound way representative of our essential nature, as it finds expression in emotion and behavior.

The most outstanding characteristic of the motor homunculus, for example—the hand, with its opposable thumb—is the defining feature of the human being. The ability to manipulate and explore characteristics of objects large and small—restricted as a general capacity to the highest primates—sets the stage for elicitation of an increased range of their properties, for their utilization as tools (for more comprehensive transformation of their infinite potential into definable actuality). The hand, used additionally to duplicate the action and function of objects, also allows first for imitation (and pointing), and then for full-blown linguistic representation.[141] Used for written language, the hand additionally enables long-distance (temporal and spatial) transfer of its ability to another (and for the elaboration and extension of exploration, during the process of writing, which is hand-mediated thinking). Even development of spoken language, the ultimate analytic motor skill, might reasonably be considered an abstract extension of the human ability to take things apart and reassemble them, in an original manner. Interplay between hand and brain has literally enabled the individual to change the structure of the world. Consideration of the structure and function of the brain must take this primary fact into account. A dolphin or whale has a large, complex brain—a highly developed nervous system—but it cannot shape its world. It is trapped, so to speak, in its streamlined test-tube-like form, specialized for oceanic life. It cannot directly alter the shape of its material environment in any complex manner. Its brain, therefore, is not likely prepared to perform any traditionally "creative" function (indeed, as one would suspect, lacks the sophisticated structuring characteristic of primate brains[142]).

It is not just the hand, however, that makes the crucial human difference, although it is the most obvious, and perhaps the most important, single factor. It is more a style or melody of adaptation that characterizes the individual human being. This style is adaptation for exploration of the unknown, within a social context. This adaptation is capacity for (speech-mediated) creation, elaboration, remembrance, description and subsequent communication of new behavior patterns, and for representation of the (frequently novel) consequences of those patterns. The hand itself was rendered more useful by the development of vertical stance, which extended visual range and freed the upper body from the demands of locomotion. The fine musculature of the face, lips and tongue—over-represented, once again, in the motor homunculus—helped render subtle communication possible. Development of explicit language extended the power of such communication immensely. Increasingly detailed exchange of information enabled the resources of all to become the resources of each, and vice versa. That process of feedback greatly extended the reach and utility of the hand, providing every hand with the ability, at least in potential, of every other hand, extant currently or previously. Evolution of the restricted central field of the eye, which has input expanded 10,000 times in the primary visual area, and is additionally represented, interhemispherically, at several higher-order cortical sites,[143] was of vital importance to development of visual language and enabled close observation, which made gathering of detailed information simpler. Combination of hand and eye enabled *Homo sapiens* to manipulate things in ways qualitatively different from those of any other animal. The individual can discover what things are like under various, voluntarily produced or accidentally encountered (yet considered) conditions—upside down, flying through the air, hit against other things, broken into pieces, heated in fire, and so on. The combination of hand and eye allowed human beings to experience and analyze the (emergent) nature of things. This ability, revolutionary as it was, was dramatically extended by application of hand-mediated, spoken (and written) language.

The human style of adaptation extends from the evidently physical to the more subtly psychological, as well. The phenomenon of consciousness, for example—arguably the defining feature of man—appears related in some unknown fashion to breadth of cellular activation in the neocortex. Bodily features with large areas of cortical representation are also therefore more thoroughly represented in consciousness (at least in potential). This can be made immediately evident to subjective awareness merely by contrasting the capacity for control and monitoring of the hand, for example, with the much less represented expanse of the back. Consciousness also evidently expands or sharpens during the course of activities designed to enhance or increase adaptive competence—during the course of creative exploration. Processing of novel or otherwise interesting sensory information, associated with the orienting complex, heightened awareness and focused concentration, activates large areas of neocortex. Similarly, increased cortical mobilization takes place during the practice phase of skill acquisition, when awareness appears required for development of control. The area of such engagement or mobilization shrinks in size as movement becomes habitual and unconscious, or when sensory information loses interest or novelty.[144] Finally, as we have noted before, intrinsic pleasure of an intense nature appears to accompany activation of the cortical systems activated during psychomotor exploratory activity, undertaken in the face of the unknown. The operation of these systems appears mediated in part by the neurotransmitter dopamine[145]—involved in producing subjective and behavioral response to cues of reward, in the form of hope, curiosity and active approach.

Human beings enjoy capacity for investigation, classification and consequent communication, which is qualitatively different from that characterizing any other animal. The material structure of *Homo sapiens* is ideal for exploration, and for the dissemination of the results thereof; spiritually—psychologically—man is characterized by the innate capacity to take true pleasure in such activity. Our physical attributes (the abilities of the hand, in combination with the other physiological specializations of man) define who we are and enable us to endlessly elicit new properties from previously stable and predictable elements of experience. The object—any object—serves us as a source of limitless possibility (or, at least, possibility limited only by the capacity for exploratory genius exhibited at any particular moment). Simple animals perform simple operations and inhabit a world whose properties are equally constrained (a world where most "information" remains "latent"). Human beings can manipulate—take apart and put together—with far more facility than any other creature. Furthermore, our capacity for communication, both verbal and nonverbal, has meant almost unbelievable facilitation of exploration, and subsequent diversity of adaptation.

Thinking might in many cases be regarded as the abstracted form of exploration—as the capacity to investigate, without the necessity of direct motoric action. Abstract analysis (verbal and nonverbal) of the unexpected or novel plays a much greater role for humans than for animals[146]—a role that generally takes primacy over action. It is only when this capacity fails partially or completely in humans—or when it plays a paradoxical role (amplifying the significance or potential danger of the unknown through definitive but "false" negative labeling)—that active exploration (or active avoidance), with its limitations and dangers, becomes necessary. Replacement of potentially dangerous exploratory action with increasingly flexible and abstracted thought means the possibility for growth of knowledge without direct exposure to danger, and constitutes one major advantage of the development of intelligence. The abstract intelligence characteristic of the human being developed in parallel with rapid evolution of the brain. We can communicate the results and interpretations of our manipulations (and the nature of the procedures that constitute that manipulation) to each other, across immense spatial and temporal barriers. This capacity for exploration, verbal elaboration and communication of such in turn dramatically heightens our capacity for exploration (as we have access to all communicated strategies and interpretive schemas, accumulated over time, generated in the course of the creative activity of others). In normal parlance, this would mean merely that we have been able to "discover" more aspects of the world. It seems to me more accurate, however, to recognize the limitations of this perspective, and to make room for the realization that new procedures and modes of interpretation literally produce new phenomena. The *word* enables differentiated thought and dramatically heightens the capacity for exploratory maneuvering. The world of human experience is constantly transformed and renewed as a consequence of such exploration. In this manner, the word constantly engenders new creation.

The capacity to create novel behaviors and categories of interpretation in response to the emergence of the unknown might be regarded as the primary hallmark of human consciousness—indeed, of human being. Our engagement in this process literally allows us to carve the world out of the undifferentiated mass of unobserved and unencountered "existence" (a form of existence that exists only hypothetically, as a necessary fiction; a form about which nothing can be experienced, and less accurately stated). We carve out the world as a consequence of our direct interactions with the unknown—most notably, with our hands, which enable us to manipulate things, to change their sensory aspects and, most importantly, to change their importance to us, to give them new, more desirable *value*. The capacity for dextrous manipulation is particularly human, and has enabled us to radically alter the nature of our experience. Equally particular, however, is our capacity for abstract exploration, which is thought about action (and its consequences), in the absence of action (and its consequences). The manner in which we conduct our abstracted exploration appears as tightly linked to the physiological structures of our brains as the manner in which we move, while exploring. In novel circumstances, our behavioral output is mediated by the systems that govern fear, and appropriate inhibition, and hope, and appropriate activation. The same things happen when we think abstractly—even when we think about how others think.[147]

Animal exploration is primarily motor in nature. An animal must move around an unfamiliar thing or situation to determine its affective relevance and sensory nature. This process of moving around experimentally appears as a consequence of the interaction between the mutually regulatory or inhibitory evaluative systems whose responsibilities are identification of potential danger or threat and potential satisfaction or promise. In the human case, each of these systems apparently comes, in the course of normal development, to dominate one of our twinned cortical hemispheres: the *right* governs response to threat (and to punishment), while the *left* controls response to promise and, perhaps (although much less clearly), to satisfaction.[148] This basically means that the right hemisphere governs our initial responses to the unknown, while the left is more suited for actions undertaken while we know what we are doing. This is in part because everything thoroughly explored has in fact been rendered either promising or satisfying (or, at least, irrelevant). If threat or punishment still lurks somewhere—that is, somewhere we must be—our behavioral adaptation is, by definition, insufficient (and the unexpected has not been vanquished). We have been unable to modify our actions to elicit from the environment—really, from the "unknown"—those consequences we wish to produce.

Richard Davidson and his colleagues have been investigating the relationship between different patterns of cortical electrical activity and mood states in adults and children. Davidson et al. have concluded that the twin hemispheres of the human brain are differentially specialized for affect, at least with regard to their frontal regions. Signs of positive affect (like genuine smiling in infants) are accompanied by heightened comparative activation of the left frontal cortex. Negative states of affect (like those occuring in .chronic depression), by contrast, are accompanied by heightened activation of the right frontal hemisphere.[149] Substantial additional evidence exists to support this general claim. To put it most fundamentally: it appears that the twin hemispheres of the brain are differentially specialized (1) for operation in unexplored territory, where the nature and valence of things remains indeterminate, and (2) for operation in explored territory, where things have been rendered either irrelevant or positive, as a consequence of previous exploration. Our brains contain two emotional systems, so to speak. One functions when we do not know what to do, and initiates the (exploratory) process that creates secure territory. The other functions when we are in fact secure. The fact of the presence of these two subsystems, but not their "locale," has been

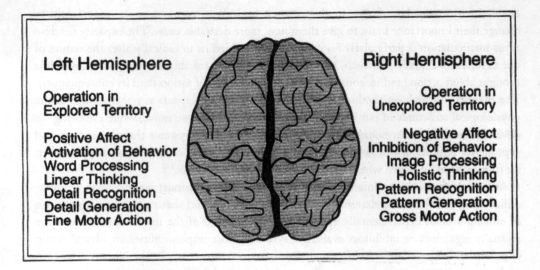

Figure 11: The Twin Cerebral Hemispheres and Their Functions

known for a good while; Maier and Schnierla[150] and Schnierla[151] hypothesized many decades ago that mechanisms of "withdrawal" and "approach" (characteristic of animals at virtually all levels of the evolutionary scale) provided the foundation for motivation as such. The nature of these two systems can best be understood by relating emotional state to motor activity, as we have done previously.

Each hemisphere, right and left, appears to have what might be described as a *family* of related functions, portrayed in *Figure 11: The Twin Cerebral Hemispheres and Their Functions*. The right hemisphere, less language-fluent than its generally more dominant twin, appears specialized for the inhibition and extinction of behavior (and, therefore, for the production of negative emotion), for generation and manipulation of complex visual (and auditory) images, for coordination of gross motor actions, and for rapid and global recognition of patterns.[152] The right hemisphere appears to come "on-line" when a particular situation is rife with uncertainty—appears particularly good at governing behavior when what *is* and *what to do* have not yet been clearly specified.[153] It might be posited, in consequence, *that this hemisphere is still under limbic control*, since the limbic system is responsible for detecting novelty and initiating exploratory behavior. This archaic control mechanism would then "drive" the processes of imagistic "hypothesis" generation that constitute the processes of abstract exploration—fantasy—we use to give determinate (and oft-bizarre) form to the unknown.

The left hemisphere, by contrast, appears particularly skilled at linguistic processing and communication, at detailed, linear thinking, at fine motor skill, and at the comprehension of wholes in terms of their constituent elements.[154] The left hemisphere—particularly its frontal or motor (sub)unit—also governs approach behavior[155] in the presence of cues of satisfaction, is integrally involved in the production of positive affect, and appears particularly good at carrying out practiced activities, applying familiar modes of apprehension. The left seems at its best when what *is* and *what should be done* are no longer questions; when tradition governs behavior, and the nature and meaning of things has been relatively fixed. The dual specialization of the left—for what has been practiced, and for what is positive—can be understood, in part, in the following manner: positive affect rules in known territory, by definition. A thing or situation has been explored most optimally (and is therefore most well known) if it has been transformed by behavioral adaptations manifested in its presence into something of determinate use (or satisfaction) or into potential for such (into promise).

The right hemisphere, in contrast to the left, appears to have remained in direct contact with, and to be specialized for encounter with, the *unknown* and its terrors, which are apperceived in the domain of instinct, motivation and affect, long before they can be classified or comprehended intellectually. The right hemisphere's capacity for inhibition and extinction of behavior (for inducing caution during exploration, for governing flight, for producing negative affect) ensures that due respect is granted the inexplicable (and therefore dangerous) when it makes its appearance. The right's aptitude for global pattern recognition (which appears as a consequence of its basic neurophysiological structure[156]) helps ensure that a provisional notion (a fantastic representation) of the unknown event (what it is like, how action should be conducted in its presence, what other things or situations it brings to mind) might be rapidly formulated. The right hemisphere appears integrally involved in the initial stages of analysis of the unexpected or novel—and its *a priori* hypothesis is always this: *this (unknown) place, this unfamiliar space, this unexplored territory is* dangerous, *and therefore partakes in the properties of all other known dangerous places and territories, and all those that remain unknown, as well.* This form of information processing—"a" is "b"—is metaphor; generation of metaphor (key to the construction of narratives—dreams, dramas, stories and myth) might well be regarded as the first stage of hypothesis construction. As situation-specific adaptive behaviors are generated as a consequence of exploration, this provisional labeling or hypothesis (or fantasy) might well undergo modification (assuming nothing actually punishing or determinately threatening occurs); such modification constitutes further and more detailed learning. Anxiety recedes, in the absence of punishment or further threat (including novelty); hope occupies the affective forefront, accompanied by the desire to move forward, and to explore (under the governance of the left hemisphere).

The right hemisphere appears capable of dealing with less determinate information. It can use forms of cognition that are more diffuse, more global[157] and more encompassing to come to terms initially with what cannot yet be understood but undeniably exists. The right hemisphere uses its capacity for massive generalization and comprehension of imagery to place the novel stimulus in an initially meaningful context, which is the *a priori* manner of appropriate categorization. This context is defined by the motivational significance of the novel thing, which is revealed first by the mere fact of novelty (which makes it both threatening and promising) and then in the course of its detailed exploration. The right hemisphere remains concerned with the question "what is this new thing like?" (meaning "what should be done in the presence of this unexpected occurrence?") and does not care "what is this thing objective-ly?" "What is the new thing like" means "is it dangerous, or threatening (first and foremost), satisfying or promising?" Categorization according to valence means that the thing is what it signifies for behavior.

The chaos that constitutes the unknown is rendered predictable—is turned into the "world"—by the generation of adaptive behaviors and modes of representation. It is the process of novelty-driven exploration that, in the individual case, produces such behaviors and strategies of classification. However, we are not only individuals. We exist in a very com-plex social environment, characterized by the constant exchange of information regarding the means and ends of "proper" adaptation. The human capacity for the generation of self-regulatory behavior and representation has been expanded immensely, in some ways beyond our own comprehension, by our capacity for verbal and nonverbal communication. We can learn through discussion and by reading—can absorb information directly even from our departed but literate ancestors. But there is more—we can learn from everyone who acts in the natural course of things or dramatically, and we can store the behaviors of individuals we come into contact with (directly, by copying them; or indirectly, through the intermediation of narrative and dramatic art forms). It is of course our ability to copy, to mimic,[158] that underlies our capacity to do things we do not necessarily "understand" (that is, cannot describe explicitly). It is for that reason, in part, that we need a "psychology."

Patterns of behavioral and representational adaptation are generated in the course of active exploration and "contact with the unknown." These patterns do not necessarily remain stable, however, once generated. They are modified and shaped, improved and made effi-cient, as a consequence of their communicative exchange. Individual "a" produces a new behavior, "b" modifies it, "c" modifies that, "d" radically changes "c's" modification and so on *ad infinitum*. The same process applies to representations (metaphors, say, or explicit con-cepts). This means that our exploratory assimilative and accommodative processes actually extend over vast periods of time and space. Some of this extension—perhaps the most obvi-ous part—is mediated by literacy. An equally complex and subtle element, however, is medi-ated by mimesis.

Patterns of behavioral adaptation and schemas of classification or representation can be derived from the observation of others (and, for that matter, from the observation of one-self). How we act in the presence of things, in their constantly shifting and generally social context, is what those things mean (or even what they are), before what they mean (or what they are) can be more abstractly (or "objectively") categorized. What a thing is, therefore, might be determined (in the absence of more useful information) by examination of how action is conducted in its presence, which is to say that if someone runs from something it is safe to presume that the thing is dangerous (the action in fact defines that presumption). The observation of action patterns undertaken by the members of any given social commu-nity, including those of the observing subject, therefore necessarily allows for the derivation and classification of provisional value schema. If you watch someone (even yourself) approach something then you can assume that the approached thing is good, at least in some determinate context, even if you don't know anything else about it. Knowing what to do, after all, is classification, before it is abstracted: classification in terms of motivational relevance, with the sensory aspects of the phenomena serving merely as a cue to recognition of that motivational relevance.[159]

It is certainly the case that many of our skills and our automatized strategies of classification are "opaque" to explicit consciousness. The fact of our multiple memory systems, and their qualitatively different modes of representation—described later—ensures that such is the case. This opaqueness means, essentially, that we "understand" more than we "know"; it is for this reason that psychologists continue to depend on notions of the "unconscious" to provide explanations for behavior. This unconsciousness—the psychoanalytic god—is our capacity for the implicit storage of information about the nature and valence of things. This information is generated in the course of active exploration, and modified, often unrecognizably, by constant, multigenerational, interpersonal communication. We live in social groups; most of our interactions are social in nature. We spend most of our time around others and, when we are alone, we still wish to understand, predict and control our personal behaviors. Our maps of the "understood part of the world" are therefore in large part *maps of patterns of actions*—of *behaviors* established as a consequence of creative exploration, and modified in the course of endless social interactions. We watch ourselves act; from this action, we draw inferences about the nature of the world (including *those acts that are part of the world*).

We know that the right hemisphere—at least its frontal portion—is specialized for response to punishment and threat. We also know that damage to the right hemisphere impairs our ability to detect patterns and to understand the meaning of stories.[160] Is it too much to suggest that the emotional, imagistic and narrative capabilities of the right hemisphere play a key role in the initial stages of transforming something novel and complex, such as the behaviors of others (or ourselves) and the valence of new things, into something thoroughly understood? When we encounter something new, after all, we generate fantasies (imagistic, verbal) about its potential nature. This means we attempt to determine how the unexpected thing might relate to something we have already mastered—or, at least, to other things that we have not yet mastered. To say "this unsolved problem appears to be like this other problem we haven't yet solved" is a step on the way to solution. To say, "here is how these (still essentially mysterious) phenomena appear to hang together" is an intuition of the sort that precedes detailed knowledge; is the capacity to see the forest, though not yet differentiating between the types of trees. Before we truly master something novel (which means, before we can effectively limit its indeterminate significance to something predictable, even irrelevant) we *imagine* what it might be. Our imaginative representations actually constitute our initial adaptations. Our fantasies comprise part of the structure that we use to inhibit our responses to the *a priori* significance of the unknown (even as such fantasies facilitate generation of more detailed and concrete information). There is no reason to presuppose that we have been able to explicitly comprehend this capacity, in part because it actually seems to serve as a *necessary or axiomatic precondition for the ability to comprehend, explicitly*.

The uniquely specialized capacities of the right hemisphere appear to allow it to derive from repeated observations of behavior images of action patterns that the verbal left can arrange, with increasingly logic and detail, into *stories*. A story is a map of meaning, a "strategy" for emotional regulation and behavioral output—a description of how to act in a circumstance, to ensure that the circumstance retains its positive motivational salience (or at least has its negative qualities reduced to the greatest possible degree). The story appears generated, in its initial stages, by the capacity for imagery and pattern recognition characteristic of the right hemisphere, which is integrally involved in narrative cognition[161] and in processes that aid or are analogous to such cognition. The right hemisphere has the ability to decode the nonverbal and melodic aspects of speech, to empathize (or to engage, more generally, in interpersonal relationships), and the capacity to comprehend imagery, metaphor and analogy.[162] The left-hemisphere "linguistic" systems "finish" the story, adding logic, proper temporal order, internal consistency, verbal representation, and possibility for rapid abstract explicit communication. In this way, our explicit knowledge of value is expanded, through the analysis of our own "dreams." Interpretations that "work"—that is, that improve our capacity to regulate our own emotions (to turn the current world into the desired world, to say it differently)—*qualify as valid*. It is in this manner that we verify the accuracy of our increasingly abstracted presumptions.

The process of creative exploration—the function of the *knower*, so to speak, who generates explored territory—has as its apparent purpose an increase in the breadth of motoric repertoire (skill) and alteration of representational schema. Each of these two purposes appears served by the construction of a specific form of knowledge, and its subsequent storage in permanent memory. The first form has been described as *knowing how*. The motor unit, charged with origination of new behavioral strategies when old strategies fail (when they produce undesired results), produces alternate action patterns, experimentally applied, to bring about the desired result. Permanent instantiation of the new behavior, undertaken if the behavior is successful, might be considered development of new *skill. Knowing how* is skill. The second type of knowing, which is representational (an *image or model of something*, rather than the thing itself) has been described as *knowing that*[163]—I prefer *knowing what*. Exploration of a novel circumstance, event or thing produces new sensory and affective input, during active or abstracted interaction of the exploring subject and the object in question. This new sensory input constitutes grounds for the construction, elaboration and update of a permanent but modifiable four-dimensional (spatial and temporal) representational model of the experiential field, in its present and potential future manifestations. This model, I would propose, is a story.

It is the hippocampal system—which, as we have seen, is an integral part of the regulation of anxiety—that is critically involved in the transfer of information from observation of ongoing activity to permanent memory,[164] and that provides the physiological basis (in concert with the higher cortical structures) for the development and elaboration of this mnestic representation. It is the right hemisphere, which is activated by the unknown, and which can generate patterns rapidly, that provides the initial imagery—the contents of fantasy—for the story. It is the left hemisphere that gives these patterns structure and communicability (as it does, for example, when it interprets a painting, a novel, a drama or a conversation). The hippocampus notes mismatch; this disinhibits the amygdala (perhaps not directly). Such disinhibition "releases" anxiety and curiosity, driving exploration. The right hemisphere, under these conditions of motivation, derives patterns relevant to encapsulation of the emergent unknown, from the information at its disposal. Much of this information can be extracted from the social environment, and the behavioral interactions and strategies of representation—emergent properties of exploration and communication—that are "embedded" in the social structure. Much of this "information" is still implicit—that is, coded in behavioral *pattern*. It is still *knowing how*, before it has been abstracted and made explicit as knowing what. The left hemisphere gets increasingly involved, as translation "up the hierarchy of abstraction" occurs.

Knowing-how information, described alternatively as *procedural*, habitual, dispositional, or skilled, and *knowing-what* information, described alternatively as *declarative*, episodic, factual, autobiographical, or representational, appear physiologically distinct in their material basis, and separable in course of phylo- and ontogenetic development.[165] Procedural knowledge develops long before declarative knowledge, in evolution and individual development, and appears represented in "unconscious" form, expressible purely in performance. Declarative knowledge, by contrast—knowledge of what—simultaneously constitutes consciously accessible and communicable episodic imagination (the world in fantasy) and subsumes even more recently developed semantic (linguistically mediated) knowledge, whose operations, in large part, allow for abstract representation and communication of the contents of the imagination. Squire and Zola-Morgan[166] have represented the relationship between these memory forms according to the schematic of *Figure 12: The Multiple Structure of Memory*.[167] The neuroanatomical basis of *knowing how* remains relatively unspecified. Skill generation appears in part as the domain of the cortical pre/motor unit; "storage" appears to involve the cerebellum. Knowing *what*, by contrast, appears dependent for its existence on the intact function of the cortical sensory unit, in interplay with the hippocampal system.[168] Much of our knowing *what*, however—our description of the world—*is about knowing how*, which is behavioral knowledge, wisdom. Much of our descriptive knowledge—representational knowledge—is representation of what constitutes wisdom (without being that wisdom, itself). We have gained our description of wisdom by watching how we act, in our culturally governed social interactions, and by representing those actions.

We know *how*, which means how to act to transform the mysterious and ever-threatening world of the present into what we desire, long before we *know how* we know how, or *why* we know how. This is to say, for example, that a child learns to *act* appropriately (assuming it does) long before it can provide abstracted explanations for or descriptions of its behavior.[169] A child can be "good" without being a moral philosopher. This idea echoes the developmental psychologist Jean Piaget's notion, with regard to child development, that adaptation at the sensorimotor level occurs prior to—and lays the groundwork for—the more abstracted forms of adaptation that characterize adulthood. Piaget regarded *imagistic representation* as

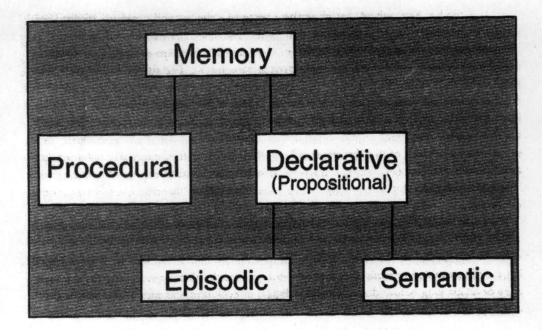

Figure 12: The Multiple Structure of Memory

an intermediary between sensorimotor intelligence and the (highest or most abstract) stage of "formal operations"; furthermore, he believed that imitation—the "acting out" of an object—served as a necessary prerequisite to such imagistic representation (portrayal in image or word, instead of behavior). The process of *play* appears as a higher-order, or more abstract form of imitation, from this perspective. Piaget presents two main theses:

The first is that in the field of play and imitation it is possible to trace the transition from sensory-motor assimilation and accommodation to the mental assimilation and accommodation which characterize the beginnings of representation.... [The second is that] the various forms of representation interact. There is representation when an absent model is imitated. There is representation in symbolic play, in imagination and even in dreams, the systems of concepts and logical relations, both in their intuitive and operational forms, implies representation.[170]

Piaget believed that imitation could be described in terms of accommodation: "If there is primacy of accommodation (matching of behavior) over assimilation (altering of schemas) ... the activity tends to become imitation."[171] This implies that the imitating child in fact *embodies* more information than he "understands" (represents). He continues: "representation ... can be seen to be a kind of interiorized imitation, and therefore a continuation of accommodation."[172] [With regard to the three-memory-system model (which Piaget is of course not directly referring to): "even if there were justification for relating the various stages of mental development to well-defined neurological levels, the fact remains that, in spite of the relative discontinuity of the structures, there is a certain functional continuity, each structure preparing for its successors while utilizing its predecessors."[173]]

What can be said of children appears true, more or less, phylogenetically: our cultures (which we absorb as children, through the processes of imitation) consist primarily of patterns of activity, undertaken in a social context. As parents are to children, cultures are to adults: we do not know how the patterns we act out (or the concepts we utilize) originated, or what precise "purposes" (what long-term "goals") they currently serve. Such patterns are in fact "emergent properties" of long-term social interactions. Furthermore, we cannot describe such patterns well, abstractly (explicitly, semantically), even though we duplicate them accurately (and unconsciously) in our behavior (and can represent them, episodically, in our literary endeavors). We do not know *why* we do what we do, or, to say the same thing, what it is that we are (all ideological theories to the contrary). We watch ourselves, and wonder; our wonder takes the shape of the story or, more fundamentally, the *myth*. Myths describing the *known*, explored territory, constitute what we know about our knowing how, before we can state, explicitly, what it is that we know how. Myth is, in part, the image of our adaptive action, as formulated by imagination, before its explicit containment in abstract language; myth is the intermediary between action and abstract linguistic representation of that action. Myth is the distilled essence of the stories we tell ourselves about the patterns of our own behavior, as they play themselves out in the social and impersonal worlds of experience. We learn the story, *which we do not understand* (which is to say, cannot make explicit), by watching. We represent the action patterns we encounter in action (*that is ritual*), image and word: we act, then represent our behavior, ever more abstractly, ever more explicitly, "consciously."

The central features of our (socially determined) behavior thus become key elements—characters—in our stories. The generation and constant refinement of these stories, told and retold over centuries, allows us to determine ever more clearly just what proper (and improper) behavior consists of, in an environment permanently characterized by the interplay between security and unpredictability. We are extremely (uncontrollably) imitative, overwhelmingly social and interminably exploratory. These characteristics allow us to generate and communicate represented images, and, simultaneously, serve as the focal point of inquiry for those images. Our capacity for creative action frees us, constantly, from the ever-shifting demands of the "environment." The ability to represent creative action—to duplicate observed creativity in our own actions, and to represent that creativity in detail and essence—allows everyone to benefit from the creative action of everyone else (at least everyone with whom communication might conceivably take place). The fact of our sociability ensures that our adaptive behaviors are structured with the social community in mind, at least in the long run, and increases our chances of exposure to creative intelligence. We observe others acting in a manner we find admirable, and duplicate their actions. In this manner, we obtain the skills of others. Our capacity for abstraction allows us to take our facility for imitation one step further, however: we can learn to imitate not only the precise behaviors that constitute adaptation, *but the process by which those behaviors were generated.* This means—we can learn not only skill, but meta-skill (can learn to mimic the pattern of behavior that generates new skills). It is the encapsulation of meta-skill in a story that makes that story great.

Our imitative proclivity, expressed in behavior, appears to find its more abstracted counterpart in the ability to admire, which is a permanent, innate or easily acquired constituent element of our intrapsychic state. This capability for awe, this desire to copy, often serves to impel further psychological and cognitive development. The worshipful attitude that small boys adopt toward their heroes, for example, constitutes the outward expression of the force that propels them toward embodying, or incarnating (or even inventing) oft ill-defined heroic qualities themselves. The capacity for imitation surfaces in more abstract guise in the human tendency to act "as-if"[174]—to identify with another—to become another, in fantasy (which means, to ritually identify with or unconsciously adopt the story of another). (This means— the ability to adopt someone else's goal, as if it were yours.[175]) The capacity to act "as if" expresses itself in admiration (ranging in intensity from the simple respect accorded a competent other, to abject worship) and, even more abstractly, in ideological possession. No independent "instinct" necessarily needs to be postulated, to account for this mimetic ability (although one may well exist): all that may be necessary is the capacity to observe that another has obtained a goal that is also valued by the observer (that observation provides the necessary motivation), and the skill to duplicate the procedures observed to lead to such fulfillment.

Mimetic propensity, expressed in imitative action, provides for tremendous expansion of behavioral competence[176]; allows the *ability* of each to become the *capability* of all. Precise duplicative facility, however, still retains pronounced limitations. Specific behaviors retain their adaptive significance only within particular, restricted environments (only within bounded frames of reference). If environmental contingencies shift (for whatever reason), the utility of strategies designed for the original circumstance (and transmitted through imitation) may become dramatically restricted or even reversed. The capacity for abstraction of imitation— which is, in the initial stages, capability for dramatic play—overcomes the specific restrictions of exact imitation, elaborating reproduction of particular acts, removing the behavior to be copied from its initial specific context, establishing its first-level declarative representation and generalization. *Play* allows for the permanent extension of competence and confidence through *pretense*, which means through metaphoric and symbolic action (which is semantic use of episodic representation), and for natural expansion of behavioral range from safe, predictable, self-defined contexts, out toward the unknown world of experience. Play creates a world in "rule-governed" fantasy—in episodic or imagistic representation—in which behavior can be rehearsed and mastered, prior to its expression in the real world, with real-world consequences. Play is another form of "as-if" behavior, that allows for experimentation with fictional narrative—pretended descriptions of the current and desired future states of the world, with plans of action appended, designed to change the former into the latter. To *play* means to set—or to fictionally transform—"fictional" goals. Such fictional goals give valence to phenomena that would, in other contexts, remain meaningless (valence that is informative, without being *serious*). Play allows us to experiment with means and ends themselves, without subjecting ourselves to the actual consequences of "real" behavior, and to benefit emotionally, in the process. The goals of play are not real; the incentive rewards, however, that accompany movement to a pretended goal—these are *real* (although bounded, like game-induced anxieties). The bounded reality of such affect accounts, at least in part, for the intrinsic interest that motivates and accompanies play (or immersion in any dramatic activity).

Play transcends imitation in that it is less context-bound; it allows for the abstraction of essential principles from specific (admirable) instances of behavior. Play allows for the initial establishment of a more general model of what constitutes allowable (or ideal) behavior. Elaboration of dramatic play into formal drama likewise *ritualizes* play, abstracting its key elements one level more, and further distills the vitally interesting aspects of behavior— which are representative (by no mere chance) of that active exploratory and communicative pattern upon which all adaptation is necessarily predicated. Theatrical ritual dramatically represents the individual and social consequences of stylized, distilled behavioral patterns, based in their expression upon different assumptions of value and expectations of outcome. Formal drama clothes potent ideas in personality, exploring different paths of directed or motivated action, playing out conflict, cathartically, offering ritual models for emulation or rejection. Dramatic *personae* embody the behavioral wisdom of history. In an analogous fashion, in a less abstract, less ritualized manner, the ongoing behavior of parents dramatizes cumulative mimetic history for children.

Emergence of narrative, which contains much more information than it explicitly presents, further disembodies the knowledge extant latently in behavioral pattern. Narrative presents semantic representation of play or drama—offers essentially abstracted episodic representations of social interaction and individual endeavor—and allows behavioral patterns contained entirely in linguistic representation to incarnate themselves in dramatic form on the private stage of individual imagination. Much of the information derived from a story is actually already contained *in episodic memory*. In a sense, it could be said that the words of the story merely act as a retrieval cue for information already in the mnestic system (of the listener), although perhaps not yet transformed into a form capable either of explicit (semantic) communication, or alteration of procedure.[177] [178] It is for this reason that Shakespeare might be viewed as a precursor to Freud (think of Hamlet): Shakespeare "knew" what Freud later "discovered," but he knew it more implicitly, more imagistically, more procedurally. (This is not say that Shakespeare was any less brilliant, just that his level of abstraction was different.) Ideas, after all, come from somewhere; they do not arise, spontaneously, from the void. Every complex psychological theory has a lengthy period of historical development (development that might not be evidently linked to the final emergence of the theory).

Interpretation of the reason for dramatic consequences, portrayed in narrative—generally left to the imagination of the audience—constitutes analysis of the *moral* of the story. Transmission of that moral—that rule for behavior, or representation—is the *purpose* of narrative, just as fascination, involuntary seizure of interest, is its (biologically predetermined) means. With development of the story, mere description of critically important (and therefore compelling) behavioral/representational patterns becomes able to promote active imitation. At this point the semantic system, activating images in episodic memory, sets the stage for the alteration of procedure itself. This means establishment of a "feedback loop," wherein information can cycle up and down "levels of consciousness"—with the social environment as necessary intermediary—transforming itself and expanding as it moves. Development of narrative means verbal abstraction of knowledge disembodied in episodic memory and embodied in behavior. It means capability to disseminate such knowledge widely and rapidly throughout a communicating population, with minimal expenditure of time and energy. Finally, it means intact preservation of such knowledge, simply and accurately, for generations to come. Narrative description of archetypal behavioral patterns and representational schemas—*myth*—appears as an essential precondition for social construction and subsequent regulation of complexly civilized individual presumption, action and desire.

It is only after behavioral (procedural) wisdom has become "represented" in episodic memory and portrayed in drama and narrative that it becomes accessible to "conscious" verbal formulation and potential modification in abstraction. Procedural knowledge is not representational, in its basic form. Knowing-how information, generated in the course of exploratory activity, can nonetheless be *transferred* from individual to individual, in the social community, through means of imitation. Piaget points out, for example, that children first act upon objects, and determine object "properties" in accordance with these actions, and then almost immediately imitate themselves, turning their own initial spontaneous actions into something to be represented and ritualized.[179] The same process occurs in interpersonal interaction, where the other person's action rapidly becomes something to be imitated and ritualized (and then abstracted and codified further). A shared rite, where each person's behavior is modified by that of the other, can therefore emerge in the absence of "consciousness" of the structure of the rite; however, once the social ritual is established, its structure can rapidly become described and codified (presuming sufficient cognitive ability and level of maturation). This process can in fact be observed during the spontaneous construction (and then codification) of children's games.[180] It is the organization of such "games"—and their elaboration, through repeated communication—that constitutes the basis for the construction of culture itself.

Behavior is imitated, then abstracted into play, formalized into drama and story, crystallized into myth and codified into religion—and only then criticized in philosophy, and provided, *post-hoc*, with rational underpinnings. Explicit philosophical statements regarding the grounds for and nature of ethical behavior, stated in a verbally comprehensible manner, were not established through rational endeavor. Their framing as such is (clearly) a secondary endeavor, as Nietzsche recognized:

What the scholars called a "rational foundation for morality" and tried to supply was, seen in the right light, merely a scholarly expression of the common faith in the prevalent morality; a new means of expression for this faith.[181]

Explicit (moral) philosophy arises from the mythos of culture, grounded in procedure, rendered progressively more abstract and episodic through ritual action and observation of that action. The process of increasing abstraction has allowed the *knowing what* "system" to generate a representation, in imagination, of the "implicit predicates" of behavior governed by the *knowing how* "system." Generation of such information was necessary to simultaneously ensure accurate prediction of the behavior of others (and of the self), and to program predictable social behavior through exchange of abstracted moral (procedural) information. Nietzsche states, further:

That individual philosophical concepts are not anything capricious or autonomously evolving, but grow up in connection and relationship with each other; that, however suddenly and arbitrarily they seem to appear in the history of thought, they nevertheless belong just as much to a system as all the members of the fauna of a continent—is betrayed in the end also by the fact that the most diverse philosophers keep filling in a definite fundamental scheme of possible philosophies. Under an invisible spell, they always revolve once more in the same orbit; however independent of each other they may feel themselves with their critical or systematic wills, something within them leads them, something impels them in a definite order, one after the other—to wit, the innate systematic structure and relationship of their concepts. Their thinking is, in fact, far less a discovery than a recognition, a remembering, a return and a homecoming to a remote, primordial, and inclusive household of the soul, out of which those concepts grew originally: philosophizing is to this extent a kind of atavism of the highest order.[182]

The knowing-what system, declarative (episodic and semantic), has developed a description of knowing-how activity, procedure, through a complex, lengthy process of abstraction. Action and imitation of action developmentally predate explicit description or discovery of the rules governing action. Adaptation through play and drama preceded development of linguistic thought, and provided the ground from which it emerged. Each developmental "stage"—action, imitation, play, ritual, drama, narrative, myth, religion, philosophy, rationality—offers an increasingly abstracted, generalized and detailed representation of the behavioral wisdom embedded in and established during the previous stage. The introduction of semantic representation to the human realm of behavior allowed for continuance and ever-increasing extension of the cognitive process originating in action, imitation, play, and drama. Language turned drama into mythic narrative, narrative into formal religion, and religion into critical philosophy, providing for exponential expansion of adaptive ability. Consider Nietzsche's words once again:

Gradually it has become clear to me what every great philosophy so far has been: namely, the personal confession of its author and a kind of involuntary and unconscious memoir; also that the moral (or immoral) intentions in every philosophy constituted the real germ of life from which the whole plant had grown.[183]

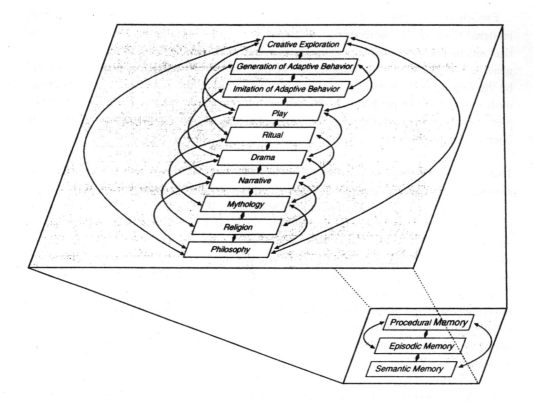

Figure 13: Abstraction of Wisdom, and the Relationship of Such Abstraction to Memory

The procedural system provides (constitutes?) memory for behavior. Such memory includes imitative representation of behaviors generated spontaneously in the course of creative individual action, whose precise circumstance of origins have been lost in the mists of history, but which have been integrated into a consistent behavioral pattern, over time (integrated into culturally determined *character*). Integration means active balance of competing subjectively grounded motivational demands within the context of the social environment, means internalization of socially regulated behavioral expression of subjective desire. Such internalization constitutes construction of a value (dominance) hierarchy; means determination of the relative contextual propriety (morality) of imitated or otherwise incorporated patterns of action. Such construction inevitably "precedes" episodic or semantic representation of the basis of the construction, although such second-order representation, once established, becomes capable (indirectly) of modifying procedure itself (as what is imagined can then be acted out). This is the loop that feeds the development of explicit "consciousness" itself: procedure is established, then represented, then altered in abstraction, then practiced; the procedure changes, as a consequence of the abstracted and practiced modification; this change in turn produces an alteration in its representation, and so on, and so on, from individual to individual, down the chain of generations. This process can occur "externally," as a consequence of social interaction, or "internally," as a consequence of word and image-mediated abstract exploratory activity ("thought"). This interactive loop and its putative relationship to underlying cognitive/memory structures are represented schematically in *Figure 13: Abstraction of Wisdom, and the Relationship of Such Abstraction to Memory*. (Only a few of the interactions between the "stages" of knowledge are indicated, for the sake of schematic simplicity.)

Behavioral knowledge is generated during the process of creative exploration. The consequences of such exploration—the adaptive behavioral patterns generated—are imitated, and represented more abstractly. Play allows for the generalization of imitated knowledge, and for the integration of behaviors garnered from different sources (one "good thing to do" may conflict in a given situation with another; "good things to do" therefore have to be ranked in terms of their context-dependent value, importance or *dominance*). Each succeeding stage of abstraction modifies all others, as our ability to speak, for example, has expanded our capacity to play. As the process of abstraction continues and information vital for survival is represented more simply and efficiently, what is represented transforms from the particulars of any given adaptive actions to the most general and broadly appropriate pattern of adaptation—that of creative exploration itself. This is to say: individual acts of voluntary and successful encounter with the unknown might be broadly imitated; might elicit spontaneous imitation. But some more essential ("prototypical"[184]) feature(s) characterize all such acts. With increasing abstraction and breadth of representation, essential features come to dominate particular features. As Eliade[185] points out, traditional (that is, nonliterate) cultures have a historical memory that may be only a few generations long—that is, as long as the oldest surviving individual is old. Events that occurred previous to this are telescoped into something akin to the aboriginal Australian's "dreamtime," the "trans-historical" period when ancestral giants walked the earth and established the behavioral patterns comprising the present mode of being. This telescoping, the "mythologization" of history, is very useful from the perspective of *efficient storage*. We learn to imitate (and to remember) not individual heroes, the "objective" historical figures of the past, but what those heroes represented: *the pattern of action that made them heroes*. That pattern is the act of voluntary and successful encounter with the unknown, the generation of wisdom through exploration. (I am not trying to imply, either, that the semantic or episodic memory systems can directly modify procedure; it is more that the operations of the semantic/episodic systems alter the world, and world alterations alter procedure. The effect of language and image on behavior is generally secondary—mediated through the environment—but is no less profound for that.)

The fact that the many "stories" we live by can be coded and transmitted at different levels of "abstraction," ranging from the purely motoric or procedural (transmitted through imitation) to the more purely semantic (transmitted through the medium of explicit ethical philosophy, say) makes comprehension of their structure and inter-relationships conceptually difficult. This difficulty is compounded by the fact that different stories have different spatial-temporal "resolutions"—that is, we may be governed at one moment by short-term, simple considerations and at the next by longer-term, more complex considerations. Someone married might think, for example, "I find my friend's spouse particularly attractive; I would like to make love to him or her"—evaluating that individual, positively—and then, immediately, correct: "My friend's spouse flirts too much for his or her own good, and looks like a lot of trouble." Perhaps both these viewpoints are valid. It is certainly not uncommon for the same "stimulus" to possess competing valences. Otherwise, as I said before, we would never have to think.

Every apprehensible phenomenon has a multitude of potential uses and significances. It is for this reason that it is possible for each of us to drown in possibility. Even something as simple as a piece of paper is not simple at all, except insofar as implicit contextual determinants make it appear so. Wittgenstein asks:

Point to a piece of paper.—And now point to its shape—now to its colour—now to its number (that sounds queer).—How did you do it?—You will say that you "meant" a different thing each time you pointed. And if I ask how that is done, you will say you concentrated your attention on the colour, the shape, etc. But I ask again: how is *that* done?[186]

A kitchen knife, for example: is it something to cut up vegetables, at dinner? Something to draw, for a still life? A toy, for mumblety-peg? A screwdriver, to fix a shelf? An implement of murder? In the first four cases, it "possesses" a positive valence. In the last case, it is negative—unless you are experiencing a frenzy of rage. How is its essential functional and affective multiplicity reduced to something singular and, therefore, useful? You can't fix the shelf and make dinner at the same time, and in the same place. You may need to do both, at some point, however, and this means that you must maintain the multiple uses and valences as possibilities. This means that you must (1) decide on one course of action, and eliminate all the rest, yet (2) retain the others for future consideration to ensure that your range of possible actions remain as broad as possible.

How is this ever-present competition to be ameliorated? How might the process of amelioration be considered, with regard to the additional complicating fact of the multilevel embodiment and abstraction of stories? So far we have considered the "ends" and the "means" of a given framework of reference (a story) as qualitatively different phenomena, echoing a dilemma that pervades ethics, as a field of study. The end or goal of a given planned sequence of behavior constitutes an image of the desired future, which serves as point of contrast, for the unbearable present. The means by which this end might be attained comprises the actual behavioral steps that might be undertaken, in pursuit of such desirable change. This seems a very reasonable perspective, in that at any given moment means and ends might be usefully distinguished. *Where we are going* is evidently different than *how we will get there*. This conceptual utility is only provisional, however—and the fact of the "means/end" distinction actually obscures more detailed and comprehensive description. Means and ends, plans and goals, are not qualitatively different, in any final sense, and can be transformed, one into the other, at any moment. Such transformation occurs, in fact, whenever a problem arises: whenever the unknown manifests itself, in the course of our ongoing behavior. It is in this manner that we switch spatial-temporal resolution (change "set" or shift our "frames of reference"), in order to re-evaluate our actions, and reconsider the propriety of our wishes.

Our stories—our frames of reference—appear to have a "nested" or hierarchical structure. At any given moment, our attention occupies only one level of that structure. This capacity for restricted attention gives us the capability to make provisional but necessary judgments about the valence and utility of phenomena. However, we can also shift levels of abstraction—*we can voluntarily focus our attention, when necessary, on stories that map out larger or smaller areas of space-time* (excuse the Einsteinian reference, but it is in fact accurate in this case, as our stories have a duration, as well as an area). "When necessary" means *depending on the status of our current operations*. For example, say you are in the kitchen, and you want to read a book in your study. An image of you reading a book in your favorite chair occupies the "ends" or "desired future" pole of your currently operational story (contrasted with the still-too-illiterate you of the present time). This "story" might have a conceived duration of, say, ten minutes; in addition, it "occupies" a universe defined by the presence of a half-dozen relevant "objects" (a reading lamp, a chair, the floor you have to walk on to get to your chair, the book itself, your reading glasses) and the limited space they occupy. You make it to your chair. Your book is at hand. You reach up to turn on the reading light—flash!—the bulb burns out. The unknown—the unexpected, in this context—has just manifested itself. You switch "set." Now your goal, still nested within the "reading a book" story, is "fix the reading lamp." You adjust your plans, find a new bulb, and place it in the lamp. Flash! It burns out again. This time you smell burnt wire. This is worrisome. The book is now forgotten—irrelevant, given the current state of affairs. Is there something wrong with the lamp (and, therefore, at a slightly more general level, with all future plans that depend on that lamp)? You explore. The lamp doesn't smell. It's the electrical outlet, in the wall! The plate covering the outlets is hot! What does that mean? You shift your apprehension up several levels of spatial-temporal resolution. Maybe something is wrong with the wiring of the house itself! The lamp is now forgotten. Ensuring that your house does not burn down has suddenly taken priority. How does this shift in attention occur?

Figure 14: Conceptual Transformation of the Means/Ends Relationship from Static to Dynamic presents a tripartite schematic, designed to take us from the state where we conceptualize means and ends as distinct, to the state where we see them as isomorphic elements, given distinct status only on a provisional basis. Subdiagram (1) is familiar and represents the "normal" story, composed of present state, desired future state, and three of the various means that might be utilized, in order to transform the former into the latter. This subdiagram is predicated on the presumption that many means might be used to get from point "a" to point "b"; in truth, however, only one means (the "most efficient" or otherwise desirable) will be employed at any one time. (We only have one motor output system, after all—and, therefore, one "consciousness"?) Subdiagram (2) is a transformed version of (1), showing that the "plans" of (1) can better be conceptualized as "stories," in and of

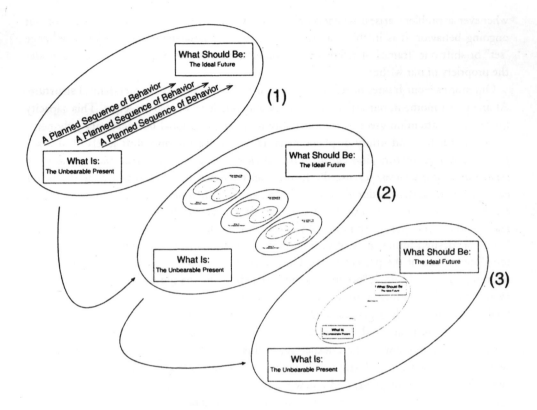

Figure 14: Conceptual Transformation of the Means/Ends Relationship from Static to Dynamic

themselves—showing that a "big" story (one that occupies a large spatial-temporal domain) is actually composed of nested "little" stories. Subdiagram (2) is still predicated on the presumption that a number of smaller stories might be used as means for a larger end. If your company is failing, you might fire half your employees, branch out into a new product line, or cut the salaries of your upper management. Each of these approaches, all designed for the same purpose, are clearly different (and complex) in their internal structure. You might do more than one thing, but if two of these multiple things conflict, one will have to be made subordinate to the other. Plans (and ends) are granted comparative importance, and organized accordingly. This state of affairs, where the relative importance of (potentially competing) plans has been fixed, is represented in subdiagram (3), which will remain our representation of choice.[187]

At any given place and time, we are considering only a fixed number of "variables" as means and ends. This is absolutely necessary, as action requires exclusion as much (or more) as inclusion.[188] However, those things we consider as "relevant variables" (and their status as relevant, or not) have to be mutable. We have to decide, yet retain the capacity to alter our decisions. Our prefrontal cortex—critical to goal-directed action[189]—appears to allow us this freedom: it does so, by "temporally sequencing" events and actions,[190] by considering contextual information and using that consideration to govern behavior,[191] and by shifting set.[192] It performs this multiplicity of operations, I submit, by considering one thing, then another, as the currently operative "consummatory reward"—as the goal toward which behavior is to be devoted, as the "desired future" against which the "unbearable present," in the form of emergent experience, is to be compared and evaluated. The structure in Figure 14, subdiagram (3), is a multilevel, nested structure, composed of the interdependent goals and plans comprising the "life story." This conceptualization helps explain the idea of a "step along the way" (a stairway or ladder to heaven, metaphorically speaking).[193]

Each step—each substory—has the same *structure* (but not the same *content*) as all those stories "above" and "below." This means that all the elements of a "good" story might be expected to mirror, in some profound manner, all the other elements: that a story, like the world itself, might be read (and read correctly) at multiple and multiply informative levels of analysis. This gives good stories their *polysemous* quality. It is for this reason that Frye can state:

One of the commonest experiences in reading is the sense of further discoveries to be made within the same structure of words. The feeling is approximately "there is more to be got out of this," or we may say, of something we particularly admire, that every time we read it we get something new out of it.[194]

A phenomenon that constitutes a goal at one level might be regarded as an incentive reward at the next, since the attainment of subsidiary goals are preconditions for the attainment of higher-level goals (this implies that most consummatory rewards will simultaneously possess an incentive aspect). The cognitive operations dependent upon the intact prefrontal cortex can move up and down these levels, so to speak, fixating at one, and allowing for determinate action, when that is deemed most appropriate (making the others implicit at that place and time); reorganizing and reconstituting the levels and their respective statuses, when that becomes necessary. *Figure 15: Bounded Revolution* sheds light on this process and, simultaneously, on the conundrum of relative novelty. How can a thing be radically new, somewhat new, somewhat familiar, or completely familiar? The simple answer is— a given phenomenon (a "thing" or "situation") can have its utility and/or meaning transformed at one level of analysis, but not at another. This means that novelty can be "bounded"; that something can be new in one manner, but remain familiar at another. This upper "familiar" level provides "walls" of security. These walls enclose a bounded territory, within which necessary change can occur, without fear of catastrophe.

Here is an exemplary "story": I am an undergraduate. I want to be a doctor. I am not sure exactly why, but that question has never become relevant (which is to say, my desire is an implicit presumption, an axiom of my behavior). I did well in high school. I have good marks in university, as a pre-med student. I take the MCAT. I fail: twentieth percentile. Suddenly and unexpectedly I realize that I am not going to be a doctor. The walls come tumbling

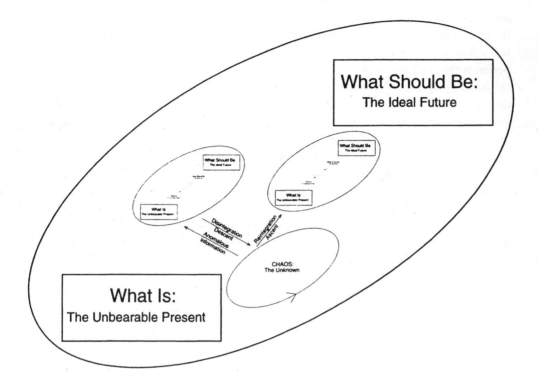

Figure 15: Bounded Revolution

down. My emotions, which were held in check by the determinate valences my ongoing story gave to experiential phenomena, now (re)emerge, viciously—in chaos. I am a depressed and anxious wreck. As I recover, I re-evaluate my life. I am disciplined and have good academic skills. I like university; I like working with people. Many of the predicates of the doctor story are still intact and do not need modification. I must journey further up the hierarchy, then!—maybe, for the first time. We do not question a story, when it is working! If it produces the desired results, it is *correct*! Why did I want to be a doctor? For monetary security. Because it was expected of me (for reasons of tradition—my father was a doctor). For reasons of status. Because I could appease the suffering of others and be a good person. So—hierarchical organization [this takes (or even is) thought]: (1) I want to help people; (2) I need some monetary security; (3) I would like to stay in the health profession; (4) perhaps status is not as important as I thought (and might therefore be "sacrificed," to appease the angry gods and restore order to the cosmos). I will become a *medical technician* or maybe even a *nurse*. I can still be a "good person" even if I'm not a doctor, and perhaps that is the most important thing of all. Reorganization completed. Utility of experiential phenomena re-established. Emotional integrity and stability reattained. Good thing I didn't do anything rash!

It is interesting and instructive to consider Eastern representations of reality (that is, of the "cosmos") in light of this conceptualization. Reality is made up of nested interpretations, that give determinate form to objects (as implements) and to the valence of those objects. Every interpretation, however, is subject to transformation, at every level. This constant (and necessary) transformation, in conjunction with the fact of at least transient (and necessary) stability, makes up the "world." Mircea Eliade describes the Indian version of the doctrine of the "eternal return"—the endlessly nested, cyclical nature of the "universe" (conceived of as the totality of experience, and not as "objective reality"):

A complete cycle, a *mahayuga*, comprises 12,000 years. It ends with a dissolution, a *pralaya*, which is repeated more drastically (*mahapralaya*, the Great Dissolution) at the end of the thousandth cycle. For the paradigmatic schema "creation-destruction-creation-etc." is reproduced *ad infinitum*. The 12,000 years of a *mahayuga* were regarded as divine years, each with a duration of 360 years, which gives a total of 4,320,000 years for a single cosmic cycle. A thousand such *mahayugas* make up a *kalpa* (form); *14 kalpas* make up a *manvantara* (so named because each *manvantara* is supposed to be ruled by Manu, the mythical Ancestor-King). A *kalpa* is equivalent to a day in the life of Brahma; a second *kalpa* to a night. One hundred of these "years" of Brahma, in other words 311,000 milliards of human years, constitute the life of Brahma. But even this duration of the god's life does not exhaust time, for the gods are not eternal and the cosmic creations and destructions succeed one another forever.[195]

Every novelty-inspired, exploration-driven "learning experience" has a revolutionary element; it is just that those reconstructions that involve stories with very limited "sizes" (that is, spatial-temporal areas) release only a proportionate amount of emotion. The "normal/ revolutionary" *dichotomy* is, therefore, not valid—it is always a matter of degree. Small-scale inconveniences require minor life-story modifications. Large-scale catastrophes, by contrast, undermine everything. The "biggest disasters" occur when the largest stories that we play out are threatened with dissolution, as a consequence of radical "environmental" transformation. Such transformation may occur in the natural course of things, when an earthquake or similar "act of God" takes place; may be generated internally, as a consequence of heretical action; or may emerge when the "foreign devils"—emissaries of chaos—threaten our explored territories (our nested stories, our cultural stability). In the latter case, we may well turn to war as an alternative deemed emotionally desirable.

Our stories are nested (one thing leads to another) and hierarchically arranged [pursuit "a" is superordinate to pursuit "b" (love is more important than money)]. Within this nested hierarchy, our consciousness—our apperception—appears to have a "natural" level of resolution, or categorization. This default resolution is reflected in the fact, as alluded to previously, of the basic object level. We "see" some things *naturally*, that is, in Roger Brown's terminology, at a level that gives us "maximal information with minimal cognitive effort."[196] I don't know what drives the mechanism that determines the appropriate level of analysis. Elements of probability and predictability must play a role. It is, after all, increasingly useless to speculate over increasingly large spatial-temporal areas, as the number of variables that must be considered increases rapidly, even exponentially (and the probability of accurate prediction, therefore, decreases). Perhaps the answer is something along the lines of "the simplest solution that does not generate additional evident problems wins," which I suppose is a variant of Occam's razor. So the simplest cognitive/exploratory maneuver that renders an unpredictable occurrence *conditionally* predictable or familiar is most likely to be adopted. This is another example of proof through utility—if a solution "works" (serves to further progress toward a given goal), then it is "right." Perhaps it is the frontal cortex that determines what might be the most parsimonious possible context, within which a given novel occurrence might be evaluated. So the notion would be that a novel occurrence initiates an exploratory procedure, part of which is devoted to determining the level of analysis most appropriate for conducting an evaluation. This would involve the shifting of stories. Also, a given stimulus is obviously not evaluated at all possible levels of analysis, simultaneously. This would constitute an impossible cognitive burden. It seems that the cortex must temporarily fixate at a chosen level, and then act "as if" that is the only relevant level. Through this maneuver, the valence of something can appear similarly fixed. It is only this arbitrary restriction of data that makes understanding—and action—possible.

Anyway: we are adapted, as biological organisms, to construe our environment as a domain with particular temporal and spatial borders—that is, as a place of a certain size, with a fixed duration. Within that "environment," conceived of as that certain size and duration, certain phenomena "leap out at us," and "cry out to be named."[197] Whenever those "natural categories" of interpretation and their associated schemas of action fail us, however, we have to look up and down the scale of spatial-temporal resolution. We do this by looking at the big picture, when we have to, or by focusing on details that may have previously escaped us. Both the details and the big picture may be considered as dwindling or trailing off into, first, the unconscious (where they exist as potential objects of cognition) and then, the unknown (where they exist as latent information or as undiscovered facts). The unconscious may then be considered as the mediator between the unknown, which surrounds us constantly, and the domain that is so familiar to us that its contents have been rendered explicit. This mediator, I would suggest, is those metaphoric, imagistic processes, dependent upon limbic-motivated right-hemispheric activity, that help us initially formulate our stories. *Figure 16: Nested Stories, Processes of Generation, and Multiple Memory Systems* helps explain the idea of this "unconscious"—the broadest span stories, which are determined by complex social interactions, are episodic (imagistic) or even procedural (manifested only in socially modified behavior) in nature. There is a very narrow window of expressible "frames of reference"—conscious stories. Just ask any young child or unsophisticated adult to describe the "rationale" for their behaviors.

Every level of analysis—that is, every definable categorization system and schema for action (every determinate story)—has been constructed, interpersonally, in the course of exploratory behavior and communication of the strategies and results thereof. Our natural levels of apprehension, the stories that most easily or by default occupy our attention, are relatively accessible to consciousness and amenable to explicit verbal/semantic formulation and

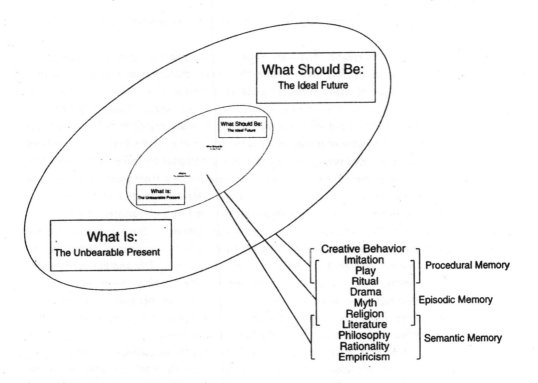

Figure 16: Nested Stories, Processes of Generation, and Multiple Memory Systems

communication. The higher-level stories, which cover a broader expanse of spatial-temporal territory, are increasingly complex and, therefore, cannot be as simply formulated. Myth steps in to fill the breach.

MYTHOLOGICAL REPRESENTATION: THE CONSTITUENT ELEMENTS OF EXPERIENCE

Myth represents the world as forum for action. The world as forum for action comprises three eternally extant constituent elements of experience and a fourth that precedes them. The unknown, the knower, and the known make up the world as place of drama; the indeterminate "precosmogonic chaos" preceding their emergence serves as the ultimate source of all things (including the three constituent elements of experience).

The precosmogonic chaos tends to take metaphorical form as the *uroboros*, the self-consuming serpent, who represents the union of matter and spirit, and the possibility of transformation. The uroboros serves as "primal source" of the mythological world parents (the *Great Mother*, nature, deity of the unknown, creative and destructive; the *Great Father*, culture, deity of the familiar, tyrannical and protective) and of their "*Divine Son*" (the *Knower*, the generative *Word*, the process of exploration).

The ancient Mesopotamian creation myth—the *Enuma elish*—provides a concrete example of the interplay of these personalities. This myth features four main characters or sets of characters; *Tiamat*, the feminine dragon of chaos, primordial goddess of creation (the uroboros and the Great Mother are conflated, as is frequently the case, in this myth); *Apsu*, Tiamat's husband and consort; the elder gods, children of Tiamat and Apsu; and *Marduk*, sun deity and mythic hero. Tiamat symbolizes the great unknown, the matrix of the world; Apsu the known, the pattern that makes regulated existence possible. The elder gods symbolize the common psychological attributes of humanity (the fragments or constituent elements of consciousness), and constitute a more thorough representation of the constituent elements of the patriarchal known; Marduk, greatest of the secondary deities, represents the process that eternally mediates between matrix and regulated existence.

The original union of Tiamat and Apsu brings the elder gods into being. These gods carelessly kill Apsu, upon whom they unconsciously depend. Tiamat reappears, with a vengeance, and decides to destroy everything she has created. Her "children" send one volunteer after another out to overpower her. All fail. Finally, Marduk offers to do battle. He is elected as king—the greatest of gods, the "determiner of destinies"—and voluntarily confronts Tiamat. He cuts her apart, and creates the cosmos from her pieces. The Mesopotamian emperor, who ritually embodies Marduk, acts out this battle during the festival of the New Year, when the "old world" is renewed.

The *Enuma elish* expresses in image and narrative the idea that the psychological function giving order to chaos (1) creates the cosmos and (2) should occupy a superordinate position, in the intrapsychic and social domains. The ideas contained in this myth are given more elaborated expression in later Egyptian works of metaphysical speculation, which more directly address the idea of the heroic renewal of culture.

The three constituent elements of experience and the fourth who precedes them can be viewed, at a higher level of resolution, as seven universal characters (who may take on any of a variety of culture-specific identities). Myth describes the interactions of these characters. The great dragon of chaos—the uroboros, the self-devouring serpent—might be conceptualized as pure (latent) information, before it is parsed into the world of the familiar, the unfamiliar, and the experiencing subject. The uroboros is the stuff of which categorical knowledge is composed, before being that knowledge; it is the primary "element" of the world, which is decomposed into cosmos, surrounding chaos, and the exploratory process which "separates" the two.

The bivalent Great Mother (second and third characters) is creation and destruction, simultaneously—the source of all new things, the benevolent bearer and lover of the hero; the destructive forces of the unknown, the source of fear itself, constantly conspiring to destroy life. The bivalent divine son (fourth and fifth) is the sun god, the hero who journeys to the underworld to rescue his incapacitated ancestors, the messianic son of the virgin mother, savior of the world—and, simultaneously, his sworn adversary, arrogant and deceitful. The bivalent Great Father (sixth and seventh) is the wise king and the tyrant, cultural protection from the terrible forces of nature, security for the weak, and wisdom for the foolish. Simultaneously, however, he is the force who devours his own offspring, who rules the kingdom with a cruel and unjust hand, and who actively suppresses any sign of dissent or difference.

Terrible, chaotic forces lurk behind the facade of the normal world. These forces are kept at bay by maintenance of social order. The reign of order is insufficient, however, because order itself becomes overbearing and deadly, if allowed unregulated or permanent expression. The actions of the hero constitute an antidote to the deadly forces of chaos, and to the tyranny of order. The hero creates order from chaos and reconstructs that order when necessary. His actions simultaneously ensure that novelty remains tolerable and that security remains flexible.

> *Mephistopheles*: Congratulations, before you part from me!
> You know the devil, that is plain to see.
> Here, take this key.
> *Faust*: That little thing? But why?
> *Mephistopheles*: First grasp it; it is nothing to decry.
> *Faust*: It glows, it shines, increases in my hand!
> *Mephistopheles*: How great its worth, you soon shall understand.
> The key will smell the right place from all others:
> Follow it down, it leads you to the Mothers! [198]

Introduction

Reasonable and informed observers—at least since the time of Frazier[199]—have established the widespread spatial and temporal dispersion of cosmogonic stories, tales of heroism and deceit, rituals of initiation, and standard imagistic representations, such as the virgin and child. These stories, tales, rituals and images often differ in detail, and temporal ordering; sometimes, however, they are simply the same. It is possible that this similarity might be the consequence of dissemination, from a single source, hundreds of centuries ago. This hypothesis, however, does not explain why standard stories are *remembered*, once disseminated, and transmitted down the generations, with little structural alteration. It is reasonable to presume that, over the long run, our species "forgets" most things that are useless: we do not forget our myths, however. Indeed, much of the activity broadly deemed "cultural" is in fact the effort to ensure that such myths are constantly represented and communicated.

Carl Jung attempted to account for the apparent universality of world interpretation with the hypothesis of the "collective unconscious." Jung believed that religious or mythological symbols sprung from a universal source, whose final point of origin was biological (and heritable). His "collective unconscious" was composed of "complexes," which he defined as heritable propensities for behavior or for classification. The Jungian position, which is almost never understood properly, has attracted more than its share of derision. Jung was not privy to our knowledge of the mechanisms of inheritance (a limitation necessarily shared by all members of his generation); the idea of "collective memories" appears impossible—

Lamarckian—from the modern perspective. Jung did not really believe that individual memories themselves could be transmitted, however—although his writings, which are very difficult, do not always make this clear. When he speaks formally of the collective unconscious, he is at pains to point out that it is the possibility of categorization that is inherited, and not the contents of memory itself. However, he frequently writes as if the contents, as well, might be inherited.

The general irritation over Jung's "heritable memory" hypothesis has blinded psychologists and others to the remarkable fact that narratives *do* appear patterned, across diverse cultures. The fact that all cultures use what are clearly and rapidly identifiable as "narratives" (or at least as "rites," which are clearly dramatic in nature) in itself strongly points to an underlying commonality of structure and purpose. It might still be objected: attempts to attribute comprehensible patterning to such narratives cannot be demonstrated, without a theory of interpretation, and that theory may be merely "reading in" patterns, where none actually "exist." The same objection can, of course, be applied—and applied validly—to literary interpretation, the study of history, dream analysis and anthropology. Cultural phenomena cannot be understood except from a cultural perspective. This fundamental problem (among others) makes *verification* of theories in the "domain of value" difficult.

Nonetheless, to live, it is necessary to act. Action presupposes belief and interpretation (implicit, if not explicit). Belief has to be grounded in faith, in the final analysis (as the criteria by which a moral theory might be evaluated have to be *chosen*, as well). There is no reason, however, why such faith cannot be informed, and critically assessed. It seems reasonable to presume that cross-cultural analysis of systems of belief, and their comparison with the essentially literary productions of the humanities, might constitute a means to attain such information. This was Jung's approach. The "causal mechanism" he constructed to account for what he found—that is, the "collective unconscious"—appears insufficiently elaborated, from the modern empirical perspective (although the idea is much more complex, and much less easily dismissable, than generally conceded). This does not mean that we should dismiss Jung's methodology or deride his otherwise valuable insights. Great modern minds, working in areas outside of psychology, have also concluded that stories have universal structures.

How can the *fact* of patterned stories—archetypal stories, if you will—be reconciled with the apparent *impossibility* of inherited memory content? We might turn our attention to the phenomenon of language, and the processes of its "storage" and transmission, to find an answer. The human linguistic ability appears to have a relatively specific biological basis. Other animals do not have language, in their natural states, and cannot be taught language at any sophisticated level. Human children, by contrast—even when severely intellectually impaired—pick up language easily and use it fluently, naturally and creatively. Language use is an intrinsic characteristic of *Homo sapiens*, and the structure of language itself appears biologically grounded. Nonetheless, human languages differ. A native Japanese speaker cannot understand a native French speaker, although it might be evident to both that the other is using language. It is possible for two phenomena to be different at one level of analysis and similar at another.

The question might be asked: upon what databank, so to speak, does a child draw, when he or she learns to talk (read, write)? The child listens to those around her. She is not explicitly "taught" how to talk, although some explicit teaching takes place. Her biological propensity encounters a cultural reality: the existence of language, in the culture. Her parents serve as primary intermediaries of culture: *they embody language in their behavior* and transmit it to her during their day-to-day activities. Nonetheless, they cannot be said to be the "creators" of language, although they may use it idiosyncratically—even creatively. It is the capability for human linguistic activity—whatever that is—that is the "creator." The cumulative consequences of this capability, expressed over centuries, have modified the behavior of all the individuals who compose a given linguistic "culture." Identifiable individuals serve as the temporary agents of embodied memory for the entire culture, at any given locale and time; nonetheless, the loss of a given individual poses no threat to the "knowledge" of the culture. This is because language is "remembered"—that is, embodied—in the behavior of all those who speak. Children pick up language by interacting with adults, who embody language. Thus, they learn to speak, and learn to know they have language, and even to observe and study the fact that they have language.

The same holds true of moral behavior and of the belief that "underlies" it. Adults embody the behavioral wisdom of their culture for their children. Children interact with adults, who serve as "cultural emissaries." Obviously, a given adult may be a better or worse representative, just as a parent may be more or less literate. However, a bad example can be as exemplary as a good example; furthermore, children are rarely limited, in their exposure, to a single "hero." If there are no other adults around in fact, they are inevitably present by proxy, in "entertainment": in ritual, drama, literature and myth. The behavioral patterns that make up our stories might therefore be regarded as "stored" in our (social) behavior. This implies that such patterns may be abstracted from that behavior, at any time. The "collective unconscious" is, from this perspective, embodied behavioral wisdom, in its most fundamental form—is the cumulative transmitted consequences of the fact of *exploration* and *culture* on *action*.

Our capacity for abstraction allows us to derive the constituent elements of successful "adaptation" itself from observation of behavioral patterns that are constantly played out in the world *as it actually exists.* The behavioral patterns that constitute adult interaction, for example, are exceedingly sophisticated and are conditioned to the last gesture by centuries of cultural work. We can extract "images" of these patterns; such images, just as sophisticated as the behaviors they represent, constitute the building blocks for our stories, and for our self-understanding. (The admirable adult, an identifiable individual, keeps her house tidy and neat, reconciles her warring brothers and learns hard moral lessons when such learning is necessary. The archetypal hero makes order out of chaos, brings peace to the world, and restructures society when it has become rigid and anachronistic.) The "collective unconscious" that constitutes the basis for shared religious mythology is in fact the behavior, the procedures, that have been generated, transmitted, imitated, and modified by everyone who has ever lived, everywhere. Images of these behaviors and of the transcendent "place" where they occur (the universe of chaos and order) constitute *metaphors*, symbolic images.

Metaphors mediate between our procedural wisdom and our explicit knowledge; they constitute the imagistic declarative point of transition between the act and the word.

We have spent hundreds of thousands of years watching ourselves act and telling stories about how we act. A good story has a universal quality, which means that it speaks a language we all understand. Any universally comprehensible language must have universal referents, and this means that a good story must speak to us about those aspects of experience that we all share. But what is it that every human being shares, regardless of place and time of birth? Is it reasonable to posit that anything might remain constant, for example, across the centuries that separate us from our Stone Age ancestors; across the ideological and religious barriers that divide the inhabitants of our modern nations? Our distant predecessors lived much closer to nature, and the problems that beset them seem far removed from our current daily struggles. The great difference between us and them seems analogous in distance, if not precisely in kind, to that obtaining between the varied cultural worlds of today—to the great gap that still separates the Hindu religious mystic, for example, from the Manhattan investment banker. It is not surprising that a world characterized by such different human lives remains rife with constant intergroup conflict, nor is it surprising that we might seem to have outgrown our traditional wisdom. But are there fundamental presuppositions we might agree upon and share despite our differences?

Most objects of experience have some properties in common, while varying with regard to others. Generally, the similarities and the differences are both significant. So it is with individuals, and with cultures. We seem peculiarly aware of our differences, however, and not of our similarities. Even groups of people who share much in common, at least from the perspective of more distant outsiders—the Irish Catholics and Protestants spring to mind—appear sufficiently conscious of those factors that make them unique in their social affiliation. I think this is in part because we are not built to focus on the predictable and familiar. Our attention gravitates naturally toward those aspects of our environments, natural and social, that contain information. The similarities of the Serb and the Croat are hidden from each other, so to speak, by a wall of habituation, but the differences stand out profoundly.

To ask the question "What is it that two or more discriminable beings or things or situations might *share*?" is really to ask "At what levels of analysis might two or more things be considered the same? And at what levels different?" It is the particulars of our individuality—our specific time and place—that differentiate us from one another. What unites us is the *fact* of those particulars, however: the fact that we each have a specific time and place, and the implications of that fact for the nature of our existence. Our lives are open to possibility, but remain eternally bounded by disease, death and subjugation to social structure. As mutable, limited social beings, we are all engaged in a massive, cooperative and competitive endeavor. We do not understand the rules that govern this endeavor, in the final analysis; we cannot state explicitly *why it is* that we do what we do. Our democratic constitutions, for example—which contain the most fundamental axioms of the "body of law" that we imitate (that governs our behavior)—are inextricably embedded in the conception of *natural rights* (which is to say, in a statement of faith: "We hold these truths to be self-evident"). We are all, in consequence, *imitating a story that we don't understand*. This story covers the broadest possible expanse of time and space (at least that expanse relevant to us), and is still implicitly "contained" in our behavior (although represented, in part, in episodic imagery and semantic description). This partially implicit containment constitutes our mythology, and our ritual, and provides the "upper-level," "unconscious" frames of reference within which our conditional and expressible individual stories retain their validity.

It is impossible to properly appreciate the nature of the categories of the mythological imagination without some understanding of the process of categorization. The act of categorization enables us to treat the mysterious and complex world we inhabit as if it were simpler—as if it were, in fact, comprehensible. We perform this act of simplification by treating objects or situations that share some aspect of structure, function or implication as if they were identical. People are very good at categorizing—so good, in fact, that the ability is taken for granted and appears simple. It is not so simple, however. Neither the "rules" that underly categorization, nor the act itself, have proved easy to describe. Roger Brown, the eminent psycholinguist, states:

Until about 1973, psychological experiments on category formation conceived of human categories on the model of a "proper set." Triangles are a proper set, which means that members of the triangle class are precisely definable in terms of a conjunction of attributes true of all members of the set, and of no nonmembers. A triangle is a closed three-sided figure. From the fact that a clear definition exists, it follows that membership in the set is not a matter of degree; one triangle is no more essentially triangular than any other. An entity either is or is not a triangle.

In retrospect, it is amazing that psychology was for so long able to think of real-life categories as proper sets. We ought to have worried more over the extreme difficulty everyone has in defining anything "natural," and natural, as used here, includes not only dogs and carrots but also artifacts like chairs, cars, and pencils. I know you can tell one when you see one, but just try listing the attributes that are true of all dogs and of no cats or wolves or hyenas, or of all the carrots and no radishes or turnips, or of all chairs and no small tables, hassocks, benches or slings.[200]

In the natural state, so to speak, human beings do not think like logicians or even like empiricists. It takes training to think like that. In the absence of such training, we still think, however; but more subjectively—like "unreasonable," idiosyncratically emotional beings, who inhabit bodies of particular size, with particular and constrained properties. Our *natural* categories, which are the groupings we generate spontaneously, do not consist solely of the consensually apprehensible properties shared by the things and situations we encounter. Neither are natural categories tightly bounded; their borders are fuzzy, and they overlap. The construction of proper sets is possible—obviously, since they exist—and the ability to construct and use such sets has proved useful, in a broad variety of manners. Nonetheless, the capability that underlies such construction appears relatively new, phylogenetically speaking, and seems dependent at least in part on the ability to think *empirically* and to regard things *objectively*. In the absence of such ability—which requires specialized training (or, at least, immersion in a culture like ours where such thinking has become commonplace) people naturally incline toward the development of what has been described (recently) as the "cognitive model." Cognitive models are characterized by a number of distinctive properties (as paraphrased, in part, from George Lakoff[201]):

1) They are *embodied* with regard to their content, which essentially means that they can be used, without necessarily being defined; are implicit in action, without necessarily being explicit in description. Two things classified within the same cognitive model are two things that evoke the same behavior, and can therefore be regarded, at least from the perspective of action, as one thing. If you are utilizing a cognitive model, and someone asks you to describe its content ("What makes a *dog*?"), you might say, "I can't say, but I know when one is around." You know that a dog is, for example, something friendly, something to be petted, and something to play with—although such knowledge does not constitute everything that makes up what you regard as *dog*. Most of the concepts you use are in fact embodied, at the most basic of levels—are habitual, procedural, motoric, behavioral. You can use them without thinking. Those that are not can only be applied slowly, with full conscious attention, and with effort.

2) They are characterized by *basic-level categorization* and *basic-level primacy*. These terms mean, respectively, that the phenomena most "naturally" apprehensible to the human mind—perceptible as a whole, or gestalt; nameable, communicable, manipulable, memorable—serve as the material for initial categorization, and that those initial categories provide the basis for the development of more abstract concepts (even for the comparison point for determining what we consider abstract). "Most naturally apprehensible" means learned and named first (generally with short names) and conceptualized at the level of distinctive action, (in association with such characteristic behaviors as petting for the category "cat" and smelling for the category "flower"). Our basic-level categories reflect our structure, as much as the structure of the external world: we most accurately conceive of those things that most simply present themselves to us. The "higher" and "lower" levels of category that surround these naturally apprehensible basic-level phenomena might be regarded, in contrast, as "achievements of the imagination," to use Roger Brown's phrase.[202] We perceive the "cat," for example, and *infer* the species that contains the cat or the subtype that makes it *Siamese*. Our basic-level categories generally occupy the middle of our conceptual hierarchies: we generalize when we move "up," and specialize when we move "down."

3) They may be used in *metonymic* or *reference-point* reasoning. Metonymic reasoning is *symbolic*, in the psychoanalytic or literary sense. *Metonymic* means *interchangeable* and more. The fact that objects in a cognitive model have metonymic properties means that any or all of those objects can stand for any or all of the others. This capacity makes sense, since all of the objects in a given category are by definition regarded as equivalent, in some nontrivial sense (most generally, in terms of *implication for action*). The human capacity for metaphor, aesthetic appreciation and allusion seems integrally related to the capacity for metonymic reasoning and the use of richly meaningful cognitive models.

4) They are characterized by *membership* and *centrality gradience*. Membership gradience implies degree of membership, which is to say that an ostrich, for example, is a bird, but perhaps not so much of a bird as a robin—because the robin has more properties that are *central* to the category *bird*. A thing can be a better or worse exemplar of its category; if it is worse, it can still be placed within that category.

5) They contain phenomena associated as a consequence of *familial resemblance*, a term used first in this context by Ludwig Wittgenstein.[203] Things with familial resemblance all share similarities with a potentially hypothetical object. The prototypical Smith brother, to use a famous example,[204] may have a dark mustache, beady eyes, balding pate, thick horn-rimmed glasses, dark beard, skinny neck, large ears and weak chin. Perhaps there are six Smith brothers, in total, none of whom has all the properties of the prototypical Smith. Morgan Smith has a weak chin, large ears, balding pate and skinny neck—but no glasses, mustache, or beard. Terry, by contrast, has the glasses, mustache and beard—but a full head of hair, small ears and a normal neck. Nelson has a receding hairline, beady eyes, and a dark beard and mustache—and so on for Lance, Randy and Lyle. None of the brothers precisely resembles another, but if you saw them in a group, you would say, "those men are all related."

6) They give rise to the phenomenon of *polysemy*, a defining characteristic of myth. A polysemic story is written and can be read validly on many levels. The phenomenon of polysemy, discussed in some detail later in this book, arises when the relationship of objects within a particular cognitive model is analogous in some sense to the relationship that obtains between cognitive models. Great works of literature are always polysemic, in this manner: the characters within the story stand in the same relationship to one another as things of more general significance stand to one another, in the broader world. The struggle of Moses against the Egyptian pharaoh, for example, to take a story we will consider later, can also be read as an allegory of the struggle of the oppressed against the oppressor or, even more generally, as the rebellion of the [world-destroying (flooding)] savior against society.

To say that two separable things belong to the same category is a tricky business. We presume, without thinking, that we group things as a consequence of something about them, rather than as a consequence of something about us. What do all *chairs* share in common, then? Any given chair may lack some of the most common chair attributes, such as legs, backs or armrests. Is a tree-stump a chair? Yes, if you can sit on it. It isn't really something about an object, considered as an independent thing, that makes it a chair: it is, rather, something about its potential for interaction with us. The category "chair" contains objects that serve a function we value. Chairs may be efficiently sat upon—at least potentially. Our action in the face of an object constitutes an elementary but fundamental form of classification (constitutes, in fact, the most fundamental of all classifications; the classification from which all abstracted divisions are derived). The category of "all things that make you want to run away when you look at them" might be considered, for example, a very basic form of construct. Closely related to this category, although slightly higher in the hierarchy of abstraction, might be the category of "all objects to be feared," or "all objects that are dangerous when approached in one fashion but beneficial when approached in another."

It is a meaningful but "irrational" classification scheme of this sort that Jung described as a *complex*—one of the constituent elements of the "collective unconscious." A complex is, in part, a group of phenomena, linked together because of shared significance [which is (essentially) *implication for action*, or *emotional equivalence*]. Jung believed that many complexes had an archetypal (or universal) basis, rooted in biology, and that this rooting had something specifically to do with memory. It appears that the truth is somewhat more complicated. We classify things according to the way they appear, the way they act, and in accordance with their significance to us, which is an indication of how to act in their presence—and may mix any or all of these attributes, irrationally (but meaningfully), in a single scheme. We categorize diverse things in similar manners, across cultures, because we share perceptual apparatus, motivational drive and emotional state, as well as structure of memory *and* physical form, manifested in observable behavior. The imagination has its natural categories, dependent for their existence on the interaction between our embodied minds and the world of shared experience; into these categories fall particular phenomena in a more or less predictable manner. Stories describe the interactions of the contents of the categories of the imagination, which take embodied form, in the shape of dramatic characters. The characters have a predictable nature, and play out their relationships in an eternally fascinating patterned fashion, time and time again, everywhere in the world.

So now we have the observation of commonality of structure, and a plausible theory to account for the presence of that commonality. Perhaps it would be reasonable, then, to describe the nature of the universal patterns in narrative—while placing a variety of additional and stringent constraints on that description, for the sake of caution (given the difficulty of verifying "interpretive theories"). First, let us make the description rationally acceptable, and internally consistent—that is, let us find a way of making sense of myth that does not conflict with the tenets of empiricism and experimental science, and that appears applicable to stories derived from many different places, and many different times. Let us further make the description simple, as a good theory should be simple—so that remembering the interpretive framework will be much easier than remembering the stories themselves. Let us make it compelling, as well, from the emotional perspective. Good theories have an affective component, sometimes described as "beauty." This beauty appears simultaneously as efficiency—the same sort of efficiency that characterizes a well-crafted tool—and as what might be described as a "window into possibility." A good theory lets you use things—things that once appeared useless—for desirable ends. In consequence, such a theory has a general sense of excitement and hope about it. A good theory about the structure of myth should let you see how a story you couldn't even understand previously might shed new and useful light on the meaning of your life. Finally, let us constrain the description by making it fit with what is known about the manner in which the brain actually operates (and which was described previously); let us ensure that the world of myth, as interpreted, is the same world apperceived by the mind.

Operation within this set of constraints allows for generation of the following straightforward hypothesis: the "partially implicit" mythic stories or fantasies that guide our adaptation, in general, appear to describe or portray or embody three permanent constituent elements of human experience: the *unknown*, or unexplored territory; the *known*, or explored territory; and the process—the *knower*—that mediates between them. These three elements constitute the cosmos—that is, the world of experience—from the narrative or mythological perspective.

No matter where an individual lives—and no matter when—he faces the same set of problems or, perhaps, the same set of metaproblems, since the details differ endlessly. He is a cultural creature, and must come to terms with the existence of that culture. He must master the domain of the *known*—explored territory—which is the set of interpretations and behavioral schemas he shares with his societal compatriots. He must understand his role within that culture—a role defined by the necessity of preservation, maintenance and transmission of tradition, as well as by capacity for revolution and radical update of that tradition, when such update becomes necessary. He must also be able to tolerate and even benefit from the existence of the transcendental *unknown*—unexplored territory—which is the aspect of experience that cannot be addressed with mere application of memorized and habitual procedures. Finally, he must adapt to the presence of himself—must face the endlessly tragic problem of the *knower*, the exploratory process, the limited, mortal subject; must serve as eternal mediator between the creative and destructive "underworld" of the unknown and the secure, oppressive patriarchal kingdom of human culture.

We cannot see the *unknown*, because we are protected from it by everything familiar and unquestioned. We are in addition habituated to what is *familiar* and *known*—by definition—and are therefore often unable to apprehend its structure (often even unable to perceive that it is there). Finally, we remain ignorant of *our own true nature*, because of its intrinsic complexity, and because we act toward others and ourselves in a socialized manner, which is to say a predictable manner—and thereby shield ourselves from our own mystery. The figures of myth, however, embody the world—"visible" and "invisible." Though the analysis of such figures, we can come to see just what meaning means, and how it reveals itself, in relationship to our actions. It is through such analysis that we can come to realize the potential breadth and depth of our own emotions, and the nature of our true being; to understand our capacity for great acts of evil—and great acts of good—and our motivations for participating in them.

Consider once again this archaic creation myth from Sumer:

So far, no cosmogonic text properly speaking has been discovered, but some allusions permit us to reconstruct the decisive moments of creation, as the Sumerians conceived it. The goddess Nammu (whose name is written with the pictograph representing the primordial sea) is presented as "the mother who gave birth to the Sky and the Earth" and the "ancestress who brought forth all the gods." The theme of the primordial waters, imagined as a totality at once cosmic and divine, is quite frequent in archaic cosmogonies. In this case too, the watery mass is identified with the original Mother, who, by parthenogenesis, gave birth to the first couple, the Sky (An) and the Earth (Ki), incarnating the male and female principles. This first couple was united, to the point of merging, in the *hieros gamos* [mystical marriage]. From their union was born En-lil, the god of the atmosphere. Another fragment informs us that the latter separated his parents.... The cosmogonic theme of the separation of sky and earth is also widely disseminated.[205]

The "sky" and "earth" of the Sumerians are categories of apprehension, characteristic of the Sumerian culture, and must not be confused with the sky and earth of modern empirical thinking. "An" and "Ki" are, instead, the dramatically represented Great Father and Great Mother of all things (including the son who "gives birth" to them). This somewhat paradoxical narrative is prototypical; mythologies of creation tend to manifest themselves in this pattern. In the *Enuma elish*, for example—the oldest written creation myth we possess—the Mesopotamian hero/deity Marduk faces the aquatic female dragon Tiamat (mother of all things, including Marduk himself), cuts her up, and creates the world from her pieces.[206] The god Marduk serves explicitly as *exemplar* for the Mesopotamian emperor,[207] whose job is to ensure that the cosmos exist and remain stable, as a consequence of his proper "moral" behavior, *defined by his imitation of Marduk*. In the Judeo-Christian tradition, it is the *Logos*[208]—the word of God—that creates order from chaos, and it is in the image of the *Logos* that man ["Let us make man in our image, after our likeness" (Genesis 1: 26)] is created. This idea has clear additional precedents in early and late Egyptian cosmology (as we shall see). In the Far East, similarly, the cosmos is imagined as composed of the interplay between *yang* and *yin*, chaos and order[209]—that is to say, unknown or unexplored territory and known or explored territory. *Tao*, from the Eastern perspective, is the pattern of behavior that mediates between them (analogous to En-lil, Marduk and the *Logos*) constantly generating, destroying, and regenerating the universe. For the Eastern individual, life in Tao is the highest good, the "way" and "meaning"; the goal toward which all other goals must remain subordinate.

Our narratives describe the world as it possesses broad but classifiable implication for motor output—as it *signifies*. We gather information about the nature of the world, as it signifies for behavior, by watching ourselves and the others who compose our social groups *act* in the world. We derive conclusions about the fundamental meanings of things by observing how we respond to them. The unknown becomes classifiable, in this manner, because we respond to its manifestation predictably. It compels our *actions*, and "makes" us *feel*. It frightens us into paralysis and entices us forward, simultaneously; it ignites our curiosity, and heightens our senses. It offers us new information and greater well-being, at the potential cost of our lives. We observe our responses, which are biologically predetermined, and draw the appropriate conclusions. The unknown is intrinsically *interesting*, in a manner that poses an endless dilemma. It promises and threatens simultaneously. It appears as the hypothetical ultimate "source" of all determinate information, and as the ultimate unity of all currently discriminable things. It surrounds all things, eternally; engenders all things and takes all things back. It can therefore be said, paradoxically, that we *know* specific things about the domain of the unknown, that we understand something about it, can act toward and represent it, even though it has not yet been explored. This paradoxical ability is a nontrivial capacity. Since the unknown constitutes an ineradicable component of the "environment," so to speak, we have to know what it is, what it signifies; we must understand its implication for behavior and its affective valence.

Explored territory is something altogether different. Habitual and familiar actions are useful there, instead of the frightened, tentative or exploratory behaviors that serve where nothing is certain. Habits and familiar actions exist, as a general rule, because they have been successful, because their implementation suffices to transform what would otherwise be unexplored territory into a safe and fruitful haven. As we have been at pains to demonstrate, the unknown does not lose its *a priori* motivational significance—promise and threat—because of the passive process of "habituation." Adaptation is *active*. "Habituation," except in the most trivial of senses, is the consequence of successful creative exploration, which means generation of behavioral patterns that turn the indeterminate meaning of something newly encountered into something positive, at best—neutral, at worst. Is fire dangerous, or beneficial? It depends on how it is approached—which is to say, fire has context-dependent potential for harm and for benefit. Which of its many "potentials" fire actually manifests depends on what behavioral strategy is undertaken in its presence. Fire heats our homes. Now and then—when we are insufficiently cautious—it burns one of them down. What fire does—which is to say, what it *is*, from the perspective of motivational significance—depends on how we treat it.

We have lost our fear of fire, not because we have habituated to it, but because we have learned how to control it. We have learned to specify and limit its "intrinsically" ambivalent affective valence, through modification of our own behavior, in its presence. Fire, insofar as we can control it, has been rendered predictable, nonthreatening—even familiar and comforting. All things we can control (which means, can bend to our own ends) have been likewise rendered predictable—by definition. The "territory" of "explored territory" is defined, at least in general, by *security*. Secure territory is that place where we know how to act. "Knowing how to act" means "being sure that our current actions will produce the results desired in the future." The affective significance of the phenomena that comprise "explored territory" have been mapped. This map takes the form of the story, which describes the valence of present occurrences, the form of the desired future, and the means that might serve usefully to transform the former into the latter. Any territories our stories serve to render beneficial constitute "home ground."

Home ground—explored territory—is that place where unfamiliar *things* do not exist. Many of the things we encounter, however, are other people. This means that "explored territory" is also that place where unfamiliar *behaviors* are not encountered. On familiar ground, we engage in those activities that are habitual, alongside others, who are doing the same thing (who are pursuing the same goals, whose emotions can be understood, whose beliefs are the same as ours, whose actions are predictable). Much of what we know how to do is behavior matched to society—individual action matched to, adapted to, modified by, the cumulative behavior of the others who surround us. "Explored" necessarily means, therefore, "where human activity has been rendered predictable," as well as "where the course of 'natural' events can be accurately determined." The maps that make territory familiar consequently consist in large part of *representations of behavior*—personal behavior, which we manifest, and the behavior of others, which we constantly encounter, and to which we have adjusted our personal actions. So, we map our own behaviors and those of others, because such behaviors constitute a large part of the world. We do not always understand what we do, however—our actions cannot be said to be explicitly comprehended. Our behavioral patterns are exceedingly complex, and psychology is a young science. The scope of our behavioral wisdom exceeds the breadth of our explicit interpretation. We act, even instruct, and yet do not understand. How can we do what we cannot explain?

We have already seen that we can represent what we do not understand—that we derive knowledge about the nature of the unknown (about the fact that it is eternally frightening and promising), by watching how we behave in its presence. We do something similar with regard to the social world and the behaviors that compose it. We watch how others act—and imitate, and learn to act, in consequence. Furthermore, we learn to *represent* the social world—explored territory, in large part—by watching the actions that take place in it; by exploring the social world itself. These representations are first patterns of actions, then stories—once the nature of the behavioral patterns have been identified, and represented in a declarative manner. A good story portrays a behavioral pattern with a large "expanse" of valid territory. It follows, therefore, that the greatest of all stories portrays the pattern of behavior with the widest conceivable territory.

We imitate and map adaptive behaviors—behaviors that efficiently reach a desired end—so that we can transform the mysterious unknown into the desirable and predictable; so that the social and nonsocial aspects of our experience remain under our control. The particular behaviors we imitate and represent, organized into a coherent unit, shared with others, constitute our cultures; constitute the manner in which we bring order to our existence. Our maps of adaptive behavior contain descriptions of the world in which that behavior is manifested—contains descriptions of explored and unexplored territory—as well as representations of the behaviors themselves. The stories mankind tells about the personal and historical past constitute expressions of the content of the declarative memory system, which is the system that knows what. Stories are generally told about animate objects, motivated, emotional beings, and might be regarded as descriptions of behavior, including antecedents, consequences and contexts. Stories contain portrayals of the outputs of the procedural system—which is the system that knows how—and inferences (explicit and implicit) about the existence and nature of factors (implicit, nonverbal, nondeclarative "presuppositions"), motivational and emotional, that guide and govern such output. The *knowing what* system therefore contains a complex sociohistorically constructed (but still somewhat "unconscious") verbal and imaginative description of the actions of the *knowing how* system. This description takes narrative form. Capacity for such representation emerges as the consequence of a complex and lengthy process of development, originating in action, culminating in the production of capacity for abstract cognition.

The episodic system, which generates representations of the experiential world, contains an elaborate model of the phenomenological world, composed in large part of encountered human behaviors generated by the other and the self, the most complex and affectively relevant phenomena in the human field of experience. This representation takes imaginative, dramatic, then narrative, mythic form as the model is constructed in fantasy, then described by the semantic system. Narrative/mythic "reality" is the world, conceived of in imagination, comprising imagistic representation of the behavioral pattern central to "morality," played out in an environment permanently characterized by the interplay of the known and the unknown. This "reality" is the world as place of action, and not as "place of objective things."

> All the world's a stage,
> And all the men and women merely players:
> They have their exits and their entrances;
> And one man in his time plays many parts.[210]

Before the emergence of empirical methodology, which allowed for methodical separation of subject and object in description, the world-model contained abstracted inferences about the nature of existence, derived primarily from observations of human behavior. This means, in essence, that pre-experimental man observed "morality" in his behavior and inferred (through the process described previously) the existence of a source or rationale for that morality in the structure of the "universe" itself. Of course, this "universe" is *the experiential field*—affect, imagination and all—and not the "objective" world constructed by the post-empirical mind. This prescientific "model of reality" primarily consisted of narrative representations of behavioral patterns (and of the contexts that surround them), and was concerned primarily with the motivational significance of events and processes. As this model became more abstract—as the semantic system analyzed the information presented in narrative format, but not understood—man generated imaginative hypotheses about the nature of the ideal human behavior, in the archetypal environment. This archetypal environment was (is) composed of three domains, which easily become three "characters":

The *unknown* is unexplored territory, nature, the unconscious, dionysian force, the *id*, the Great Mother goddess, the queen, the matrix, the matriarch, the container, the object to be fertilized, the source of all things, the strange, the unconscious, the sensual, the foreigner, the place of return and rest, the maw of the earth, the belly of the beast, the dragon, the evil stepmother, the deep, the fecund, the pregnant, the valley, the cleft, the cave, hell, death and the grave, the moon (ruler of the night and the mysterious dark), uncontrollable emotion, matter and the earth.[211] Any story that makes allusion to any of these phenomena instantly involves all of them. The grave and the cave, for example, connote the destructive aspect of the maternal—pain, grief and loss, deep water, and the dark woods; the fountain in the forest (water and woods in their alternative aspect), by contrast, brings to mind sanctuary, peace, rebirth and replenishment.

The *knower* is the creative explorer, the ego, the I, the eye, the phallus, the plow, the subject, consciousness, the illuminated or *enlightened* one, the trickster, the fool, the hero, the coward; spirit (as opposed to matter, as opposed to dogma); the sun, son of the unknown and the known (son of the Great Mother and the Great Father).[212] The central character in a story must play the role of hero or deceiver; must represent the sun (or, alternatively, the adversary—the power that eternally opposes the "dominion of the light").

The *known* is explored territory, culture, Apollinian control, superego, the conscience, the rational, the king, the patriarch, the wise old man and the tyrant, the giant, the ogre, the cyclops, order and authority and the crushing weight of tradition, dogma, the day sky, the countryman, the island, the heights, the ancestral spirits and the activity of the dead.[213] Authority and its danger play central roles in interesting tales, because human society is hier- archical, and because the organized social world is omnipresent. Authority and power mani- fest themselves, implicitly or explicitly, in all human relationships; we have never lived—cannot live—without others. The fact of power relationships and authority consti- tutes an eternally challenging and necessary constant of the human domain of experience.

The unknown is *yang*, cold, dark and feminine; the known, *yin*, warm, bright and mascu- line; the knower is the man living in *Tao*, on the razor's edge, on the straight and narrow path, on the proper road, in meaning, in the kingdom of heaven, on the mountaintop, cruci- fied on the branches of the world-tree—is the individual who voluntarily carves out the space between nature and culture. The interpretation of words in relationship to these proto- types (unknown, knower, known) is complicated by the fact of shifting meaning: *earth*, for example, is unknown (feminine) in relationship to sky, but known (masculine) in relation- ship to water; *dragon* is feminine, masculine and subject simultaneously. This capacity for meanings to shift is not illogical, it is just not "proper."[214] Meaning transforms itself endless- ly with shift in interpretive context—is determined in part by that context (that frame of ref- erence, that story). The same word in two sentences—one ironical, for example, the other straightforward—can have two entirely different, even opposite, meanings. Likewise, the sentence taken out of the context of the paragraph may be interpreted in some fashion entirely foreign to the intent of the author. Admission of the property of context-dependent meaning is neither illogical nor indicative of sloppy reasoning, nor primitive—it is merely recognition that context determines significance. The fact of context-dependence, however, makes interpretation of a given symbol difficult—particularly when it has been removed from its culturally constructed surroundings or milieu.

The unknown, the known and the knower share tremendous affective bivalence: the domain of nature, the Creat Mother, contains everything creative and destructive, because creation and destruction are integrally linked. The old must be destroyed to give way to the new; the mysterious source of all things (that is, the unknown) is also their final destination. Likewise, the domain of culture, the Great Father, is simultaneously and unceasingly tyranny and order, because security of person and property is always obtained at the cost of absolute freedom. The eternal subject, man, the knower, is equally at odds: the little god of earth is also mortal worm, courageous and craven, heroic and deceitful, possessed of great and dan- gerous potential, knowing good and evil. The unknown cannot be described, by definition. The known is too complicated to be understood. The knower—the conscious individual human being—likewise defies his own capacity for comprehension. The interplay between these ultimately unfathomable "forces" nonetheless constitutes the world in which we act, to which we must adapt. We have configured our behavior, accordingly; the natural categories[215] we use to apprehend the world reflect that configuration.

> The Tao existed before its name,
> and from its name, the opposites evolved,
> giving rise to three divisions,
> and then to names abundant.
> These things embrace receptively,
> achieving inner harmony,
> and by their unity create
> the inner world of man.[216]

The mythological world—which is the world as drama, story, forum for action—appears to be composed of three constituent elements and a "fourth" that precedes, follows and surrounds those three. These elements, in what is perhaps their most fundamental pattern of inter-relationship, are portrayed in *Figure 17: The Constituent Elements of Experience*. This figure might be conceptualized as three disks, stacked one on top of another, "resting" on an amorphous background. That background—chaos, the ultimate source and destination of all things—envelops the "world" and comprises everything that is now separate and identifiable: subject and object; past, present and future; "conscious" and "unconscious"; matter and spirit. The Great Mother and Father—the world parents (unexplored and explored territory, respectively; nature and culture)—can be usefully regarded as the primordial "offspring" of primeval chaos. The Great Mother—the unknown, as it manifests itself in experience—is the feminine deity who gives birth to and devours all. She is the unpredictable as it is encountered, and is therefore characterized, simultaneously, by extreme positive and extreme negative valence. The Great Father is order, placed against chaos; civilization erected against nature, with nature's aid. He is the benevolent force that protects individuals from catastrophic encounter with what is not yet understood; is the walls that surrounded the maturing Buddha and that encapsulated the Hebrew Eden. Conversely, however, the Great Father is the tyrant who forbids the emergence (or even the hypothetical existence) of anything new. The Archetypal Son is the child of order and chaos—culture and nature—and is therefore clearly their product. Paradoxically, however, as the deity who separates the earth (mother) from the sky (father), he is also the process that gives rise to his parents. This paradoxical situation arises because the existence of defined order and the unexplored territory defined in opposition to that order can come into being only in the light of consciousness, which is the faculty that knows (and does not know). The Archetypal Son, like his parents, has a positive aspect and a negative aspect. The positive aspect continually reconstructs defined territory as a consequence of the "assimilation" of the unknown [as a consequence of "incestuous" (that is, "sexual"—read *creative*) union with the Great Mother]. The negative aspect rejects or destroys anything it does not or will not understand.

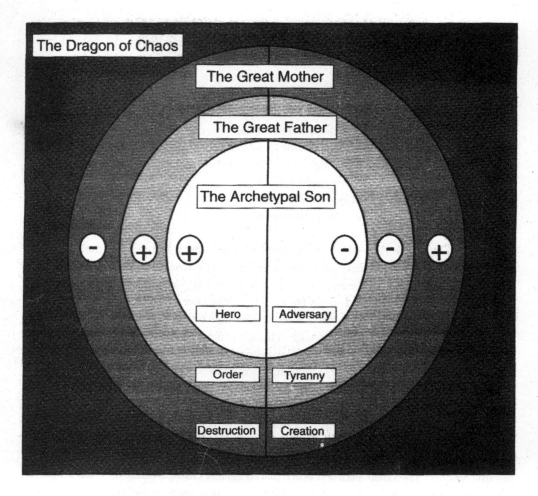

Figure 17: The Constituent Elements of Experience

Figure 18: The Positive Constituent Elements of Experience, Personified[217] portrays the "Vierge Ouvrante," a fifteenth century French sculpture, which represents the "constituent elements of the world" in personified, and solely positive form. Personification of this sort is the rule; categorical exclusion or inclusion in accordance with valence (all "bad" elements; all "good" elements) is almost equally common. All *positive* things are, after all, reasonably apprehended as similar, or identical—likewise, all *negative* things. It is for this reason, in part, that the terror of the unknown, the tyranny of the state, and the evil aspect of man are "contaminated" with one another—for this reason that the devil and the stranger are easily perceived as one. The "Vierge Ouvrante" is a strange work, from the standard Christian perspective, as it portrays Mary, the "mother of God," as superordinate to God the Father and

Figure 18: The Positive Constituent Elements of Experience, Personified

Christ the son. That superordinate position is perfectly valid, however, from the more general mythological perspective (although not *exclusively* valid). Each "constituent element of experience" can be regarded as progenitor, or as offspring, with regard to any other (as the world parents give birth to the divine son; as the divine son separates the world parents; as order is a derivative of chaos; as chaos is defined by order). So the most familiar Christian "sequence of generation" (which might be God ⇒ Mary ⇒ Christ) is only one of many "valid" configurations (and is not even the only one that characterizes Christianity).

The world of experience is composed of the known—explored territory—in paradoxical juxtaposition with the unknown—unexplored territory. Archaic notions of "reality" presuppose that the familiar world is a sacred space, surrounded by chaos (populated, variously, by demons, reptiles, spirits and barbarians—none of whom is really *distinguishable*). The world of order and chaos might be regarded as the stage, for man—for the twin aspects of man, more accurately: for the aspect that inquires, explores and transforms (which voluntarily expands the domain and structure of order, culture) and for the aspect that opposes that inquiry, exploration and transformation. The great story is, therefore, good *vs.* evil, played out against the endless flux of being, as it signifies. The forces of "good" have an eternal character (in the same way that Platonic objects are represented, eternally, in supracelestial space); unfortunately, so do the forces of evil. This eternality exists because all members of the species *Homo sapiens* are essentially equivalent, equal before God: we find ourselves vulnerable, mortal creatures, thrown into a universe bent on our creation and protection—and our transformation and destruction. Our "attitude" toward this ambivalent universe can only take one of two prototypical forms: positive or negative. The precise nature of these two forms (which can only be regarded as complex "personalities") and of the background against which they work constitutes the central subject matter of myth (and, dare it be said, the proper subject matter of the humanities and fine arts).

Analysis of a series of myths—the series which, I would argue, underlies Western civilization itself—should make these points painfully self-evident. We will begin with a discussion of the *Enuma elish*. This Mesopotamian creation story, which was elaborated in detail and complexity over the course of numerous centuries, is the most ancient complete cosmogonic myth at our disposal. We turn from the Sumerians to ancient Egyptian cosmology; then, from these specific examples to a more general discussion of mythological representation.

The Enuma elish: A Comprehensive Exemplar of Narrative Categorization

Creation myths are generally considered primitive or superstitious attempts to perform the magic of modern science. We assume that our ancestors were trying to do the same thing we do when we construct our cosmological theories and describe the generation of the objective world. This presumption is wrong. Our ancestors were not as simple-minded as we think they were, and their theories of the generation of the cosmos were not merely primitive science. Archaic theories of creation attempted to account for the existence of the world, as experienced in totality (which means, including *meaning*), and not for the isolated fact of the material world. The world as experienced in totality is made up of the material things we are familiar with, and the valences we consider epiphenomenal; of the objects of experience, and the fact of the subject, who does the experiencing. The world brought into being in archaic myths of creation is phenomenological, rather than material—it includes all aspects of experience, including those things we now regard as purely subjective. The archaic mind had not yet learned how to forget what was important. Ancient stories of the generation of the world therefore focus on all of reality, rather than on those distant and abstracted aspects we regard as purely objective.

Science might be considered "description of the world with regard to those aspects that are consensually apprehensible" or "specification of the most effective mode of reaching an end (given a defined end)." Narrative—myth, most fundamentally—can be more accurately regarded as description of the world as it *signifies* (for *action*). The mythic universe is *a place to act*, not *a place to perceive*. Myth therefore describes things in terms of their unique or shared affective valence, their value, their motivational significance. If we can tell (or act out) a story about something, we can be said to have mapped that thing, at least in part. We tell stories about the unknown, and the knower, and the known, and can therefore be said, somewhat paradoxically, to have adapted to the unpredictable, to the *fact* that we can adapt to the unpredictable, and to explored territory itself, where everything has been rendered secure. Although the unknown is truly unknown, it can be regarded as possessed of stable characteristics, in a broad sense. These characteristics are revealed in the actions we undertake in response to the appearance of unexpected things.

The world as experienced is composed of all the things we are familiar with and have classified in accordance with their relevance, and all the things we are unfamiliar with, which have a relevance all of their own, and of the process that mediates between the two, which turns the unfamiliar into the familiar and, sometimes, makes the predictable strange. The domain of the unfamiliar might be considered the ultimate source of all things, since we generate all of our determinate knowledge as a consequence of exploring what we do not understand. Equally, however, the process of exploration must be regarded as seminal, since nothing familiar can be generated from the unpredictable in the absence of exploratory action and conception. The domain of the known—created in the process of exploration—is the familiar world, firm ground, separated from the maternal sea of chaos. These three domains comprise the fundamental building blocks of the archaic world of myth. We briefly discussed an archaic Sumerian creation myth, previously, describing the "world" as the consequence of the separation of the cosmic parents, *An* (Sky) from *Ki* (Earth) by *En-lil*, their son and god of the atmosphere. The ancient Egyptians regarded the situation similarly:

Like so many other traditions, the Egyptian cosmogony begins with the emergence of a mound in the primordial waters. The appearance of this "First Place" above the aquatic immensity signifies the emergence of the earth, but also the beginning of light, life, and consciousness. At Heliopolis, the place named the "Hill of Sand," which formed part of the temple of the sun, was identified with the primordial hill. Hermopolis was famous for its lake, from which the cosmogonic lotus emerged. But other localities took advantage of the same privilege. Indeed, each city, each sanctuary, was considered to be a "center of the world," the place where the Creation had begun. The initial mound sometimes became the cosmic mountain up which the pharaoh climbed to meet the sun god.

Other versions tell of the primordial egg, which contained the "Bird of Light" . . . , or of the original lotus that bore the Child Sun, or, finally, of the primitive serpent, first and last image of the god

Atum. (And in fact chapter 175 of the *Book of the Dead* prophesies that when the world returns to the state of chaos, Atum will become the new serpent. In Atum we may recognize the supreme and hidden God, whereas Re, the Sun, is above all the manifest God....) The stages of creation—cosmogony, theogony, creation of living beings, etc.—are variously presented. According to the solar theology of Heliopolis, a city situated at the apex of the Delta, the god Re-Atum-Khepri [three forms of the sun, noontime, setting, and rising, respectively] created a first divine couple, Shu (the Atmosphere) and Tefnut, who became parents of the god Geb (the Earth) and of the goddess Nut (the Sky). The demiurge performed the act of creation by masturbating himself or by spitting. The expressions are naively coarse, but their meaning is clear: the divinities are born from the very substance of the supreme god. Just as in the Sumerian tradition, Sky and Earth were united in an uninterrupted *hieros gamos* until the moment when they were separated by Shu, the god of the atmosphere [in other similar traditions, Ptah]. From their union were born Osiris and Isis, Seth and Nephthys [who will be discussed later].[218]

Primordial myths of creation tend to portray the origin of things as the consequence of at least one of two related events. The universe was symbolically born into being, for example, as a result of the action of a primeval hermaphroditic deity. Alternatively, it arose from the interaction of somewhat more differentiated masculine and feminine spirits or principles (often the offspring of the most primordial god)—emerged, for example, from the interplay of the sky, associated (most frequently) with the father, and the earth (generally but not invariably granted a female character). Imagery of the latter sort remains latently embedded in the oldest (Jahwist) creation myth in the familiar Old Testament book of Genesis. The Jahwist story begins in the fourth stanza of the second chapter of Genesis and describes the masculine God breathing life (spirit) into the *adamah*, mother earth, thereby creating the original (hermaphroditic) man, Adam.[219] In alternative, more actively dramatic accounts—such as that of the *Enuma elish*, the Babylonian creation myth—the creative demiurge slays a dragon, or a serpent, and constructs the universe out of the body parts. The two forms of story, very different on the surface, share deep grammatical structure, so to speak; utilize metaphors that are closely associated, psychologically and historically, to drive their fundamental message home:

In the Babylonian creation hymn *Enuma elish* [("when above"[220]), circa 650 B.C., in its only extant form; derived from a tradition at least two thousand years older] the god of the fresh-water sea, Apsu, was killed and his widow Tiamat, goddess of the "bitter" or salt waters, threatened the gods with destruction. Marduk, the champion of the gods, killed her and split her in two, creating heaven out of one half and earth out of the other. Similarly, the creation in Genesis begins with a "firmament" separating the waters above from the waters below, but succeeding a world that was waste (*tohu*) and void, with darkness on the face of the deep (*tehom*). The Hebrew words are said to be etymologically cognate with Tiamat, and there are many other allusions in the Old Testament to the creation as a killing of a dragon or monster.[221]

It is easy, or, at least appears easy, to understand why the pre-experimental mind might frequently have associated the creation of everything with femininity—with the source of new life through birth (most evidently, the cause and concrete origin of all living things). The role of the male in the original creation—the part played by the "masculine principle," more precisely—is comparatively difficult to comprehend, just as the male role in procreation is less obvious. Nonetheless, the most widely disseminated of creation myths—and, arguably, the most potent and influential—essentially reverses the standard pattern of mythic origin, and places particular emphasis on the masculine element. In the Judeo-Christian tradition, creation depends on the existence and action of *Logos*, mythically masculine discriminant consciousness or exploratory spirit, associated inextricably with linguistic ability—with the Word, as St. John states (in what was perhaps designed to form the opening statement of the New Testament, structurally paralleled with the beginning of Genesis[222]):

In the beginning was the Word, and the Word was with God, and the Word was God.

The same was in the beginning with God. All things were made by him; and without him was not any thing made that was made.

In him was life; and the life was the light of men. And the light shineth in darkness; and the darkness comprehended it not. (John 1:1–4)

The explicit stress placed by the Judeo-Christian tradition on the primacy of the word and its metaphorical equivalents makes it somewhat unique in the pantheon of creation myths. The early Jews were perhaps the first to clearly posit that activity in the mythically masculine domain of spirit was linked in some integral manner to the construction and establishment of experience as such. It is impossible to understand why the Judeo-Christian tradition has had such immense power—or to comprehend the nature of the relationship between the psyche and the world—without analyzing the network of meaning that makes up the doctrine of the Word.

There exists clear psychological precedent for the philosophy of the early Jews (and the later Christians) in the Mesopotamian and Egyptian schools of metaphysical speculation—in their rituals, images and acts of abstract verbal representation. The Mesopotamian creation myth, which we will consider first—the *Enuma elish*—portrays the emergence of the earliest world as the consequence of the (sexual, generative, creative) union of the primal deities *Apsu* and *Tiamat*. Apsu, masculine, served as the *begetter* of heaven and earth, prior to their identification as such (before they were named). Tiamat, "she who gave birth to them all,"[223] was his consort. Initially, Apsu and Tiamat existed (?) indistinguishably from one another, "still mingled their waters together"[224] when "no pasture land had been formed, and not even a reed marsh was to be seen; when none of the other gods had been brought into being, when they had not yet been called by their name, and their destinies had not yet been fixed."[225] Their *uroboric* union served as the source from which more differentiated but still fundamental structures and processes or spirits issued: "at that time, were the gods created

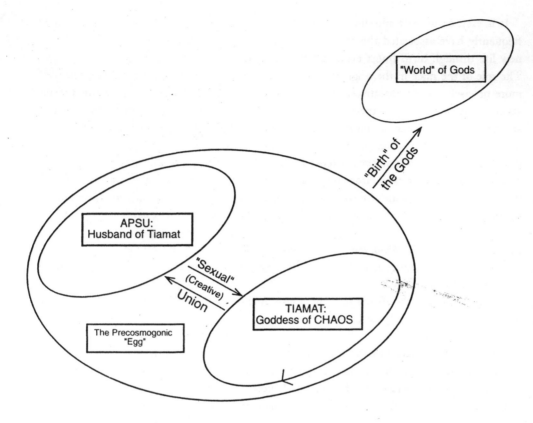

Figure 19: The Birth of the World of Gods

with them."[226] The "precosmogonic egg" "inhabited" by Tiamat and Apsu gave rise to the initial "world of gods." This process is portrayed schematically in *Figure 19: The Birth of the World of Gods*.

The Mesopotamian gods—like deities everywhere—present somewhat of a mystery to the modern mind. Archaic cultures are rife with deities. We seem unable to locate them now. They do not seem part of the objective external world. It is tempting, therefore, to regard such beings as imaginative constructions, as personifications of subjective affective or emotional states or drives, the incarnated form of subjective experience. The term "personification," however, implies a voluntary act—connotes the conscious use of metaphor, on the part of the individual driven to represent and knowing that he is representing. There is no indication, however, that it is *an act of conscious creation* that gives rise to the pre-experimental deity; in fact, the opposite appears more true: it is the "action of the deity" that gives rise to creative endeavor, as such, on the part of the creating subject. The god must therefore be more than the subject; more than the subject's original narrative conception of himself.

The phenomena that we would now describe as emotions or motive forces, from the perspective of our modern, comparatively differentiated and acute self-consciousness, do not appear to have been experienced precisely as "internal" in their original form. Rather, they made their appearance as part and parcel of the experience (the event, or sequence of events) that gave rise to them, and adopted initial representational form in imaginative embodiment. The modern idea of the "stimulus" might be regarded as a vestigial remnant of this form of thinking—a form that grants the power of affective and behavioral control to the object (or which cannot distinguish between that which elicits a response, and the response itself). We no longer think "animistically" as adults, except in our weaker or more playful moments, because we attribute motivation and emotion to our own agency, and not (generally) to the stimulus that gives proximal rise to them. We can separate the thing from the implication of the thing, because we are students and beneficiaries of empirical thinking and experimental method. We can remove attribution of motive and affective power from the "object," and leave it standing in its purely sensory and consensual aspect; can distinguish between *what is us* and *what is world*. The pre-experimental mind could not (cannot) do this, at least not consistently; could not reliably discriminate between *the object* and *its effect on behavior*. It is that object and effect which, in totality, constitute a god (more accurately, it is a *class* of objects and their effects that constitute a god).

A god, so considered—more specifically, a potent and powerful god, one with a history—constitutes the manner in which a group or family of stimuli of isomorphic motivational significance reveals itself to or grips the collective (communicated) imagination of a given culture. Such a representation is a peculiar mix (from the later, empirical viewpoint) of psychological and sociological phenomena and objective "fact"—an undifferentiated mix of subject and object (of emotion and sensory experience), transpersonal in nature (as it is historically elaborated "construction" and shared imaginative experience). The primitive deity nonetheless serves as accurate representation of the ground of being, however, because it is affect and subjectivity as well as pure object (before the two are properly distilled or separated)—because it is primordial *experience*, rather than the mere primordial *thing*.

The original "children of Tiamat and Apsu"—the "elder gods"—should therefore be regarded as embodiments of the archaic transpersonal intrapsychic phenomena that give rise to human motivation, as well as those aspects of the objective world that activate those intrapsychic systems. The Sumerians considered themselves destined to "clothe and feed" such gods, because they viewed themselves as the servants, in a sense, of what we would call instinctive forces, "elicited" by the "environment." Such forces can be reasonably regarded as the Sumerians regarded them—as deities inhabiting a "supracelestial place," extant prior to the dawn of humanity. Erotic attraction, for example—a powerful god—has a developmental history that predates the emergence of humanity, is associated with relatively "innate" releasing "stimuli" (those that characterize erotic beauty), is of terrible power, and has an existence "transcending" that of any individual who is currently "possessed." *Pan*, the Greek god of nature, produced/represented fear (produced "panic"); *Ares* or the Roman *Mars*, warlike fury and aggression. We no longer personify such "instincts," except for the purposes of literary embellishment, so we don't think of them "existing" in a "place" (like heaven, for example).

But the idea that such instincts inhabit a space—and that wars occur in that space—is a metaphor of exceeding power and explanatory utility. Transpersonal motive forces do wage war with one another over vast spans of time; are each forced to come to terms with their powerful "opponents" in the intrapsychic hierarchy. The battles between the different "ways of life" (or different philosophies) that eternally characterize human societies can usefully be visualized as combat undertaken by different standards of value (and, therefore, by different hierarchies of motivation). The "forces" involved in such wars do not die, as they are "immortal": the human beings acting as "pawns of the gods" during such times are not so fortunate.

Back to the *Enuma elish*: The secondary/patriarchal deities of the Mesopotamian celestial pantheon—including the couples *Lahmu* and *Lahamu* and *Kishar* and *Anshar*—arose as a direct consequence of the interactions of the original sexualized "unity" of Tiamat and Apsu, the most primal of couples. This undifferentiated precosmogonic egg (a common metaphor in other creation myths) "contains" an alloy of "order" (the "masculine" principle") and "chaos" (the "feminine" principle). This alloy is the "world parents," locked in "creative embrace" (is spirit and matter, conceived alternatively, still "one thing"). Tiamat and Apsu's union gives rise to children—the primordial instincts or forces of life who, in turn, engender more individualized beings. The *Enuma elish* itself does not spend much time fleshing out the specific characteristics of these forces of life, as it is concerned with more general issues. Lahmu and Lahamu and Kishar and Anshar are incidental characters, serving only as intermediaries between the real protagonists of the drama—Marduk, a late-born individual-like god, and Tiamat, his turncoat mother. Kishar and Anshar therefore serve only as progenitors of *Anu*, who in turn "begot *Ea*,[227] his likeness," "the master of his fathers,"[228] "broad of understanding, wise, mighty in strength, much stronger than his grandfather, Anshar,"[229] without "rival among the gods his brothers."[230]

The elder gods serve merely to reproduce and to noisily act. Their incessant racket and movement upsets the divine parents; disturbs "the inner parts of Tiamat."[231] So Tiamat and Apsu conspire to "devour" their children. This is a common mythological occurrence; one echoed later in the story of Yahweh, Noah and the Flood. The gods give birth to the cosmos, but ceaselessly attempt to destroy it.

Ea catches wind of his parents' plot, however, however, and slays Apsu—adding insult to injury by building a house on his remains (and by naming that house Apsu, in mockery or remembrance). Into this house he brings his bride, *Damkina*, who soon gives birth to Marduk, the hero of the story, "the wisest of the wise, the wisest of the gods,"[232] filled with "awe-inspiring majesty."[233] When Ea saw his son:

> He rejoiced, he beamed, his heart was filled with joy.
> He distinguished him and conferred upon him double equality with the gods,
> So that he was highly exalted and surpassed them in everything.
> Artfully arranged beyond comprehension were his members,
> Not fit for human understanding, hard to look upon.
> Four were his eyes, four were his ears.

When his lips moved, fire blazed forth.

Each of his four ears grew large,

And likewise his eyes, to see everything.

He was exalted among the gods, surpassing was his form;

His members were gigantic, he was surpassing in height.

Mariyutu, Mariyutu:

Son of the sun-god, the sun-god of the gods![234]

Marduk is characterized by the metaphoric associates of consciousness. He has exaggerated sensory capacities; his very words are characterized by creative and destructive power (by the transformative capacity of fire). He is the "sun-god," above all, which means that he is assimilated to (or, more accurately, occupies the same "categorical space") as "sight," "vision," "illumination," "enlightenment," "dawn," the "elimination of darkness" and the "death of the night."

In the midst of all this action—war plans, death, birth—*Anu* (Marduk's grandfather, Ea's father) busies himself with the generation of the four winds. His work raises waves upon the surface of the waters occupied by Tiamat, and the (previously unidentified) primary/matriarchal subdeities that (apparently) accompany her there. This new intrusion troubles her beyond tolerance, upset as she was already at the noise of her offspring and death of her husband. She decides to rid the universe of the (secondary/patriarchal) elder gods, once and for all, and begins to produce horrible "soldiers" to aid her in battle:

. . . bearing monster serpents

Sharp of tooth and not sparing the fang.

With poison instead of blood she filled their bodies.

Ferocious dragons she clothed with terror,

She crowned them with fear-inspiring glory and made them like gods,

So that he who would look upon them should perish from terror.[235]

The angry Tiamat—the unknown, chaos, in its terrible or destructive aspect—produces eleven species of monsters to aid her in her battle, including the viper, the dragon, the great lion, the rabid dog, the scorpion-man and the storm-demon. She elects the firstborn, *Kingu* by name, to reign over them all, giving him "the tablet of destinies"[236] to signify his ascension and dominion. The story continues:

After Tiamat had made strong preparations,

She made ready to join battle with the gods her offspring.

To avenge Apsu, Tiamat did this evil.

How she got ready for the attack was revealed to Ea.

When Ea heard of this matter,

He became benumbed with fear and sat in silent gloom.

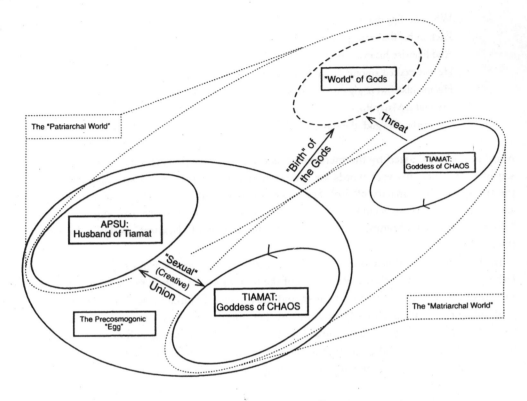

Figure 20: The "Death" of Apsu, and the (Re)Emergence of Tiamat as Threat

> After he had reflected on the matter and his wrath had subsided,
> He went to Anshar, his [great-]grandfather.
> And when he had come into the presence of Anshar, his [great-]grandfather,
> He communicated to him all that Tiamat had planned.[237]

Now, although Apsu isn't well described in the *Enuma elish*, it is clear that he is the masculine consort of Tiamat. The "masculine" consort of the "goddess of the unknown" is inevitably the "god of the known" (or his "progenitor" and dependent, the *knower*). It is the "known" that serves as protection from the unknown, whether this is understood or not. Ea kills Apsu, which means that he unconsciously strips himself of protection.

Ea might therefore be reasonably regarded as representative of that part of humanity eternally (and ignorantly) contemptuous of tradition and willing to undermine or destroy the past without understanding its necessity or nature. Those "unconsciously" protected from the outside world by the walls of culture may become irritated by the limitations such walls represent, and incautiously pull them down. This act of destruction, disguised as a blow for freedom, lets the terrible unknown flood back in. The Great Mother is a terrible force, in the absence of patriarchal protection. The *Enuma elish* makes this vital point, implicitly. This state of affairs is represented schematically in *Figure 20: The "Death" of Apsu, and the (Re)Emergence of Tiamat as Threat.*

Anshar is terribly upset by the news of Tiamat's anger. He asks Ea to stand against Tiamat. Ea fails, deservedly, and Anshar sends Anu in his stead. He is routed as well, and returns, overcome by terror. In desperation and final hope, Anshar and Ea call on Marduk, the young sun-god:

> Ea called Marduk to his private room;
> He advised him, telling him the plan of his heart:
> "Marduk, consider my idea, hearken to thy father.
> Thou art he, my son, who relieves his heart;
> Draw nigh into the presence of Anshar, ready for battle;
> Speak and stand forth; when he sees thee, he will be at rest."
> [Marduk] was glad at the word of his father;
> He drew nigh and stood before Anshar.
> When Anshar saw him, his heart was filled with joy;
> He kissed his lips, his fear was removed.
> "Anshar, be not silent, but open thy lips;
> I will go and accomplish all that is in thy heart!
> What man is it who has brought battle against thee?
> Tiamat, who is a woman, is coming against thee with arms!
> My father, creator, be glad and rejoice;
> Soon thou shalt trample upon the neck of Tiamat!
> Yea, my father, creator, be glad and rejoice;
> Soon thou shalt trample upon the neck of Tiamat!"

Anshar answers:

> "My son, who knowest all wisdom,
> Quiet Tiamat with thy holy incantation."[238]

Marduk's "magic words" (remember, he speaks fire) are clearly and reasonably portrayed as one of the most powerful weapons in the battle against the forces of chaos. Anshar continues:

> "On the storm chariot quickly pursue the way!
> [...] ... turn her back!"
> The lord was glad at the word of his father;
> His heart exulted, and he said to his father:
> "Lord of the gods, destiny of the great gods,

If I am indeed to be your avenger,
To vanquish Tiamat and to keep you alive,
Convene the assembly and *proclaim my lot supreme* [emphasis added].
When ye are joyfully seated together in the Court of Assembly,
May I through the utterance of my mouth determine the destinies, instead of you.
Whatever I create shall remain unaltered,
The command of my lips shall not return, ... it shall not be changed."[239]

Alexander Heidel, who provided the translation of the *Enuma elish* cited here, comments:

Marduk demands supreme and undisputed authority as the price for risking his life in combat with Tiamat. When therefore the gods, at the New Year's festival [see discussion below], convened in the Court of Assembly, "they reverently waited" on Marduk, the "king of the gods of heaven and earth," and in that spirit they decided the destinies. The gods, indeed, "continue to determine destinies long after Marduk has received the powers he here desires";[240] but the final decision rested with Marduk, so that in the last analysis it was he who decided the fates.[241]

This is an example of the "hierarchical organization of the gods," a concept frequently encapsulated in mythology, and one we shall return to later. All the original children of Tiamat are potent and impersonal elder gods, "psychological forces"—the "deities" that eternally rule or constitute human motivation and affect. The question of the proper ordering of those forces ("*who*, or *what*, should rule?") is the central problem of morality, and the primary problem facing human individuals and social organizations. The Sumerian "solution" to this problem was the elevation of Marduk—the sun-god who voluntarily faces chaos—to the position of "king" (and the subjugation of the other gods to that "king"):

Anshar opened his mouth
And addressed these words to Kaka, his vizier:
"Kaka, my vizier, who gladdenest my heart,
Unto Lahmu and Lahamu I will send thee;
Thou knowest how to discern and art able to relate.
Cause the gods my fathers to be brought before me.
Let them bring all the gods to me!
Let them converse and sit down to banquet.
Let them eat bread and prepare wine.
For Marduk, their avenger, let them decree the destiny.
Set out, O Kaka, go, and stand thou before them.
What I am about to tell thee repeat unto them.
Anshar, your son, has sent me.
The command of his heart he has charged me to convey,
Saying: Tiamat, our bearer, hates us.

She held a meeting and raged furiously.

All the gods went over to her;

Even those who ye have created march at her side.

They separated themselves and went over to the side of Tiamat;

They were angry, they plotted, not resting day or night;

They took up the fight, fuming and raging;

They held a meeting and planned the conflict.

Mother Hubur [Tiamat], who fashions all things,

Added thereto irresistible weapons, bearing monster serpents

Sharp of tooth and not sparing the fang.

With poison instead of blood she filled their bodies.

Ferocious dragons she clothed with terror,

She crowned them with fear-inspiring glory and made them like gods.

So that they might cause him who would look upon them to perish from terror,

So that their bodies might leap forward and none turn back their breasts.

She set up the viper, the dragon, and the lahumum

The great lion, the mad dog, and the scorpion-man,

Driving storm demons, the dragonfly, and the bison,

Bearing unsparing weapons, unafraid of battle.

Powerful are her decrees, irresistible are they.

Altogether eleven kinds of monsters of this sort she brought into being.

Of those among the gods, her first-born, who formed her assembly,

She exalted Kingu; in their midst she made him great.

To march at the head of the army, to direct the forces,

To raise the weapons for the engagement, to launch the attack,

The high command of the battle,

She intrusted to his hand; she caused him to sit in the assembly, saying:

I have cast the spell for thee, I have made thee great in the assembly of the gods.

The dominion over all the gods I have given into thy hand.

Mayest thou be highly exalted, thou my unique spouse!

May thy names become greater than those of the Anunnaki!

She gave him the tablet of destinies, she fastened it upon his breast, saying

"As for thee, thy command shall not be changed, the word of thy mouth shall be
    dependable!"

Now when Kingu had been exalted and had received supreme dominion,

They decreed the destinies of the gods, her sons, saying:

"May the opening of our mouths quiet the fire-god!

May thy overpowering poison vanquish the opposing night!"

I sent Anu, but he could not face her.

Ea also was afraid and turned back.

Then Marduk, the wisest of the gods, your son, came forward.

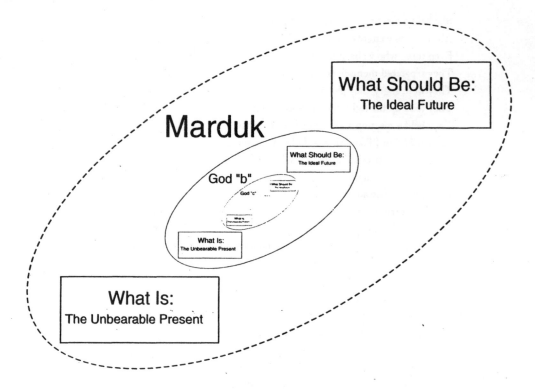

Figure 21: "World" of Gods: Hierarchical Organization

His heart prompted him to face Tiamat.

He opened his mouth and said to me:

"If I am indeed to be your avenger,

To vanquish Tiamat and to keep you alive,

Convene the assembly and proclaim my lot supreme.

When ye are joyfully seated together in the Court of Assembly,

May I through the utterance of my mouth determine the destinies, instead of
　　　　you.

Whatever I create shall remain unaltered,

The command of my lips shall not return void, it shall not be changed."

Hasten to me then and speedily fix for him your destiny,

That he may go to meet your powerful enemy!"[242]

The "hierarchical organization of the gods" is represented schematically in *Figure 21: "World" of Gods: Hierarchical Organization*, which portrays Marduk as the superordinate personality or pattern of action, "designed" to transform the unbearable present into the desired future. The *Enuma elish* states, essentially: "When things are normal, any god might rule. However, in the case of a true crisis, everyone turns to the sun-god (the embodiment of "consciousness"). Perhaps it is reasonable to presume, therefore, that he should always reign supreme." The "formulation" of this "hypothesis" was a work of unsurpassed genius, and a decisive move in the history of the Western mind.

The vizier Kaka goes on his way, as commanded, and spreads the word among the (secondary/patriarchal) elder deities, who assemble to contemplate the upcoming battle:

> They entered into the presence of Anshar and filled the Court of Assembly;
> They kissed one another as they came together in the assembly;
> They conversed and sat down to a banquet.
> They ate bread and prepared wine.
> The sweet wine dispelled their fears;
> Their bodies swelled as they drank the strong drink.
> Exceedingly carefree were they, their spirit was exalted;
> For Marduk, their avenger, they decreed the destiny.
> They erected for him a lordly throne-dais,
> And he took his place before his fathers to receive sovereignty.
> "Thou art the most important among the great gods,
> Thy destiny is unequaled, thy command is like that of Anu.
> Marduk, thou art the most important among the great gods.
> Thy destiny is unequaled, thy command is like that of Anu.
> From this day onward thy command shall not be changed.
> To exalt and abase—this shall be thy power!
> Dependable shall be the utterance of thy mouth, thy command shall not prove vain.
> None among the gods shall infringe upon thy prerogative."[243]

The gods place "the starry garment of the night sky"[244] in their midst. At the command of Marduk's mouth—on his word—it appears; at his command, it disappears, "as the night sky on the passage of the sun." [245] Marduk is clearly part of the pantheon who eternally vanquish the dragon of the night. The story continues:

> When the gods his fathers beheld the power of his word,
> They were glad and did homage, saying: "Marduk is king!"
> They bestowed upon him the scepter, the throne, and the royal robe;
> They gave him an irresistible weapon smiting the enemy, saying:
> "Go and cut off the life of Tiamat.
> May the winds carry her blood to out-of-the-way places."
> After the gods his fathers had determined the destiny of [Marduk],
> They set him on the road—the way to success and attainment.[246]

Marduk gathers his armaments—bow, club and lightning—sets himself ablaze, and fashions a net to enclose Tiamat. He is a master of fire and armaments—which is to say, a master of the technology that serves most fundamentally to transform the unknown and terrifying world into the comforting, productive and familiar. He is able to bind the unknown; to limit its sphere of action, and to bring it under control. He raises the winds, and the storm to aid him, using the forces of nature against nature itself. He dresses himself in a terrifying coat of mail, and wears "terror-inspiring splendor" on his head. Prepared carefully in this manner, and fortified against poison, he takes the "direct route" to Tiamat. He confronts (re-emergent) novelty voluntarily, at a time of his choosing, after careful preparation, and without avoidance. His mere appearance strikes terror into the heart of *Kingu* and his legion of monsters (just as Christ, much later, terrifies the Devil and his minions). Marduk confronts Tiamat, accuses her of treachery, and challenges her to battle.

> When Tiamat heard this,
> She became like one in a frenzy and lost her reason.
> Tiamat cried out loud and furiously,
> To the very roots her two legs shook back and forth.
> She recites an incantation, repeatedly casting her spell;
> As for the gods of battle, they sharpen their weapons.
> Tiamat and Marduk, the wisest of the gods, advanced against one another;
> They pressed on to single combat, they approached for battle.[247]

Marduk fills Tiamat with "an evil wind," which distends her belly. When she opens her mouth to devour him, he lets an arrow fly, which tears her interior, and splits her heart. He subdues her, completely, casts down her carcass, and stands upon it. His voluntarily encounter with the forces of the unknown produces a decisive victory. He rounds up her subordinates—including Kingu, whom he deprives of the tablet of destinies—and encapsulates them with netting. Then he returns to Tiamat:

> The lord trod upon the hinder part of Tiamat,
> And with his unsparing club he split her skull.
> He cut the arteries of her blood
> And caused the north wind to carry it to out-of-the-way places....
> He split her open like a mussel into two parts;
> Half of her he set in place and formed the sky (therewith) as a roof.
> He fixed the crossbar and posted guards;
> He commanded them not to let her waters escape.
> He crossed the heavens and examined the regions.
> He placed himself opposite the Apsu, the dwelling of Ea.
> The lord measured the dimensions of the Apsu,
> And a great structure, he established, namely Esharra [Earth]."[248]

Marduk then constructs the heavenly order, fashioning the year, defining the twelve-sign zodiac, determining the movement of the stars, the planets and the moon.[249] Finally, he deigns to create man (out of Kingu, the greatest and most guilty of Tiamat's allies), so that "upon him shall the services of the gods be imposed that they may be at rest";[250] then he returns the gods allied with him to their appropriate celestial abodes. Grateful, they deliver him a present:

> Now, O Lord, who hast established our freedom from compulsory service,
> What shall be the sign of our gratitude before thee?
> Come, let us make something whose name shall be called "Sanctuary."
> It shall be a dwelling for our rest at night; come, let us repose therein![251]

The dwelling is Babylon, center of civilization, mythic sacred space, dedicated in perpetuity to Marduk.

The mythic tale of the *Enuma elish* describes the nature of the eternal relationship between the (unknowable) source of all things, the "gods" who rule human life, and the subject or process who constructs determinate experience, through voluntary encounter with the unknown. The "full story" presented in the Sumerian creation myth is presented, schematically, in *Figure 22: The Enuma elish in Schematic Representation*. Tiamat is portrayed, simultaneously, as the thing that breeds everything (as the mother of all the gods); as the thing that destroys all things; as the consort of a patriarchal spiritual principle, upon who creation also depends (Apsu); and, finally, as the thing that is cut into pieces by the hero who constructs the world. Marduk, the last-born "child" of instinct, is the hero who voluntarily faces the creative/destructive power that constitutes the "place" from which all things emerge. He is the martial deity, role model for the culture of the West, who violently carves the unknown into pieces, and makes the predictable world from those pieces.

This tale contains within it a complex and sophisticated notion of causality. None of its elements exists in contradiction with any other, even though each lays stress on different aspects of the same process. Something must exist, prior to the construction of identifiable things (something that cannot be imagined, in the absence of a subject). That thing might usefully be portrayed as the "all-devouring mother of everything." The particular, discriminable, familiar elements of human experience exist as they do, however, because the conscious subject can detect, construct and transform them. The "son-hero's" role in the "birth" of things is therefore as primal as the mother's, although this part is somewhat more difficult to comprehend. Nonetheless, the Sumerians manage the representation, in narrative form. It is a relatively small step from this dramatic/imagistic portrayal of the hero to the most explicit Christian doctrine of *Logos*—the creative Word (and from there to our notion of "consciousness").

The mythic tale of Marduk and Tiamat refers to the capacity of the individual to explore, voluntarily, and to bring things into being as a consequence. The hero cuts the world of the unpredictable—unexplored territory, signified by Tiamat—into its distinguishable elements;

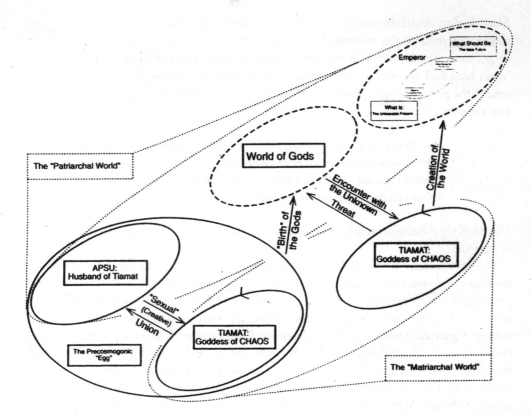

Figure 22: The Enuma elish in Schematic Representation

weaves a net of determinate meaning, capable of encompassing the vast unknown; embodies the divine "masculine" essence, which has as its most significant feature the capacity to transform chaos into order. The killing of an all-embracing monster and the construction of the universe from its body parts is symbolic (metaphorical) representation of the central, adaptive process of heroic encounter with the undifferentiated unknown, and the construction or generation of differentiated order as a consequence. It is this process, emulated by the emperor of Mesopotamia (who ritually embodied Marduk), that served as the basis for his authority—and, indeed, *that serves to undergird the idea of lawful authority to the present day.* The Mesopotamian emperor's identification with the most divine of all the deities (according to the judgment and election of those selfsame powers) lent him power, and served to maintain social and psychological order among his people. Furthermore, the Mesopotamian emperor stood in the same relationship to his people as Marduk stood to him: as ritual model for emulation, as the personality whose actions served as pattern for all actions undertaken in the kingdom—as the personality that was the state, *insofar as the state defined and brought order to interpersonal interactions* (which, after all, was and is its primary function). Babylon was therefore conceptualized as "the kingdom of god on earth"—that is, as a profane imitation of heaven. The emperor served this "imitated heaven" as the "imitator of Marduk," at least insofar as he was conservative, just, courageous and creative. Eliade comments on the sacrality of the Mesopotamian sovereign, and describes the rituals designed to maintain that sacrality:

At Babylon the *Enuma elish* was recited in the temple on the fourth day of the New Year festival. This festival, named *zagmuk* ("beginning of the year") in Sumerian and *akitu* in Akkadian [note: the Sumerians and Akkadians united to form Babylon], took place during the first twelve days of the month of Nisan. It comprised several sequences, of which we will mention the most important: (1) a day of expiation for the king, corresponding to Marduk's "captivity"; (2) the freeing of Marduk; (3) ritual combats and a triumphal procession, led by the king, to the *Bit Akitu* (the house of the New Year festival), where a banquet was held; (4) the *hieros gamos* [mystical marriage] of the king with a hierodule [ritual slave/prostitute] personifying the goddess; and (5) the determination of destinies by the gods.[252]

The meaning of some terminology, and the nature of the latter two sequences, must be clarified here:

First [with regard to (4)], it should be noted that *hieros gamos* means *mystical marriage*—the marriage of the king, and the queen or goddess. This marriage provides dramatic representation of the union of the exploratory tendency (incarnated by the king) with the positive aspect of the unknown, incarnated by the *hierodule*. Marduk (the king) is originally "shut up," signifying his temporary disappearance (see the description of Osiris, below) during the normal or routinized operations of the state. He is freed, to meet Tiamat; he does so, in sexual union. This sexual (read: creative) union—the juxtaposition of the process of knowing, embodied by the king (Marduk), with the unknown, embodied by Tiamat (incarnated by the *hierodule*)—is what gives rise to the generation of new information and patterns of adaptation. The process of generating knowledge is therefore assimilated to the domain of sexual union, as the primordial creative process. The deity of chaos, or the unknown, appears most generally as feminine (and as half-negative, and half-positive) once the initial division between order and chaos has been established. The *attribution* of femininity to this deity, so to speak, occurs most fundamentally because the unknown serves as the matrix from which determinate forms are borne. The negative attribution (Tiamat serves as example) exists because the unknown has a destructive aspect; the positive (the *hierodule* here, *Isis* in the Egyptian myth of *Osiris*, *Mary* in Christianity) because the unknown is also creative or generative.

Second [with regard to (5)], it should be noted that the king (in his incarnation as god) served to "determine destinies" because he was both hero—ritual model for emulation—and absolute ruler. As such, he literally controlled individual destinies, serving as he did, practically and in representation, as the most powerful individual in society, and the most dominant "strategy" in the hierarchy of behavioral adaptation. What he could not determine, by

law, however, he was to provide by creative example (since the "body of laws," as embodiment of past wisdom, is insufficient to deal with the challenges of the present). This idea is
developed much more explicitly by the Egyptians, as we shall see. Back to Eliade's story:

The first sequence of this mythico-ritual scenario—the king's humiliation and Marduk's captivity—
indicates the regression of the world to the precosmogonic chaos. In the sanctuary of Marduk the high
priest stripped the king of his emblems (scepter, ring, scimitar and crown) and struck him in the face.
Then, on his knees, the king uttered a declaration of innocence: "I have not sinned, O lord of the lands,
I have not been negligent regarding thy divinity." The high priest, speaking in Marduk's name, replied:
"Do not fear. . . . Marduk will hear thy prayer. He will increase thy dominion."

   During this time the people sought for Marduk, supposed to be "shut up in the mountain". (a formu-
la indicating the "death" of a divinity) . . . [in consequence of a descent] "far from the sun and light." . . .

[When the world "regresses" to "precosmogonic chaos," it is always the case that the hero is
missing. The hero is, after all, incarnation of the process by which chaos is transformed into
order. If chaos has the upper hand, it is by definition because of a current paucity of heroism.
It can be said, therefore, that the reappearance of the Great Mother, in her terrible guise, the
death of the Great Father (who serves as protection from his creative and destructive wife),
and the absence of the hero (who turns chaos into order) all represent different ways of
telling the same story—the story which describes a *life-threatening imbalance in the powers of
the constituent elements of experience*. Eliade continues, describing the "rediscovery" or "re-
emergence" of Marduk.]

. . . Finally, he was delivered, and the gods assembled (that is, their statues were brought together) to
determine the destinies. (This episode corresponds, in the *Enuma elish*, to Marduk's advancement to the
rank of supreme god.) The king led the procession to the Bit Akitu, a building situated outside of the
city [*outside the domain of civilization, or order*]. The procession represented the army of the gods advanc-
ing against Tiamat. According to an inscription of Sennacherib, we may suppose that the primordial
battle was mimed, the king personifying Assur (the god who had replaced Marduk). The *hieros gamos*
took place after the return from the banquet at the Bit Akitu. The last act consisted in the determination
of the destinies for each month of the year. By "determining" it, the year was ritually *created*, that is, the
good fortune, fertility, and richness of the new world that had just been born were insured. . . .

   The role of the king in the *akitu* is inadequately known. His "humiliation" corresponds to the regression of the world to chaos and to Marduk's captivity in the mountain. The king personifies the god in
the battle against Tiamat and in the *hieros gamos* with a hierodule. But identification with the god is
not always indicated; as we have seen, during his humiliation the king addresses Marduk. Nevertheless,
the sacrality of the Mesopotamian sovereign is amply documented. . . .

   Though the king recognized his earthly begetting, he was considered a "son of god." . . . This twofold
descent made him supremely the intermediary between gods and men. The sovereign represented the
people before the gods, and it was he who expiated the sins of his subjects. Sometimes he had to suffer
death for his people's crimes; this is why the Assyrians had a "substitute for the king." The texts pro-

claim that the king had lived in fellowship with the gods in the fabulous garden that contains the Tree of Life and the Water of Life.... The king is the "envoy" of the gods, the "shepherd of the people," named by god to establish justice and peace on earth....

It could be said that the king shared in the divine modality, but without becoming a god. He *represented* the god, and this, on the archaic levels of culture, also implied that he *was*, in a way, he whom he personified. In any case, as mediator between the world of men and the world of the gods, the Mesopotamian king effected, in his own person, a ritual union between the two modalities of existence, the divine and the human. It was by virtue of this twofold nature that the king was considered, at least metaphorically, to be the creator of life and fertility.[253]

Marduk, in his manifestation as *Namtillaku*, was also "the god who restores to life,"[254] who can restore all "ruined gods, as though they were his own creation; The lord who by holy incantation restore[s] the dead gods to life."[255] This idea echoes through ancient Egyptian theology, as described below. Marduk was *Namshub*, as well, "the bright god who brightens our way"[256]—which once again assimilates him to the sun—and *Asaru*, the god of resurrection, who "causes the green herb to spring up."[257] Whatever Marduk represents was also considered central to creation of rich abundance,[258] to mercy,[259] and justice,[260] to familial love,[261] and, most interestingly, to the "creation of ingenious things" from the "conflict with Tiamat."[262] [!!!] He was in fact addressed by fifty names by the Mesopotamians. Each name signified an independent valuable attribute or property (likely at one time separate gods), now regarded as clearly dependent for its existence upon him. It seems evident that the attribution of these fifty names to Marduk parallels the movement toward monotheism described in the *Enuma elish* itself (with all the gods organizing themselves voluntarily under Marduk's dominion) and occurring in Mesopotamian society, at the human and historical level. It might be said that the Mesopotamians "came to realize" (in ritual and image, at least) that all the life-sustaining processes that they worshiped in representation were secondary aspects of the exploratory/creative/rejuvenating process embodied by Marduk.

A similar pattern of ritual and secondary conceptualization characterized ancient Egyptian society. In the earliest Egyptian cosmology (circa 2700 B.C.), the god *Ptah*, a spiritualized manifestation of *Atum*, the all-encircling serpent, creates "by his mind (his 'heart') and his word (his 'tongue')."[263] Eliade states:

Ptah is proclaimed the greatest god, Atum being considered only the author of the first divine couple. It is Ptah "who made the gods exist." ...

In short, the theogony and the cosmogony are effected by the creative power of the thought and word of a single god. We here certainly have the highest expression of Egyptian metaphysical speculation. As John Wilson observes,[264] it is *at the beginning* of Egyptian history that we find a doctrine that can be compared with the Christian theology of the Logos [or Word].[265]

The Egyptians "realized" that consciousness and linguistic ability were vital to the existence of things—precisely as vital as the unknowable matrix of their being. This idea still has not fully permeated our explicit understanding (since we attribute the existence of things purely to their material "substrate"), despite its centrality to Christian thinking. The Egyptians viewed Ptah—the spermatic word—as the original, or primordial (read "heavenly" king). As in Mesopotamia, essentially, he ceded this power, in the earthly domain, to his successor, the pharaoh [his "actual" or "literal" son, from the Egyptian viewpoint (as the Pharaoh was viewed *as* god)]. The creative power thus transferred was literally defined by the Egyptians as the ability to put order (*ma^{c}at*) "in the place of Chaos."[266] Eliade comments:

It is these same terms that are used of Tut-ankh-Amon when he restored order after the "heresy" of Akh-en-Aton, or of Pepi II: "He put *ma^{c}at* in the place of falsehood (of disorder)." Similarly, the verb *khay*, "to shine," is used indifferently to depict the emergence of the sun at the moment of creation or at each dawn and the appearance of the pharaoh at the coronation ceremony, at festivals, or at the privy council.

The pharaoh is the incarnation of *ma^{c}at*, a term translated by "truth" but whose general meaning is "good order" and hence "right," "justice." *Ma^{c}at* belongs to the original creation; hence it reflects the perfection of the Golden Age. Since it constitutes the very foundation of the cosmos and life, *ma^{c}at* can be known by each individual separately. In texts of different origins and periods, there are such declarations as these: "Incite your heart to know *ma^{c}at*"; "I make thee to know the thing of *ma^{c}at* in thy heart; mayest thou do what is right for thee!" Or: "I was a man who loved *ma^{c}at* and hated sin. For I knew that (sin) is an abomination to God." And in fact it is God who bestows the necessary knowledge. A prince is defined as "one who knows truth (*ma^{c}at*) and whom God teaches." The author of a prayer to Re cries: "Mayest Thou give me *ma^{c}at* in my heart!"

As incarnating *ma^{c}at*, the pharaoh constitutes the paradigmatic example for all his subjects. As the vizier Rekh-mi-Re expresses it: "He is a god who makes us live by his acts." The work of the pharaoh insures the stability of the cosmos and the state and hence the continuity of life. And indeed the cosmogony is repeated every morning, when the solar god "repels" the serpent Apophis, though without being able to destroy him; for chaos (= the original darkness) represents virtuality; hence it is indestructible. *The pharaoh's political activity repeats Re's exploit: he too "repels" Apophis, in other words he sees to it that the world does not return to chaos. When enemies appear at the frontiers, they will be assimilated to Apophis* [the god of primordial chaos], *and the pharaoh's victory will reproduce Re's triumph* [emphasis added].[267]

The ideas of kingship, creativity and renewal are given a different and more sophisticated slant in the central myth of Osiris, which served as an alternate basis for Egyptian theology.

The story of Osiris and his son *Horus* is much more complex, in some ways, than the Mesopotamian creation myth, or the story of *Re*, and describes the interactions between the "constituent elements of experience" in exceedingly compressed form. Osiris was a primeval king, a legendary ancestral figure, who ruled Egypt wisely and fairly. His evil brother, *Seth*— whom he did not understand[268]—rose up against him. *Figure 23: The Battle Between Osiris and Seth in the Domain of Order* portrays this conflict as a "war" in the "(heavenly) domain of order." Seth kills Osiris (that is, sends him to the underworld) and dismembers his body, so

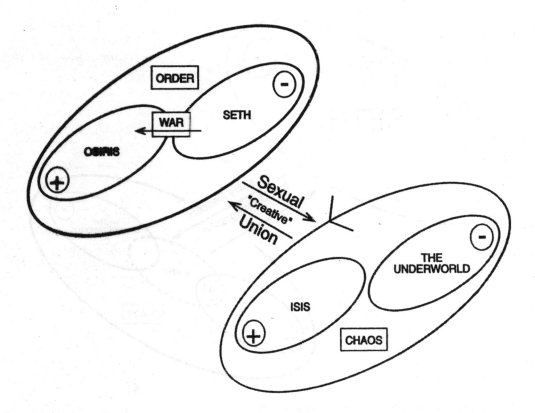

Figure 23: The Battle Between Osiris and Seth in the Domain of Order

that it can never be "found." *Figure 24: The Involuntary Descent and Disintegration of Osiris* portrays Osiris' "involuntary descent and disintegration," and his quasi-"existence" in the underworld of chaos.

The death of Osiris signifies two important things: (1) the tendency of a (static) ruling idea, system of valuation, or particular story—no matter how initially magnificent or appropriate—to become increasingly irrelevant with time; and (2) the dangers that necessarily accrue to a state that "forgets" or refuses to admit to the existence of the immortal deity of evil. Seth, the king's brother and opposite, represents the mythic "hostile twin" or "adversary" who eternally opposes the process of creative encounter with the unknown; signifies, alternatively speaking, a pattern of adaptation characterized by absolute opposition to establishment of divine order. When this principle gains control—that is, usurps the throne—the "rightful king" and his kingdom are necessarily doomed. Seth, and figures like him—often represented in narrative by the corrupt "righthand man" or "adviser to the once-great king"—view human existence itself with contempt. Such figures are motivated only to protect or advance their position in the power hierarchy, even when the prevailing order is clearly

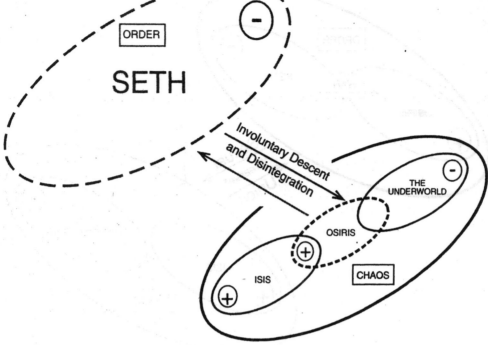

Figure 24: The Involuntary Descent and Disintegration of Osiris

counterproductive. Their actions necessarily speed the process of decay, endemic to all structures. Osiris, although great, was naïve in some profound sense—blind, at least, to the existence of "immortal" evil. This blindness, and its resultant incaution, brings about (or at least hastens) Osiris' demise.

Osiris has a wife, as befits the "king of order." Isis, as Osiris' mythic counterpart, is representative of the positive aspect of the unknown (like the *hierodule* in the Mesopotamian New Year's ritual). She is possessed of great magical powers, as might be expected, given her status. She gathers up Osiris' scattered pieces and makes herself pregnant with the use of his dismembered phallus. This story makes a profound point: the degeneration of the state or domain of order and its descent into chaos serves merely to fructify that domain and to make it "pregnant." In chaos lurks great potential. When a great organization disintegrates, falls into pieces, the pieces might still usefully be fashioned into, or give rise to, something else (perhaps something more vital, and still greater). Isis therefore gives birth to a son, Horus, who returns to his rightful kingdom to confront his evil uncle. This process is schematically represented in *Figure 25: The Birth and Return of Horus, Divine Son of Order and Chaos*.

Horus fights a difficult battle with Seth—as the forces of evil are difficult to overcome—

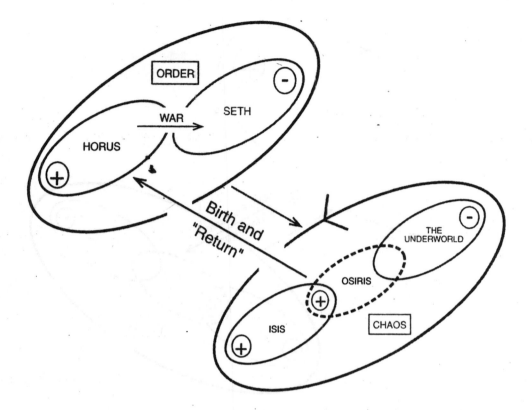

Figure 25: The Birth and Return of Horus, Divine Son of Order and Chaos

and loses an eye in the process. Seth is overcome, nonetheless; Horus recovers his eye. The story could stop there, narrative integrity intact, with the now-whole and victorious Horus' well-deserved ascension to the throne. However, Horus does the unexpected, descending *voluntarily* to the underworld to find his father (as portrayed schematically in *Figure 26: Voluntary Encounter with the Underworld*). It is representation of this move—reminiscent of Marduk's voluntary journey to the "underworld" of Tiamat—that constitutes the brilliant and original contribution of Egyptian theology.

Horus discovers Osiris, extant in a state of torpor. He offers his recovered eye to his father—so that Osiris can "see," once again. They return, united and victorious, and establish a revivified kingdom. The kingdom of the "son and father" is an improvement over that of the father or the son alone, as it unites the hard-won wisdom of the past (that is, of the dead) with the adaptive capacity of the present (that is, of the living). The (re)establishment and improvement of the domain of order is schematically represented in *Figure 27: Ascent, and Reintegration of the Father*.

In the story of Osiris, the senescence/death of the father (presented as a consequence of

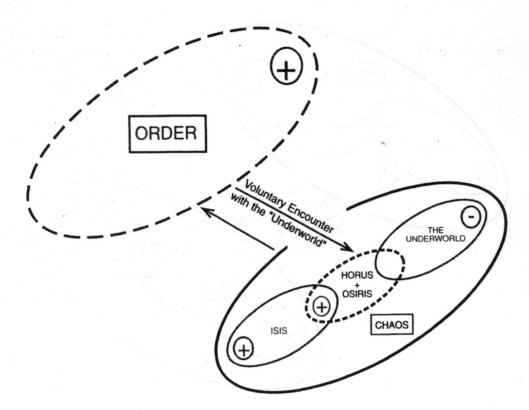

Figure 26: Voluntary Encounter with the Underworld

the treachery of Seth) is overcome by the mythic son, the hero who (temporarily) defeats the power of evil, and who rejuvenates the father. Marduk, the Mesopotamian supreme god, is by comparison a straightforward hero: he carves the familiar world from the unfamiliar. Horus, equally brave, is more complete, and more sophisticated. He cannot remain content with his own ascension, feeling himself incomplete without his father. He therefore journeys voluntarily into the underworld, releases the disintegrated forces of tradition trapped there, and makes them part of himself. This pattern of behavior constitutes an elaboration of that represented by Marduk—or of Re, the Egyptian sun-god.

Marduk creates order from chaos. That capacity, which is theoretically embodied in the form of the Mesopotamian emperor, lends temporal authority its rightful power. The same idea, elaborated substantially, applies in Egypt. Osiris constitutes the old state, once great, but dangerously anachronistic. Horus partakes of the essence of tradition (he is the son of his father), but is vivified by an infusion of "new information" (his mother, after all, is "the positive aspect of the unknown"). As an updated version of his father, he is capable of dealing with the problems of the present (that is, with the emergent evil represented by his uncle). Victorious over his uncle, he is nonetheless incomplete, as his youthful spirit lacks the

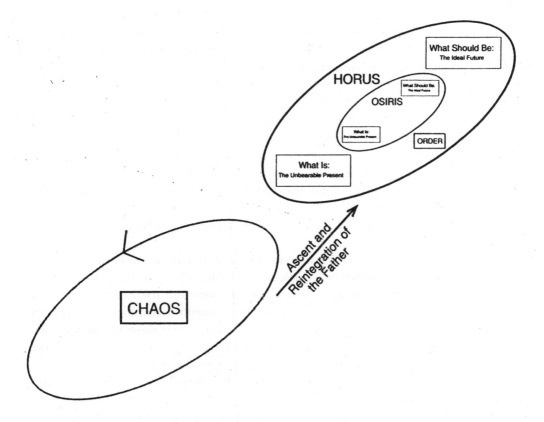

Figure 27: Ascent, and Reintegration of the Father

wisdom of the past. So he journeys into the unknown, where his father rests, "lifeless"—that is, uncomprehended; without embodiment or incarnation (in action) in the present. Horus unites himself with his father and becomes the ideal ruler—the consciousness of present youthful life, conjoined with the wisdom of tradition.

The "dead" Egyptian pharaoh—that is, the ruler whose death preceded the ascension of the current pharaoh—was assimilated to (occupied the same categorical space as) Osiris. That meant he was regarded as equivalent to "the spirit that founded the state"—the archetypal creator-god or legendary ancestor whose courageous actions had cosmogonic significance. The current ruler (who depended for much of his power on the traditions of his predecessors, modified when necessary) was regarded as equivalent to Horus and to Re, the sun-god. The ruling pharaoh was therefore the power that generated order from chaos (as Re), and the power that rejuvenated order, once it had degenerated into unthinking authoritarianism or too-rigid (and blind) tradition. Furthermore, however, he *was the rejuvenated Osiris* (who was the "dead pharaoh")—so he was tradition, given sight. The sophistication of this idea of reputable leadership—creative power, regenerative power and revivified tradition—can hardly be regarded as anything but remarkable. It is also of overwhelming historical interest and modern relevance that the Egyptians increasingly came to regard Osiris-Horus as an examplar, not just of the pharaoh, *but of every individual in the kingdom*. Eliade states, with regard to later Egyptian burial practice:

The texts formerly inscribed on the walls of the hidden chambers in the pyramids erected for the pharaohs are now reproduced inside the coffins of the nobility and even of totally unprivileged people. Osiris becomes the model for all those who hope to conquer death. A *Coffin Text* proclaims: "Thou art now the son of a king, a prince, as long as thy heart (i.e., spirit) shall be with thee." Following Osiris' example, and with his help, the dead are able to transform themselves into "souls," that is, into perfectly integrated and hence indestructible spiritual beings. Murdered and dismembered, Osiris was "reconstituted" by Isis and reanimated by Horus. In this way he inaugurated a new mode of existence: from a powerless shade, he became a "person" who "knows," a duly initiated spiritual being.[269]

This development might also be regarded as an illustration of the increasing psychologization, abstraction and internalization of religious ideation: in the earliest stages of representation, deities are viewed as pluralistic, and as individualistic and fractious members of a supracelestial (that is, transpersonal and immortal) community. Later, they are integrated into a hierarchy, as the culture becomes more integrated, more sure about relative valuation and moral virtue—and a single god, with a multitude of related features, comes to dominate. *Development of monotheism thus parallels intrapsychic and intracultural moral integration.* As the average citizen identifies more and more clearly with this monotheistic, integrated pattern, its external nature, as an attribute of the gods, recedes. It becomes more clearly an attribute of the individual human being, and more like what we would conceive of as a psychological trait. The god's subjective aspect—his or her intrapsychic quality—becomes more evident, at least to the most sophisticated of intuitions, and the possibility of "personal relationship" with the deity emerges as a prospect at the conceptual level of analysis. The process is just begining, in abstraction, in Mesopotamia and Egypt; the ancient Israelites bring it most clearly to fruition, with potent and lasting effect. It does not seem unreasonable to regard this development as a precursor to the Christian revolution—which granted every individual the status of "son of god"—and as implicitly akin to our modern notion of the intrinsic "human right."

The Egyptian pharaoh, like the Mesopotamian king, served as material incarnation of the process that separates order from chaos; simultaneously, the pharaoh/king literally embodied the state. Finally, the pharoah/king was the rejuvenator of his own "father." The "ideal" pharaoh/king was therefore the exploratory process that gave rise to the state, the state itself, and the revivifying (exploratory) process that updated the state when it was in danger of too-conservative ossification. This massively complex and sophisticated conceptualization is given added breadth and depth by consideration of its psychological element. The state is not merely cultural; it is also "spiritual." As custom and tradition is established, it is inculcated into each individual, and becomes part of their intrapsychic structure. The state is therefore personality and social organization, simultaneously—personality and social order conjoined in the effort to keep the terror of chaos at bay (or, better still, united in the effort to make something positively useful of it). This means that the hero/king who establishes, embodies and updates the social world is also the same force that establishes, embodies and updates the intrapsychic world, the *personality*—and that one act of update cannot necessarily or reasonably be distinguished from the other. In "improving" the world, the hero improves himself; in improving himself, he sets an example for the world.

Initially, the "personality of state" was in fact a ritual human model (a hero) to observe and imitate (an entity represented in behavioral pattern); then a story about such ritual models (an entity represented in imagination), and, finally—and only much later—an abstract construction of rules describing the explicit rights and responsibilities of the citizenry—(an entity of words, the "body" of law). This increasingly abstract and detailed construction develops from imitation to abstract representation, and comprises rules and schemas of interpretations useful for maintaining stability of interpersonal interaction. The establishment of these rules and schemas gives determinate meaning to human experience, by bringing predictability to all social situations (to all things encountered interpersonally). The same thing might be said from the psychological perspective. It is incorporation of the "personality of state," dominated by the figure of the hero, that brings order to the inner community of necessity and desire, to the generative chaos of the soul.

The Mesopotamian culture-hero/deity Marduk represents the capacity of the process of exploration to generate the world of experience; the Egyptian gods Horus-Osiris represent the extended version of that capacity, which means not only generation of the world from the unknown, but transformation of the pattern of adaptation which constitutes the known, when such transformation becomes necessary.

Sometimes "adaptation" is merely a matter of the adjustment of the means to an end. More rarely, but equally necessarily, adaptation is reconceptualization of "what is known" (unbearable present, desirable future and means to attain such) because what is is known is out of date, and therefore deadly. It is the sum of these processes that manifests itself in the Judeo-Christian tradition as the mythic Word of God (and which is embodied in Christ, the Christian culture-hero). This is the force that generates subject and object from the primordial chaos (and, therefore, which "predates" the existence of both); the force that engenders the tradition that makes vulnerable existence possible, in the face of constant mortal threat; and the force that updates protective tradition when it has become untenable and tyrannical on account of its age.

The Sumerian and Egyptian myths portray ideas of exceeding complexity, in ritual, drama, and imagistic form. This form is not purposeful mystification, but the manner in which ideas emerge, before they are sufficiently developed to be explicitly comprehensible. We acted out and provisionally formulated complete, "impressionistic" models of the world of experience (which was the world we always had to understand) long before the "contents" of such models could be understood in the way we currently conceive of understanding.

Brief analysis of the Sumerian and Egyptian theologies, and of the relationship of those

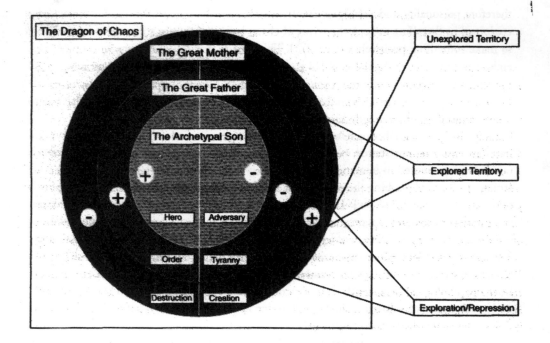

Figure 28: The Constituent Elements of Experience as Personality, Territory and Process

theologies to political action, shed substantial light on the manner in which many of our most important modern ideas developed (and on what those ideas actually mean). This understanding, derived from two or three specific examples, can be further enhanced through more general discussion. We therefore turn our attention from analysis of complete stories—which have as their advantage a more compelling nature—to detailed description of the mythological characters whose essence and interactions constitute the world. The totality of the world, which includes the significance of experienced things, as well as the things themselves, is composed of what has been explored and rendered familiar; what has yet to be encountered, and is therefore unpredictable; and the process that mediates between the two. One final element must be additionally considered: the state of being that includes or precedes the division of everything into these three constituent elements. This state might be regarded as the true source of all things, subjects and objects—the single ancestor and final destination of all. The complete mythological "world of experience" is portrayed schematically in *Figure 28: The Constituent Elements of Experience as Personality, Territory and Process.* Our discussion turns first to the diverse nature of representations of the original, undifferentiated state (the condition of primordial chaos) and then to a more elaborated description of its "children"—the divine parents, nature and culture, and the divine son, simultaneously child, primal creator and eternal adversary.

The Dragon of Primordial Chaos

> The source of things is the boundless. From whence they arise, thence
> they must also of necessity return. For they do penance and make com-
> pensation to one another for their injustice in the order of time. [270]

It might seem futile to speculate about the nature of that which existed prior to any experience, or that which has not yet been explored. Futile or not, such speculation has occupied a
good portion of man's time, as he attempted to understand the mystery of his emergence and
of the world he found himself occupying. It seems impossible to determine what it *is* that was
before anything was; myth attempts that task, despite its impossibility. It does so using the
tool of metaphor. The metaphorical statements of myth work because unknown or partially
known things inevitably share characteristics of importance with somewhat more thoroughly
investigated, comprehended and familiar things. Two or more objects or situations come to
occupy the same mythological or categorical space, therefore, because they share similar form,
function or capacity to induce affect and compel behavior. A mandrake root, for example, has
the nature of a man, symbolically speaking, because it has the *shape* of a man; Mars is a warlike planet because it is red, and red, the color of blood, is associated indelibly with aggression;
the metal mercury (and the "spirit" that inhabits it) is akin to seawater because both may serve
as solvents or agents of transformation; the dark and the animal of the forest are the same,
because they are both unfamiliar—because they both inhibit ongoing behavior, when they
make their appearance; because they both cause *fear*. Metaphor links thing to thing, situation
to situation, concentrating on the phenomenological, affective, functional and motivational
features the linked situations share. Through such linkage, what might otherwise remain
entirely mysterious can begin to become comprehended.

Myths of the origin metaphorically portray the nature of the infinite potential that characterized being prior to the dawn of experience. This general symbolic construction takes
many particular forms, each of which might be said to constitute a partial attempt to represent the unrepresentable whole. These particular forms range in nature from the specific and
concrete to the general and abstract, and are influenced in their development by the environmental and cultural conditions obtaining at the time of their emergence. The process of
metaphorical representation provides a bridge—and an increasingly communicable bridge—
between what can be directly explored, experienced and "comprehended," and what remains
eternally unknown.

Mythic symbols of the chaos of the beginning are imaginative pictures, whose purpose is
representation of a paradoxical totality, a "state" (which is already to say something too determinate) self-contained, uniform and complete, where everything now distinct resides in
union: a state where being and nonbeing, beginning and end, matter and energy, spirit and

body, consciousness and unconsciousness, femininity and masculinity, night and day remain compounded, prior to their discrimination into the separable elements of experience. In this "state," all conceivable pairs of opposites and contradictory forces exist together, within the all-encompassing embrace of an omniscient, omnipresent, omnipotent and altogether mysterious God. This "paradisal" precondition, lacking nothing, characterized by absolute completion, exists in contradistinction to the profane world, imperfect and partial, suspended unbearably in time and in space; it surrounds that world completely, like the night surrounds the day, comprising the beginning of things, the fountainhead for everything and, similarly, the resting place and destination point for all. William James turned to poetry in his attempt to conceptualize this "place":

> No verbiage can give it, because the verbiage is other,
> Incoherent, coherent—same.
> And it fades! And it's infinite! AND it's infinite! ...
> Don't you see the difference, don't you see the identity?
> Constantly opposites united!
> The same me telling you to write and not to write!
> Extreme—extreme, extreme! ...
> Something, and other than that thing!
> Intoxication, and otherness than intoxication.
> Every attempt at betterment,—every attempt at otherment
> —is a—
> It fades forever and forever as we move.[271]

This state—"the totality of all things"—might be regarded as the objective world, in the absence of the subject, although this conceptualization is too narrow, as primordial chaos also contains that which evolves into the subject, when it is differentiated. What might be regarded as the standard objective viewpoint is predicated on the idea that "things" as they are perceived exist regardless of the perceiver. From a certain perspective, this is true. Things have a nature that appears independent of subjective will, and follow their own laws of being and development—despite our wishes. However, the job of determining what a thing is in the absence of the subject is much more difficult than might initially be imagined. It is certainly the case—as we have seen—that the *value* of an object can shift with shifts in frame of reference. It appears to be true, however, that what an *object* is "is and of itself" is also subject to such shift. Any given object—a table, say—exists *as a table* because it is apprehended only in a very limited and restrained manner. Something is a table at a particular and isolated level of analysis, specified by the nature of the observer. In the absence of this observer, one might ask, what is it that is being apprehended? Is the proper level of analysis and specification subatomic, atomic or molecular (or all three at once)? Should the table be considered an indistinguishable element of the earth upon which it rests, or of the solar system, which contains the earth, or of the galaxy itself? The same problem obtains from the perspective of temporality. What is now table was once tree; before that, earth—before that, rock; before that, star. What is now table also has before it an equally complex and lengthy developmental history waiting in "front" of it; it will be, perhaps, ash, then earth, then—far enough in the future—part of the sun again (when the sun finally re-envelops the earth). The table is what it "is" only at a very narrow span of spatial and temporal resolution (the span that precisely characterizes our consciousness). So what is the table as an "independent object"— "free," that is, of the restrictions that characterize the evidently limited human viewpoint? What is it that can be conceptualized at all spatial and temporal levels of analysis simultaneously? Does the "existence" of the thing include its interactions with everything it influences, and is influenced by, gravitationally and electromagnetically? Is that "thing" everything it once was, everything it is, and everything it will be, all at the same time? Where then are its borders? How can it be distinguished from other things? And without such distinction, in what manner can it be said to exist?

Question: what is an object, in the absence of a frame of reference? Answer: it is everything conceivable, at once—is something that constitutes the union of all currently discriminable opposites (and something that cannot, therefore, be easily distinguished from nothing).

I am not saying that there are no such things as "things"—that would of course be patently absurd. It is also fully apparent that the things we apprehend are rule-governed—the cosmos as we experience it is orderly and rationally comprehensible. What I am claiming is that "objective" things are in fact the product of an interaction between whatever constitutes our limited consciousness and whatever constitutes the unlimited "background" that makes up the world, in the absence of a subject. This is a stance informed by mythology—in particular, by myths of the origin.

Archaic myths describing the ultimate origin concern themselves with representation of the source, not of *objects*, in the modern sense, but of *subjects and the experience of those subjects* (some part of which can be regarded as objects). Such myths typically describe the genesis of the world of experience by relating the existence of a primordial god, portraying the division of this god into the world-parents, and detailing the separation of those parents by their own "son." This is the division of the hermaphroditic, all-encompassing, self-devouring and nourishing serpent of chaos into earth/matter and sky/spirit, and the subsequent discrimination of those "primordial opposing forces" into identifiable aspects of being. The Indo-European myth of *Indra* and *Vṛtra* provides a representative example:

The central myth of Indra, which is, furthermore, the most important myth in the Rig Veda, narrates his victorious battle against Vṛtra, the gigantic dragon who held back the waters in the "hollow of the mountains." Strengthened by *soma*, Indra lays the serpent low with his *vajra* ("thunderbolt"), the weapon forged by Tvaṣṭr, splits open his head, and frees the waters, which pour into the sea "like bellowing cows." (RV 1.32)

The battle of a god against an ophidian or marine monster is well known to constitute a widespread mythological theme. We need only remember the struggle between Re and Apophis, between the Sumerian god Ninurta and Asag, Marduk and Tiamat, the Hittite storm god and the serpent

Illuyankas, Zeus and Typhon, the Iranian hero Thraëtona and the three-headed dragon Azhi-dahâka. In certain cases (Marduk-Tiamat, for example) the god's victory constitutes the preliminary condition for the cosmogony. In other cases the stake is the inauguration of a new era or the establishment of a new sovereignty (cf. Zeus-Typhon, Baal-Yam). In short, it is by the slaying of an ophidian monster—symbol of the virtual, of "chaos," but also of the "autochthonous"—that a new cosmic or institutional "situation" comes into existence. A characteristic feature, and one common to all these myths, is the fright, or a first defeat, of the champion (Marduk and Re hesitate before fighting; at the onset, the serpent Illyunakas succeeds in mutilating the god; Typhon succeeds in cutting and carrying off Zeus's tendons). According to the *Śatapatha Brāhmaṇa* (1.6.3–17), Indra, on first seeing Vṛtra, runs away as far as possible, and the *Mārkaṇḍeya Purāṇa* describes him as "sick with fear" and hoping for peace.[272]

It would serve no purpose to dwell on the naturalistic interpretations of this myth; the victory over Vṛtra has been seen either as rain brought on by a thunderstorm or as the freeing of the mountain waters (Oldenberg) or as the triumphs of the sun over the cold that had "imprisoned" the waters by freezing them (Hillebrandt). Certainly, naturalistic elements are present, since the myth is multivalent; Indra's victory is equivalent, among other things, to the triumph of life over the sterility and death resulting from the immobilization of the waters by Vṛtra. But the structure of the myth is cosmogonic. In Rig Veda 1.33.4 it is said that, by his victory, the god created the sun, the sky, and dawn. According to another hymn (RV 10.113.4–6) Indra, as soon as he was born, separated the Sky from the Earth, fixed the celestial vault, and hurling the *vajra*, tore apart Vṛtra, who was holding the waters captive in the darkness. Now, Sky and Earth are the parents of the gods (1.185.6); Indra is the youngest (3.38.1) and also the last god to be born, because he put an end to the hierogamy [mystical union] of Sky and Earth: "By his strength, he spread out these two worlds, Sky and Earth, and caused the sun to shine." (8.3.6). After this demiurgic feat, Indra appointed Varuṇa cosmocrator and guardian of *ṛta* (which had remained concealed in the world below; 1.62.1)....

There are other types of Indian cosmogonies that explain the creation of the world from a *materia prima*. This is not the case with the myth we have just summarized, for here a certain type of "world" already existed. For Sky and Earth were formed and had engendered the gods. Indra only separated the cosmic parents, and, by hurling the *vajra* at Vṛtra, he put an end to the immobility, or even the "virtuality," symbolized by the dragon's mode of being. [Indra comes across Vṛtra "not divided, not awake, plunged in the deepest sleep, stretched out" (RV 4.19.3).] According to certain traditions, the "fashioner" of the gods, Tvaṣṭṛ, whose role is not clear in the Rig Veda, had built himself a house and created Vṛtra as a sort of roof, but also as walls, for his habitation. Inside this dwelling, encircled by Vṛtra, Sky, Earth and the Waters existed. Indra bursts asunder this primordial monad by breaking the "resistance" and inertia of Vṛtra. In other words, the world and life could not come to birth except by the slaying of an amorphous Being. In countless variants, this myth is quite widespread.[273]

The primordial theriomorphic serpent-god is endless *potential*; is whatever being is prior to the emergence of the capacity for experience. This potential has been represented as the self-devouring dragon (most commonly) because this image (portrayed in *Figure 29: The Uroboros—Precosmogonic Dragon of Chaos*[274] aptly symbolizes *the union of incommensurate opposites*. The uroboros is simultaneously representative of two antithetical primordial ele-

Figure 29: The Uroboros—Precosmogonic Dragon of Chaos

ments. As a snake, the uroboros is a creature of the ground, of *matter;* as a bird (a winged animal), it is a creature of the air, the sky, *spirit.* The uroboros symbolizes the union of *known* (associated with spirit) and *unknown* (associated with matter), explored and unexplored; symbolizes the juxtaposition of the "masculine" principles of security, tyranny and order with the "feminine" principles of darkness, dissolution, creativity and chaos. Furthermore, as a snake, the uroboros has the capacity to shed its skin—to be "reborn." Thus, it also represents the possibility of transformation, and stands for the *knower,* who can transform chaos into order, and order into chaos. The uroboros stands for, or comprises, everything that is as of yet unencountered, prior to its differentiation as a consequence of active exploration and classification. It is the source of all the information that makes up the determinate world of experience and is, simultaneously, the birthplace of the experiencing subject.

The uroboros is one thing, as everything that has not yet been explored is one thing; it exists everywhere, and at all times. It is completely self-contained, completely self-referential: it feeds, fertilizes and engulfs itself. It unites the beginning and the end, being and becoming, in the endless circle of its existence. It serves as symbol for the ground of reality itself. It is the "set of all things that are not yet things," the primal origin and ultimate point of return for every discriminable object and every independent subject. It serves as progenitor of all we know, all that we don't know, and of the spirit that constitutes our capacity to know and not know. It is the mystery that constantly emerges when solutions to old problems cause new problems; is the sea of chaos surrounding man's island of knowledge—and the source of that knowledge, as well. It is all new experience generated by time, which incessantly works to transform the temporarily predictable once again into the unknown. It has served mankind as the most ubiquitous and potent of primordial gods:

This is the ancient Egyptian symbol of which it is said, "*Draco interfecit se ipsum, maritat se ipsum, impraegnat se ipsum.*" It slays, weds, and impregnates itself. It is man and woman, begetting and conceiving, devouring and giving birth, active and passive, above and below, at once.

As the Heavenly Serpent, the uroboros was known in ancient Babylon; in later times, in the same area, it was often depicted by the Mandaeans; its origin is ascribed by Macrobius to the Phoenicians. It is the archetype of the εντόπαν, the All One, appearing as Leviathan and as Aion, as Oceanus, and also as the Primal Being that says "I am Alpha and Omega." As the Kneph of antiquity it is the Primal Snake, the "most ancient deity of the prehistoric world." The uroboros can be traced in the Revelation of St. John and among the Gnostics as well as among the Roman syncretists; there are pictures of it in the sand paintings of the Navajo Indians and in Giotto; it is found in Egypt, Africa, Mexico, and India, among the gypsies as an amulet, and in the alchemical texts.[275]

The uroboros is Tiamat, the dragon who inhabits the deep, transformed by Marduk into the world; Apophis, the serpent who nightly devours the sun; and *Rahab*, the leviathan, slain by Yahweh in the course of the creation of the cosmos:

> Canst thou draw out leviathan with an hook? or his tongue with a cord which thou
>   lettest down?
> Canst thou put an hook into his nose? or bore his jaw through with a thorn?
> Will he make many supplications unto thee? will he speak soft words unto thee?
> Will he make a covenant with thee? wilt thou take him for a servant for ever?
> Wilt thou play with him as with a bird? or wilt thou bind him for thy maidens?
> Shall the companions make a banquet of him? shall they part him among the merchants?
> Canst thou fill his skin with barbed irons? or his head with fish spears?
> Lay thine hand upon him, remember the battle, do no more.
> Behold, the hope of him is in vain: shall not one be cast down even at the sight of him?
> None is so fierce that dare stir him up: who then is able to stand before me?
> Who hath prevented me, that I should repay him? whatsoever is under the whole
>   heaven is mine.
> I will not conceal his parts, nor his power, nor his comely proportion.
> Who can discover the face of his garment? or who can come to him with his double
>   bridle?
> Who can open the doors of his face? his teeth are terrible round about.

His scales are his pride, shut up together as with a close seal.

One is so near to another, that no air can come between them.

They are joined one to another, they stick together, that they cannot be sundered.

By his neesings a light doth shine, and his eyes are like the eyelids of the morning.

Out of his mouth go burning lamps, and sparks of fire leap out.

Out of his nostrils goeth smoke, as out of a seething pot or caldron.

His breath kindleth coals, and a flame goeth out of his mouth.

In his neck remaineth strength, and sorrow is turned into joy before him.

The flakes of his flesh are joined together: they are firm in themselves; they cannot be
    moved.

His heart is as firm as a stone; yea, as hard as a piece of the nether millstone.

When he raiseth up himself, the mighty are afraid: by reason of breakings they purify
    themselves.

The sword of him that layeth at him cannot hold: the spear, the dart, nor the habergeon.

He esteemeth iron as straw, and brass as rotten wood.

The arrow cannot make him flee: slingstones are turned with him into stubble.

Darts are counted as stubble: he laugheth at the shaking of a spear.

Sharp stones are under him: he spreadeth sharp pointed things upon the mire.

He maketh the deep to boil like a pot: he maketh the sea like a pot of ointment.

He maketh a path to shine after him; one would think the deep to be hoary.

Upon earth there is not his like, who is made without fear.

He beholdeth all high things: he is a king over all the children of pride. (Job 41:1–34)

The uroboros is that which exists as pure unqualified potential, prior to the manifestation of such potential, in the experience of the limited subject; is the infinite possibility for sudden dramatic unpredictability that still resides in the most thoroughly explored and familiar of objects (things, other people, ourselves). That unpredictability is not mere material possibility or potential; it is also meaning. The domain of chaos—which is where what to do has not yet been specified—is a "place" characterized by the presence of potent emotions, discouragement, depression, fear, rootlessness, loss and disorientation. It is the affective aspect of chaos that constitutes what is most clearly known about chaos. It is "darkness, drought, the suspension of norms, and death."[276] It is the terror of the dark of night, which fills itself with demons of the imagination, yet exerts an uncanny fascination; it is the fire that magically reduces one determinate thing to another; it is the horror and curiosity engendered by the stranger and foreigner.

The uroboros—the primordial matrix—contains in "embryonic" form everything that can in principle possibly be experienced, and the thing that does the experiencing. The great serpent (the matrix) is therefore consciousness—*spirit*, before it manifests itself—and *matter*, before it is separated from spirit. This great mythological idea finds its echo in certain modern theories of the development of the subject; most particularly, among those entitled *constructivist*. The famous Swiss developmental psychologist Jean Piaget claimed, for example, that the experiencing subject constructs himself in infancy, as a consequence of his exploratory activity.[277] He acts, and observes himself acting; then imitates the action, forming a primordial representation of himself—later, formulates a more abstracted model of his own actions. *Thus the subject is created from the information generated in the course of exploratory activity.* Contemporaneously, the world comes into being:

> Thou brakest the heads of leviathan in pieces, *and* gavest him *to be* meat to the
>   people inhabiting the wilderness.
> Thou didst cleave the fountain and the flood: thou driedst up mighty rivers.
> The day *is* thine, the night also *is* thine: thou has prepared the light and the sun.
> Thou hast set all the borders of the earth: thou hast made summer and winter.
>   (Psalms 74:14–17).

Actions have consequences. The consequences of actions constitute the world—the familiar world, when they are predictable; the world of the unexpected, when they are not.

The state of the origin has been represented most abstractly as a circle, the most perfect of geometric forms, or as a sphere, without beginning or end, symmetrical across all axes. Plato, in the *Timaeus*, described the primary source as the round, there at the beginning.[278] In the Orient, the world and its meaning springs from the encircled interplay and union of the light, spiritual, masculine *yang* and the dark, material, feminine *yin*.[279] According to the adepts of medieval alchemy, discernible objects of experience (and the subjects who experienced them) emerged from the round chaos, which was a spherical container of the primordial element.[280] The God of Islam, Judaism and Christianity, "Alpha and Omega, the beginning and the end, the first and the last" (Revelations 22:13), places himself outside of or beyond worldly change, and unites the temporal opposites within the great circle of his being. The assimilation of the origin to a circle finds narrative echo in myths describing heaven as the end to which life is, or should be, devoted (at least from the perspective of the "immortal soul.") The Kingdom of God, promised by Christ, is in fact re-establishment of Paradise (although a Paradise characterized by reconciliation of opposing forces, and not regressive dissolution into preconscious unity). Such re-establishment closes the circle of temporal being.

The uroboric initial state is the "place" where all opposite things were (will be) united; the great self-devouring dragon whose division into constituent elements constitutes the precondition for experience itself. This initial state is a "place" free of problems, and has a paradisal aspect, in consequence; however, the price that must be paid for uroboric paradise is being itself. It is not until the original unity of all things is broken up—until the most primordial of gods is murdered—that existence itself springs into being. The emergence of things, however, brings with it the problem of conflict—a problem that must be solved, optimally, without eliminating the fact of existence itself.

The uroboros is the unified parent of the known, the Great Father (explored territory and the familiar), and of the unknown, the Great Mother (anomalous information and the

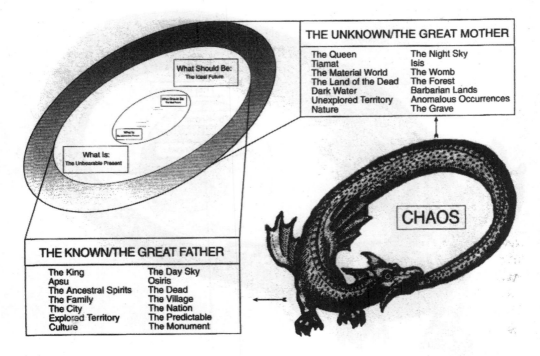

The Unknown/The Great Mother

The Queen	The Night Sky
Tiamat	Isis
The Material World	The Womb
The Land of the Dead	The Forest
Dark Water	Barbarian Lands
Unexplored Territory	Anomalous Occurrences
Nature	The Grave

CHAOS

The Known/The Great Father

The King	The Day Sky
Apsu	Osiris
The Ancestral Spirits	The Dead
The Family	The Village
The City	The Nation
Explored Territory	The Predictable
Culture	The Monument

Figure 30: The Birth of the World Parents

unpredictable). It might be regarded, as well, as the single androgynous grandparent of the hero, son of the night and the day, mediator between the known and unknown, whose being constitutes a necessary precondition for the existence of differentiated things (and who can, therefore, also be regarded as a *causa prima*). The world parents, Earth and Sky, emerge when the uroboric dragon undergoes a first division. *Figure 30: The Birth of the World Parents* presents the "birth of the world" in schematic form, insofar as it has been conceptualized by the mythic imagination. The chaos that constitues totality divides itself into what has been explored, and what has yet to be explored.

From the mythic perspective, this division is equivalent to the emergence of the cosmos—and, therefore, to creation or genesis itself. One thing is missing—the fact of the explorer, and the nature of his relationship with what is known and what has yet to be known. With the "birth" of the explorer—with his construction from the interplay between culture and nature—the entire "world" comes into being. This "emergence of experience" is portrayed in *Figure 31: The Constituent Elements of the World, in Dynamic Relationship*. The "knower" is simultaneously child of nature and culture, creator of culture (as a consequence of his encounter with nature or the unknown world) and the "person" for whom the unknown is a reality.

It is almost impossible to overestimate the degree to which the "world parent" schema of

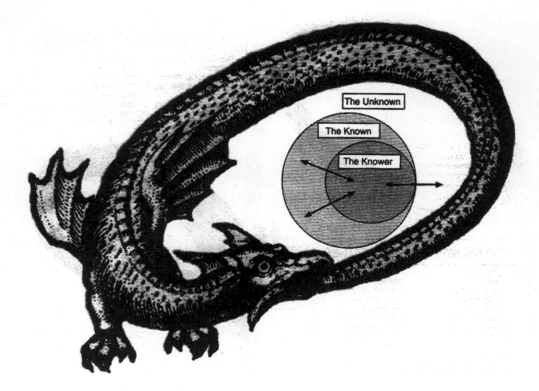

Figure 31: The Constituent Elements of the World, in Dynamic Relationship

categorization colors (or, alternatively, has been derived from) fundamental human presumption and activity. The "world" is explored territory, surrounded by mystery; that mystery is experienced as undifferentiated but oft-menacing chaos. Everything that "occupies" such chaos is *directly perceived* as (not abstractly conceptualized as) *identical* to it—is directly perceived as unknown and anxiety-provoking. The foreigner, therefore—the occupant of the "habitation of dragons" (Isaiah 34:13)—is *naturally* apprehended as an agent of formless chaos. Eliade states:

One of the outstanding characteristics of traditional societies is the opposition that they assume between their inhabited world and the unknown and indeterminate space that surrounds it. The former is the world (more precisely, our world), the cosmos; everything outside it is no longer a cosmos but a sort of "other world," a foreign, chaotic space, peopled by ghosts, demons, "foreigners" (who are assimilated to [undistinguished from, more accurately] the demons and the souls of the dead).[281]

Everything *outside* occupies the same categorical space as the dragon of chaos, or the terrible mother. The early Indo-Europeans equated the destruction of enemies in battle to the slaying of Vṛtra by Indra;[282] the ancient Egyptians regarded the Hyksos, "barbarians," as equivalent to Apophis, the serpent who nightly devours the sun;[283] and the archaic Iranians (Zoroastrians) equated the mythic struggle of King Faridun against a foreign usurper—the dragon Azdahak—with the cosmogonic fight of the hero Thraëtona against Azhi Dahâka, the primordial serpent of chaos.[284] The enemies of the Old Testament Hebrews also suffer the same fate: they are regarded as equivalent to Rahab, or Leviathan, the serpent overcome by Yahweh in his battle to establish the world ["Speak, and say, Thus saith the Lord GOD; Behold, I am against thee, Pharaoh king of Egypt, the great dragon that lieth in the midst of his rivers, which hath said, My river is mine own, and I have made it for myself." (Ezekiel 29:3); also, "Nebuchadnezzar the king of Babylon hath devoured me, he hath crushed me, he hath made me an empty vessel, he hath swallowed me up like a dragon, he hath filled his belly with my delicates, he hath cast me out." (Jeremiah 51:34)]. Eliade continues:

At first sight this cleavage in space appears to be due to the opposition between an inhabited and organized—hence cosmicized—territory and the unknown space that extends beyond its frontiers; on one side there is a cosmos, on the other a chaos. But we shall see that if every inhabited territory is a cosmos, this is precisely because it was first consecrated, because, in one way or another, it is the work of the gods or is in communication with the world of the gods. The world (that is, our world) is a universe within which the sacred has already manifested itself, in which, consequently, the break-through from plane to plane has become possible and repeatable. It is not difficult to see why the religious moment implies the cosmogonic moment. The sacred reveals absolute reality and at the same time makes orientation possible; hence it *founds the world* in the sense that it fixes the limits and establishes the order of the world.

All this appears very clearly from the Vedic ritual for taking possession of a territory; possession becomes legally valid through the erection of a fire altar consecrated to Agni. "One says that one is installed when one has built a fire altar (*garhapatya*) and all those who build the fire altar are legally established." (*Shatapatha Brahmana*, VII, 1,1,1–4). By the erection of a fire altar Agni is made present, and communication with the world of the gods is ensured; the space of the altar becomes a sacred space. But the meaning of the ritual is far more complex, and if we consider all of its ramifications, we shall understand why consecrating a territory is equivalent to making it a cosmos, to *cosmicizing* it. For, in fact, the erection of an altar to Agni is nothing but the reproduction—on the microcosmic scale—of the Creation. The water in which the clay is mixed is assimilated to the primordial water; the clay that forms the base of the altar symbolizes the earth; the lateral walls represent the atmosphere, and so on. And the building of the altar is accompanied by songs that proclaim which cosmic region has just been created (*Shatapatha Brahmana* I, 9, 2, 29, etc.). Hence the erection of a fire altar—which alone validates taking possession of a new territory—is equivalent to a cosmogony.

An unknown, foreign and unoccupied territory (which often means "unoccupied by our people") still shares in the fluid and larval modality of chaos. By occupying it and, above all, by settling in it, man symbolically transforms it into a cosmos through a ritual repetition of the cosmogony. What is to become "our world" must first be "created," and every creation has a paradigmatic model—the creation of the universe by the gods. When the Scandinavian colonists took possession of Iceland (*landnama*) and cleared it, they regarded the enterprise neither as an original undertaking nor as human and profane work. For them, their labor was only repetition of a primordial act, the transformation of chaos into cosmos by the divine act of creation. When they tilled the desert soil, they were in fact repeating the act of the gods who had organized chaos by giving it a structure, forms, and norms.

Whether it is a case of clearing uncultivated ground or of conquering and occupying a territory already inhabited by "other" human beings, ritual taking possession must always repeat the cosmogony. For in the view of archaic societies everything that is not "our world" is not yet a world. A territory can be made ours only by creating it anew, that is, by consecrating it. This religious behavior in respect to unknown lands continued, even in the West, down to the dawn of modern times [and was reflected recently in the "planting of the flag" on the moon, by the American astronauts]. The Spanish and Portuguese conquistadors, discovering and conquering territories, took possession of them in the name of Jesus Christ [the world-creating *Logos*].[285]

A similar form of ritual and ideation dominates processes even as "simple" as the establishment of a new building. In India,

Before a single stone is laid, "The astrologer shows what spot in the foundation is exactly above the head of the snake that supports the world. The mason fashions a little wooden peg from the wood of the Khadira tree, and with a coconut drives the peg into the ground at this particular spot, in such a way as to peg the head of the snake securely down.... If this snake should ever shake its head really violently, it would shake the world to pieces."[286] A foundation stone is placed above the peg. The cornerstone is thus situated exactly at the "center of the world." But the act of foundation at the same time repeats the cosmogonic act, for to "secure" the snake's head, to drive the peg into it, is to imitate the primordial gesture of Soma (*Rig Veda* II, 12, 1) or of Indra when the latter "smote the Serpent in his lair" (*Rig Veda*, VI, 17, 9), when his thunderbolt "cut off his head" (*Rig Veda* I, 52, 10).[287]

Order—explored territory—is constructed out of chaos and exists, simultaneously, in opposition to that chaos (to the "new" chaos, more accurately: to the unknown *now defined in opposition to explored territory*). Everything that is not order—that is, not predictable, not usable—is, by default (by definition) *chaos*. The foreigner—whose behaviors cannot be predicted, who is not kin, either by blood or by custom, who is not an inhabitant of the "cosmos," whose existence and domain has not been sacralized—is *equivalent* to chaos (and not merely metaphorically equated with chaos). As such, his appearance means threat, as his action patterns and beliefs have the capacity to upset society itself, to dissolve and flood the world, and to reinstitute the dominion of the uroboros.

The Great Mother: Images of the Unknown, or Unexplored Territory

The Mother of Songs, the mother of our whole seed, bore us in the beginning. She is the mother of all races of men and the mother of all tribes. She is the mother of the thunder, the mother of the rivers, the

mother of trees and of all kinds of things. She is the mother of songs
and dances. She is the mother of the older brother stones. She is the
mother of the grain and the mother of all things. She is the mother of
the younger brother Frenchmen and of the strangers. She is the
mother of the dance paraphernalia and of all temples, and the only
mother we have. She is the mother of the animals, the only one, and
the mother of the Milky Way. It was the mother herself who began to
baptize. She gave us the limestone coca dish. She is the mother of the
rain, the only one we have. She alone is the mother of all things, she
alone. And the mother has left a memory in all the temples. With her
sons, the saviors, she left songs and dances as a reminder. Thus the
priests, the fathers, and the older brothers have reported.[288]

Representation of culture, the known, is simple, comparatively; it is second-order abstraction, depiction of that which has already been made subject to order. Representation of culture is encapsulation of that to which behavioral adaptation has previously occurred; of those
things or situations whose sensory properties, affective implications and motivational significances have been and are presently specified. Representation of the knower, the human subject, is also depiction of that which is constantly encountered, in all interpersonal
interactions, and in all self-conscious states: is portrayal of those aspects of an infinitely complex set of data which have at least been experienced, if not exhausted. Representation of the
unknown, however, appears impossible, a contradiction in terms. How can what has not yet
been encountered be comprehended, understood, embodied, faced or adapted to? But what
has not been encountered must be comprehended. The range of our experience continually
supersedes the domain of our determinate knowledge. We are therefore prone to constant
contact with the unknown. It appears every time we make an error; every time our presumptions are wrong—every time our behaviors do not produce the consequences we expect and
desire. The absence of specific depiction, appropriate to inexplicable circumstance, does not
alleviate the necessity of appropriate action—even though the nature of that action cannot
yet be specified. This means that the nature of the unknown, as such, must become represented, in order to design action patterns, *which are broadly suited for response to what cannot
yet (and cannot eternally) be predicted or controlled.* We are in fact capable of a set of paradoxical abilities: we know what to do, when we do not know what to do; we know how to represent what to do, when we do not know what to do; finally, we know how to represent what
we have not yet encountered. These adaptive capacities—impossible, at first glance—
immensely further our capacity to behave, successfully, in the face of our mysterious experience, and to communicate and broaden that capacity.

If an error in judgment, interpretation or behavior occurs, and something unexpected
appears, that unexpected thing has identifiable properties: it is dangerous, and promising.
The danger is potential for punishment, frustration, disappointment, social isolation, physical damage—even death. Every moment of threat, however, is simultaneously a moment of

opportunity. The change that upsets the presently predictable and orderly also means potential for advancement into a more promising future. The unexpected is information itself, information necessary for the constant expansion of adaptive competence. Such information comes packaged in danger and promise. To gain the information promised, the danger must be overcome. This process of necessary eternal overcoming constantly constructs and transforms our behavioral repertoires and representational schemas.

Everything presently known about the subject and objects of human experience was at one time merely the undifferentiated unknown—which was far more than what yet remained to be discovered about the collectively apprehensible sensory qualities of the world. The unknown may manifest itself in the consensually validatable empirical realm, as an aspect of the material world; likewise, it may appear as new significance, where none was evident before. What is known and familiar poses no threat, but offers no possibility beyond that which has been previously determined. The explored thing or situation has been associated with behaviors that render it beneficial, in the ideal, or at least irrelevant. The omnipresent unknown, by contrast, presents threat and promise infinite in scope, impossible to encapsulate, equally impossible to ignore. The unknown, unexpected, or unpredictable is the source of all conditional knowledge—and the place that such knowledge "returns" to, so to speak, when it is no longer useful. Everything we know, we know because someone explored something they did not understand—explored something they were afraid of, in awe of. Everything we know, we know because someone generated something valuable in the course of an encounter with the unexpected.

"Civilization advances by extending the number of important operations which we can perform without thinking about them."[289] All things that we know no longer demand our attention. To know something is to do it automatically, without thinking, to categorize it at a glance (or less than a glance), or to ignore it entirely. The nervous system is "designed" to eliminate predictability from consideration, and to focus limited analytical resources where focus would produce useful results. We attend to the places where change is occurring; where something is happening that has not yet been modeled, where something is happening that has not yet had behaviors erected around it—where something is happening that is not yet understood. Consciousness itself might be considered as that organ which specializes in the analysis and classification of unpredictable events. Attention and concentration naturally gravitate to those elements in the experiential field that contain the highest concentration of novelty, or that are the least expected, prior to what might normally be considered higher cognitive processing. The nervous system responds to irregular change and eliminates regularity. There is limited information, positive and negative, in the predictable. The novel occurrence, by contrast, might be considered a window into the "transcendent space" where reward and punishment exist in eternal and unlimited potential.

The unknown or unexpected or novel appears when plans go wrong: when behavioral adaptation or interpretive schema fails to produce what is desired or to predict what occurs. The appearance of the unexpected or unpredictable inhibits ongoing goal-directed activity in the absence of conscious volition. Concurrently with this inhibition of activity comes inexorable redirection of attention toward the unexpected event. The unexpected grips behavior and spontaneously generates antithetical affects, varying in intensity with the improbability of the occurrence, creating heightened interest, fear, intense curiosity or outright terror. This motivational significance appears to have been experienced as an intrinsic feature of the unknown, prior to the strict formal modern division of experiential world into empirical object and subjective observer—and is still fundamentally experienced in that manner today. Rudolf Otto, in his seminal investigation into the nature of religious experience, described such experience as *numinous*,[290] involuntarily gripping, indicative of significance beyond the normal and average. The "numinous" experience has two aspects: *mysterium tremendum*, which is capacity to invoke trembling and fear; and *mysterium fascinans*, capacity to powerfully attract, fascinate and compel. This numinous power, divine import, is extreme affective relevance and concomitant direction of behavior by the (unknown) object. This "power" is commonly considered by those subject to it as a manifestation of God, personification of the unknown, and ultimate source of all conditional knowledge:

The feeling of it may at times come sweeping like a gentle tide, pervading the mind with a tranquil mood of deepest worship. It may pass over into a more set and lasting attitude of the soul, continuing, as it were, thrillingly vibrant and resonant, until at last it dies away and the soul resumes its "profane," non-religious mood of everyday experience. It may burst in sudden eruptions up from the depths of the soul with spasms and convulsions, or lead to the strangest excitements, to intoxicated frenzy, to transport, and to ecstasy. It has its wild and demonic forms and can sink to an almost grisly horror and shuddering. It has its crude, barbaric antecedents and early manifestations, and again it may be developed into something beautiful and pure and glorious. It may become the hushed, trembling, and speechless humility of the creature in the presence of—whom or what? In the presence of that which is a *mystery* inexpressible and above all creatures.[291]

Nothing that is not represented can be said to be understood—not as we normally use that term. Nonetheless, understanding of the unknown—which cannot, in theory, be represented—is vital to continued survival. Desire to represent the unknown, to capture its essence, is in consequence potent enough to drive the construction of culture, the net that constrains the unknowable source of all things. The impetus for representation of the domain of the unexpected arose (and arises) as consequence of the intrinsic, biologically determined affective or emotional significance of the unknown or novel world. Representations of the unknown constitute attempts to elaborate upon its nature, to illuminate its emotional and motivational significance (to illuminate its *being*, from the prescientific or mythic perspective). This is categorization of all that has not yet been explored and represented, in the service of adaptation to that which has not yet been understood. This is the attempt to formulate a conception of "the category of all as-of-yet uncategorized things," so that a useful stance might be adopted, with regard to that category.

The novel ceaselessly inspires thought and allows itself to be entangled, yet inevitably transcends all attempts at final classification. The unknown therefore provides a constant powerful source of "energy" for exploration and the generation of new information. Desire to formulate a representation of that which supersedes final classification and remains eternally motivating might well be understood as a prepotent and irresistible drive. That drive constitutes what might be regarded as the most fundamental religious impulse—constitutes the culturally universal attempt to define and establish a relationship with God—and underlies the establishment of civilized historical order. The product of this drive, the culturally constructed complex, extant in fantasy—the *symbol*, composed of communicable representation of all things constantly threatening and promising to man—affects and structures the experience of each individual, yet remains impersonal, distinct and separate:

The living symbol formulates an essential unconscious factor, and the more widespread this factor is, the more general is the effect of the symbol, for it touches a corresponding chord in every psyche. Since, for a given epoch, it is the best possible expression for what is still unknown, it must be the product of the most complex and differentiated minds of that age. But in order to have an effect at all, it must embrace what is common to a large group of men. This can never be what is most differentiated, the highest attainable, for only a very few attain to that or understand it. The common factor must be something that is still so primitive that its ubiquity cannot be doubted. Only when the symbol embraces that and expresses it in the highest possible form is it of general efficacy. Herein lies the potency of the living, social symbol and its redeeming power.[292]

This dynamic representation might form part of the subjective experience of a myriad of people, and therefore have its "own" biologically grounded, culturally determined existence, independent of any given person at any given time—even to follow its own intrinsic rules of development—yet fail to exist "objectively" as the objective is currently understood.

Ritualized, dramatic or mythic representations of the unknown—the domain that emerges when error is committed—appear to have provided the initial material for the most primordial and fundamental aspects of formalized religions. Appreciation of the nature of the unknown as a category developed as a consequence of observation of our inherent response to what we did not expect, manifested as predictable pattern of affect and behavior: fear and curiosity, terror and hope, inhibition of ongoing activity and cautious exploration, "habituation" and generation of novel and situation-specific appropriate behavioral strategies. Two things are the same, from the empirical viewpoint, if they share collectively apprehended sensory features. Two things are the same, from the metaphoric, dramatic, or mythical perspective—from the perspective of the natural category—if they produce the same subjective state of being (affect or motivation), or have the same functional status (which is implication for behavior). Experiences that share affective tone appear categorizable in single complexes, symbolic in nature (from the standpoint of abstract cognition)—appear as products of culture, which evolved in the social environment characteristic of ancestral *Homo sapiens* and later disappeared. Such complexes might play a useful role, in the promotion of general adaptive behavior, in the face of feared and promising objects, in the absence of detailed exploration-generated information, regarding the explicit nature of these objects.

These representations might be considered the consequence of first-level representation—of imitation, as Piaget pointed out—and then, later, the consequence of more abstracted second-order representation (of symbolic understanding). Understanding can be reached at the most inclusive, yet primary level, through ritual and mimesis. An unknown phenomenon, gripping but incomprehensible, can yet be represented ritually, can be *acted out*. Secondary representation of this "acting out" constitutes the initial form of abstract representation. To understand the lion, for example—or the hunted beast—it is first necessary to "become" the lion or the hunted beast—to mimic, physically, and later to represent the mimicry in imagination. It is in this manner that the son imitates the father, whom he will later become. A child's embodiment of the parent means his incorporation of the knowledge of the parent, at least insofar as that knowledge is action. The child acts out his father, without understanding him and without understanding the reasons for his acting out. It could be said, metaphorically, that the imitating child is possessed by the *spirit of the father*, as the father was possessed in his own childhood. The "spirit of the father" may be conceived, in this representational schema, as an entity independent of the particular father, or the particular son—as something that manifests itself in imagination and in possession of behavior generation after generation, in more or less constant and traditional form. Similarly, the unknown, which might be considered object and subject simultaneously—which manifests itself in the perceptible world, in affect, and which grips behavior—might well be regarded as (or manifest itself in imagination as) a transpersonal entity (or as the result of the actions of a transpersonal entity). The ancestral "primordial hunter," terrified by something unknown in the bush, portrays his encounter with what frightened him by acting out the unknown demon when he returns to the village. This acting out is simultaneously embodiment and representation; it is basic-level hypothesis regarding the nature of the unknown, as such. Alternatively, perhaps, he fashions an image, an idol, of the thing—and gives concrete form to what until then is merely behavioral compulsion. The unknown first appears symbolically as an independent personality, when it cannot be conceived of in any other fashion, and later appears *as if* it were a personality (in evidently metaphoric guise). Evidence for the adoption of "personality" by representational or quasi-representational "complexes" is plentiful.[293] Such "complexes" may "construct themselves" over the course of many centuries, as a consequence of the exploratory and creative endeavors of many disparate individuals, united within the communicative network of culture.

It is in this manner, over vast stretches of time, that the "transpersonal" domain of the imagination becomes populated with "spirits." Jung described the "space" occupied by such "spirits" as the *pleroma* (a gnostic term).[294] The pleroma might be described as the subjective world of experience, in remembrance—the episodic world, perhaps, from the perspective of modern memory theory—although representations apparently collectively apprehensible under certain peculiar circumstances (like those of the Virgin Mary, in Yugoslavia, prior to the devastating Serbian-Bosnian-Muslim war, or those of "alien spaceships" [UFOs] during the Cold War) also make their "home" there. The pleroma is the "space" in which heaven and hell have their existence, the place where Plato's "supracelestial" ideals reside, the ground of dream and fantasy. It appears to have a four-dimensional structure, like that of objective space-time (and of memory),[295] but is characterized by a tremendous vagueness with regards to category and temporality. The "spirits" which inhabit the pleroma, in its "natural" condition, are *deities*—undifferentiated mixes of subject and object, motivational significance and sensory aspect, elaborated into personified representations by the efforts of many. This is merely to say that a representation is a *social construct*, with *historical (even biological) roots*—like any idea—and that the spirit who inhabits the imagination is not necessarily a figment created by the person who "has" that imagination. The devil is not the product of the particular Christian. It is more accurate to note that the figure of the devil—or of Christ, for that matter—inhabits the mind of the Christian (and of all Christians), and that such habitation occurs as a consequence of transpersonal social and historical processes operating almost completely beyond the realm of individual control. [296] The child, similarly, cannot be said to create the monsters that live in his imagination. They grow there, so to speak, and are then subjectively observed—are fed by casual statements on the part of adults, by action patterns the child observes but cannot explain, by emotions and motivational states that emerge suddenly and unpredictably, by the fantasies in books, on TV and in the theater.

Events or experiences that remain beyond the reach of exploration, assimilation and accommodation stay firmly entrenched in or automatically ascribed to the domain of the unknown, threatening and promising. The category of *all events that cannot yet be categorized* can nonetheless be modeled, through metaphoric application of partially comprehensible yet affect-inducing occurrences whose emotional relevance in some way matches that of the unknown. Each of the specific things that signifies danger, for example—or, alternatively, the enhancement of life—appears easily associated with every other specific thing, characterized by the same property, as well as with novelty itself, which produces fear and hope as part of its (subjectively) intrinsic nature. These experiences appear interassociated on the basis of the similar affective or behavioral states they inspire—the motivational effects they engender, prior to development of "habituation" in course of exploratory behavior.[297] The archaic "limbic system" has its own method of classification, so to speak, experienced privately as emotion—or as behavior spontaneously undertaken—manifested outside the realm of conditional abstract culturally determined presumption.[298] Everything novel encountered, avoided because of involuntary or willfully manifested fear or ignorance, is potentially or actively linked with all that remains outside of individual competence and/or cultural classification. Everything that produces fear may be subjectively considered one aspect of the same (subterranean) thing. What is that thing?

The unknown, as such, surrounds all things, but exists only in a hypothetical state, and finds representation in symbolic form as the uroboros, as we have seen. The disintegration or division of the uroboros gives rise to all things, including the disorder or unpredictability *that is defined in opposition to what has been explored.* This more narrowly defined domain of disorder or unpredictability—which is the unknown as it is actually experienced (rather than as a hypothetical entity)—tends to be portrayed as something distinctly feminine, as the daughter of the great serpent, as the matrix of all determinate being. It is useful to regard the Great

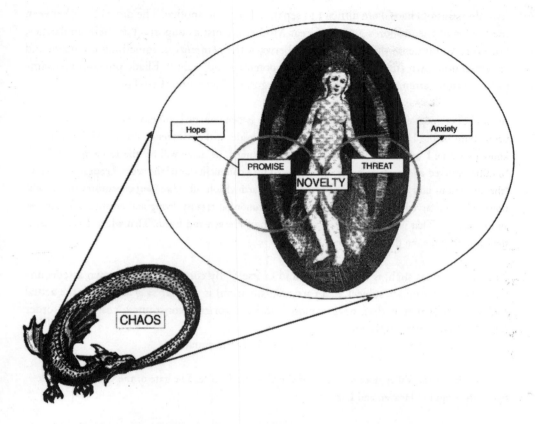

Figure 32: Novelty, the Great Mother, as Daughter of the Uroboros

Mother as the primary agent of the serpent of chaos—as the serpent's representative, so to speak, in the profane domain. The serpent of chaos can be seen lurking "behind" the Great Mother, as we shall see, and she often "adopts" reptilian (material) or birdlike (spiritual) features. This relationship is schematically represented in *Figure 32: Novelty, the Great Mother,*[299] *as Daughter of the Uroboros*. In the incarnation depicted, the Great Mother is Venus, goddess of fertility and love. As the winged mother—bird and matter—she is "spirit" and "earth" at once; the wings might just as easily be replaced by the icon of a snake, which would tie her figure more closely to the earth (and to the idea of transformation). The capsule that surrounds her, for example—frequently found enveloping Christ (as son of the Divine Mother) or Mary (the Divine Mother herself) in late medieval and early Renaissance art—is the *mandorla*, or *vesica pisces*, the "fishes' bladder," which appears to have served as sexual/symbolic representation of the source of all things since well before written history began.[300]

The uroboros and the figure of the Great Mother commonly overlap because the "chaos comprising the original state" is hard to distinguish from the "chaos defined in opposition to established order." Two things that have no distinguishable features (as is the case for the "two domains of chaos") are difficult to separate from one another. The distinctions between the figures of the uroboros and the Great Mother are just as important as their similarities, however. An immense difference obtains between the *possibility* of something unknown and an actual unknown (the difference between potential and reality). Eliade provides an example of a careful attempt to disentangle the categories, drawn from Lao-Tzu:

In another cosmogonic fragment (chap. 25), the Tao is denominated "an undifferentiated and perfect being, born before Heaven and Earth.... We can consider it the Mother of this world, but I do not know its name; I will call it Tao; and if it must be named, its name will be: the Immense (*ta*)." The "undifferentiated and perfect" being is interpreted by a commentator of the second century B.C. thus: "the mysterious unity [*Hung-t'ung*] of Heaven and Earth chaotically [*hun-tun*] constitutes [the condition] of the uncarved block." Hence the Tao is a primordial totality, living and creative but formless and nameless. "That which is nameless is the origin of Heaven and Earth. That which has a name is the Mother of the ten thousand beings."[301]

The unknown, as such, is the thing "in and of itself." By contrast, the unknown as encountered (by a determinate subject, in a particular situation) is the *matrix of all being*—the actual source of information that, once explored and categorized, constitutes "cosmos" or order (and, for that matter, *exploring agent*).

Lao-Tzu also says, in an attempt to further clarify the situation:

The divinity of the Valley does not die: it is the Obscure Female. The gate of the Obscure Female— that is the origin of Heaven and Earth.[302]

The unknown appears to be generally conceptualized or symbolically represented as female primarily because the female genitalia—hidden, private, unexplored, productive—serve as "gateway" or "portal" to the "(divine) unknown world or source of creation," and therefore easily come to stand for that "place." Novelty and femininity share analogical or categorical identity, from this perspective: both constitute a window, so to speak, into the world "beyond." Woman, insofar as she is subject to natural demands, is not merely a model for nature—she is divine nature, in imagination and actuality. She literally *embodies* the matrix of biological being, and provides, as such, an appropriate figure for the metaphoric modeling of the ground of everything. The female body constitutes the *border* between normal experience and the totality from which all forms emerge. Infants come from mothers; this hypothesis, based upon direct observation, accounts for the provisional source of particular individuals. The origin, *per se*, partakes of the same essential ineffable nature—partakes of whatever is characteristic of the (experienceable) mother, and other identifiable points of origin, which cannot be described or comprehended so easily (such as the caves where ores "grow and mature," or the ground where crops thrive). The matrix of all things is something *feminine*, like the mothers of experience; it is something with an endlessly fecund and renewed (maternal and virginal) nature—something that defines fertility and, therefore, femininity itself. Things come from somewhere; all things have their birthplace. The relationship of man *writ large* to nature, eternal mother, endlessly mimics that of the particular child to his personal mother—or, to be more accurate, the child and the mother mimic life and the world.

The unknown—as it can be encountered—is female, with paradoxical qualities. The Great and Terrible Mother of All Things promises endlessly; she also threatens, absolutely. The outcome of an encounter with the unknown—which constitutes the necessary precondition for the generation of new information (for generation of the "cosmos" and of the experiencing subject)—cannot be specified beforehand; something new might benefit, or destroy. Femininity shares emotional valence with novelty and threat, furthering the utility of the female as metaphoric grist, because of the union that exists within experience between creation of one thing, and destruction and transformation of another.[303] The processes of embryogenesis itself require that blood change form, as the fetus thrives on the "blood" of its mother. The act of birth itself is traumatic, painful, dangerous and frightening, recapitulating the natural theme of creation, transformation and destruction. Nourishment is linked integrally with death and terror, even from the beginning, when the metamorphosis of blood into milk transforms the mother into food for the infant. Nature is feminine, in addition, because of the isomorphic relationship that exists between childhood dependency on maternal beneficience and caprice, and adult subjugation to biological reality. Human infants are prepared, instinctively, to establish relationship with the mother, and to respond with vitality to manifestation of maternal interest. Every individual's primordial world-experience is experience of mother, who is the world itself, in initial developmental stages (insofar as the world has any motivational significance, whatsoever). (Indeed, for individuals who are sufficiently stunted in their psychological development, the "world" never evolves into anything other beyond "mother."[304]) Furthermore, the ontogenesis of the individual, and the mother-child symbiosis, is comparable to the phylogenesis of humanity, and the relationship of that humanity with—or its dependence upon—earth and sea. The archetypal infantile situation, which extends back into time, prior to the establishment of culture itself, is recapitulated in adulthood, with the maternal object of fear and respect, hope, love and gratitude abstracted into experience itself.

The threatening aspects of the Great Mother gather metaphoric representation as chimeras of anxiety-producing places, animals, gestures, expressions and things. These elements—diverse from the objective perspective (from the standpoint of the "proper set")—nonetheless unite to produce an image of the ever-present potential danger inherent in anything unpredictable. The Great Mother—unexplored territory—is the dark, the chaos of the night, the insect, ophidian and reptilian worlds, the damaged body, the mask of anger or terror: the entire panoply of fear-inducing experiences, commonly encountered (and imagined) by *Homo sapiens*. A dynamic complex of such objects appears as the most subtle and exact representation of the unknown imaginable—something capable, simultaneously, of characterizing the active bite of the snake, the life of fire, the sting of the scorpion, the trap of the spider—the most suitable embodiment of the manifest desire of nature's vital transformative forces, generative of death, dissolution, destruction and endless creation. Feared

Figure 33: The Spontaneous Personification of Unexplored Territory

experiences, grounded in the inexplicable, acquire representation in fantasy, as fear-producing spirits. These spirits, clothed in particular anxiety-provoking occurrences, give form to
aspects of experience that otherwise remain inexplicable—beyond understanding, from the
perspective of conditional adaptation, action and abstract thought, but impossible to ignore,
from the standpoint of affect. The "personality" of such beings constitutes the embodiment
of incomprehensible, and often intolerable, motivational significance—comprises representation of the ground of violent emotional experience, capable of inducing cognitive and
behavioral possession, impossible to incorporate into the domain of normal, culturally established being. *Figure 33: The Spontaneous Personification of Unexplored Territory* presents one
such figure, and its process of development, in comical form.[305] Equivalent but more serious
dynamic representations of this type are deities, gods borne of human experience, possessed
of quasi-objective transpersonal status—like the Word—manifestations of the unfamiliar,
the other, the unknown and the unpredictable.

What can now be calmly described as an archaic symbol or god from the past may also reasonably be considered as the manifestation of a primeval "independent" personality—the unified "embodiment" in ritual or imagination of some set of phenomena united by their affective or functional equivalence. These personalities—deities—have with time lost affective and conceptual relevance, as a consequence of the constant expansion of human adaptive capacity, and have become "broken down" into less complex, more determinate aspects of experience. In their original form, however, these "representational personalities" revealed themselves within the creative, compensatory experience of exceptional individuals, beset by their own incomprehensible (although not purely idiosyncratic) personal tragedy. Concrete realization of such manifestation—transformation into an artistic production or potent story, for example—involuntarily seized the attention of peers and inspired a sense of fascination and awe. Centuries-long cultural elaboration of such production gave rise to the elaborated "existence" of transpersonal beings, of transcendent power, who inhabited the "space" defined by the collective imagination of mankind, and who behaved in accordance with the dictates of their own irrational, myth-predicated souls. These "representations" served as active images, detailing to everyone what was as of yet explicitly unknown, only partially known; they pointed the way toward aspects of experience beyond the grasp of "conscious" abstract apprehension, but dangerous to ignore.

It is no simple manner for the limited subject to formulate an accurate representation of the unlimited unknown, of nature, the ground of existence. The unknown is the matrix of everything, the source of all birth and the final place of rest. *It hides behind our personal identity and our culture; it constantly threatens and engenders all that we do, all that we understand, and all that we are.* It can never be eliminated permanently from consideration, since every solution merely provides the breeding place for a host of new problems. The unknown is *Homo sapiens'* everlasting enemy and greatest friend, constantly challenging individual facility for adaptation and representation, constantly pushing men and women to greater depths and more profound heights. The unknown as Nature appears as paradoxical formidable overwhelming power, applied simultaneously in one direction and its opposite. Hunger, the will to self-preservation, drives living creatures to devour each other rapaciously, and the hunters have no mercy for the hunted. Sexuality bends the individual will inexorably and often tragically to the demand of the species, and existence maintains itself in endless suffering, transformation and death. Life generates and destroys itself in a pitiless cycle, and the individual remains constantly subject to forces beyond understanding or control. The desire to exist permeates all that lives, and expresses itself in terrible fashion, in uncontrollable impulse, in an endless counterpoint of fecundity and decay. The most basic, fundamental and necessary aspects of experience are at the same time most dangerous and unacceptable.

Empirical (classical) "objects" are either one thing or another. Nature, by contrast—the great unknown—is one thing and its (affective) opposite at the same time, and in the same place. The novel, primeval experience was (and remains) much too complex to be gripped, initially, by rational understanding, as understood in the present day. Mythic imagination, "willing" to sacrifice discriminatory clarity for inclusive phenomenological accuracy, provided the necessary developmental bridge. The earliest embodiments of nature are therefore symbolic combinations of rationally irreconcilable attributes; monsters, essentially feminine, who represent animal and human, creation and destruction, birth and cessation of experience. The analytical psychologist Erich Neumann, who wrote a definitive, comprehensive and useful book on the symbolism of the feminine, states:

In the early phases of consciousness, the numinosity [that is, the emotional valence] of the archetype ... exceeds man's power of representation, so much so that at first no form can be given to it. And when later the primordial archetype takes form in the imagination of man, its representations are often monstrous and inhuman. This is the phase of the chimerical creatures composed of different animals or of animal and man—the griffins, sphinxes, harpies, for example—and also of such monstrosities as phallic and bearded mothers. It is only when consciousness learns to look at phenomena from a certain distance, to react more subtly, to differentiate and distinguish [this is a function of exploration and its related abstract processes], that the mixture of symbols prevailing in the primordial archetype separates into the groups of symbols characteristic of a single archetype or of a group of related archetypes; in short, that they became recognizable.[306]

The terrible aspects of the primordial Great Mother have been represented, symbolized, in variety of manners, but her underlying reality and essential ideation remain immediately recognizable. Neumann states:

These figures are gruesomely alike. Their sheer frightfulness makes us hesitate, whether they represent a skull, the head of a snake or hippopotamus, a face showing human likeness, or a head consisting of two stone knives borne by a body pieced together from parts of snakes, panthers, lions, crocodiles, and human beings. So great is the inhuman, extrahuman, and superhuman quality in this experience of dread that man can visualize it only through phantoms. But all this—and it should not be forgotten— is an image not only of the Feminine but particularly and specifically of the Maternal. For in a profound way life and birth are always bound up with death and destruction. That is why this Terrible Mother is "Great," and this name is also given to Ta-Urt, the gravid monster, which is hippopotamus and crocodile, lioness and woman, in one. She too is deadly and protective. There is a frightening likeness to Hathor, the good cow goddess, who in the form of a hippopotamus is the goddess of the underworld. She has a positive aspect, and at the same time she is the goddess of war and death.

In the course of the later[307] development of patriarchal values, i.e., of the male deities of the sun and light, the negative aspect of the Feminine was submerged. Today it is discernible only as a content of the primordial age, or of the unconscious. Thus the terrible Ta-Urt, as well as the terrible Hathor, Isis, Neith, and others, can be reconstituted from their pictures that have been "painted over," but cannot be viewed directly. Only the monster Am-mit or Aman, which devours the souls condemned at the judgment of the dead, points by its parallelism to the terrible aspect of Ta-Urt. Am-mit was described as follows: "Her forepart (is that of) crocodiles, her hinderpart (is that of) hippopotamus, and her middle (is that of a) lion." The feminine, animal-mother character of this many-breasted creature is evident as is that of the monster wielding the terrible knife, which guards one of the underworld gates through which the souls of the departed must pass.

Am-mit devours the souls that have not withstood the midnight judgment of the dead in the underworld. But her role has become subordinate, for the religion of Osiris and Horus with its mysteries has now promised rebirth and resurrection to all human souls, and not only, as originally, to the soul of Pharaoh. The certainty of magical success in following the path of the sun, which is communicated to each man after death by the priests, has overlaid the primordial fear represented by Am-mit. But originally she was the terrible ancestral spirit of the matriarchal culture, in which the Feminine takes back what has been born of it—just as among the primitive inhabitants of the Melanesian island of Melekula or in the high culture of Mexico.[308]

The Terrible Mother challenges and threatens the individual, absolutely. She is goddess of anxiety, depression and psychological chaos—goddess of the possibility of pain and death. She is horror, insofar as horror can be imagined, and is the ground of that horror, beyond. She exposes and turns to her advantage constant mortal vulnerability. She barters, paradoxically, offering continuance of life for sacrificial death. She demands reconciliation, without offering the certainty of survival. She embodies the potential for salvation, and the central problem of life; impels the individual, involuntarily, toward further expansion of consciousness, or induces involuntary contraction, leading to death.[309] The Great Mother impels—pushes (with certainty of mortality) and pulls (with possibility of redemption)—development of consciousness and of self-consciousness. The identity of death with the unknown has permanently and incurably destroyed any possibility of final habituation to—adaptation to, more accurately—the world of experience. Man is in consequence the (incurably) anxious animal:

Thus the womb of the earth becomes the deadly devouring maw of the underworld, and beside the fecundated womb and the protective cave of earth and mountain gapes the abyss of hell, the dark hole of the depths, the devouring womb of the grave and death, and darkness without light, of nothingness. For this woman who generates life and all living things on earth is the same who takes them back into herself, who pursues her victims and captures them with snare and net. Disease, hunger, hardship, war above all, are her helpers, and among all peoples the goddesses of war and the hunt express man's experience of life as a female exacting blood. This Terrible Mother is the hungry earth, which devours its own children and fattens on their corpses; it is the tiger and the vulture, the vulture and the coffin, the flesh-eating sarcophagus voraciously licking up the blood seed of men and beasts and, once fecundated and sated, casting it out again in new birth, hurling it to death, and over and over again to death.[310]

The terrible feminine has been represented by figures such as the chimera, the sphinx, the griffin and the gorgon, which combined and unified the most disparate, yet related, aspects of nature (those aspects which, individually, intrinsically, inspire terror and deference). Gorgon-like figures and their "sisters" appear commonly throughout the world.[311] As the Aztec *Coatlicue*, whose gruesome headdress was composed of skulls, the Terrible Mother was goddess of death and dismemberment, object of sacrificial homage. As Goddess of the Snake, she was sacred in ancient Crete, and worshiped by the Romans. Her modern equivalents remain extant in Bali and India. *Kali*, Hindu goddess—portrayed in *Figure 34*

Figure 34: Unexplored Territory as Destructive Mother

Unexplored Territory as Destructive Mother [312]—is eight-armed, like a spider, and sits within a web of fire. Each of her arms bears a tool of creation or weapon of destruction. She wears a tiara of skulls, has pointed, phallic breasts, and aggressive, staring eyes. A snake, symbol of ancient, impersonal power, transformation and rebirth, is coiled around her waist. She simultaneously devours, and gives birth, to a full-grown man. Medusa, Greek monster, with her coif of snakes, manifests a visage so terrible that a single exposure turns strong men to stone—paralyzes them, permanently, with fear. This gorgon is a late "vestigial" remnant, so to speak, of an early goddess, who simultaneously embodied nature's incredible productive fecundity and callous disregard for life.

A neuropsychological description of the brain's response to the unexpected—such as we encountered earlier—is one thing; the mythological representation is another. Consideration of the figure of the Great and Terrible Mother is salutary; helps breed understanding of just what it is that our cultures—that is, our ritual identification with the dead—protects us from. We are shielded from the terrors of our imagination (and from the things that breed such terror) by the overlay of familiarity granted by shared frameworks of action and interpretation. These "walls" serve their purpose so well that it is easy for us to forget our mortal vulnerability; indeed, we generated those walls to aid that forgetting. But it is impossible to understand why we are so motivated to maintain our cultures—our beliefs, and associated patterns of action—without gazing at and appreciating the horrible figures generated by our ancestors.

The Great Mother, in her negative guise, is the force that induces the child to cry in the absence of her parents. She is the branches that claw at the night traveler, in the depths of the forest. She is the terrible force that motivates the commission of atrocity—planned rape and painful slaughter—during the waging of war. She is aggression, without the inhibition of fear and guilt; sexuality in the absence of responsibility, dominance without compassion, greed without empathy. She is the Freudian id, unconsciousness contaminated with the unknown and mortal terror, and the flies in the corpse of a kitten. She is everything that jumps in the night, that scratches and bites, that screeches and howls; she is paralyzing dismay, horror and the screams that accompany madness. The Great Mother aborts children, and is the dead fetus; breeds pestilence, and is the plague; she makes of the skull something gruesomely compelling, and is all skulls herself. To unveil her is to risk madness, to gaze over the abyss, to lose the way, to remember the repressed trauma. She is the molester of children, the golem, the bogey-man, the monster in the swamp, the rotting cadaverous zombie who threatens the living. She is progenitor of the devil, the "strange son of chaos." She is the serpent, and Eve, the temptress; she is the *femme fatale*, the insect in the ointment, the hidden cancer, the chronic sickness, the plague of locusts, the cause of drought, the poisoned water. She uses erotic pleasure as bait to keep the world alive and breeding; she is a gothic monster who feeds on the blood of the living. She is the water that washes menacingly over the ridge of the crumbling dam; the shark in the depths, the wide-eyed creature of the deep forests, the cry of the unknown animal, the claws of the grizzly and the smile of the criminally insane. The Great and Terrible Mother stars in every horror movie, every black comedy; she lies in wait for the purposefully ignorant like a crocodile waits in the bog. She is the mystery of life that can never be mastered; she grows more menacing with every retreat.

I dreamed I saw my maternal grandmother sitting by the bank of a swimming pool, which was also a river. In real life, she had been a victim of Alzheimer's disease and had regressed to a semi-conscious state. In the dream, as well, she had lost her capacity for self-control. Her genital region was exposed, dimly; it had the appearance of a thick mat of hair. She was stroking herself, absentmindedly. She walked over to me, with a handful of pubic hair compacted into something resembling a large artist's paintbrush. She pushed this at my face. I raised my arm, several times, to deflect her hand; finally, unwilling to hurt her, or interfere with her any further, I let her have her way. She stroked my face with the brush, gently, and said, like a child, "Isn't it soft?" I looked at her ruined face and said, "Yes, Grandma, it's soft."

Out from behind her stepped an old white bear. It stood to her right, to my left. We were all beside the pool. The bear was old, like little dogs get old. It could not see very well, seemed miserable and behaved unpredictably. It started to growl and wave its head at me—just like little mean dogs growl and look just before they bite you. It grabbed my left hand in its jaws. We both fell into the pool, which was by this time more like a river. I was pushing the bear away with my free hand. I yelled, "Dad, what should I do?" I took an axe and hit the bear behind the head, hard, a number of times, killing it. It went limp in the water. I tried to lift its body onto the bank. Some people came to help me. I yelled, "I have to do this alone!" Finally I forced it out of the water. I walked away, down the bank. My father joined me and put his arm around my shoulder. I felt exhausted but satisfied.

The unknown never disappears; *it is a permanent constituent element of experience.* The ability to represent the terrible aspects of the unknown allow us to conceptualize what has not yet been encountered, and to practice adopting the proper attitude toward what we do not understand.

> For I am the first and the last.
> I am the honored one and the scorned one.
> I am the whore and the holy one.
> I am the wife and the virgin.[313]

The positive aspect of the matrix of all being—the "twin sister" of Kali, so to speak—stands in marked contrast to the Terrible Mother. The beneficial unknown is the source of eternal plenitude and comfort. It is "positive femininity," metaphorically speaking, that constitutes the ground for hope itself—for the faith and belief in the essential goodness of things necessary to voluntary maintenance of life and culture. The beneficial "sister" has in consequence acquired breadth and depth of metaphoric mythic representation equivalent to that of the Terrible Mother. The beneficent aspect of the matrix of all things—the eternally fecund "virgin" (because eternally renewed), the mother of the savior—is the embodiment of the helpful source, a constant aid to painful travail, tragic suffering and existential concern. Redemptive knowledge itself springs from the generative encounter with the unknown, from exploration of aspects of novel things and novel situations; is part of the potential of things, implicit in them, intrinsic to their nature. This redemptive knowledge is *wisdom*, knowledge of how to act, generated as a consequence of proper relationship established with the positive aspect of the unknown, the source of all things:

Wisdom is radiant and unfading,
and she is easily discerned by those who love her,
and is found by those who seek her.
She hastens to make herself known to those who desire her.
He who rises early to seek her will have no difficulty,
for he will find her sitting at his gates.
To fix one's thought on her is perfect understanding,
and he who is vigilant on her account will
soon be free from care,
because she goes about seeking those worthy of her,
and she graciously appears to them in their paths,
and meets them in every thought.
The beginning of wisdom is the most sincere
desire for instruction,
and concern for instruction is love of her,
and love of her is the keeping of her laws,
and giving heed to her laws is assurance of immortality,
and immortality brings one near to God;
so the desire for wisdom leads to a kingdom. [Wisdom of Solomon (of the
     Apocrypha), RSV 6:12–20]

also:

Therefore I prayed, and understanding was given me;
I called upon God, and the spirit of wisdom came to me.
I preferred her to scepters and thrones,
and I accounted wealth as nothing in comparison with her.
Neither did I liken to her any priceless gem,
because all gold is but a little sand in her sight,
and silver will be accounted as clay before her.
I loved her more than health and beauty,
and I chose to have her rather than light,
because her radiance never ceases.
All good things came to me along with her,
and in her hands uncounted wealth.
I rejoiced in them all, because wisdom leads them;
but I did not know that she was their mother.
I learned without guile and I impart without grudging;
I do not hide her wealth,
for it is an unfailing treasure for men;

those who get it obtain friendship with God,
commended for the gifts that come from instruction.
May God grant that I speak with judgment
and have thought worthy of what I have received,
for he is the guide even of wisdom
and the corrector of the wise. (Wisdom 7:7–15)

Wisdom may be personified as a *spirit who eternally gives*, who provides to her adherents unfailing riches. She is to be valued higher than status or material possessions, as the source of all things. With the categorical inexactitude characteristic of metaphoric thought and its attendant richness of connotation, *the act of valuing this spirit is also Wisdom*. So the matrix itself becomes conflated with—that is, grouped into the same category as—the attitude that makes of that matrix something beneficial. This conflation occurs because primal generative capacity characterizes both the "source of all things" *and* the exploratory/hopeful attitudes and actions that make of that source determinate things. We would only regard the latter— the "subjective stance"—as something clearly psychological (as something akin to "wisdom" in the modern sense). The former is more likely to be considered "external," from our per-spective—something beyond subjective intervention. But it is the case that without the appropriate attitude (Ask, and it shall be given you; seek, and ye shall find; knock, and it shall be opened unto you: For every one that asketh receiveth; and he that seeketh findeth; and to him that knocketh it shall be opened" [Matthew 7:7–8].) the unknown is a sterile wasteland.[314] Expectation and faith determine the "response" of the unknown (as courageous approach eliminates anticipatory anxiety, and exploration makes the unexpected something valuable). So the indiscriminate categorization characterizing these passages has its worth.

We are motivated to protect the products of our exploration, our familiar territories, because unexplored phenomena are intrinsically meaningful, and that meaning is apt to show itself as threat. The probability that the meaning of unexplored territory will be threat, however, *appears to be a function of the interpretive context within which it makes its appearance*. If the unknown is approached voluntarily (which is to say, "as if" it is beneficial), then its promising aspect is likely to appear more salient. If the unknown makes its appearance despite our desire, then it is likely to appear more purely in its aspect of threat. This means that if we are willing to admit to the existence of those things that we do not understand, those things are more likely to adopt a positive face. Rejection of the unknown, conversely, increases the likelihood that it will wear a terrifying visage when it inevitably manifests itself. It seems to me that this is one of the essential messages of the New Testament, with its express (although difficult-to-interpret) insistence that God should be regarded as all-good.

The beneficial aspect of the unknown is something unavailable to the "unworthy," some-thing eternal and pure; something that enters into relationship with those who are willing, from age to age; and something that makes friends of God. The unknown is also something that may be conceptualized using sexual symbolism: something that may be "known," in the biblical sense. Joined with, as with a "bride," she produces all things that are good:

I learned both what is secret and what is manifest,

for wisdom, the fashioner of all things, taught me.

For in her there is a spirit that is intelligent, holy,

unique, manifold, subtle,

mobile, clear, unpolluted,

distinct, invulnerable, loving the good, keen,

irresistible,

beneficient, humane,

steadfast, sure, free from anxiety,

all-powerful, overseeing all,

and penetrating through all spirits

that are intelligent and pure and most subtle.

For wisdom is more mobile than any motion;

because of her pureness she pervades and penetrates all things.

For she is a breath of the power of God,

and a pure emanation of the glory of the Almighty;

therefore nothing defiled gains entrance into her.

For she is a reflection of eternal light,

a spotless mirror of the working of God,

and an image of his goodness.

Though she is but one, she can do all things,

and while remaining in herself, she renews all things;

in every generation she passes into holy souls

and makes them friends of God, and prophets;

for God loves nothing so much as the man who lives with wisdom.

For she is more beautiful than the sun,

and excels every constellation of the stars.

Compared with the light she is found to be superior,

for it is succeeded by the night,

but against wisdom evil does not prevail.

She reaches mightily from one end of the earth to the other,

and she orders all things well.

I loved her and sought her from my youth,

and I desired to take her for my bride,

and I became enamored of her beauty.

She glorifies her noble birth by living with God,

and the Lord of all loves her.

For she is an initiate in the knowledge of God,

and an associate in his works.

If riches are a desirable possession in life,

what is richer than wisdom who effects all things?

Figure 35: Unexplored Territory as Creative Mother

And if understanding is effective,
who more than she is fashioner of what exists? (Wisdom 7:22–8:6)[315]

The terrible unknown compels representation; likewise, the beneficial unknown. We are driven to represent the fact that possibility resides in every uncertain event, that promise beckons from the depths of every mystery. Transformation, attendant upon the emergence of change, means the death of everything old and decayed—means the death of everything whose continued existence would merely mean additional suffering on the part of those still striving to survive. The terrible unknown, which paralyzes when it appears, is also succour for the suffering, calm for the troubled, peace for the warrior, insight and discovery for the perplexed and curious—is the redemptive jewel in the head of the toad or in the lair of the fire-belching dragon. The unknown is the fire that burns and protects, the endlessly mysterious transcendent object that simultaneously gives and takes away. The positive aspect of the unknown, incarnated as the many-breasted Greco-Roman Goddess *Diana* or *Artemis*, mistress of the animals, is portrayed in *Figure 35: Unexplored Territory as Creative Mother*.[316]

Everything that *contains, shelters* and *produces* exists as source for the symbolic representation of—occupies the same category as—this promising element. Fruit distinctive for its seed-bearing properties, such as the pomegranate or poppy, provides appropriate motif for gravid containment. The pig stands as representative of fertility, and the cow—the holy beast of India—as embodiment of principle of nourishment. Shellfish "stand for" generation and fertility because of their vulvalike shape. Inanimate items such as boxes, sacks and troughs contain and shelter, while similar objects, such as the bed, the cradle and the nest are characterized by protective and therefore "maternal" function.[317] Humanized representations—statuettes of nude goddesses, among the most ancient objects of representation known[318]—appear to represent fecundity and the productive countenance of nature, in anthropomorphic form. The creation and subsequent appreciation of such figures perhaps aided individuals and societies in their efforts to clarify the nature of the human relationship to the protective aspect of existence. Makers of such statuettes placed great emphasis on collective, impersonal features of generation, such as breasts, genitals and hips (features whose functions remain largely outside voluntary control), but devoted little attention to features defining self-conscious individuality—like those of the face. Such figures apparently represented the vessel of life, and were rendered in the image of woman, whose body generated human life, and nourishment for that life. The body-vessel represented beneficial nature itself:

All the basic vital functions occur in this vessel-body schema, whose "inside" is an unknown. Its entrance and exit zones are of special significance. Food and drink are put into this unknown vessel, while in all creative functions, from the elimination of waste and the emission of seed to the giving forth of breath and the word, something is "born" out of it. *All* body openings—eyes, ears, nose, mouth (navel), rectum, genital zone—as well as the skin, have, as places of exchange between inside and outside, a numinous accent for early man. They are therefore distinguished as "ornamental" and protected zones, and in man's artistic self-representation they play a special role as idols.[319]

The unknown, source of all determinate information, is simultaneously destructive and creative. The terrible aspect of the Great Mother threatens everything with dissolution. Her positive sister is the generative aspect of being. *Figure 36: The "Heavenly Genealogy" of the Destructive and Creative Mothers* portrays the relationship between the two "discriminable" sisters, their derivation from the unified but ambivalent unknown, and their ultimate "descent" from the "dragon of chaos."

The ability to "restrict the appearance of the Terrible Mother," and "foster the realization of her Benevolent Sister" (that is, the ability to decrease threat, and maximize promise and satisfaction) might well be regarded as the secret of successful adaptation. The existence of representations of the twin aspects of the unknown allowed for *practice in adaptation* in the face of such representations, allowed for exposure of the individual, in imagination and action, in controlled fashion, to potently constructed representations of those things that he or she was destined to fear most, was necessarily most vulnerable to, but which could not be forever avoided. Similar "rituals" underly every form of successful modern psychotherapy.

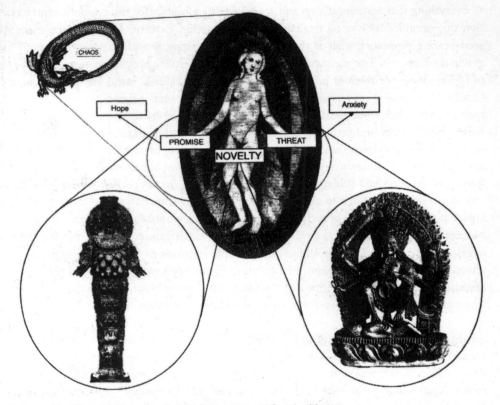

Figure 36: The "Heavenly Genealogy" of the Destructive and Creative Mothers

Modern treatment for disorders of anxiety, to take a specific example—"desensitization"—involves exposing an individual, "ritualistically" (that is, under circumstances rendered predictable by authority), to novel or otherwise threatening stimuli (with appropriate reaction modeled by that authority).[320] Such desensitization theoretically induces "habituation"; what is actually happening is that *guided exploration, in the course of behavior therapy, produces reclassification and behavioral adjustment* [such that the once terrifying thing or once again terrifying thing is turned (back) into something controllable, familiar, and *known*]. Voluntary exposure additionally teaches the previously anxiety-ridden individual the nontrivial lesson *that he or she is capable of facing the "place of fear" and prevailing*. The process of guided voluntary exposure appears to produce therapeutic benefits even when the "thing being avoided" is traumatic[321]—when it might appear cruel, from a superficially "empathic" perspective, to insist upon exposure and "processing."

The ritual of voluntary exposure *fosters mimetic identification with the hero* (whether this is explicitly recognized or not); teaches the individual that the courageous exploratory spirit can eternally prevail over threat. It is this mimetic identification and its abstracted equivalents and consequences that account for the increased general confidence and capabilities that tend to accompany exposure training. The reclassification and behavioral adjustment, attendant upon therapeutic exposure, places the previously terrible forces of the unknown back under the dominion of knowledge, into the domain of the known—expands "explored territory" into chaos—places the "Great Mother" under the strictures of her "consort," culture, the Great Father. This is exploration-predicated "creation of the cosmos" from the precosmogonic chaos, and the fostering of implicit identification with the *Logos*, the creative and redemptive Word.

Analysis of the much more dramatic, very widespread, but metaphorically equivalent phenomena of the sacrifical ritual—a rite whose very existence compelled one insightful author to argue for the essential insanity of man[322]—provides additional insight into the nature of the ability to transform threat into promise. We have already discussed the fact that the valence of an object switches with context of interpretation. It is knowledge of this idea that allows for comprehension of the meaning of the sacrificial attitude. The beautiful countenance of the beneficial mother is the face the unknown adopts *when approached from the proper perspective*. Everything unknown is simultaneously horrifying and promising; it is courage and genius (and the grace of God) that determines which aspect dominates. The uncontrollable strength and bloodlust of the bull is the power which, when domesticated, serves to foster, protect and engender the herd. The devastating power of sudden explosive combustion is reliable and efficient transportation when appropriately harnessed. The gorgon of Nature is helpmistress when approached by the brave, the honest and the humble.

Primary religious rituals, serving a key adaptive purpose, "predicated" upon knowledge of proper approach mechanisms, evolved to suit the space surrounding the primary deity, embodiment of the unknown. The ubiquitous drama of human sacrifice, (proto)typical of primordial religious practice, enacted the idea that the essence of man was something to be offered up voluntarily to the ravages of nature—something to be juxtaposed into creative encounter with the terrible unknown. The offering, in ritual, was often devoured, in reality or symbolically, as aid to embodiment of the immortal human spirit, as aid to incorporation of the heroic process. Such rituals were abstracted and altered, as they developed—with the nature of the sacrificial entity changing (with constancy of underlying "ideation").

As late as 1871, in India the festival honoring the Great Mother, in the guise of Durga[323] or Kali, was accompanied by the daily slaughter of twenty buffalo, two hundred and fifty goats, and two hundred and fifty pigs. The blood-drenched sand in the sacrificial pits was replaced twice a day—removed and buried in the earth to ensure fertility. The slaughter of animals is a relatively late development from the psychohistorical viewpoint, and is generally preceded by, and stands in place of, the ritual sacrifice of human victims. The indologist Heinrich Zimmer states:

In her "hideous aspect" (*ghora-rupa*) the goddess, as Kali, "the dark one," raises the skull full of seething blood to her lips; her devotional image shows her dressed in blood red, standing in a boat floating on a sea of blood: in the midst of the life flood, the sacrificial sap, which she requires that she may, in her gracious manifestation *(sundara-murti)* as the world mother *(jagad-amba)*, bestow existence upon new living forms in a process of unceasing procreation, that as world nurse *(jagad-dhatri)* she may suckle them at her breasts and give them the food that is "full of nourishment" *(anna-purna)*. An ancient conception, extending back as far as the Stone Age: Nature must at every step be given a helping hand; even she can accomplish nothing by herself. She is no more self-sufficient than man. Nothing takes place of itself, either in the cosmos or in human beings. Man must perform clamorous rites in order to liberate the moon from the clutches of the eclipse, to dispel its demons; and if the sun is to be released from its winter feebleness and rise ever higher with the rising year, a young girl, symbolizing the sun, must swing higher and higher into the sky. In order to bear fruit and nurture life, the earth mother demands to be fertilized and strengthened by potations of blood, the vital fluid.[324]

The mysterious and seemingly irrational "sacrificial" ritual actually dramatizes or acts out two critically important and related ideas: first, that the essence of man—that is, the divine aspect—must constantly be "offered up" to the unknown, must present itself voluntarily to the destructive/creative power that constitutes the Great Mother, incarnation of the unpredictable (as we have seen); and second, *that the "thing that is loved best" must be destroyed—that is, sacrificed—in order for the positive aspect of the unknown to manifest itself*.

The former idea is "predicated" on the notion that the unknown must be encountered, voluntarily, for new information to be generated, for new behavioral patterns to be constructed; the latter idea is "predicated" on the observation that an improper or outdated or otherwise invalid attachment—such as the attachment to an inappropriate pattern of behavior or belief—turns the world into waste, by interfering with the process of adaptation itself. Rigid, inflexible attachment to "inappropriate things of value"—indicative of dominance by a pathological hierarchy of values (a "dead god")—is tantamount to denial of the hero. Someone miserable and useless in the midst of plenty—just for the sake of illustration—is unhappy because of his or her attachments to the wrong "things." Unhappiness is frequently the consequence of immature or rigid thinking—a consequence of the overvaluation of phenomena that are in fact trivial. The neurotic clings to the things that make her unhappy, while devaluing the processes, opportunities and ideas that would free her, if she adopted them. The sacrifice of the "thing loved best" to "appease the gods" is the embodiment in procedure of the idea that the benevolent aspect of the unknown will return if the present schema of adaptation (the "ruling king") is sufficiently altered (that is, destroyed and regenerated). An individual stripped of his "identification" with what he previously valued is simultaneously someone facing the unknown—and is, therefore, someone "unconsciously" imitating the hero. The voluntary "stripping" of such identity makes the supplicant into a "new man"—at least if the sacrifice was genuine. This is not to say that such ideas cannot degenerate into meaningless, empty and cruel ritual.

The intimate relationship between clinging to the past, rejection of heroism, and denial of the unknown is most frequently explicated in narrative form (perhaps because the association is so complex that it has not yet been made explicit). The following fairy tale—a "wake-up call," from the psychoanalytic "unconscious"—may serve as useful exemplar. It occurred spontaneously to me, in a single piece, while I was trying to help a man I knew, who was undergoing a psychological crisis. His attachment to the unnecessary and superfluous was putting his future in serious danger, but he would not admit this. I wanted to warn him that he would eventually pay a great price for his short-sightedness. He ignored the story, however, at least in the short term—with predictable results.

Cock-a-doodle-doo

Once upon a time there was a man who had a long hard journey ahead of him. He was trudging along the way, over the boulders and through the brushes, when he saw a little shiny gnome with big white teeth and a black toupee sitting by the side of the road. He was drumming on a log with two white bones, and humming oddly to himself. The little gnome said,

"John—why work so hard? Why walk so fast? Who knows if you'll ever get there anyway? Come over here. I have something to show you."

So John walked off the road. He was sick of walking, anyhow, because people kept throwing sticks and stones at him. The little gnome said,

"I have a shiny red jewel I would like to sell you. Cheap. Here it is," and from beneath his cloak he pulled the biggest ruby that the man had ever seen. It must have weighed a hundred pounds, and it shone like the sun.

The gnome said, "Do you like it? It is an enchanted stone. What will you offer me for it?" and the man said, "I don't have much—much money. But I will give you everything I have." The gnome looked displeased, so John added: "I could pay some more monthly."

So the gnome accepted: "Fair enough! Buy now, pay later. Sounds good to me. I'm all for the installment plan."

So the man gave the gnome all his money, and promised to pay the rest later. And the gnome walked back into the bush by the road, clacking his teeth and giggling and twitching.

The more the man thought about his ruby, and the great deal he got, the happier he became. He started back on the road, with a light heart, but soon discovered that he couldn't make much progress, because a hundred pounds was a lot to carry. He said to himself:

"Why continue, anyway? I have what I want. Why don't I just stand here, holding my ruby—and when people walk by, they can see how well I have already done!"

So he stopped. A little while later, one of his friends came along, and saw him standing there. His friend said,

"John, why don't you come along with me? I have just opened a new business, and I could really use some help! Come along quick! It will be opening soon!"

John thought that sounded good, but his friend was in a hurry. Besides, couldn't he see the ruby? How could he speed along beside him? Where would he put his jewel? So he said, "Thanks, but I have to take care of my jewel. Maybe I'll see you later."

His friend looked at him like he was crazy—but he was trying to get somewhere quick. So he just shrugged a bit and said, "Okay, John. See you later," and he sped on down the road.

A little while later, another friend came by, and he said, "John! Nice to see you! I am going back to school! There are lots of wonderful things to learn! Lots of great things to do! The world is full of unsolved problems! I could use some company! Would you like to come along?" John thought that sounded pretty good—but this friend, too, looked like he was in a hurry. Besides, standing beside the road, holding the jewel, was tiring, and he needed all the energy he had for that. So he said to his friend, "Thanks, but I have to take care of my jewel. Isn't it beautiful? Maybe I'll see you later."

His friend looked at him like he was crazy—but he was trying to get somewhere quick. So he just shrugged and said, "Hope everything goes all right with you. See you later."

Many friends came and went, and the years went by. The jewel got heavier and heavier, but the man got more and more attached to it. The only thing was, nobody seemed to notice how beautiful it was. People would rush by, and talk about their plans

and nobody had a ruby as big

and nobody seemed likely to get a ruby as big

so you'd think that someone might have said something

something, at least, like

"Nice ruby, John. Sure wish I had one like that."

But it never happened.

Then one day someone new came down the road. He was bent over, and he was thin, and his hair was gray, although he did not look that old. He was carrying a big, dirty rock carefully in his arms, and he was not making much progress.

The strange figure approached and glanced up at John. Then he grinned and said,

"Why are you standing there stupidly, with a big ugly rock in your tired old hands? You look pretty daft. I bet you wish you had a big ruby, like the one I am carrying!" and John thought, "This poor man is deluded. He is carrying a rock—it is I who have the ruby!" so he said, "Excuse me sir, but you are sadly mistaken. I am the one with the jewel. I met a little gnome by the side of the road, and he sold it to me. I am still paying for it—although not so very much! You are carrying a rock!"

The tired stranger looked annoyed. He said, "I don't know what game you are playing, mister. You have a rock. I have a jewel. The little gnome you described sold it to me—and he said it was the only one! I have been carrying it for twenty years, and I will never let it go!"and John said, "But I have been carrying mine for twenty years, too. It can't be just a rock!"

Rock or jewel? On and on they argued.

Suddenly, out stepped the little gnome, as if he had never left! Only this time, he wasn't so little. He was bigger, and redder, and menacing, and his laugh sounded like the rattling of chains.

"Quit arguing, you two! I've never seen a sight quite so pathetic. You're carrying rocks—both of you. And if you ever would have had the sense to put them down for a second or two, you would have seen that!

"Oh well, at least you were diligent. And I played a mean trick. I feel bad.

"So, I'm going to give you what you really deserve. Do you want what you really deserve?" and John and the thin stranger nodded eagerly. Finally, they thought.

"You haven't seen anything yet. Throw down your rocks!"

So John and the thin stranger obeyed. Each rock split down the middle, when it hit the ground. Out flowed a river of ravenous white worms, which rushed toward the men and devoured them whole, while they thrashed about and screamed.

Soon, nothing was left except a leg bone from each. The little gnome picked them up, and walked off the road. He sat down by a hollow log, and started to drum.

He drummed, and he waited, and he hummed an odd little tune:

"A picture of food

feeds the whole hungry clan

the image of good

makes the whole healthy man

Why walk the mile?

Why do the work?

Just smile the smile!

success

after all

is a quirk!

Life isn't real

that's the message I give

It's easy that way

plus

who wants to live?"

It is ideas of the "necessity of sacrifice" that underlie, for example, the well-known but explicitly incomprehensible ritual of Christian communion (more accurately, the ritual of the Christian communion serves as one behavioral precursor for these explicit ideas). The Christian hero—Christ—is the spirit who offers himself voluntarily to the cross, to the grave, to suffering and death, to the terrible mother. Such a spirit is, above all, "humble"—which is a very paradoxical term, in this context. Arrogance is belief in personal omniscience. Heroic humility, set against such arrogance, means recognition of constant personal error, conjoined with belief in the ability to *transcend* that error (to face the unknown, and to update fallible belief, in consequence). "Humble" therefore means, "greater than dogma" (as the spirit of man is a "higher power" than the laws which govern his behavior). Christ's body (represented, in the communion ritual, by the "ever-resurrecting" wheaten wafer), is the container of the incarnate spirit of the dying, reborn and redemptive deity. This "body" is ritually devoured—that is, incorporated—to aid the ritual participants in their identification with Christ, the eternally dying and resurrecting (sun) god. Construction of this awful ritual meant furtherance of the abstract conceptualization of a permanent structural aspect of (every) human psyche—the heroic aspect, the *Word*—as active, individually doomed, yet mythically eternal, destined to tragic contact with threat and promise of unknown, yet constant participant in the creative adaptive redemptive process.

The ritual act of exposure is held simultaneously to placate or minimize the cruel aspect of nature, and to allow for establishment of contact with the beneficient. From the modern perspective, it might be said (much more abstractly) that voluntary cautious, careful, exploratory encounter with the threatening and unknown constitutes the precondition for transformation of that unknown into the promising (or at least the mundane), as a consequence of shift in behavior or interpretation. We moderns interpret this "change in experience" as alteration in subjective state. The pre-experimental mind, less capable of clearly differentiating subject from object, more concerned with the motivational significance of the experience, observes instead *that the fear-inducing character of the object has receded* (as a consequence of the courage of the explorer, or the benevolence of the thing in question).

Ritual sacrifice was an early (pre-abstract behavioral) variant of the "idea" of heroism, of belief in individual power—the acting out of the idea that voluntary exposure to the unknown (or dissolution of the most favored thing) constituted a necessary precondition (1) for the emergence of the beneficial "goddess" and (2) for continued successful adaptation. Incorporation of the sacrificial individual, in actuality (in ritual cannibalism) or in religious ceremony (in the mass, for example) meant assimilation of the culture-hero. Such incorporation was a "preconscious" attempt to embody the heroic essence, to fortify the constituent elements of the community against paralyzing fear of death and darkness—to fortify the individual and the social group against fear of the unknown itself. The sacrificial ritual was *acting out* of the hero, before such "acting out" could be represented in abstraction, in drama, in story. More abstract narrative representation of the target of the "heroic sacrifice" then came to portray the emergence of the beneficient goddess, capable of showering reward upon man, her eternal lover and child.

The spirit forever willing to risk personal (more abstractly, intrapsychic) destruction to gain redemptive knowledge might be considered *the archetypal representative of the adaptive process as such*. The pre-experimental mind considered traumatic union of this "masculine" representative with the destructive and procreative feminine unknown a necessary precedent to continual renewal and rebirth of the individual and community. This is an idea precisely as magnificent as that contained in the Osiris/Horus myth; an idea which adds additional depth to the brilliant "moral hypotheses" contained in that myth. The exploratory hero, divine son of the known and unknown, courageously faces the unknown, unites with it creatively—abandoning all pretense of pre-existent "absolute knowledge"—garners new information, returns to the community, and revitalizes his tradition. It is to this more complete story that we now turn our attention.

The Divine Son: Images of the Knower, the Exploratory Process

> Awake, awake, put on strength, O arm of the Lord; awake, as in the ancient days, in the generations of old. Art thou not it that hath cut Rahab, and wounded the dragon? (Isaiah 51:9)

The great androgynous dragon of chaos is also the mythic figure who guards a great treasure, hidden in the depths of a mountain, or who conceals a virgin princess in his lair. He is the fire-breathing winged serpent of transformation—the undescribable union of everything now discriminated, who constantly schemes to take back what he produced. The Great and Terrible Mother, daughter of chaos, destroys those who approach her accidentally, incautiously, or with the inappropriate attitude, but showers upon those who love her (and who act appropriately) all good things. The Great and Terrible Father, son of chaos, gives rise to sons of his own, but then attempts to crush, or even to devour them: he is precondition for existence, but impediment to its successful elaboration. What might possibly constitute "the appropriate pattern of action" in the face of such permanent and multifarious contradiction?

The fundamental act of creativity in the human realm, in the concrete case, is the construction of a pattern of behavior which produces emotionally desirable results in a situation that previously reeked of unpredictability, danger and promise. Creative acts, despite their unique particulars, have an eternally identifiable structure, because they always takes place under the same conditions: what is known is "extracted," eternally, from the unknown. In consequence, it is perpetually possible to derive and re-derive the central features of the metapattern of behavior which always and necessarily means human advancement. Human beings are curious about the structure and function of everything, not least themselves; our capacity to tell stories reflects our ability to describe ourselves. It has been said that Freud merely recapitulated Shakespeare. But it was Freud's genius, despite his manifold errors, to bring what Shakespeare portrayed dramatically *up one level of abstraction*, toward the philosophical (or even the empirical). Freud moved information about behavior from the implicit narrative to the explicit theory (or, at least, to the more explicit theory). Shakespeare performed a similar maneuver, like all storytellers, at a more "basic" level—*he abstracted from what was still behavioral, from what had not even yet been captured effectively in drama.*

During exploration, behavior and representational schema are modified in an experimental fashion, in the hopes of bringing about by ingenious means whatever outcome is currently envisioned. Such exploration also produces alteration of the sensory world—since that world changes with shift in motor output and physical locale. Exploration produces transformation in *assumption guiding behavior*, and in *expectation of behavioral outcome*: produces learning in *knowing how and knowing what* mode. Most generally, new learning means the application of a new means to the same end, which means that the pattern of presumptions underlying the internal model of the present and the desired future remain essentially intact. This form of readaptation might be described as *normal* creativity, and constitutes the bulk of human thought. However, on rare occasions, ongoing activity (specifically goal-directed or exploratory) produces more profound and unsettling mismatch. This is more stressful (and more promising), and necessitates more radical update of modeling—necessitates exploration-guided reprogramming of fundamental behavioral assumption and associated episodic or semantic representation. Such reprogramming also constitutes creativity, but of the *revolutionary* type, generally associated with genius. Exploration is therefore creation *and* re-creation of the world. The generation of new information from contact with the unknown means the construction of experience itself; the destruction of previous modes of adaptation and representation (previous "worlds") means return of "explored territory" to the unexplored condition that preceded it, and then its restructuring, in more comprehensive form. This is encounter with the Great and Terrible Mother, and death and resurrection of the Son and the Father.

A new manner of dealing with (that is, *behaving with regard to* or *classifying*) an emergent unknown is *the gift of the hero*. This gift demands to be given; *compels* communication—either directly (say, in the form of immediate imitation), or indirectly (in the form of abstract description, or narrative). There is no real qualitative distinction between transformation of means and transformation of ends (as we have seen): what constitutes "ends" at a lower level of analysis becomes "means" at a higher level. It follows that the "gift of the hero" constitutes normal and revolutionary adaptation, *simultaneously*—normal adaptation, as schemas of action and representation are extended, such that the unknown is rendered beneficial; revolution, as the old is restructured, to allow place for the new. This restructuring is equivalent to the establishment of *peace*—the peace characterizing the mythic paradise where the lion lays down with the lamb. Such peace emerges as a consequence of the hierarchical organization of the "gods of tradition" under the dominion of the hero. This means that the creative exploratory hero is also peacemaker, in his complete manifestation:

I dreamed that I was standing in the grassy yard of a stone cathedral, on a bright sunny day. The yard was unblemished, a large, well-kept green expanse. As I stood there, I saw a slab of grass pull back under the earth, like a sliding door. Underneath the "door" was a rectangular hole that was clearly a grave. I was standing on an ancient graveyard, whose existence had been forgotten. A medieval king, dressed in solid armor, rose out of the grave, and stood at attention at the head of his burial site. Similar slabs slid back, one after another, in numerous places. Out of each rose a king, each from a different period of time.

The kings were all powerful, in their own right. Now, however, they occupied the same territory. They became concerned that they would fight, and they asked me how this might be prevented. I told them the meaning of the Christian wedding ceremony—a ritual designed to subjugate the two central participants to the superordinate authority of Christ, the Christian hero, and said that this was the way to peace.

If all the great kings would bow, voluntarily, to the figure of the hero, there would be no more reason for war.

Every unmapped territory—that is, every place where what to do has not been specified—also constitutes the battleground for ancestral kings. The learned patterns of action and interpretation that vie for application when a new situation arises can be usefully regarded, metaphorically, as the current embodiments of adaptive strategies formulated as a consequence of past exploratory behavior—as adaptive strategies invented and constructed by the heroes of the past, "unconsciously" mimicked and duplicated by those currently alive.

Adaptation to new territory—that is, to the unexpected—therefore also means successful mediation of archaic or habitual strategies competing, in the new situation, for dominance over behavioral output. Rank-ordering of these "warring" strategies—construction of a context-specific behavioral "dominance hierarchy" (which corresponds to the nested narrative model proposed earlier)—therefore constitutes adaptation, just as much as creation of new situation-specific behaviors or modes of interpretation (which are inevitably composed, anyway, of bits and pieces of the past). The process of exploration, including its assimilative and accommodative aspects, is therefore inevitably entangled with the process of peacemaking. Exploration, in a given situation, can hardly be regarded as complete until the tendencies and theories that struggle for predominance in that situation have been organized to make internal (or externalized) conflict and emotional upheaval cease.

The exploratory hero, mankind's savior, cuts the primordial chaos into pieces and makes the world; rescues his dead father from the underworld, and revivifies him; and organizes the "nobles" occupying his kingdom into an effective, flexible and dynamic hierarchy. There is no categorical difference between the individual who explores and the individual who reconstructs "society," as a consequence of that exploration. Accomodation to new information is an integral part of the exploratory process: an anomaly has not been processed until the preexistent interpretive schemas extant prior to its emergence have been reconfigured to take its presence into account. Every explorer is therefore, by necessity, a revolutionary—and every *successful* revolutionary is a peacemaker.

We act appropriately before we understand how we act—just as children learn to behave before they can describe the reasons for their behavior. It is only through the observation of our actions, accumulated and distilled over centuries, that we come to understand our own motivations, and the patterns of behavior that characterize our cultures (and these are changing as we model them). *Active adaptation precedes abstracted comprehension of the basis for such adaptation.* This is necessarily the case, because we are more complex than we can understand, as is the world to which we must adjust ourselves.

First we act. Afterward, we envision the pattern that constitutes our actions. Then we use that pattern to guide our actions. It is establishment of conscious (declarative) connection between behavior and consequences of that behavior (which means establishment of a new feedback process) that enables us to abstractly posit a desired future, to act in such a way as to bring that future about, and to judge the relevance of emergent phenomena themselves on the basis of their apparent relevance to that future. This ability appears to be predicated on some developmental leap—at least insofar as the "guiding story" has become conscious (or represented in episodic or semantic memory, as opposed to remaining implicitly embedded in behavior)—and appears unlikely to characterize very young children (or animals, for that matter). Jean Piaget solved the problem of the "goal-like" behavior in creatures not yet capable of abstract conceptualization by presuming that "goals" are initially embedded in sensorimotor reflex operations, which are instinctive. This essentially means that what is later story is at first *pattern*—the pattern of socially modified behavior that constitutes human being. It is only later, as "higher-order" (episodic or semantic) cognitive systems become activated, that goals become explicitly imagined (and that they can be considered, abstractly, before their enaction). So this means that it is possible to act in a manner that looks as if it were goal-directed, before goals as such have manifested themselves. Rychlak describes Piaget's observation: "Children do not appear to be logicians at birth, conceptually interacting by constructing schema from the outset. The initial constructions are being done biologically, and only at some time later does the child schematize the reflexive patterns already underway...."[325]

First comes the action pattern, guided by instinct, shaped without conscious realization by the consequences of socially mediated "rewards" and "punishments" (determined in their "structure and locale" by the current social mores, products of historical forces). Then comes the capacity to imagine the end toward which behavior "should" be directed. Information generated from the observation of behavior provides the basis for constructing fantasies about such ends. Actions that satisfy emotions have a pattern; abstraction allows us to represent and duplicate that pattern, as an end. The highest-level abstractions therefore allow us to represent the most universally applicable behavioral pattern: that characterizing the hero, who eternally turns the unknown into something secure and beneficial; who eternally reconstructs the secure and beneficial, when it has degenerated into tyranny.

The myth of the hero has come to represent the essential nature of human possibility, as manifested in adaptive behavior, as a consequence of observation and rerepresentation of such behavior, conducted cumulatively over the course of thousands of years. The hero myth provides the structure that governs, but does not determine, the general course of history; expresses one fundamental preconception in a thousand different ways. This idea (analogous in structure to the modern hypothesis, although not explicitly formulated, nor rationally constructed in the same manner) renders individual creativity socially acceptable and provides the precondition for change. The most fundamental presumption of the myth of the hero is that the nature of human experience can be (should be) improved by voluntary alteration in individual human attitude and action. This statement—*the historical hypothesis*—is an expression of faith in human possibility itself and constitutes the truly revolutionary idea of historical man.

All specific adaptive behaviors (which are acts that restrict the destructive or enhance the beneficial potential of the unknown) follow a general pattern. This "pattern"—which *at least* produces the results intended (and therefore desired)—inevitably attracts social interest. "Interesting" or "admirable" behaviors engender imitation and description. Such imitation and description might first be of *an interesting or admirable behavior,* but is later of *the class of interesting and admirable behaviors.* The class is then imitated as a general guide to specific actions; is redescribed, redistilled and imitated once again. The image of the hero, step by step, becomes ever clearer, and ever more broadly applicable. The pattern of behavior characteristic of the hero—that is, voluntary advance in the face of the dangerous and promising unknown, generation of something of value as a consequence and, simultaneously, dissolution and reconstruction of current knowledge, of current *morality*—comes to form the kernel for the *good story,* cross-culturally. That story—which is what to do, when you no longer know what to do—defines the central pattern of behavior embedded in all genuinely religious systems (furthermore, provides the basis for the "respect due the individual" undergirding our conception of natural rights). Representations of the uroboros, the dragon of chaos, and his daughter, the Great Mother, are symbolic portrayals of the unknown. Mythological representation of the hero and his cultural construction are, by contrast, examination and portrayal of *who or what it is that knows*, and of *what it is that is known*. The creative and destructive feminine is the personality manifested in mythology by everything unknown, threatening and promising, about and within existence. Myth tends to portray the generative individual consciousness eternally willing to face this unknown power as masculine, in essence—in contradistinction to unconscious, impersonal, and unpredictable femininity, and in light of its "seminal," active, "fructifying" nature.

The earliest "stages" of the development of the figure of the hero take the form of mythic representations of the infant or adolescent, fully or partially dominated by potent maternal force.[326] This infant or adolescent is the specific individual, under the sway of the particular mother—and *Homo sapiens*, the species, subject to nature. "Generative individual consciousness" as "eternal son of the virginal mother" is represented in *Figure 37: The Exploratory Hero as Son of the Heavenly Mother*.[327] In his more mature form, the hero—formerly "son of the heavenly mother"—can be portrayed as "lover of the Great Mother" [the mother whose body he "enters" into, in creative (sexual) union, to die and reincarnate (to fertilize and impregnate)]. The Great Mother is the holy prostitute, the whore of Babylon, as well as the Virgin Mother, a maiden forever renewed, forever young, belonging to all men, but to no one man. Myth commonly utilizes the (symbolically sexual) motif of *heavenly incest*—the image of devouring or engulfing encounter, rife with creative potential—to represent union with the primordial feminine, to portray *act of creative (or destructive) encounter between the hero and the possibilities of life itself*. This is "knowledge" as sexual, creative act: the "voluntary generative union" of consciousness and chaos produces—or revives—order and cosmos.

The mythology of the hero, in *toto*, depicts the development and establishment of a personality capable of facing the most extreme conditions of existence. The hero's quest or journey has been represented in mythology and ritual in numerous ways, but the manifold representations appear in accordance with the myth of the way, as previously described: a harmonious community or way of life, predictable and stable in structure and function, is unexpectedly threatened by the emergence of (previously harnessed) unknown and dangerous forces. An individual of humble and princely origins rises, by free choice, to counter this threat. This individual is exposed to great personal trials and risks or experiences physical and psychological dissolution. Nonetheless, he overcomes the threat, is magically restored (frequently improved) and receives a great reward, in consequence. He returns to his community with the reward, and (re)establishes social order (sometimes after a crisis engendered by his return).

This most fundamental of stories is portrayed schematically in *Figure 38: The Metamythology of the Way, Revisited*.[328] Chaos breeds novelty, promising and threatening; the hero leaves his community, voluntarily, to face this chaos. His exploratory/creative act quells the

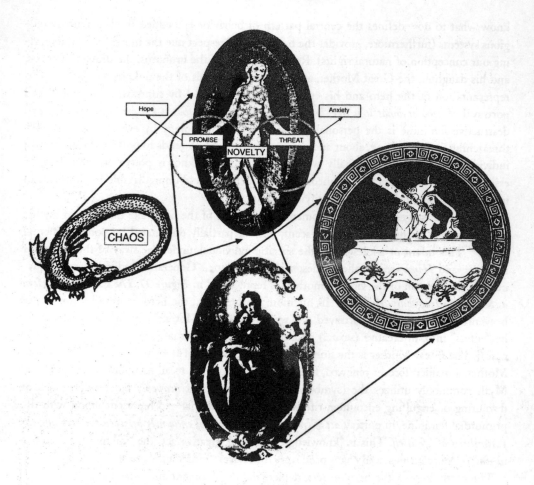

Figure 37: The Exploratory Hero as Son of the Heavenly Mother

threat embedded in chaos, and frees what is promising from its grip. Incorporation of this freed promise (this "redemptive" information)—symbolized by union with the virgin, or discovery of the treasure—transforms the hero. His transformed (enriched) behavior then serves his community as model. The group is therefore transformed and restabilized in turn.

The ultimate or archetypal representation of the original "threatened" state is the unselfconscious (but "incomplete") paradise that existed prior to the "fall" of humanity. More prosaically, that state is the innocence and potential of childhood, the glory of the past, the strength of the well-ruled kingdom, the power of the city, the stability, wealth and happiness of the family. The most primordial threat is the sudden (re)appearance or discovery of one of the manifestations of the Terrible Mother: a flood, an earthquake, a war, a monster (some type of dragon), a fish, a whale—anything unpredictable or unexpected that destroys,

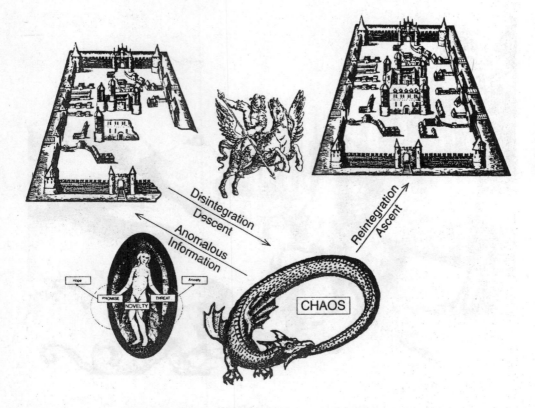

Figure 38: The Metamythology of the Way, Revisited

devours, traps, engulfs, dismembers, tortures, terrifies, weakens, mystifies, entrances, smothers or poisons (this is a partial list). The hero, product of divine parentage and miraculous birth, survivor of a dangerous childhood, faces the Terrible Mother in single combat and is devoured. He is swallowed by a great fish, or snake, or whale, and spends time underground, in the dark, in the winter, in the kingdom of the dead, or in hell; faces a dragon, a gorgon, a witch or temptress; is inundated by water, by fire, by storm, by dangerous animals; is tormented, buried alive, mesmerized, dismembered, disemboweled and deluded. He defeats the monster, freeing those who had been previously defeated, and gains or regains a lost or previously undiscovered object of value, a (virginal) woman or a treasure. Much older, much wiser, he returns home, transformed in character, bearing what he has gained, and reunites himself triumphantly with his community, which is much enriched—or even utterly transformed—by his fortune.[329]

The battle of the hero is a frequent motif in mythologically inspired sculpture, drawing and painting. A representative example is presented in *Figure 39: Castle, Hero, Serpent and Virgin: St. George and the Dragon.*[330] All of the elements of the "meta-myth" are portrayed in

Figure 39: Castle, Hero, Serpent and Virgin: St. George and the Dragon

this drawing: the threatened community, represented by the walled city or castle; the winged dragon, who has emerged from the underworld (and whose lair is surrounded by the bones of the dead); the hero, armed with the sword, who "cuts" the leviathan into pieces, and makes the world; and the virgin, freed from the dragon's clutches, who represents the benevolent, creative and fruitful aspect of the unknown. (The city is commonly portrayed on a mountain, in such representations—the serpent in a valley, or across a river. The battle takes place at sundown [when the sun deity encounters the dragon of the night].[331])

Solar myths portray the journey of the hero, utilizing simultaneously the motifs of the dragon fight and the "night sea journey." In the typical solar myth, the hero is identified with the sun, bearer of the light of consciousness, who is devoured nightly by the water serpent of the West. In the night, he battles terribly with this monster, and emerges victorious in the morning, rising renewed in the East:

In this sequence of danger, battle, and victory, the light—whose significance for consciousness we have repeatedly stressed—is the central symbol of the hero's reality. The hero is always a light-bringer and emissary of the light. At the nethermost point of the night sea journey, when the sun hero journeys through the underworld and must survive the fight with the dragon, the new sun is kindled at mid-night and the hero conquers the darkness. At this same lowest point of the year Christ is born as the shining Redeemer, as the light of the year and light of the world, and is worshipped with the Christmas tree at the winter solstice. The new light and the victory are symbolized by the illumination of the head, crowned and decked with an aureole.[332]

The Mesopotamian emperors and the pharaohs of Egypt were solar gods, representatives of the incarnated sun deity, eternal victor of the unending battle between order and chaos, light and darkness, known and unknown. In an allegorical sense, they might be considered the first true individuals—at least from the perspective of the Western historical tradition. The Egyptian people devoted their entire cultural endeavor to glorification of their rulers—motivated, unconsciously, by their participation in (their imitative identification with) the essential god-stature of the pharaoh. This idea was developed (abstracted and generalized) further by the Greeks, who attributed to each male Greek a soul, and taken to its logical conclusion by the Jews and the Christians, who granted every person absolute and inviolable individual worth before (or [potential] identity with) God.

The Great Mother is embodiment of the unknown, of the novel. The hero—her son and lover, offspring of the mystical marriage—is dramatic (first concrete behavioral, then imita-tive/imagistic, then verbal) representation of the pattern of action capable of making creative use of that unknown. The potential for expression of (and admiration for, or representation of) that pattern constitutes a heritable characteristic of the human psyche, expressed constantly in behavior during the course of human cultural activity. Containment of this pattern in dynamic image, in myth, follows centuries of observation, and generation of hypotheses, regarding the core nature of *Homo sapiens*, the historical animal. The development of such containment followed a complex path of increasingly abstracted description and redescription of self and other.

The hero is a *pattern* of action, designed to make sense of the unknown; he emerges, nec-essarily, wherever human beings are successful. Adherence to this central pattern ensures that respect for the process of exploration (and the necessary reconfiguration of belief, attendant upon that process) always remains *superordinate to all other considerations, including that of the maintenance of stable belief.* This is why Christ, the defining hero of the Western ethical tradi-tion, is able to say, "I am the way, the truth, and the life: no man cometh unto the Father, but by me" (John 14:6); why adherence to the Eastern way (Tao)—extant on the border between chaos (yin) and order (yang)—ensures that the "cosmos" will continue to endure. *Figure 40: The Process of Exploration and Update, as the Meta-Goal of Existence* schematically presents the "highest goal" of life, conceptualized from such a perspective: identification with the process of constructing and updating contingent and environment-specific goals is in this schema given necessary precedence over identification with any particular, concretized goal. Spirit is thus elevated over dogma.

We use stories to regulate our emotions and govern our behavior. They provide the present we inhabit with a determinate point of reference—the desired future. The optimal "desired

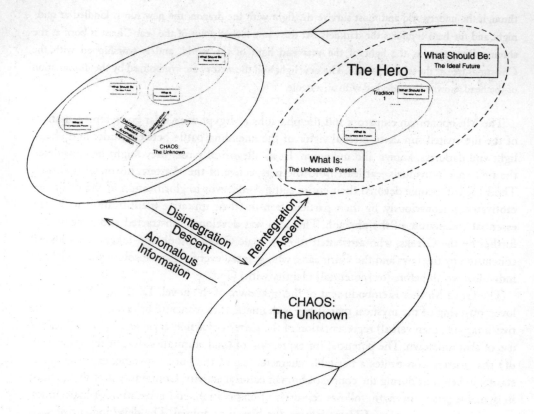

Figure 40: The Process of Exploration and Update, as the Meta-Goal of Existence

future" is not a state, however, but a process: the (intrinsically compelling) process of mediating between order and chaos; the process of the incarnation of *Logos*—the Word—which is the world-creating principle.[333] Identification with this process, rather than with any of its determinate outcomes (that is, with any "idols" or *fixed frames of reference* or *ideologies*) ensures that emotion will stay optimally regulated and action remain possible no matter how the environment shifts, and no matter when. In consequence of such identification, respect for belief comes to take second place to respect for the process by which belief is generated.

The hero is narrative representation of the individual eternally willing to take creative action, endlessly capable of originating new behavioral patterns, eternally specialized to render harmless or positively beneficial something previously threatening or unknown. It is declarative representation of the pattern of behavior characteristic of the hero that eventually comes to approximate the story of the *savior*. Behind every particular (that is, *historical*) adventurer, explorer, creator, revolutionary and peacemaker lurks the image of the "son of god," who sets his impeccable character against tyranny and the unknown. The archetypic or ultimate example of the savior is the world redeemer, the Messiah—world-creating and -redeeming hero, social revolutionary and great reconciliator. It is the sum total of the activity of the Messiah, accumulated over the course of time, that constitutes culture, the Great Father, order itself—explored territory, the domain of the known. In the "meta-stable" society, however, the Father, though healthy, is subordinate to the Son: all fixed values necessarily remain subject to the pattern of being represented by the hero. In the "City of God"— that is, the archetypal human kingdom—the Messiah eternally rules:

I saw in the night visions, and, behold, *one* like the Son of man came with the clouds of heaven, and came to the Ancient of days, and they brought him near before him.

And there was given him dominion, and glory, and a kingdom, that all people, nations, and languages, should serve him: his dominion is an everlasting dominion, which shall not pass away, and his kingdom that which shall not be destroyed. (Daniel 7:13–14)

The Great Father: Images of the Known, or Explored Territory

All particular adaptive behaviors (and interpretive schemas—schemas of *value*) are generated over the course of time by the eternal pattern of behavior described in mythic language as characteristic of the archetypal hero, the sun god. These behaviors and schemas accumulate over the centuries (as a consequence of imitation and other forms of memory communication), but do not necessarily *agree*, are not necessarily commensurate. Our hard-won adaptive methods struggle for predominance, often violently, within a given individual, between individuals within societies, and between societies. The problem of *organization* therefore arises. How do you arrange your possibilities, once you have originated them or copied them from someone else? How is it possible to make sense of the historical accretion of knowledge and wisdom? After all, multiple opportunities for behavioral output exist, in any given situation; furthermore, the possibility of interpretation makes even the "situation" mutable. How can competing possibilities—the multiplicity of potential choices—be amalgamated into some sort of unity; the kind of unity that makes mutual coexistence (and mutually beneficial coexistence) possible? How, in brief, is it possible to construct and maintain a society?

Procedural knowledge, generated in the course of heroic behavior, is not organized and integrated within the group and the individual as a consequence of simple *accumulation*. Procedure "a," appropriate in situation one, and procedure "b," appropriate in situation two, may clash in mutual violent opposition in situation three. Under such conditions intrapsychic or interpersonal conflict necessarily emerges. When such antagonism arises, moral revaluation becomes necessary. As a consequence of such revaluation, behavioral options are brutally rank-ordered, or, less frequently, entire moral systems are devastated, reorganized and replaced. This organization and reorganization occurs as a consequence of "war," in its concrete, abstract, intrapsychic and interpersonal variants. In the most basic case, an individual is rendered subject to an intolerable conflict, as a consequence of the perceived (affective) incompatibility of two or more apprehended outcomes of a given behavioral procedure. In the purely intrapsychic sphere, such conflict often emerges when attainment of what is desired presently necessarily interferes with attainment of what is desired (or avoidance of what is feared) in the future. Permanent satisfactory resolution of such conflict (between temptation and "moral purity," for example) requires the construction of an abstract moral system, *powerful enough to allow what an occurrence signifies for the future to govern reaction to what it signifies now*. Even that construction, however, is necessarily incomplete when considered only as an "intrapsychic" phenomena. The individual, once capable of coherently integrating competing motivational demands in the private sphere, nonetheless remains destined for conflict with the other, in the course of the inevitable transformations of personal experience. This means that the person who has come to terms with him- or herself—at least in principle—is still subject to the affective dysregulation inevitably produced by interpersonal interaction. It is also the case that such subjugation is actually indicative of insufficient "intrapsychic" organization, as many basic "needs" can only be satisfied through the cooperation of others.

The problems posed by the "future self," whose still-potential existence has to be taken into account and used to govern action in the present, are very similar to those posed by the existence of others, whose affective responses are equally hypothetical (as they cannot be experienced directly, but only inferred). The properly socialized individual has been trained to grant this "abstract other" (future self and other person) ontological status equivalent to the experienced self, however—has been trained to use the existence of that other as a guide to "proper action and interpretation" in the present. This means that for the social being all individual actions come to be evaluated with regard to their likely current and future consequences, for the self and for the others likely affected. Such evaluation may take place directly—that is, as a matter of "conscious deliberation"; alternatively, the well-socialized individual may act "as if" he or she thought the matter through, by remaining in well-trodden moral pathways (which have been established under the cumulative historical pressure produced by the necessity of maintaining intrapsychic and social order). The more implicit information extant in the latter case is "placed there" as a consequence of the exchange of emotional information, attendant upon given action patterns, in the potential absence of explicit rationale: someone is informed by subtle scornful gesture, for example, that a given (theoretically pleasurable and even evidently harmless) behavior is "just not done," which means that it is regarded by custom as harmful, to the self and others, in some manner not easily observed but still important. It is such arbitrary rules that constitute the implicit information coded in societal structure—information not necessarily placed there by rational means, not necessarily "comprehended" in any declarative sense; but information that is nonetheless transmissible and representable as a consequence of extended-term pattern recognition and analysis.

The "stories" by which individuals live (which comprise their schemas of interpretation, which guide their actions, which regulate their emotions) are therefore emergent structures shaped by the necessity of organizing competing internal biological demands, over variable spans of time, in the presence of others, faced with the same fate. This similarity of *demand* (constrained by physiological structure) and *context* (constrained by social reality) produces

similarity of *response*. It is this similarity of response, in turn, that is at the base of the emergent "shared moral viewpoint" that accounts for cross-cultural similarity in myth. This
means, by the way, that such "shared viewpoints" refer to something *real*, at least insofar as
emergent properties are granted reality (and most of the things that we regard without question as real are precisely such emergent properties).

The reactions of a hypothetical firstborn child to his or her newborn sibling may serve as
concrete illustration of the interactions between the individual, the interpersonal and the
social. The elder sibling may be drawn positively to the newborn by natural affiliative tendencies and curiosity. At the same time, however, the new arrival may be receiving a substantial amount of parental attention, sometimes in preference to the older child. This shift of
parental care often produces frustration, manifested in aggressive behavior, on the part of the
supplanted sibling. The older child will therefore become conflicted, internally, in consequence of his affection for the new family member, curiosity about its nature, and irritation
at the creature's existence, demands and influence on the (once) predictable interactions of
the familial social unit. The protective attitude of the parents, who restrict aggression on the
part of the elder child, further complicates things—draws the additional requirements of the
social unit into the already difficult situation.

How is the child to resolve his conflicts? He must build himself a *personality* to deal with
his new sibling (must become a proper big brother). This means that he might subordinate
his aggression to the fear, guilt and shame produced by parental adjudication on behalf of the
baby. This will mean that he will at least "act like a human being" around the baby, in the
direct presence of his parents. He might also learn to act *as if* the aggressive reaction motivated by his shift in status is less desirable, in total, than the affiliative response. His *as if*
stance may easily be bolstered by intelligent shift in interpretation: he may reasonably gain
from his younger sibling some of the attention he is no longer paid by parents—if he is diligent and genuine in his attempts to be friendly. He might also develop some more independent interests, suitable to his new position as relatively mature family member. In the former,
simplest case (when he subordinates his aggression to fear), the child rank-orders his motivational states, as manifested in behavior. In the latter, revolutionary situation, the child
restructures the implicit presumptions that originally gave rise to the conflict. Either way, the
situation is resolved (restoried) in the course of what might reasonably be described as an
internal war—accompanied, inevitably, by intense outbursts of pain, fear and rage. The personality that emerges as a consequence of such a war is, at least in the revolutionary case,
something "more like the hero" than the personality that existed prior to the change in environmental circumstances.

The situation of marriage provides an additional illustrative example, relevant to the adult
situation. In marriage, the desire for individual self-expression is necessarily limited by the
desire for maintenance of the intimate interpersonal relationship, and for adoption of the
"respectable" social role that constitutes such maintenance. The male, bachelor no more, may
attempt to carry on his premarital mode of activity, more purely dependent on personal
desire and whim, limited by whatever minimal necessary social obligations he may have

acquired previously. Soon he will discover, if he has taken an appropriately assertive partner, that his (heretofore individualistic) wishes and desires produce conflict in his married life, manifested in interpersonal strife and consequent emotional dysregulation.

The clashes that commonly accompany establishment of a permanent affiliative relationship arise as a result of the incompatibility of (implicit and explicit) individual moral presumptions and propositions, in the interpersonal sphere (arise as a result of an interpersonal "war of implicit gods"). Such clashes may be resolved in a variety of manners. One partner may, through judicious application of physical or psychological punishment, render the other impotent, so to speak, and subordinate—permanently frustrated, miserable, anxious and hostile. The marriage may thus lose much of its value or may dissolve altogether. This does not constitute a "solution"—merely regression, in the face of emergent anomaly, to the pre-existent "single personality." Alternatively, each partner may determine to take "the other" into serious consideration, and rearrange personal behavior (and emergent value) accordingly. This process will not occur without capacity to engage in open conflict (to exchange often distressing information, realistically speaking) or without the courage to voluntarily submit to the experience of negative emotion [including anxiety, guilt and shame, as previously "unconscious" (implicit) faults and insufficiencies come to light]. The mythic subjugation of the partners in a marriage to the higher authority of Christ, the culture hero, ritually represented in the Christian marriage ceremony, constitutes symbolic aid to this process.[334]

Voluntary subordination of the personal wishes of both individuals to the higher moral order embodied in the action patterns of the Christian savior, for example, means implicit agreement about the nature of transcendent principles that can be referred to when mediation between incompatible desires and presuppositions becomes necessary. This means that the "personality" constituted by the "mystical union" of both partners in the marriage is supposed to approximate Christ—to stand as an entity superordinate to the "less complete" individuals who compose the "married couple." This process of voluntary subordination "to a higher deity" parallels the extended transpersonal historical process described in the *Enuma elish*, with regard to the ascendance of Marduk. Through conflict (and cooperation), within the "container" of the marriage, new moralities are created—new patterns of behavior (and assumption and expectation) manifested and internally represented. This process may be guided to a healthy outcome through mutual participation in community-sponsored religious ritual. Alternatively, individuals may succeed, or fail, in isolation.

Motivational states compete for predominance in the present, in the purely subjective and interpersonal spheres, and also compete across time. What is fear-provoking now may be tolerated because it means less punishment (or less fear, or more pleasure, or more hope) in the future, insofar as intelligence or custom can make that judgment; similarly, the social group and the additional pressure it produces is tolerated because the group constitutes the most effective currently imaginable solution to the problem of adaptation. This group, the current embodiment of human custom, is the consequence of a battle between various ways of being fought across generations.

Although the "battle for predominance" that characterizes exchange of morally relevant information can easily be imagined as a war (and is often fought out in the guise of genuine war), it is more frequently the case that it manifests itself as a struggle between "beliefs." In the latter case, it loss of *faith*, rather than life, that determines the outcome of the battle. Human beings can substitute loss of faith for death partly because they are capable of abstractly constructing their "territories" (making beliefs out of them) and of abstractly abandoning those territories once they are no longer tenable. Animals, less capable of abstraction, are also able to lose face, rather than life, although they "act out" this loss, in behavioral routines, rather than in verbal or imagistic battles (rather than through argument). It is the capacity to "symbolically capitulate" and to "symbolically destroy" that in large part underlies the ability of individual animals to organize themselves into social groups (which require a hierarchical organization) and to maintain and update those groups once established. Much the same can be said for human beings (who also engage in abstract war, at the procedural level, as well as in real war and argumentation).

Strong ideas produce profound displays of faith, or, alternatively put: unshakeable displays of faith are indicative of the strength of an idea. The strength of an integrative idea, or its preabstract procedural equivalent, might be considered reasonably measured by its capacity to inhibit competing impulses—especially those motivated by fear. Dominance displays in groups of primates and other complex higher-order social animals provide a useful example of this. Most dominance disputes are settled before escalation into physical aggression. It is the animal most capable of holding its ground in the face of challenge—despite threat, regardless of fear—that is the likely victor in the case of such a dispute. The capacity to maintain territorial position when challenged is therefore indicative of the degree to which intrapsychic state is integrated with regard to current motivation (which means, indicative of how "convinced" a given animal is that it can [should] hold its ground). This integration constitutes power—*charisma*, in the human realm—made most evident in behavioral display. The certainty with which a position is held (whether it is a territorial position, dominance hierarchy niche, or abstract notion)—insofar as this can be inferred from observable behavior, such as absence of fear—constitutes a valid indication of the potential integrative potency of that position; constitutes an indication of how much the creature who is holding the position believes in the rightness (justice, goodness) of his or her stance. The integrative strength of beliefs of this type can be determined, accurately, through challenge (since the capacity to withstand challenge is dependent upon that strength). This means that the ability of those who hold an idea to withstand challenge without wavering constitutes one [nonempirical (?)] *affective* criterion for determination of the truth of that idea—or at least of its intrapsychic utility. Hence the power of the martyr, and the unwillingness of even modern totalitarians to allow their enemies to make public sacrifices of themselves.

Rank-ordering of behavior in terms of comparative utility is (procedural, episodic or semantic) judgment rendered upon value. Such judgment constitutes a decision about the "nature of good and evil," from the mythic or narrative viewpoint. Such determinations of value are decisions whose function is organization of future-oriented present individual behavior, manifested in the (inevitably) social context, in accordance with the wisdom of past

experience. The content of mythically transmitted behavioral schemas and their value-predi-
cated arrangements generally remains implicit, outside the domain of descriptive compre-
hensibility, because of their exceedingly complex structure, which evolved through the action
of primarily nondeclarative evolutionary processes. The emotional upheaval caused by simultaneous application of noncommensurate behavioral or interpretive strategies provides the
impetus for the organization of those strategies. Such organization emerges as a result of the
"struggle for dominion," intrapsychically or interpersonally—emerges in consequence of a
quasi-Darwinian struggle for survival.

Over the course of centuries, the actions of ancestral heroes, imitated directly and then represented in myth, become transformed, simplified, streamlined and quickened—reduced as it
were ever more precisely to their "Platonic" forms. Culture is therefore the sum total of surviving historically determined hierarchically arranged behaviors and second- and third-order
abstract representations, and more: it is the integration of these, in the course of endless social
and intrapsychic conflict, into a single pattern of behavior—a single system of morality,
simultaneously governing personal conduct, interpersonal interaction and imagistic/semantic
description of such. This pattern is the "corporeal ideal" of the culture, its mode of transforming the unbearable present into the desired future, its guiding force, its central personality.
This personality, expressed in behavior, is first embodied in the king or emperor, socially
(where it forms the basis for "sovereignty"). Abstractly represented—imitated, played, ritualized, and storied—it becomes something ever more psychological. This embodied and represented "cultural character" is transmitted through the generations, transmuting in form, but
not in essence—transmitted by direct instruction, through imitation, and as a consequence of
the human ability to incorporate personality features temporarily disembodied in narrative.

The "integrative conflict" of complex ideas, giving rise to the "central character of culture,"
appears as a process extending over untold centuries. This process represents itself, in
mythology, as the "battle of the gods in heaven," which Eliade has described as the "conflict
between divine generations."[335] Eliade discusses Hittite/Hurrian and Canaanite mythology
(*circa* 1740–1200 B.C.) and its relationship to similar myths in ancient Phoenicia and elsewhere. In the Hittite theogony, the relative sovereignty of the gods was determined by war
between them:

The initial episode, "Kingship in Heaven," explains the succession of the first gods. In the beginning,
Alalu was king, and Anu, the most important of the gods, bowed before him and served him. But after
nine years Anu attacked and vanquished him. Then Alalu took refuge in the subterranean world, and
Kumarbi became the new sovereign's servant. Nine years passed, and Kumarbi in his turn attacked
Anu. The latter fled, flying into the sky, but Kumarbi pursued him, caught him by the feet, and threw
him to the ground, after biting his "loins." Since he was laughing and rejoicing over his exploit, Anu
told him that he had been impregnated. Kumarbi spat out what was still in his mouth, but a part of
Anu's virility entered his body, and he became big with three gods. The rest of the text is badly mutilated, but it is presumed that Anu's "children," with Teshub, the storm god leading them, make war on
Kumarbi and dethrone him.[336]

Eliade continues, drawing upon Philo of Byblos' archaic *Phoenician History*:

The first sovereign [Phoenician] god was Elioun (in Greek, Hypistos, "The Most High"), corresponding in the Hurrian/Hittite mythology to Alalu. From his union with Bruth there came into the world
Uranus (corresponding to Anu) and Ge (Gaea). In their turn, these two engendered four sons, the first
of whom, El (or Kronos), corresponds to Kumarbi. As the result of a quarrel with his wife, Uranus tries
to destroy his progeny, but El forges a saw (or lance?) for himself, drives out his father, and becomes
the sovereign. Finally, Baal (representing the fourth generation and corresponding to Teshub and
Zeus) obtains the sovereignty; exceptionally, he obtains it without combat.

It is important to emphasize at once the "specialized" and at the same time syncretistic character of
this myth, and not only in its Hurrian/Hittite version (in which, besides, there are a number of
Sumero-Akkadian elements). The *Enuma elish* [337] likewise presents (1) a series of divine generations,
(2) the battle of the young gods against the old gods, and (3) the victory of Marduk, who thus assumes
the sovereignty.

To sum up: all the myths that recount the conflicts between successive generations of gods for the
conquest of universal sovereignty justify, on the one hand, the exalted position of the last conquering
god and, on the other hand, explain the present structure of the world and the actual condition of
humanity. [338]

The "gods" are transpersonal forces, "instinctive" and socially modified, comprising universal
elements of human experience. The organization of these gods, as a consequence of combat,
is an abstracted and poetic description of the manner in which emergent behavioral patterns
and interpretive schemas—moral positions, so to speak—fight for predominance, and therefore organize themselves, over the course of time.

The manner in which a given society has come to organize its behavioral hierarchies is
implicit in its mode of attributing to, or perceiving value in, "objects" (which is to say, *implicit in its mode of restricting the meaning manifested by objects to an acceptable range and magnitude*). The brutally organized consequence of "the battle of the gods" constitutes the tradition
that structures the intrapsychic hierarchy of values, regulates interpersonal interaction, and
keeps individual emotion in check (as the consequences of individual and social behavior,
when guided by tradition, remain predictable). A given behavior, manifested in the *absence* of
another being, does not necessarily produce the same outcome when it plays itself out in the
presence of others. Two children and one toy is not the same situation as one child and one
toy (because, in a sense, the *toy* is not the same—not from the phenomenological perspective). The behavioral tendencies of individuals undergo constant modification in the social
situation, because the fact of the society in the situation changes the motivational relevance
of all the objects in the situation. Two children with one toy have to come to an agreement,
which is mutual modification of behavior, before the toy can be what it is when encountered
alone—which is fun, rather than trouble.

The behavioral tendencies of individuals are mimicked action patterns which were originally established as a consequence of heroic behavior. The mutual interplay of action pat-

terns in the social world, however, results in their inevitable modification. Patterns of behavior—those motivated by aggression, for example, or love or fear—have a transpersonal basis, which accounts, in part, for their personification as gods (or, for their *existence* as gods, from a more liberal interpretive perspective). It is the constant clash of these gods that allows for their mutual coexistence and their social organization. A number of "gods" might operate simultaneously in the domain of a disputed toy, for example (in the unknown territory brought about by the fact of something desirable but singular in a social environment). The "god of war" (Ares, say, for the sake of argument) might emerge "within" one child, or both—in which case a fight will ensue. The winner, assuming there is one, may then be more likely to be warlike, in the future, in a social situation characterized by ambiguity. The loser might have other thoughts [may come, for example, to be dominated by *Pan(ic)* when faced with emergent toy-conflict with a stranger (may come to cry and withdraw)]. Alternatively, in the optimistic case, one or both children may *negotiate* a fair settlement, so both are satisfied, and neither hurt. The "negotiation" of a "fair settlement" presupposes that each child treats the other as an "object of value"—that is, as one who must be taken into account in the course of behavioral decisions. This *taking into account of others* is recognition of their implicit worth—their "basic human rights"—as (mythologically equivalent) community members. Such recognition, acted out before it is understood, provides the basis for the organization of societies, on a foundation other than that of force. Despite the lack of explicit understanding, however, the fact of negotiation is indicative of identity with the hero (the eternal "means to peace") as the hero is divine peacemaker, in one of his many guises. The emergence of negotiation, during time of dispute, is therefore both "spontaneous incarnation of the savior," and source of information for the derivation of stories about the nature of the hero (which are useful for future reference).

In the case of children engaged in a dispute over toys: a parent who allows the stronger child preferential access to the desired object is making the moral claim that the thing—and the aggressive desire for the thing, which may well be conflated with the thing—is something of higher value than the emotional state or physical well-being of the defeatable other. Alternatively, the parent may require the children to mediate between their competing demands without reverting to "might makes right," and to construct for themselves a hierarchy of value governing behavior in the chaotic situation defined by the mutually desirable but singular toy. It is the sum total of such interactions, conducted in once unexplored territory, hierarchically organized, that come to compose culture.

In the case of broader society: the "meaning" of an object—that is, the significance of that object for emotional regulation and behavioral output—is determined by the social consequences of behaviors undertaken and inferences drawn in its presence. Thus internal motivational forces vie for predominance under the influence of social control. The valence of erotic advances made by a given woman, for example—which is to say, whether her behavior invokes the "goddess of love" or the "god of fear"—will depend on her current position in a given social hierarchy. If she is single and acting in context, she may be considered desirable;

if she is the intoxicated wife of a large and dangerous man, by contrast, she may be placed in the category of "something best run away from quickly."

When exploration culminates in punishment, to take another example, the exploratory tendency matched to that situation will come under the inhibitory control of fear. When this subordination occurs as a consequence of the investigation of a natural object, the interpretation would be that something has been learned about the nature of the world (about that part of it which is dangerous, at any rate). The process is extended complexly in the social sphere. A motivated pattern of action (even the motivated state itself) may come under the inhibitory control of fear, because its behavioral expression within the social community results in social rejection (or other interpersonally mediated punishment). Thus it could be said that the structure of the internal motivational state reflects the consequences of behavior undertaken in the nature and social worlds—or, more particularly, *that there is an isomorphic relationship between the state of the internal representation of motivational states and the external, social world*. It is for this reason that a political state and a psychological state can be in some sense regarded as identical (and why individuals come so easily to identify with their social groupings).

The culturally determined meaning of an object—apprehended, originally, as an aspect of the object—is in fact in large part implicit information about the nature of the current dominance hierarchy, which has been partially transformed into an abstract hypothesis about the relative value of things (including the self and others). *Who owns what*, for example, determines *what things signify*, and *who owns what* is dominance-hierarchy dependent. What an object signifies is determined by the value placed upon it, manifested in terms of the (socially determined) system of promises, rewards, threats and punishments associated with exposure to, contact with, and use or misuse of that object. This is in turn determined by the affective significance of the object (its relevance, or lack thereof, to the attainment of a particular goal), in combination with its scarcity or prevalence, and the power (or lack thereof) of those who judge its nature. In keeping with this observation, the existentialist psychotherapist Ludwig Binswanger states:

All "metamorphoses of the egoistic in social instincts" and thus, properly said, all metamorphoses of evil into good drives and dispositions, occur, according to Freud, under compulsion. "Originally, i.e., in human history [such transformations occurred] only under external compulsion, but they [occurred] through the bringing into the world of hereditary dispositions for such transformations and also through their perpetuation and reinforcement 'during the life of the individual himself.'" Indeed, this whole "development" takes the direction in which external compulsion is introjected, and which, in the case of the human Super-ego, is completely absorbed. This transformation occurs, as we know, "by the admixture of erotic components": "we learn to value being loved as an advantage by virtue of which we may do without other advantages." Culture is thus "attained through the renunciation of instinctual gratifications and furthered by every new development which serves the purposes of renunciation."

In all this, we stand before the pure specimen of *homo natura*: bodily instinct, the gaining of pleasure (sacrificing a lesser for a greater gain), inhibition because of compulsion or pressures from society (the prototype being the family), a developmental history in the sense of ontogenetic and phylogenetic transformations of outer into inner compulsions, and the inheritance of these transformations.[339]

Whether a particular behavioral strategy (planned or exploratory) produces a positive or negative outcome in a particular situation depends, for social animals, on the nature of the social environment in which it is manifested. Any given "object" capable of eliciting behavior is *necessarily part of a social context, among social animals; that social context plays an important role in determining the value of the object*. It is social determination of value that helps make an object neutral, dangerous, promising or satisfying—in large part, independently of the "objective" properties of the item in question. The socially determined affective significance of the object is "naturally" experienced as an aspect of the object—which is to say that the charisma radiating from an Elvis Presley guitar is "part" of the guitar. This means that the meaning of objects in a social context *is actually information about the structure of that social context* [as well as "part" of the object (its "magic") from the mythological or narrative perspective].

Identification of the context-dependent meaning of objects in the social environment, which is determination of the behavioral patterns whose manifestation is appropriate in that situation, means encounter with cultural structure designed to bring predictability to the ongoing flow of events. Participation in the processes and representations comprising that structure (that is, *adoption of social identity*) means heightened capacity to predict behavior of self and other—and, therefore, capacity to regulate emotion through the ebb and flow of life. *Much potential unpredictability remains "constrained" by the shared identity constituting culture.* This social identity, which is a story about how things are and how they should be—"things" including the self and the other—provides the framework that constrains the otherwise unbearable *a priori* motivational significance of the ultimately unknowable experiential object. The unknown surrounds the individual, like the ocean surrounds an island, and produces affect, compels behavior, whenever it shows its terrible but promising face. Culture is constructed in spite of (in cooperation with, in deference to) this omnipresent force, and serves as a barrier, quelling emotion, providing protection against exposure to the unbearable face of God.

It is the *conservative* aspect of society that ensures that the past, as presently *reincarnated* and *remembered*, continues to serve as ultimate source of moral virtue and emotional protection. This remembered past is the mythical Father, echoed more abstractly in one "person" of the Christian Trinity. The power of the past is given due recognition in the ritual of ancestor worship, for example, which is motivated by desire to remain "in communication" with the dead (to retain the wisdom, protective power and guiding hand of the dead). Such motivation comprised a force sufficient to give impetus to the construction of megaliths—massive stone "testaments to the past"—in a geographical zone stretching from western and northern Europe, through the Middle East, into Tibet and Korea, from 4000 B.C. to the present day.[340] The megaliths, like the modern necropolises or cemeteries, are sites of the dead, monuments and aid to memory and the continuity of culture. Eliade states:

Megaliths have a relation to certain ideas concerning existence after death. The majority of them are built in the course of ceremonies intended to defend the soul during its journey into the beyond; but they also insure an internal postexistence, both to those who raise them during their own lifetime and to those for whom they are built after death. In addition, megaliths constitute the unrivaled connection between the living and the dead; they are believed to perpetuate the magical virtues of those who constructed them or for whom they were constructed, thus insuring the fertility of men, cattle, and harvests.[341]

also

By virtue of the megalithic constructions, the dead enjoy an exceptional power; however, since communication with the ancestors is ritually assured, this power can be shared by the living.... What characterizes the megalithic religions is the fact that the ideas of *perenniality* and of *continuity between life and death* are apprehended through the *exaltation of the ancestors as identified, or associated, with the stones.*[342]

What is cast in stone, so to speak, is *remembered*, and what is remembered (in the absence of permanent literate means of communication) is the value of culture, the significance of the discoveries of all those whose existence preceded the present time. The past, made metaphorically present in the form of stone, is the mythical ancestor-hero—is Osiris, the founder of the community. In traditional communities, awe-inspired imitation of the actions of that primary personage, modified by time and abstracted representation, retains primary and potent force (even in revolutionary cultures such as our own). The action of the pre-experimental man consists of ritual duplication and simultaneous observation of taboo—action bounded by custom. When such a man endeavors to produce a particular end, he follows an exemplary pattern. This pattern was established by his ancestral progenitors in a time subsuming all time, and in a "divine" (actually, communitarian-intrapsychic) space. His tradition, after all, is not merely the force of the past—it is that force, *as it is exists and is represented in the present*. What is remembered takes on representation as a pattern—as that pattern of behavior characteristic of the culture-creating "supernatural beings" who lived prior to living recollection. This pattern is traditional behavior, as established and organized by those who were capable of originating adaptation—or, it could be said, as established and organized by the immortal and central human spirit who constantly battles the fear of death and creates the conditions that promote life:

For the man of traditional societies everything significant—that is, everything creative and powerful—that has ever happened took place in the beginning, in the Time of myths.

In one sense it could almost be said that for the man of archaic societies history is "closed"; that it exhausted itself in the few stupendous events of the beginning. By revealing the different modes of deep-sea fishing to the Polynesians at the beginning of Time, the mythical Hero exhausted all the possible forms of that activity at a single stroke; since then, whenever they go fishing, the Polynesians repeat the exemplary gesture of the mythical Hero, that is, they imitate a transhuman model.

But, properly considered, this history preserved in the myths is closed only in appearance. If the man of primitive societies had contented himself with forever imitating the few exemplary gestures revealed by the myths, there would be no explaining the countless innovations he has accepted during the course of Time. No such thing as an absolutely closed primitive society exists. We know of none that has not borrowed some cultural elements from outside; none that, as the result of these borrowings, has not changed at least some aspects of its institutions; none that, in short, has had no history. But, in contrast to modern society, primitive societies have accepted all innovations as so many "revelations," hence as having a superhuman origin. The objects or weapons that were borrowed, the behavior patterns and institutions that were imitated, the myths or beliefs that were assimilated, were believed to be charged with magico-religious power; indeed, it was for this reason that they had been noticed and the effort made to acquire them. Nor is this all. These elements were adopted because it was believed that the Ancestors had received the first cultural revelations from Supernatural Beings. And since traditional societies have no historical memory in the strict sense, it took only a few generations, sometimes even less, for a recent innovation to be invested with all the prestige of the primordial revelations.

In the last analysis we could say that, though they are "open" to history, traditional societies tend to project every new acquisition into the primordial Time, to telescope all events in the same atemporal horizon of the mythical beginning.[343]

The social structure that emerges, over time, as a consequence of the "battle of the gods," might be most accurately likened to a personality (to the personality *adopted by all who share the same culture*). It is in fact the personality of "the dead heroes of the past" (the "hero as previously realized") and is most frequently symbolized by the figure of the Great Father, simultaneous personification of order and tyranny. Culture binds nature. The archetypal Great Father protects his children from chaos; holds back the precosmogonic water from which everything was derived, to which everything will return; and serves as progenitor of the hero. The protective capacity of benevolent tradition, embodied in the form of political order, constitutes a common mythological/narrative theme. This may be illustrated for our purposes through consideration and analysis of a Polish folktale: *The Jolly Tailor Who Became King*.[344] Nitechka, the hero of the story, is a simple tailor. He courageously aids a wounded gypsy—that is, acts humanely toward an outsider, a stranger, a personified "emissary of chaos." In return, the gypsy provides him with "redemptive" information—informs him that if he walks westward, he will become king. He acquires a scarecrow—"the Count"—as a companion, and has a number of adventures with him. Finally, the two travelers arrive at the town of Pacanow and observe the proceedings there—in great astonishment:

All around the town it was sunshiny and pleasant; but over Pacanow the rain poured from the sky as from a bucket.

"I won't go in there," said the Scarecrow, "because my hat will get wet."

"And even I do not wish to become King of such a wet kingdom," said the Tailor.

Just then the townspeople spied them and rushed toward them, led by the Burgomaster riding on a shod goat.

"Dear Sirs," they said, "perhaps you can help us."

"And what has happened to you?" asked Nitechka.

"Deluge and destruction threaten us. Our King died a week ago, and since that time a terrible rain has come down upon our gorgeous town. We can't even make fires in our houses, because so much water runs through the chimneys. We will perish, honorable Sirs!"

"It is too bad," said Nitechka very wisely.

"Oh, very bad! And we are most sorry for the late King's daughter, as the poor thing can't stop crying and this causes even more water."

"That makes it still worse," replied Nitechka, still more wisely.

"Help us, help us!" continued the Burgomaster. "Do you know the immeasurable reward the Princess promised to the one who stops the rain? She promised to marry him and then he will become King."

The basic plot is established. The tailor—he who clothes, mends and ties—is the hero. Although simple (poor in outward appearance, humble, willing to take risks, helpful and kind), he has the capacity to become King. He journeys to a town threatened by a deluge (by chaos, in the guise of "return of the primordial waters"). This deluge began after the recent death of the King. The king's daughter—benevolent (young, beautiful, good) counterpart to the forces of the negative feminine (the unstoppable rain)—appears willing to unite with whomever saves the kingdom. She represents the potential embedded in voluntarily confronted chaos (yet is assimilated to her primordial partner, the Great Mother, by her "rainlike" tears).

Nitechka realizes that he must bring back "pleasant weather." He ponders the situation for three long days. Finally, he is granted a revelation:

"I know where the rain comes from!"

"Where from?"

"From the sky" [that is, from "heaven"].

"Eh!" grumbled the Scarecrow. "I know that too. Surely it doesn't fall from the bottom to the top, but the other way around."

"Yes," said Nitechka, "but why does it fall over the town only, and not elsewhere?"

"Because elsewhere is nice weather."

"You're stupid, Mr. Count," said the Tailor. "But tell me, how long has it rained?"

"They say since the King died."

"So you see! Now I know everything! The King was so great and mighty that when he died and went to Heaven he made a huge hole in the sky."

"Oh, oh, true!"

The death of the King—who is the ritual model for emulation, the figure who brings order or predictability to interpersonal interaction undertaken among his subjects—means potential dissolution of security and protection. The King's death (his "return to heaven," or to the kingdom of the dead) is equivalent to the fracturing of a protective wall. The unknown, from which his subjects were protected, pours through the breached wall. The kingdom risks inundation:

"Through the hole the rain poured and it will pour until the *end of the world* [emphasis added] if the hole isn't sewed up!"

Count Scarecrow looked at him in amazement.

"In all my life I have never seen such a wise Tailor," he said.

Nitechka orders the townspeople to bring "all the ladders in the town," to "tie them together," and to "lean them against the sky." He ascends the ladder, with a hundred needles, threading one:

Count Scarecrow stayed at the bottom and unwound the spool on which there was a hundred miles of thread.

When Nitechka got to the very top he saw that there was a huge hole in the sky, a hole as big as the town. A torn piece of the sky hung down, and through this hole the water poured.

This narrative fragment is particularly interesting, as it is apparent that the water is coming, somehow, from "behind" the sky. The sky is utilized in mythology, in general, as a "masculine" symbol (at least the day sky) and tends to be assimilated to the same natural category as "the king." It appears to be damage to the *general structure* of the "masculine" sky, produced by the death of a specific king, that constitutes the breach through which *precosmogonic material* (in the form of water) is able to pour through. The "death of the king" and the "breach in the sky" is equivalent in meaning to the death of Apsu, in the *Enuma elish*—the death that heralded the reappearance of Tiamat. In this tale, however, Nitechka "repairs the structure of the sky" (an act equivalent to the reconstitution of Osiris) instead of directly battling the "dragon of chaos":

So he went to work and sewed and sewed for two days. His fingers grew stiff and he became very tired but he did not stop. When he had finished sewing he pressed out the sky with the iron and then, exhausted, went down the ladders.

Once more the sun shone over Pacanow. Count Scarecrow almost went mad with joy, as did all the other inhabitants of the town. The Princess wiped her eyes that were almost cried out, and throwing herself on Nitechka's neck, kissed him affectionately.

The "creative union" of the hero with the "benevolent aspect of the unknown" is evidently approaching.

Nitechka was very happy. He looked around, and there were the Burgomaster and Councilmen bringing him a golden scepter and a gorgeous crown and shouting:

"Long live King Nitechka! Long live he! Long live he! And let him be the Princess' husband and let him reign happily!"

So the merry little Tailor reigned happily for a long time, and the rain never fell in his kingdom.

This fairy tale constitutes a specific example of a more general type of story: that is, the story of the "god who binds."[345] The god who binds might be Marduk, who encloses Tiamat in a net given to him by his father, Anu—in which case the binding is clearly benevolent (even "world-engendering"). Binding may also be conceptualized as the prerogative of the sovereign, who binds his "enemies"—that is, those who threaten the stability of the kingdom—with cords, ropes and legal strictures. Binding brings order, in short, but too much order can be dangerous. The closing line of *The Jolly Tailor* informs us that rain never falls in the newly established kingdom. While this might sound like a happy ending to those who have been recently inundated with water, it isn't so suitable a trick if it engenders a drought. We may turn to another literary example, to illustrate this point.

In the famous children's novel *A Wrinkle in Time*, a small boy with magical powers becomes inhabited by a powerful patriarchal extraterrestrial spirit, while trying to rescue his father from "dark powers" threatening the universe. While possessed, this boy, Charles Wallace, remarks to his sister:

"You've got to stop fighting and relax. Relax and be happy. Oh, Meg, if you'd just relax you'd realize that all our troubles are over. You don't understand what a wonderful place we've come to. You see, on this planet everything is in perfect order because everybody has learned to relax, to give in, to submit. All you have to do is look quietly and steadily into the eyes of our good friend, here, for he is our friend, dear sister, and he will take you in as he has taken me."[346]

Everyone who inhabits the state dominated by "the good friend" behaves in a programmatic and identical manner. Anyone who differs is "adjusted," painfully, or eliminated. There is no space for disorder of any type:

Charles Wallace's strange, monotonous voice ground against her ears. "Meg, you're supposed to have *some* mind. Why do you think we have wars at home? Why do you think people get confused and unhappy? Because they all live their own, separate, individual lives. I've been trying to explain to you in the simplest possible way that [in this state] individuals have been done away with.... [here there is] ONE mind. It's IT. And that's why everybody is so happy and efficient...."

"Nobody suffers here," Charles intoned. "Nobody is ever unhappy."[347]

The (necessary) meaning-constraint typical of a given culture is a consequence of uniformity of behavior, imposed by that culture, toward objects and situations. The push toward uniformity is a primary characteristic of the "patriarchal" state (as everyone who acts in the same situation-specific manner has been rendered comfortably "predictable"). The state becomes increasingly tyrannical, however, as the pressure for uniformity increases. As the drive toward similarity becomes extreme, everyone becomes the "same" person—that is, imitation of the past becomes total. All behavioral and conceptual variability is thereby forced from the body politic. The state then becomes truly *static:* paralyzed or deadened, turned to stone, in mythological language. Lack of variability in action and ideation renders society and the individuals who compose it increasingly vulnerable to precipitous "environmental" transformation (that is, to an involuntary influx of "chaotic" changes). It is possible to engender a complete social collapse by constantly resisting incremental change. It is in this manner that the gods become displeased with their creation, man—and his willful stupidity—and wash away the world. The necessity for interchange of information between "known" and "unknown" means that the *state risks its own death by requiring an excess of uniformity.* This risk is commonly given narrative representation as "the senescence and frailty of the old King," or as "the King's mortal illness, brought on by lack of 'water' (which is "precosmogonic chaos," in its positive aspect)." Such "ideas" are well illustrated in the Brothers Grimm fairy tale *The Water of Life*: [348]

There once was a king who was so ill that it was thought impossible his life could be saved. He had three sons, and they were all in great distress on his account, and they went into the castle gardens and wept at the thought that he must die. An old man came up to them and asked the cause of their grief. They told him that their father was dying, and nothing could save him.

The old man said, "There is only one remedy which I know. It is the Water of Life. If he drinks of it he will recover, but it is very difficult to find."

The two eldest sons determine to seek out the Water of Life, one after the other, after gaining their father's reluctant permission. They both encounter a dwarf, at the beginning of their journeys, and speak rudely to him. The dwarf places a curse on them, and they each end up stuck fast in a mountain gorge.

The "youngest son" then sets out. He is humble and has the "right attitude" toward what he does not understand. When he encounters the dwarf, therefore—who plays out the same role as the gypsy woman in *The Jolly Tailor*—he receives some valuable information:

"As you have spoken pleasantly to me, and not been haughty like your false brothers, I will help you and tell you how to find the Water of Life. It flows from a fountain in the courtyard of an enchanted castle.[349] But you will never get in unless I give you an iron rod and two loaves of bread. With the rod strike three times on the iron gate of the castle and it will spring open. Inside you will find two lions with wide-open jaws, but if you throw a loaf to each they will be quiet. Then you must make haste to fetch the Water of Life before it strikes twelve, or the gates of the castle will close, and you will be shut in."

The story is making a point: when you don't know where you are going, it is counterproductive to assume that you know how to get there. This point is a specific example of a more general moral: Arrogant ("prideful") individuals presume they know who and what is important. This makes them too haughty to pay attention when they are in trouble—too haughty, in particular, to attend to those things or people whom they habitually hold in contempt. The "drying up of the environment" or the "senescence of the king" is a consequence of a too rigid, too arrogant value hierarchy. ("What or who can reasonably be ignored" is as much a part of such a hierarchy as "who or what must be attended too.") When trouble arrives, the traditional value hierarchy must be revised. This means that the formerly humble and despised may suddenly hold the secret to continued life[350]—and that those who refuse to admit to their error, like the "elder brothers," will inevitably encounter trouble. The story continues:

The Prince thanked him, took the rod and the loaves, and set off. When he reached the castle all was just as the dwarf had said. At the third knock the gates flew open, and when he had pacified the lions with their loaves, he walked into the castle. In the great hall he found several enchanted princes, and he took the rings from their fingers. He also took a sword and a loaf which were lying by them.

The enchanted princes might be regarded as equivalent, in an important sense, to Osiris, the "ancestral hero" whose potential lay unutilized in the underworld after his dismemberment by Seth. The enchanted princes are ancestral forces with magical powers (like the "dead kings" in the churchyard dream we discussed earlier). The young prince's voyage into the "enchanted castle" is equivalent to a voluntary descent into the dangerous kingdom of the dead. His "encounter with the dead ancestors" allows him access to some of their power (in the guise of their tools and other belongings). The young prince also encounters the "benevolent aspect of the unknown" in the underworld, as might well be expected, in her typical personification:

On passing into the next room he found a beautiful maiden, who rejoiced at his coming. She embraced him and said that he had saved her, and if he would come back in a year she would marry him. She also told him where to find the fountain with the enchanted water, but she said he must make haste to get out of the castle before the clock struck twelve.

Then he went on and came to a room where there was a beautiful bed freshly made, and as he was very tired he thought he would take a little rest. So he lay down and fell asleep. When he woke it was a quarter to twelve. He sprang up in a fright, and ran to the fountain and took some of the water in a cup which was lying nearby, and then hurried away. The clock struck just as he reached the iron gate, and it banged so quickly that it took off a bit of his heel.

He rejoiced at having got some of the Water of Life, and hastened on his homeward journey. He again passed the dwarf, who said when he saw the sword and the loaf, "Those things will be of much service to you. You will be able to strike down whole armies with the sword, and the loaf will never come to an end."

The sword and the loaf are the concrete forms taken by the "possibility" released during the prince's heroic journey into the terrible unknown. The sword is a tool which might find its use in the battle with negative forces. The loaf is magical, in the same manner as the loaves and fishes in the story of Christ's miraculous provendor:

In those days, when again a great crowd had gathered, and they had nothing to eat, he called his disciples to him, and said to them,

"I have compassion on the crowd, because they have been with me now three days, and have nothing to eat;

and if I send them away hungry to their homes, they will faint on the way; and some of them have come a long way."

And his disciples answered him, "How can one feed these men with bread here in the desert?"

And he asked them, "How many loaves have you?" They said, "Seven."

And he commanded the crowd to sit down on the ground; and he took the seven loaves, and having given thanks he broke them and gave them to his disciples to set before the people; and they set them before the crowd.

And they had a few small fish; and having blessed them, he commanded that these also should be set before them.

And they ate, and were satisfied; and they took up the broken pieces left over, seven baskets full.

And there were about four thousand people.

And he sent them away; and immediately he got into the boat with his disciples, and went to the district of Dalmanu'tha.

The Pharisees came and began to argue with him, seeking from him a sign from heaven, to test him.

And he sighed deeply in his spirit, and said, "Why does this generation seek a sign? Truly, I say to you, no sign shall be given to this generation."

And he left them, and getting into the boat again he departed to the other side.

Now they had forgotten to bring bread; and they had only one loaf with them in the boat.

And he cautioned them, saying, "Take heed, beware of the leaven of the Pharisees and the leaven of Herod."

And they discussed it with one another, saying, "We have no bread."

And being aware of it, Jesus said to them, "Why do you discuss the fact that you have no bread? Do you not yet perceive or understand? Are your hearts hardened?

Having eyes do you not see, and having ears do you not hear? And do you not remember?

When I broke the five loaves for the five thousand, how many baskets full of broken pieces did you take up?" They said to him, "Twelve."

"And the seven for the four thousand, how many baskets full of broken pieces did you take up?" And they said to him, "Seven."

And he said to them, "Do you not yet understand?" (Mark 8:1–21, RSV)

The hero provides "food that never ends."

Back to the story: the dwarf tells the prince where his brothers can be found—warning him that they have bad hearts and should be left to their fate. The young prince seeks them out, nevertheless, rescues them, and tells them everything that has happened.

Then they rode away together and came to a land where famine and war were raging. The King thought he would be utterly ruined, so great was the destitution.

The Prince went to him and gave him the loaf, and with it he fed and satisfied his whole kingdom. The Prince also gave him his sword, and he smote the whole army of his enemies with it, and then he was able to live in peace and quiet. Then the Prince took back his sword and his loaf, and the three brothers rode on.

But later they had to pass through two more countries where war and famine were raging, and each time the Prince gave his sword and his loaf to the King and in this way he saved three kingdoms.

The tale takes this diversion to help drive home the general utility of what has been rescued from the "enchanted kingdom, where the princess dwells." The treasures released from that kingdom have a powerful, protective, revitalizing capacity, no matter where they are applied.

The brothers continue homeward, but the older two deceive the younger on the voyage, exchanging the true Water of Life for salt sea water (the "arrogant elder brothers" replace the "benevolent aspect of the Great Mother" with her "destructive counterpart"). When he arrives at home, the younger son unwittingly gives this poisonous water to his father, making him sicker. The older brothers then heal the poisoned king with the genuine but stolen water, masking their evil souls with the appearance of benevolence, and arrange to have their unfortunate sibling banished and killed. The huntsman assigned to do the killing cannot bring himself to do it, however, and allows the young prince to escape. Then the tide starts to turn. The previous generous exploits of the young prince are revealed, and the old king repents:

After a time three wagonloads of gold and precious stones came to the King for his youngest son. They were sent by the kings who had been saved by the Prince's sword and miraculous loaf, and who now wished to show their gratitude.

Then the old King thought, "What if my son really was innocent?" And he said to his people, "If only he were still alive! How sorry I am that I ordered him to be killed."

"He is still alive," said the huntsman. "I could not find it in my heart to carry out your commands." And he told the King what had taken place.

A load fell from the King's heart on hearing the good news, and he sent out a great proclamation to all parts of his kingdom that his son was to come home, where he would be received with great favor.

In the meantime, the princess is preparing for the return of the prince. She

had caused a road to be made of pure shining gold leading straight to her castle, and told her people that whoever came riding straight along it would be her true bridegroom, and they were to admit him. But anyone who came either on one side of the road or the other would not be the right one, and he was not to be let in.

When the year had almost passed, the eldest Prince thought that he would hurry to the Princess, and by giving himself out as her deliverer would gain a wife and a kingdom as well. So he rode away, and when he saw the beautiful golden road he thought it would be a thousand pities to ride upon it, so he turned aside and rode to the right of it. But when he reached the gate the people told him that he was not the right bridegroom, and he had to go away.

Soon after the second Prince came, and when he saw the golden road he thought it would be a thousand pities for his horse to tread upon it, so he turned and rode up on the left of it. But when he reached the gate he also was told that he was not the true bridegroom, and like his brother was turned away.

The two elder princes are too bound up in their traditional thoughts of power, wealth and glory to concentrate on what is of true importance. Because of their "great respect" for the gold that makes up the road, they miss a great opportunity. Their overarching admiration for material goods blinds them to the possibility of establishing a relationship with the source of all good things—in the guise of the princess (playing a "part" similar to that of the Wisdom of Solomon). The youngest son makes no such mistake:

When the year had quite come to an end, the third Prince came out of the wood to ride to his beloved, and through her to forget all his past sorrows. So on he went, thinking only of her and wishing to be with her, and he never even saw the golden road. His horse cantered right along the middle of it, and when he reached the gate it was flung open and the Princess received him joyfully, and called him her deliverer and the lord of her kingdom. Their marriage was celebrated without delay and with much rejoicing. When it was over, she told him that his father had called him back and forgiven him. So he went to him and told him everything: how his brothers had deceived him, and how they had forced him to keep silence. The old King wanted to punish them, but they had taken a ship and sailed away over the sea, and never came back as long as they lived.

The old king is dying for lack of water. He has two elder sons, who could rescue him, but they are narrow-minded, traditional, materialistic, selfish and rigid. They lack proper "spirit" for the quest. The youngest son, a proper hero, pays attention to what the "sensible" ignore, makes a voyage into the unknown, and brings back what is needed. It is the journey of the hero that revitalizes the king. Osiris languishes in the underworld—regardless of past greatness—without Horus.

It was the emergence of the heroic stance, mythically represented by man as equal in divinity to the unknown or Nature, that provided the precondition for the generation of concrete behavioral adaptations to the world of experience. Emergence of heroism meant construction of culture: historically determined procedural knowledge and communicable description thereof. Construction of culture is creation of the mythic Great and Terrible Father, tyrant and wise king, as intermediary between the vulnerable individual and the overwhelming natural world. This Father is the *consequence* of voluntary heroic action—temporally summed and integrated effect of creative exploratory behavior—as well as *progenitor* of those who take heroic action. This paradoxical child and father of the hero is *primarily* "personality" (procedure) and only *secondarily* abstracted first- and second-order representation thereof (and is most certainly not cumulative description of the "objective" world). That this is so can be seen, even today, when the members of totalitarian cultures such as the modern North Korean collapse into genuine hysteria as a consequence of the death of their leader, who is embodiment of order and determinate meaning. Such tendencies are not restricted to those dominated by the totalitarian, either. Frye states:

The function of the king is primarily to represent, for his subjects, the unity of their society in an individual form. Even yet Elizabeth II can draw crowds wherever she appears, not because there is anything remarkable about her appearance, but because she dramatizes the metaphor of society as a single "body." Other societies have other figures, but there seems to be a special symbolic eloquence, even a pathos, about the *de jure* monarch, whose position has been acquired by the pure accident of birth, and who has no executive power. At the same time most societies have done away with monarchical figures; "charismatic" leaders, dictators, and the like are almost invariably sinister and regressive; the mystique of royalty that Shakespeare's plays take for granted means little to us now; and theologians talking about the "sovereignty" of God risk alienating their readers by trying to assimilate the religious life to the metaphors of a barbaric and outmoded form of social organization. It is natural that our news media should employ the royal metaphor so incessantly in telling us about what France or Japan or Mexico "is" doing, as though they were individual beings. But the same figure was used in my younger days, to my own great annoyance, to boost the prestige of dictators: "Hitler is building roads across Germany," "Mussolini is draining the marshes in Italy," and the like. Those who employed this figure were often democratic people who simply could not stop themselves from using the royal metaphor. It seems as though the sovereign may be either the most attractive of icons or the most dangerous of idols.[351]

The Great Father is a product of history—or, is history itself, insofar as it is acted out and spontaneously remembered—intrapsychically instantiated during the course of socialization, and embedded in the social interactions and specific object-meanings that make up a given culture. This culturally determined structure—this inhibitory network, this *intrapsychic representative of the social unit*—provides experiential phenomena with determinate significance. This determinate significance is restricted meaning—reduced from the general meaning of the unknown, *per se*, to the particular—and not relevance or import added to a neutral background. The unknown manifests itself in an intrinsically meaningful manner: a manner composed of threat and promise. The specific meaning of objects discriminated from the unknown consists of restrictions of that general significance (often, of restrictions to zero—to irrelevance). Such restriction is, however, purely conditional, and remains intact only as long as the culturally determined model of meaning itself maintains its functional utility (including credibility). "Maintains its functional utility" means insofar as the culture posits a reasonable current description, a believable end goal, and a workable mode of transforming the former into the latter (workable for the individual, and for the maintenance and expansion of the culture itself).

Figure 41: Order, the Great Father, as Son of the Uroboros [352] schematically portrays the Great Father as masculine offspring of precosmogonic chaos; as embodiment of the known, the predictable, the familiar; as security and tyranny simultaneously. The Great Father is

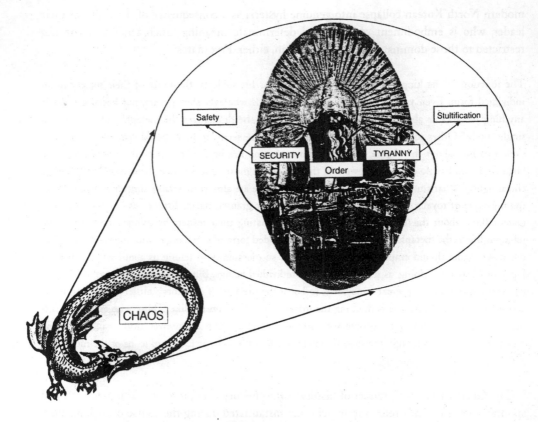

Figure 41: Order, the Great Father, as Son of the Uroboros

patriarchal society, tradition, pomp and circumstance, military-industrial complex, and superego: demanding, rigid, unjust, dangerous and necessary. He is ambivalent in precisely the same manner as the Great Mother, his "wife." In the guise of literal father, he is protection for children, who are too immature and vulnerable to deal with the unknown. More abstractly, he is the pattern of behavior the father represents, that becomes "internalized" during maturation. The Great Father takes the infinite possibility of spirit that the infant represents and forges it into something limited but actual. He is manner incarnate, ruling all social interactions.

Figure 42: Explored Territory as Orderly, Protective Father [353] presents the Great Father as wise king, *security*. The wise king maintains stability, not because he is afraid of the unknown, but because nothing new can be built without a strong foundation. He is the adaptive routine, developed by the heroes of the past, whose adoption by those in the present allows for control and safety. He is a house with doors; a structure that shelters, but does not stifle; a master who teaches and disciplines but does not indoctrinate or crush. He represents the tradition fostering cooperation among people whose shared culture makes trust possible,

Figure 42: Explored Territory as Orderly, Protective Father

even easy. The Great Father as Wise King keeps one foot on the Terrible Mother; the monsters of chaos are locked up in his dungeon or banished to the nether regions of the kingdom. He is the personality of dead heroes (that is, the action patterns and hierarchies of value established through exploration in the past) organized according to the principle of "respect for the intrinsic value of the living." This makes him the king who takes advice from his subjects—who is willing to enter into creative interchange with those he "dominates" legally—and to benefit from this advice from the "unworthy."

Knowledge of the necessity for such interplay between strong and weak emerged into explicit Western consciousness not least through the actions of the ancient Hebrew prophets. The philosopher of religion Huston Smith draws two examples from the Bible to illustrate this point:

One is the story of Naboth who, because he refused to turn over his family vineyard to King Ahab, was framed on false charges of blasphemy and subversion and then stoned; as blasphemy was a capital crime, his property then reverted to the throne. When news of this travesty reached Elijah, the word of the Lord came to him, saying,

["Arise, go down to meet Ahab king of Israel, who is in Sama'ria; behold, he is in the vineyard of Naboth, where he has gone to take possession.

And you shall say to him, 'Thus says the LORD, "Have you killed, and also taken possession?"' and you shall say to him, 'Thus says the LORD: "In the place where dogs licked up the blood of Naboth shall dogs lick your own blood."'" (1 Kings 21:18, 19 RSV)]

The story carries revolutionary significance for human history, for it is the story of how someone without official position took the side of a wronged man and denounced a king to his face on grounds of injustice. One will search the annals of history in vain for its parallel. Elijah was not a priest. He had no formal authority for the terrible judgment he delivered. The normal pattern of the day would have called for him to be struck down by bodyguards on the spot. But the fact that he was "speaking for" an authority not his own was so transparent that the king accepted Elijah's pronouncement as just.

The same striking sequence recurred in the incident of David and Bathsheba. From the top of his roof David glimpsed Bathsheba bathing and wanted her. There was an obstacle, however: she was married. To the royalty of those days this was a small matter; David simply moved to get rid of her husband. Uriah was ordered to the front lines, carrying instructions that he be placed in the thick of the fighting and support withdrawn so he would be killed. Everything went as planned; indeed, the procedure seemed routine until Nathan the prophet got wind of it. Sensing immediately that "the thing that David had done displeased the Lord," he went straight to the king, who had absolute power over his life, and said to him:

[Thus says the LORD, the God of Israel, "I anointed you king over Israel, and I delivered you out of the hand of Saul;

and I gave you your master's house, and your master's wives into your bosom, and gave you the house of Israel and of Judah; and if this were too little, I would add to you as much more.

Why have you despised the word of the LORD, to do what is evil in his sight? You have smitten Uri'ah the Hittite with the sword, and have taken his wife to be your wife, and have slain him with the sword of the Ammonites.

Now therefore the sword shall never depart from your house, because you have despised me, and have taken the wife of Uri'ah the Hittite to be your wife."

Thus says the LORD, "Behold, I will raise up evil against you out of your own house; and I will take your wives before your eyes, and give them to your neighbor, and he shall lie with your wives in the sight of this sun.

For you did it secretly; but I will do this thing before all Israel, and before the sun."

David said to Nathan, "I have sinned against the LORD." And Nathan said to David, "The LORD also has put away your sin; you shall not die.

Nevertheless, because by this deed you have utterly scorned the LORD, the child that is born to you shall die." (2 Samuel 12:7–14)]

The surprising point in each of these accounts is not what the kings do, for they were merely exercising the universally accepted prerogatives of royalty in their day. The revolutionary and unprecedented fact is the way the prophets challenged their actions.[354]

Smith concludes:

Stated abstractly, the Prophetic Principle can be put as follows: The prerequisite of political stability is social justice, for it is in the nature of things that injustice will not endure. Stated theologically, this point reads: God has high standards. Divinity will not put up forever with exploitation, corruption and mediocrity.[355]

The initially "undeclarable" constraint of "respect for the weaker" provides the precondition for the emergence of abstract and statable principles of social justice. Societies that lack such constraint or that come, over time, to forget the necessity of such constraint risk the "vengeance of God":

Thus says the LORD: "For three transgressions of Moab, and for four, I will not revoke the punishment; because he burned to lime the bones of the king of Edom.

So I will send a fire upon Moab, and it shall devour the strongholds of Ker'ioth, and Moab shall die amid uproar, amid shouting and the sound of the trumpet;

I will cut off the ruler from its midst, and will slay all its princes with him," says the LORD.

Thus says the LORD: "For three transgressions of Judah, and for four, I will not revoke the punishment; because they have rejected the law of the LORD, and have not kept his statutes, but their lies have led them astray, after which their fathers walked.

So I will send a fire upon Judah, and it shall devour the strongholds of Jerusalem."

Thus says the LORD: "For three transgressions of Israel, and for four, I will not revoke the punishment; because they sell the righteous for silver, and the needy for a pair of shoes—

they that trample the head of the poor into the dust of the earth, and turn aside the way of the afflicted; a man and his father go in to the same maiden, so that my holy name is profaned;

they lay themselves down beside every altar upon garments taken in pledge; and in the house of their God they drink the wine of those who have been fined.

"Yet I destroyed the Amorite before them, whose height was like the height of the cedars, and who was as strong as the oaks; I destroyed his fruit above, and his roots beneath.

Also I brought you up out of the land of Egypt, and led you forty years in the wilderness, to possess the land of the Amorite.

And I raised up some of your sons for prophets, and some of your young men for Nazirites. Is it not indeed so, O people of Israel?" says the LORD.

"But you made the Nazirites drink wine, and commanded the prophets, saying, 'You shall not prophesy.'

"Behold, I will press you down in your place, as a cart full of sheaves presses down.

Flight shall perish from the swift, and the strong shall not retain his strength, nor shall the mighty save his life;

Figure 43: Explored Territory as Tyrannical Father

he who handles the bow shall not stand, and he who is swift of foot shall not save himself, nor shall he who rides the horse save his life; and he who is stout of heart among the mighty shall flee away naked in that day," says the LORD. (Amos 2:1–16 RSV)

Such societies are *tyrannical*. Tyrannical societies violate the implicit principles upon which society itself is founded. This renders them inevitably self-defeating.[356]

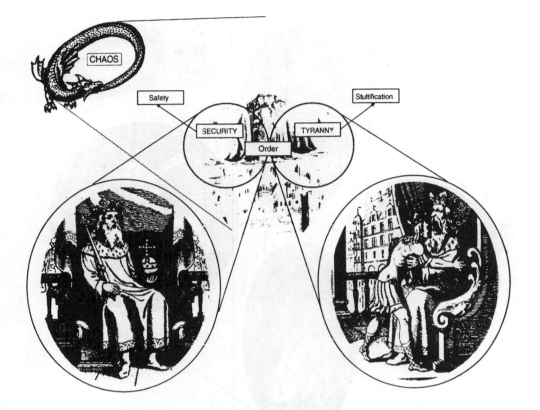

Figure 44: The "Heavenly Genealogy" of the Tyrannical and Protective Fathers

Figure 43: Explored Territory as Tyrannical Father [357] presents the forces of tradition as "son-devouring king." The conservative tendency of any culture, striving to maintain itself, can easily transform into the deadening weight of absolute authority. The Great Father as tyrant destroys what he once was and undermines what he still depends upon. The tyrant is the force of everything that has been, including everything that once was good, against everything that could be. This is the aspect of the Great Father that motivates adolescent rebellion and gives rise to ideological narratives attributing to society everything that produces the negative in man. It is the Tyrannical Father who consumes his own children, and who walls up the virgin princess in an inaccessible place. The Tyrannical Father rules absolutely, while the kingdom withers or becomes paralyzed; his decrepitude and age are matched only by his arrogance, inflexibility and blindness to evil. He is the personification of the authoritarian or totalitarian state, whose "goal" is reduction of all who are currently living to manifestation of a single dead "past" personality. When everyone is the same, everything is predictable; all things are of strictly determinable value, and everything unknown (and fear-provoking) is hidden

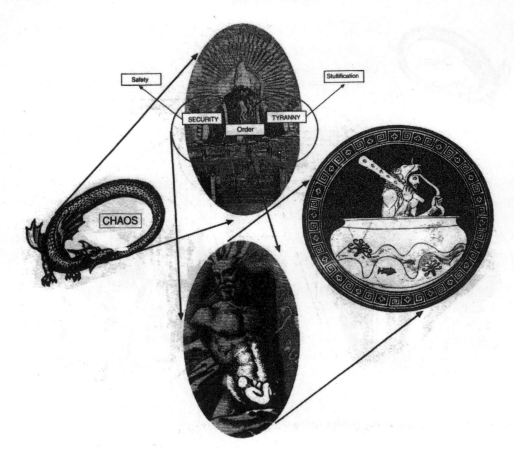

Figure 45: The Exploratory Hero as Son of the Great Father

from view. Unfortunately, of course, every unpredictable and fear-provoking thing is also informative, and new information is vital to continued successful adjustment.

The Great Father in his dual guise is the taboo, the barrier set up against the intrusion of the dangerously unpredictable, the floodgate that controls the ocean behind. He is protection for fools, and impediment to genius, and precondition for genius, and punishment for fools. His ambivalence is unavoidable, and should be recognized, for such recognition serves as effective antidote to naive ideologically-motivated utopian thought. Anything that protects and fosters (and that is therefore predictable and powerful) necessarily has the capacity to smother and oppress (and may manifest those capacities, unpredictably, in any given situation). No static political utopia is therefore possible—and the kingdom of God remains spiritual, not worldly. Recognition of the essentially ambivalent nature of the predictable—stultifying but secure—means discarding simplistic theories which attribute the existence of human suffering and evil purely to the state, or which presume that the state is all that is good, and that the individual should exist merely as subordinate or slave. The king is a wall. Walls provide a barrier to the sudden influx of the unknown, and block progress forward. One function presupposes the other (although either may certainly come to dominate). *Figure 44: The "Heavenly Genealogy" of the Tyrannical and Protective Fathers* portrays the relationship between the two discriminable aspects of the known, their derivation from the unified but ambivalent known, and their original "descent" from the "dragon of chaos."

The Great Father is order, vs. chaos; the past, vs. the present; the old, vs. the young. He is the ancestral spirit whose force extends beyond the grave, who must be kept at bay with potent and humble ritual. He is the single personality composed of the consequences of the eternal war between all the great heroes of the past, and he stands over the developing individual, in the guise of the actual father, like a god. The Great Father is the old emperor, dangerously out of date—a powerful warrior in his youth, now under the spell of a hostile force. He is the eternal impediment to the virgin bride, the tyrannical father who wishes to keep his fruitful daughter firmly under his control. He is the authoritarian who rules the land ravaged by drought; keeper of the castle in which everything has been brought to a standstill.

The Great Father is protection and necessary aid to growth, but absolute identification with his personality and force ultimately destroys the spirit. Culture, career and role do not sufficiently exhaust the possibilities of the individual. *Figure 45* therefore portrays *The Exploratory Hero—scion of chaos and order—as Son of the Great Father.* [358]

3

APPRENTICESHIP AND ENCULTURATION

Adoption of a Shared Map

———————————◯———————————

Ideologies may be regarded as incomplete myths—as partial stories, whose compelling nature is a consequence of the appropriation of mythological ideas. The philosophy attributing individual evil to the pathology of social force constitutes one such partial story. Although society, the Great Father, has a tyrannical aspect, he also shelters, protects, trains and disciplines the developing individual—and places necessary constraints on his thought, emotion and behavior.

Subjugation to lawful authority might more reasonably be considered in light of the metaphor of the apprenticeship. Childhood dependency must be replaced by group membership, prior to the development of full maturity. Such membership provides society with another individual to utilize as a "tool," and provides the maturing but still vulnerable individual with necessary protection (with a group-fostered "identity"). The capacity to abide by social rules, regardless of the specifics of the discipline, can therefore be regarded as a necessary transitional stage in the movement from childhood to adulthood.

Discipline should therefore be regarded as a skill that may be developed through adherence to strict ritual, or by immersion within a strict belief system or hierarchy of values. Once such discipline has been attained, it may escape the bounds of its developmental precursor. It is in this manner that true freedom is attained. It is at this level of analysis that all genuine religious and cultural traditions and dogmas are equivalent, regardless of content: they are all masters whose service may culminate in the development of self-mastery, and consequent transcendence of tradition and dogma.

Apprenticeship is necessary, but should not on that account be glamorized. Dogmatic systems make harsh and unreasonable masters. Systems of belief and moral action—and those people who are identified with them—are concerned above all with self-maintenance and preservation of predictability and order. The (necessarily) conservative tendencies of great systems makes them tyrannical, and more than willing to crush the spirit of those they "serve." Apprenticeship is a precursor to freedom, however, and nothing necessary and worthwhile is without danger.

Adoption of this analytic standpoint allows for a certain moral relativism, conjoined with an absolutist higher-order morality. The particulars of a disciplinary system may be somewhat unimportant. The fact that adherence to such a system is necessary, however, cannot be disregarded.

We are all familiar with the story of benevolent nature, threatened by the rapacious forces of the corrupt individual and the society of the machine. The plot is solid, the characters believable, but Mother Nature is also malarial mosquitoes, parasitical worms, cancer and Sudden Infant Death Syndrome. The story of peaceful and orderly tradition, undermined by the incautious and decadent (with the ever-present threat of chaos lurking in the background) is also familiar, and compelling, and true—except that the forces of tradition, however protective, tend to be blind, and to concern themselves more with their own stability than with the well-being of those subject to them. We have all heard and identified with the story of the brave pioneer, additionally—plough in hand, determined to wrest the good life and the stable state from the intransigent forces of nature—although we may be sporadically aware that the "intransigent forces" shaped so heroically included the decimated original inhabitants of our once-foreign landscape. We all know, finally, the story of the benevolent individual, genuine and innocent, denied access to the nourishing forces of the true and natural world, corrupted by the unreasonable strictures of society. This tale has its adherents, as well—not least because it is reassuring to believe that everything "bad" stems from without, rather than within.

These stories are all *ideologies* (and there are many more of them). Ideologies are attractive, not least to the educated modern mind—credulous, despite its skepticism—particularly if those who embody or otherwise promote them allow the listener every opportunity to identify with the creative and positive characters of the story, and to deny their association with the negative. Ideologies are powerful and dangerous. Their power stems from their incomplete but effective appropriation of mythological ideas. Their danger stems from their attractiveness, in combination with their incompleteness. Ideologies tell only part of the story, but tell that part as if it were complete. This means that they do not take into account vast domains of the world. It is incautious to act in the world as if only a set of its constituent elements exist. The ignored elements conspire, so to speak, as a consequence of their repression, and make their existence known, inevitably, in some undesirable manner.

Knowledge of the grammar of mythology might well constitute an antidote to ideological gullibility. Genuine myths are capable of representing the totality of conflicting forces, operating in any given situation. Every positive force has its omnipresent and eternal "enemy." The beneficial aspect of the "natural environment" is therefore properly viewed in light of its capacity to arbitrarily inflict suffering and death. The protective and sheltering capacity of society is therefore understood in light of its potent tendency to tyranny and the elimination of necessary diversity. The heroic aspect of the individual is regarded in light of the ever-lurking figure of the adversary: arrogant, cowardly and cruel. A story accounting for all of these "constituent elements of reality" is balanced and stable, in contrast to an ideology—and far less likely to produce an outburst of social psychopathology. But the forces that make up the world as a forum for action constantly war in opposition. How is it possible to lay a path between them, so to speak—to configure a mode of being that takes "all things" into account, without being destroyed in the process? A developmental account of the relationship between "the forces of the individual, society and chaos" might aid in the comprehension of their proper interplay.

I counseled an immature thirty-something-year-old man at one point during my service as a psychological intern. He was always working at cross-purposes to himself, placing obstacles in his path and then tripping over them. (This was the literal truth, upon occasion. He was living with his mother, after the failure of his marriage. I suggested that he start cleaning up his life by cleaning up his room—which is a more difficult step than might be casually presupposed for someone habitually and philosophically undisciplined. He placed a vacuum cleaner in the doorway of his bedroom, after getting about half-way through the task. For a week he had to step over it, but he didn't move it, and he didn't finish the job. That situation could reasonably be regarded as a polysemic sample of his life.) This person had sought help because his disintegrated marriage had produced a son, whom he loved (or at least wanted to love). He came to therapy because he didn't want his child to grow up badly, as he had. I tried to scare him into behaving properly, because I believed (and believe) that terror is a great and underutilized motivator. (Anxiety—which is ineradicable—can work against you, or for you). We spent a long time outlining, in great detail, the consequences of his undisciplined behavior, to that point in his life (no successful career, no intimate relationship, an infant son thrust into a broken family) and the likely long-term future results (increasing self-disgust, cynicism about life, increased cruelty and revenge-seeking, hopelessness and despair). We also discussed the necessity for discipline—that is, for adherence to a coherent and difficult moral code—for himself and for his son.

Of course, he worried that any attempt on his part to shape the behavior of his son would interfere with the natural development and flowering of the child's innate potential. So it might be said, using Jung's terminology, that he was an "unconscious exponent"[359] of the philosophy of Rousseau:

With what simplicity I should have demonstrated that man is by nature good, and that only our institutions have made him bad![360]

That is—the Rousseau who repeatedly placed his own children in foundling asylums because their existence was inconvenient to him (and, we must presuppose, detrimental to the flowering of his intrinsic goodness). Anyway, the fervent hope of every undisciplined person (even an undisciplined genius) is that his current worthlessness and stupidity is someone else's fault. If—in the best of cases—it is *society's* fault, then society can be made to pay. This sleight-of-hand maneuver transforms the undisciplined into the admirable rebel, at least in his own eyes, and allows him to seek unjustified revenge in the disguise of the revolutionary hero. A more absurd parody of heroic behavior can hardly be imagined.

One time my client came to me with a dream:

My son was asleep in his crib inside a small house. Lightning came in through his window, and bounced around inside the house. The lightning was powerful, and beautiful, but I was afraid it would burn the house down.

Dream interpretation is a difficult and uncertain business, but I believed that this image was interpretable within the context of our ongoing discussions. The lightning represented the potential implicit in the infant. This potential was an exceedingly strong and useful force—like electricity. But electricity is only useful when harnessed. Otherwise it burns down houses.

I can't say much about the outcome of this particular case, as internship contact with those seeking psychological help tends to be restricted in time. My client seemed, at least, more negatively affected by his immature behavior, which struck me as a reasonable start; furthermore, he understood (at least explicitly, although not yet procedurally) that discipline could be the father of the hero, and not just his enemy. The dawning of such understanding meant the beginnings of a mature and healthy philosophy of life, on his part. Such a philosophy was outlined in explicit detail by Friedrich Nietzsche, despite his theoretically "antidogmatic" stance.

Nietzsche has been casually regarded as a great enemy of Christianity. I believe, however, that he was consciously salutary in that role. When the structure of an institution has become corrupt—particularly according to its own principles—it is the act of a friend to criticize it. Nietzsche is also viewed as fervid individualist and social revolutionary—as the prophet of the superman, and the ultimate destroyer of tradition. He was, however, much more sophisticated and complex than that. He viewed the "intolerable discipline" of the Christian church, which he "despised," as a necessary and admirable precondition to the freedom of the European spirit, which he regarded as not yet fully realized:

Every morality is, as opposed to *laisser aller*, a bit of tyranny against "nature"; also against "reason"; but this in itself is no objection, as long as we do not have some other morality which permits us to decree that every kind of tyranny and unreason is impermissible. What is essential and inestimable in every morality is that it constitutes a long compulsion: to understand Stoicism or Port-Royal or Puritanism, one should recall the compulsion under which every language so far has achieved strength and freedom—the metrical compulsion of rhyme and rhythm.

How much trouble the poets and orators of all peoples have taken—not excepting a few prose writers today in whose ear there dwells an inexorable conscience—"for the sake of some foolishness," as utilitarian dolts say, feeling smart—"submitting abjectly to capricious laws," as anarchists say, feeling "free," even "free-spirited." But the curious fact is that all there is or has been on earth of freedom, subtlety, boldness, dance, and masterly sureness, whether in thought itself or in government, or in rhetoric and persuasion, in the arts just as in ethics, has developed only owing to the "tyranny of such capricious laws"; and in all seriousness, the probability is by no means small that precisely this is "nature" and "natural"—and *not* that *laisser aller*.

Every artist knows how far from any feeling of letting himself go his "most natural" state is—the free ordering, placing, disposing, giving form in the moment of "inspiration"—and how strictly and subtly he obeys thousandfold laws precisely then, laws that precisely on account of their hardness and determination defy all formulation through concepts (even the firmest concept is, compared with them, not free of fluctuation, multiplicity and ambiguity).

What is essential "in heaven and on earth" seems to be, to say it once more, that there should be *obedience* over a long period of time and in a *single* direction: given that, something always develops, and has developed, for whose sake it is worthwhile to live on earth; for example, virtue, art, music, dance, reason, spirituality—something transfiguring, subtle, mad, and divine. The long unfreedom of the spirit, the mistrustful constraint in the communicability of thoughts, the discipline thinkers imposed on themselves to think within the directions laid down by a church or court, or under Aristotelian presuppositions, the long spiritual will to interpret all events under a Christian schema and to rediscover and justify the Christian god in every accident—all this, however forced, capricious, hard, gruesome, and antirational, has shown itself to be the means through which the European spirit has been trained to strength, ruthless curiosity, and subtle mobility, though admittedly in the process an irreplaceable amount of strength and spirit had to be crushed, stifled, and ruined (for here, as everywhere, "nature" manifests herself as she is, in all her prodigal and indifferent magnificence, which is outrageous but noble).

That for thousands of years European thinkers thought merely in order to prove something—today, conversely, we suspect every thinker who "wants to prove something"—that the conclusions that *ought* to be the result of their most rigorous reflection were always settled from the start, just as it used to be with Asiatic astrology, and still is today with the innocuous Christian-moral interpretation of our most intimate personal experiences "for the glory of God" and "for the salvation of the soul"—this tyranny, this caprice, this rigorous and grandiose stupidity has *educated* the spirit. Slavery is, as it seems, both in the cruder and in the more subtle sense, the indispensable means of spiritual discipline and cultivation, too. Consider any morality with this in mind: what there is in it of "nature" teaches hatred of the *laisser aller*, of any all-too-great freedom, and implants the need for limited horizons and the nearest tasks—teaching the *narrowing of our perspective*, and thus in a certain sense stupidity, as a condition of life and growth.

"You shall obey—someone and for a long time: *else* you will perish and lose the last respect for yourself"—this appears to me to be the categorical imperative of nature which, to be sure, is neither "categorical" as the old Kant would have it (hence the "else") nor addressed to the individual (what do individuals matter to her?), but to peoples, races, ages, classes—but above all to the whole human animal, to *man*.[361]

This is the *philosophy of apprenticeship*—useful for conceptualizing the necessary relationship between subordination to a potent historically constructed social institution and the eventual development of true freedom.

A child cannot live on its own. Alone, it drowns in possibility. The unknown supersedes individual adaptive capacity, in the beginning. It is only the transmission of historically determined behavioral patterns—and, secondarily, their concomitant descriptions—that enables survival in youth. These patterns of behavior and hierarchies of value—which children mimic and then learn expressly—give secure structure to uncertain being. It is the group, initially in parental guise, that stands between the child and certain psychological catastrophe. The depression, anxiety and physical breakdown that is characteristic of too early childhood separation from parents is the result of exposure to "too much unknown" and incorporation of "too little cultural structure." The long period of human dependency must be met with the provision of a stable social environment—with predictable social interactions, which meet individual motivational demands; with the provision of behavioral patterns and schemas of value capable of transforming the unpredictable and frightening unknown into its beneficial equivalent. This means that transformation of childhood dependency entails adoption of ritual behavior (even regular meal-and-bed-times are rituals) and incorporation of a morality (a framework of reference) with an inevitably metaphysical foundation.

Successful transition from childhood to adolescence means *identification with the group*, rather than continued dependency upon the parents. Identification with the group provides the individual with an alternative, generalized, nonparental source of protection from the unknown, and provides the group with the resources of another soul. The group constitutes a historically validated pattern of adaptation (specific behaviors, descriptions of behavior and general descriptions). The individual's identification with this pattern strengthens him when he needs to separate from his parents and take a step toward adulthood, and it strengthens the group, insofar as it now has access to his individual abilities. The individual's identification with this pattern bolsters his still-maturing ability to stand on his own two feet—supports his determination to move away from the all-encompassing and too secure maternal-dependent world. Identity with the group therefore comes to replace recourse to parental authority as "way of being in the face of the unknown." It provides structure for social relationships (with self and others), determines the meaning of objects, provides desirable end as ideal, and establishes acceptable procedure (acceptable mode for the "attainment of earthly paradise").

Personal identification with the group means socialization, individual embodiment of the valuations of the group—primarily, as expressed in behavior. Group values constitute cumulative historical judgment rendered on the relative importance of particular states of motivation, with due regard for intensity, as expressed in individual action, in the social context. All societies are composed of individuals whose actions constitute embodiment of the creative past. That creative past can be conceptualized as the synthesis of all culture-creating exploratory communicative activity, including the act of synthesis itself.

Myth comprises description of procedural knowledge; constitutes episodic/semantic representation of cumulative behavioral wisdom, in increasingly abstracted form. Introduction of the previously dependent individual at adolescence to the world of ancestral behavior and myth constitutes *transmission* of culture—inculcation of the Great Father, historically determined personality and representation of such—as adaptation to, explanation of and protection against the unknown, the Great and Terrible Mother. This introduction reaches its culmination with initiation, the primary ritual signifying cultural transmission—the event which destroys the "unconscious" union between child and biological mother.

The child is born in a state of abject dependence. The caring mother is simultaneously individual force and embodiment of impersonal biological beneficence—the eternal mythic virgin mother, material consort of God. The infant comes equipped with the ability to respond to this innately nurturing presence, to develop a symbiotic relationship with his or her caregiver, and to grow increasingly strong. The maturation of creative exploratory capacity, which constitutes the basis for mature self-reliance, appears dependent for its proper genesis upon the manifestation of maternal solicitude: upon love, balanced promotion of individual ability and protection from harm. Tender touch and care seduce the infant to life, to expansion of independence, to potential for individual strength and ability.[362] The absence of such regard means failure to thrive, depression and intrapsychic damage, even death.[363]

The maturing individual necessarily (tragically, heroically) expands past the domain of paradisal maternal protection, in the course of development; necessarily attains an apprehension whose desire for danger and need for life exceeds the capability of maternal shelter. This means that the growing child eventually comes to face problems—how to get along with peers in peer-only play groups; how to select a mate from among a myriad of potential mates—that cannot be solved (indeed, may be made more difficult) by involvement of the beneficial maternal. Such problems might be regarded as emergent consequences of the process of maturation itself; of the increased possibility for action and comprehension necessarily attendant upon maturation. A four-year-old, making the transition to kindergarten, cannot use a three-year-old's habits and schemas of representation to make his way in the novel social world. A thirteen-year-old cannot use a seven-year-old's personality—no matter how healthy—to solve the problems endemic to adolescence. The group steps in—most evidently, at the point of adolescence—and provides "permeable" protective shelter to the child too old for the mother but not old enough to stand alone. The universally disseminated rituals of initiation—induced "spiritual" death and subsequent rebirth—catalyzes the development of adult personality; follows the fundamental pattern of the cyclic, circular cosmogonic myth of the way. The culturally determined rites and biological processes associated with initiation constitute absolute destruction of childhood personality, of childhood dependence—initial unselfconscious "paradisal" stability—for necessary catalysis of group identification. Such rituals tend to be more complex and far-reaching for males than for females. This is perhaps in part because male development seems more easily led astray, in a socially harmful manner, than female (adolescent males are more delinquent and aggressive[364]) and in part because female transition to adulthood is catalyzed "by nature" in the form of comparatively rapid maturation and the naturally dramatic onset of menstruation.

The group to which the initiate is introduced consists of a complex interweaving of behavioral patterns established and subsequently organized in the past, as a consequence of voluntary creative communicative exploration. The group is the current expression of a pattern of behavior developed over the course of hundreds of thousands of years. This pattern is constructed of behaviors established initially by creative heroes—by individuals who were able and willing to do and to think something that no one had been able to do or to think before. Integration of these behaviors into a stable hierarchy, and abstract representation of them, in the course of a process beginning with imitation and ending in semantic description, produces a procedural and declarable structure whose *incorporation* dramatically increases the individual's behavioral repertoire and his or her descriptive, predictive and representational ability. This incorporation—which is primarily implicit, and therefore invisible—is *identification with the group*. Identification with the group means the provision of determinate meaning, as the antidote to excruciating ignorance and exposure to chaos.

A multitude of (specific) rituals have evolved to catalyze such identification. Catalysis often appears necessary, as the movement to adolescence is vitally important but psychologically challenging, involving as it does "voluntary" sacrifice of childhood dependency (which is a valid form of adaptation, but predicated upon [nondeclarative] assumptions suitable only to the childhood state). Such transitional rituals are generally predicated upon enaction of the fundamental narrative structure—the Way—previously presented. Ritual initiation, for example—a ubiquitous formal feature of pre-experimental culture[365]—takes place at or about the onset of puberty, when it is critical for further psychological development and continued tribal security that boys transcend their dependency upon their mothers. This separation often takes place under purposefully frightening and violent conditions. In the general initation pattern, the men, acting as a unit (as the *embodiment of social history* [366]), separate the initiates from their mothers, who offer a certain amount of more or less dramatized resistance and some genuine sorrow (at the "death" of their children).

The boys know that they are to be introduced to some monstrous power who exists in the night, in the forest or cave, in the depths of the unknown. This power, capable of devouring them, serves as the mysterious deity of the initiation. Once removed from their mothers, the boys begin their ritual. This generally involves some mixture of induced regression of personality—reduction to the state of "precosmogonic chaos," extant prior even to earliest childhood—and induction of overwhelming fear, accompanied by severe physical or spiritual hardship or torture. The initiates are often forbidden to talk, and may be fed by the men. They may be circumcised, mutilated or interred alive—required to undergo intense punishment, subjected to intense dread. They symbolically pass into the maw of the Terrible Mother and are reborn as men, as adult members of the "tribe," which is the historical cumulation of the consequences of adaptive behavior. (Initiates often actually pass, literally, through the body of some constructed beast, aided by the elders of the tribe, who serve as the agents of this deity.[367]) When the rite is successfully completed, the initiated are no longer children, dependent upon the arbitrary beneficience of nature—in the guise of their mothers—but are members of the tribe of men, active standard-bearers of their particular culture, who have had their previously personality destroyed, so to speak, by fire. They have successfully faced the worst trial they are likely ever to encounter in their lives.

The terror induced by ritual exposure to the forces of the unknown appears to put the brain into a state characterized by enhanced suggestibility—or, at least, by dramatically heightened need for order, by need for coherent and meaningful *narrative*. The person who is in a "state" where he no longer knows what to do or what to expect is highly motivated to escape that state, by whatever means necessary. The stripping away of a former mode of adaptation, engendered by dramatic shift of social locale (of "context"), produces within the psyche of those so treated a state of acute apprehension, and intense desire for the re-establishment of predictability and sense. This acute apprehension is, as we have seen, the consequence of the "renovelization" of the environment: sufficient challenge posed to the integrity

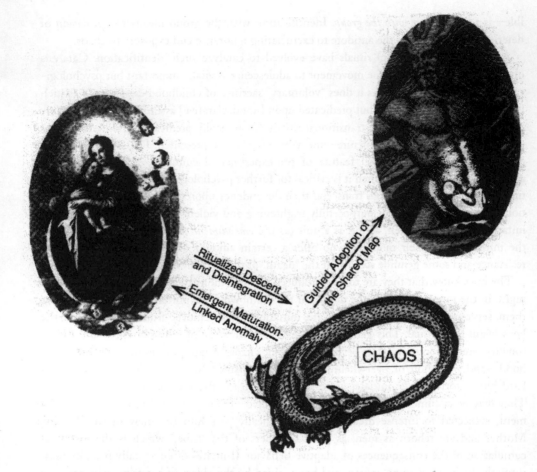

Figure 46: The "Death" and "Rebirth" of the Adolescent Initiate

of a previous personality disrupts its structure, "freeing" phenomena previously adapted to from the grasp of familiar action and valuation. The phenomena, thus "free," then once again "possess" sufficient "energy" to motivate their reconceptualization (that is, to make of that process of reconceptualization something sufficiently vital and important to stamp itself into memory—into permanent incarnation as "personality").

The ritually "reduced" and terrified initiates, unable to rely on the adaptive strategies utilized during their childhoods, desperately need new explanations and new patterns of behavior to survive in what is, after all, a new environment. That new environment is the society of men, where women are sexual partners and equals instead of sources of dependent comfort; where the provision of food and shelter is a responsibility, and not a given; where security—final authority, in the form of parent—no longer exists. As the childhood "personality" is destroyed, the adult personality—a manifestation of transmitted culture—is inculcated. The general initiatory "narrative" or ritual is presented schematically in *Figure 46: The "Death" and "Rebirth" of the Adolescent Initiate.*

The comparatively more abstracted rite of baptism is predicated upon similar principles. Baptism is the dramatic or episodic representation of the act or ritual of initiation—or, at least, stands midway between the entirely "unconscious" or procedural forms of initiation and their semantically abstracted symbolic equivalents. Baptism is spiritual birth (rebirth), as opposed to birth of the flesh. The font of the church, which contains the baptismal water, is a symbolic analog of the uterus[368] (the *uterus ecclesiastiae*), which is the "original" place that transforms precosmogonic chaos into spirit-embodied matter (into "personality"). When the initiate is plunged into (now sprinkled with) baptismal water, he or she is symbolically reduced, from insufficient stability to chaos; is drowned as a profane being, and then resurrected; is reunited (incestuously, mythically speaking) with the Great Mother, then reborn formally into the community of the spirit.[369] Such abstracted reductions to "death" and symbolic reconstructions constitute ritualization and representation of the processes endlessly necessary to revitalization of the individual personality and the social group. Eliade states:

The majority of initiatory ordeals more or less clearly imply a ritual death followed by resurrection or a new birth. The central moment of every initiation is represented by the ceremony symbolizing the death of the novice and his return to the fellowship of the living. But he returns to life a new man, assuming another mode of being. Initiatory death signifies the end at once of childhood, of ignorance, and of the profane condition....

All the rites of rebirth or resurrection, and the symbols that they imply, indicate that the novice has attained to another mode of existence, inaccessible to those who have not undergone the initiatory ordeals, who have not tasted death. We must note this characteristic of the archaic mentality: the belief that a state cannot be changed without first being annihilated—in the present instance, without the child's dying to childhood. It is impossible to exaggerate the importance of this obsession with beginnings, which, in sum, is the obsession with the absolute beginning, the cosmogony. For a thing to be well done, it must be done as it was the first time. But the first time, the thing—this class of objects, this animal, this particular behavior—did not exist: when, in the beginning, this object, this animal, this institution, came into existence, it was as if, through the power of the Gods, being arose from nonbeing.

Initiatory death is indispensable for the beginning of spiritual life. Its function must be understood in relation to what it prepares: birth to a higher mode of being.... [I]nitiatory death is often symbolized, for example, by darkness, by cosmic night, by the telluric womb, the hut, the belly of a monster. All these images express regression to a preformal state, to a latent mode of being (complementary to the precosmogonic Chaos), rather than total annihilation (in the sense in which, for example, a member of the modern societies conceives death). These images and symbols of ritual death are inextricably connected with germination, with embryology; they already indicate a new life in course of preparation....

For archaic thought, then, man is made—he does not make himself all by himself. It is the old initiates, the spiritual masters, who make him. But these masters apply what was revealed to them at the beginning of Time by Supernatural Beings. They are only the representatives of these Beings; indeed, in many cases, they incarnate them. This is as much as to say that in order to become a man, it is necessary to resemble a mythical model.[370]

Groups are individuals, uniform in their acceptance of a collective historically determined behavioral pattern and schema of value. Internalization of this pattern, and the description thereof (the myths—and philosophies, in more abstracted cultures—which accompany it), simultaneously produces ability to act in a given (social) environment, to predict the outcomes of such action, and to determine the meaning of general events (*meaning* inextricably associated with behavioral outcome). Such internalization culminates in the erection of implicit procedural and explicit declarable structures of "personality," which are more or less isomorphic in nature, which simultaneously constitute habit and moral knowledge. Habit is a way of being, a general strategy for "redemption" in the "natural" and "cultural" spheres, shaped by the social exchange of affect-laden information, mastered to the point of "unconscious" automaticity. Moral knowledge is fixed representation of the (previously) "unknown"; is generation of capacity to predict the behavior of objects, other people and the self. The sum total of accurate behaviorally linked representation of the world as forum for action constitutes the structure which reduces the manifold meaning of the experiential plenum to a restricted and therefore manageable domain. The manifold meaning is anxiety, on first contact (or under uncontrolled, overwhelming or involuntary conditions of exposure)—anxiety, which would otherwise be generated in response to everything. Interference with adolescent initiation-catalyzed group incarnation is therefore disruption of or failure to (re)generate the structure providing for respite from unbearable existential anxiety.

A society "works" to the degree that it provides its members with the capacity to predict and control the events in their experiential field—to the degree that it provides a barrier, protection from the unknown or unexpected. Culture provides a ritual model for behavioral emulation, and heuristics for desire and prediction—active procedures for behavior in the social and nonsocial worlds, plus description of processes in the social and nonsocial worlds, including behavioral processes. Incorporation of culture therefore means fixed adaptation to the unknown; means, simultaneously, inhibition of novelty-induced fear, regulation of interpersonal behavior, and provision of redemptive mode of being. The group is the historical structure that humanity has erected between the individual and the terrible unknown. Intrapsychic representation of culture—establishment of group identity—protects individuals from overwhelming fear of their own experience; from contact with the *a priori* meaning of things and situations. This is the intercession of the mythic Great Father against the terrible world of the Great Mother. This intercession is provision of a specific goal schema, allowing for the transformation of the vagaries of individual experience into positive events, within a social context, in the presence of protection against the unbearable unknown.

This historically determined cultural structure is constructed of courageously engineered and creatively integrated responses to situations that arise typically in the course of human experience, arranged in terms of their relative importance, organized simultaneously to minimize intrapsychic motivational and external interpersonal conflict, and to allow for continued adaptation. This (primarily nonverbal) socially transmitted structure of assumption, expectation and behavior is very stable, under most circumstances. It has seen everything, and done everything, so to speak, and cannot be easily undermined. In most situations, it efficiently governs social interaction, general expectation, and organization of goal-directed behavior. In its implicit imitative, dramatic, narrative form, it is exceptionally durable, and highly resistant to naïve social revolution.[371] However, such stability is advantageous only in stable times. Under exceptional circumstances—when the environment shifts rapidly, for reasons independent of or dependent upon human activity—the historical "personality" must be altered or even qualitatively reconfigured to allow equally rapid adaptation to take place. This process of rearrangement is necessarily predicated upon disruption (death) of the old order. Dissolution of the old order means (potential) return of the determinate meaning of experiential objects to their preclassified state of chaos—simultaneously unbearably threatening and, secondarily, infinitely promising. Apprehension of the inevitability of such dissolution, however vague, constitutes one potent barrier to the process of creative readaptation.

The historical structure "protects itself" and its structure in two related manners. First, it inhibits intrinsically rewarding but "antisocial" behaviors (those which might upset the stability of the group culture) by associating them with certain punishment (or at least with the threat thereof). This punishment might include actual application of undesirable penalties or, more "subtly"—removal of "right to serve as recognized representative of the social structure." This means, in the latter case, forced individual forfeit of identification with (imitation of, internalization of) said social structure (at least for the once-socialized), and induction of overwhelming guilt or anxiety, as a consequence of goal loss, value dissolution, and subsequent re-exposure to the novelty of decontextualized experience. It is the potential for such an affectively unbearable state that constitutes the power of banishment—which can be used "consciously" by societies to punish wrongdoers—or that can be experienced as a self-induced state, by individuals careless, arrogant or ignorant enough to "kill" what supports them.[372]

The culturally-determined historical structure protects and maintains itself secondly by actively promoting individual participation in behavioral strategies that satisfy individual demand, and that simultaneously increase the stability of the group. The socially constructed way of a *profession*, for example, allows the individual who incarnates that profession opportunity for meaningful activity in a manner that supports or at least does not undermine the stability of the historically determined structure which regulates the function of his or her threat-response system. Adoption of a socially sanctioned "professional personality" therefore provides the initiated and identified individual with peer-approved opportunity for intrinsic goal-derived pleasure, and with relative freedom from punishment, shame and guilt. Potentially upsetting competition between socially sanctified ways of being, within a given social group, is also subject to cultural minimization. Each of the many professions whose union comprises a functioning complex society is the consequence of the heroic past activities which established the profession, modified by the equally heroic activities that allowed for its maintenance and "update" (in the presence of other competing activities and ever-changing

"environmental" demand). "Lawyer" and "physician," for example, are two embodied ideologies, nested within more complex overarching narrative schemas, whose domains of
activity, knowledge and competence have been delimited, one against the other, until
both can occupy the same "territory" without emergence of destructive and counterproductive conflict. This is the "organization of dead kings," so to speak, under the dominion
of the "hero": doctors and lawyers are both subject to "higher-order (legal) principles"
which govern their behavior such that one group can tolerate—at least within reason—
the presence of the other.

The properly structured patriarchal system fulfills the needs of the present while "taking
into account" those of the future; simultaneously, it balances the demands of the self with
those of the other. The suitability of the "cultural solution" is judged by individual affective
response. This grounding of verification in universally constant affect, in combination with
the additional constraints of stability and adaptability, means inevitable construction of
human groups and human moral systems with centrally identifiable features and processes of
generation. The construction of a successful group, the most difficult of feats, means establishment of a society composed of individuals who act in their own interest (at least enough
to render their life bearable) and who, in doing so, simultaneously maintain and advance
their culture. The "demand to satisfy, protect and adapt, individually and socially"—and to
do so over vast and variable stretches of time—places severe intrinsic constraints on the
manner in which successful human societies can operate. It might be said that such constraints provide universal boundaries for acceptable human morality. The nature of what
constitutes such acceptability fosters direct conflict or debate, in terms of the details, but the
broad picture is necessarily clear. That picture is presented and represented in ritual, mythology and narrative, which eternally depict intrinsically meaningful themes, playing themselves out, in eternally fascinating fashion. Nietzsche states:

That individual philosophical concepts are not anything capricious or autonomously evolving, but grow
up in connection and relationship with each other; that, however suddenly and arbitrarily they seem to
appear in the history of thought, they nevertheless belong just as much to a system as all the members of
the fauna of a continent—is betrayed in the end also by the fact that the most diverse philosophers keep
filling in a definite fundamental scheme of possible philosophies. Under an invisible spell, they always
revolve once more in the same orbit; however independent of each other they may feel themselves with
their critical or systematic wills, something within them leads them, something impels them in a definite order, one after the other—to wit, the innate systematic structure and relationship of their concepts.
Their thinking is, in fact, far less a discovery than a recognition, a remembering, a return and a homecoming to a remote, primordial, and inclusive household of the soul, out of which those concepts grew
originally: philosophizing is to this extent a kind of atavism of the highest order.[373]

Adoption of a particular way of being allows, concurrently, for determination of the meaning of objects, and the morality of behaviors. Objects attain significance according to their
perceived utility—with regard to their capacity to further movement away from the unbear-

able present toward the ideal future; likewise, moral behavior is seen as furthering and immoral behavior as impeding or undermining such movement. Of course, identification of what constitutes the basis for establishing the nature of morality or the comparative value of objects is no simple matter. In fact, such judgment comprises the constant central demand of adaptation. No fixed answer solution to this problem can be offered—this question, "the nature of the highest ideal" or "the nature of the highest good"—because the environment posing the query, so to speak, constantly shifts, as time progresses (that shift constitutes, in fact, time's progression). The constant fact of eternal change does not eliminate the utility of all "moral" answers, however, as such answers must be formulated, before any action or interpretation can take place. Time merely makes eternal nonsense of the offer of fixed structure as solution—fixed structure, that is, as opposed to *process* (in this case, the patterned creative communicative process of generating adaptive structure).

Conflict, on the individual and social planes, constitutes dispute about the comparative value of experiences, objects and behaviors. Nondeclarative presumption "a," upon which behavior "a" is (hypothetically) predicated, becomes subjugated to presumption "b," "b" to "c," and so on, in accordance with some implicit scheme or notion of ultimate value which firsts manifests itself in behavior, and in behavioral conflict, long before it can be represented episodically or semantically. It might be said that the emergence of a scheme of ultimate value is an inevitable consequence of the social and exploratory evolution of man. Cultural structure, incarnated intrapsychically, originates in creative action, imitation of such action, integration of action and imitated action—constitutes adaptive action and representation of integrated pattern of action. Procedures may be mapped in episodic memory and abstracted in essence by the semantic system. This process results in construction of a story, or narrative. Any narrative contains, implicit in it, a set of moral assumptions. Representation of this (primarily social) moral code in form of episodic memory constitutes the basis for myth; provides the ground and material for eventual linguistically mediated development of religious dogma or codified morality. Advantages of such codification are the advantages granted by abstraction *per se*—ease of communication, facilitation of transformation—and formal declaration of (historically sanctified) principles useful in mediation of emergent value-centered dispute. Disadvantages—more subtle, and more easily unrecognized—include premature closure of creative endeavor, and dogmatic reliance on wisdom of the (dead) past.

Human beings, as social animals, act "as if" motivated by a (limited) system of more or less internally consistent and integrated set of moral virtues, even in the absence of the explicit (declarative) representation of this system. The nature of these virtues, embodied in behavior, in their origin, have become more and more conscious (more represented in declarative thinking and remembering) over the course of socially mediated human cognitive evolution. Nonetheless it is very difficult to determine and explicitly state just what virtuous behavior consists of; to describe, with accuracy, how it is that people should (and do) act—to identify those ends toward which behavior should be devoted, and to provide explicit and rigorous justification for such claims. A culture is, to a large degree, a shared moral code, and deviations from that code are generally easily identified, at least *post-hoc*. It is still the case, however, that description of the domain of morality tends to exceed the capability of declarative thought, and that the nature of much of what we think of as moral behavior is still, therefore, embedded in unconscious procedure. As a consequence, it is easy for us to become confused about the nature of morality, and to draw inappropriate, untimely and dangerous "fixed" conclusions.

The conservative worships his culture, appropriately, as the creation of that which deserves primary allegiance, remembrance and respect. This creation is the concrete solution to the problem of adaptation: "how to behave?" (and how can that be represented and communicated?). It is very easy, in consequence, to err in attribution of value, and to worship the specific solution itself, rather than the source of that solution. Hence the biblical injunction:

Thou shalt have no other gods before me.
Thou shalt not make unto thee any graven image, or any likeness *of any thing* that *is* in heaven above, or that *is* in the earth beneath, or that *is* in the water under the earth:
Thou shalt not bow down thyself to them, nor serve them: for I the Lord thy God *am* a jealous God, visiting the iniquity of the fathers upon the children unto the third and fourth *generation* of them that hate me. (Exodus 20:3–5)

This "arbitrary" injunction exists in large part because much less explicit attention is generally paid (*can* be paid, in the initial stages of abstract representation) to the more fundamental, but more abstract and difficult, *meta-problem* of adaptation: "how is (or was) how to behave determined?" or "what is the nature of the behavioral procedure that leads to the establishment of and rank-ordering of valid forms of how to behave? (that leads to successful adaptation, as such?)" and "how can that be represented and communicated?" The answer to the question "what constitutes the highest value?" or "what is the highest good?" is in fact the solution *to the meta-problem*, not the problem, although solutions to the latter have been and are at present constantly confused with solutions to the former—to the constant (often mortal) detriment of those attempting to address the former.

The precise nature of that which constitutes morality still eludes declarative exposition. The moral structure, encoded in behavior, is too complex to completely consciously formulate. Nevertheless, that structure remains an integrated system (essentially, a historically determined personality, and representation thereof), a product of determined efforts (procedural and declarative) devoted toward integrated adaptation, and not a merely random or otherwise incomprehensible compilation of rituals and beliefs. Culture is a structure aimed toward the attainment of certain (affectively-grounded) ends, in the immediate present and over the longer course of time. As such, a given cultural structure necessarily must meet a number of stringent and severely constrained requirements: (1) it must be self-maintaining (in that it promotes activities that allow it to retain its central form); (2) it must be sufficiently flexible to allow for constant adaptation to constantly shifting environmental circumstances; and (3) it must acquire the allegiance of the individuals who compose it.

The first requirement is so fundamental, even in the short term, that it appears self-evident. A culture must promote activities that allow for its own maintenance, or it will devour itself. The second requirement—flexibility—is more difficult to fulfill, particularly in combination with the first (self-maintenance). A culture must promote activity that supports itself, but must simultaneously allow for enough innovation so that essentially unpredictable alteration in "environmental" circumstance can be met with appropriate change in behavioral activity. Cultures that attempt to maintain themselves through promotion of absolute adherence to traditional principles tend rapidly to fail the second requirement, and to collapse precipitously. Cultures that allow for unrestricted change, by contrast, tend to fail the first, and collapse equally rapidly. The third requirement (allegiance of the populace) might be considered a prerequisite for the first two. A culture that lasts must be supported (voluntarily) by those who compose it. This means, in the final analysis, that its mode of operation must remain verified by the sum total of individual affect; means that those who constitute the group must remain satisfied by its operation—must derive sufficient reward, protection from punishment, provision of hope, and alleviation of threat to render the demands of group maintenance bearable. Furthermore, the group solution must *appear* ideal—in comparison to any or all actual or imaginable alternatives. The compelling attractiveness of simplistic utopian ideologies, even in the "skeptical" twentieth century, is evidence for the stringent difficulty of this final requirement.

In suboptimal circumstances, the problem of "protection for the developing individual" and "maintenance of the protective, uniform social structure" is solved by the permanent sacrifice of individual diversity to the stability and identity of the group. This solution banishes fear effectively, in the short term, but also eliminates necessary potential and the capacity for "adaptive" transformation. The suboptimal solution to the problem of authoritarian or totalitarian danger, in turn, is denigration of the role of society, attribution of evil to its effects, and degeneration of traditional skills and learning. This is sacrifice of the Terrible Father, without recognition of the need for his resuscitation—and is, therefore, an invitation to the intrusion of chaos. The solution to the problem of the necessity for group identification is, by contrast, to be found in the philosophy of the apprenticeship: each individual must voluntarily subjugate him- or herself to a master—a "wise king"—whose goal is not so much maintenance and protection of his own identity and status as it is construction of an individual (a "son") capable of transcending the restrictions of the group.

The optimal "wise king" to whom subordination might be regarded as necessary must therefore either be an individual whose "identity" is nested within a hierarchy whose outermost territory is occupied by the exploratory hero, or a group about which the same might be said. So the ideal "group" or master might be conceptualized, once again, as Osiris (the traditions of the past) nested within Horus/Re (the process that originally created those traditions, and which presently updates them). This means that the "meta-problem" of adaptation—*"what is the nature of the behavioral procedure that leads to the establishment of and rank-ordering of valid forms of how to behave? (that leads to successful adaptation, as such?)"*—has been answered by groups who ensure that their traditions, admired and imitated, are nonetheless subordinate to the final authority of the creative hero. So the "highest good" becomes "imitation (worship) of the process represented by the hero," who, as the ancient Sumerians stated, restores all "ruined gods, as though they were his own creation."[374]

Human morality is exploratory activity (and allowance for such), undertaken in a sufficiently stable social context, operating within stringent limitations, embodied in action, secondarily represented, communicated and abstractly elaborated in episodic and semantic memory. Such morality—act and thought—is nonarbitrary in structure and specifically goal-directed. It is predicated upon conceptualization of the highest good (which, in its highest form, is stable social organization allowing for manifestation of the process of creative adaptation), imagined in comparison to the represented present. Such conceptual activity allows for determination of acceptable behavior, and for constraint placed upon the meaning of objects (considered, always, in terms of their functional utility as tools, in a sense, for the attainment of a desired end).

The pathological state takes imitation of the "body of the laws" to an extreme, and attempts to govern every detail of individual life. This "total imitation" reduces the behavioral flexibility of the state, and renders society increasingly vulnerable to devastation through environmental transformation (through the influx of "chaotic change"). Thus the state suffers, for lack of "the water of life," until it is suddenly flooded, and swept away. The healthy state, by contrast, compels imitation more in the form of voluntary affiliation (until the establishment of individual competence and discipline). Following the successful "apprenticeship," the individual is competent to serve as his own master—to serve as an autonomous incarnation of the hero. This means that the individual's capacity for "cultural imitation"—that is, his capacity for subservience to traditional order—has been rendered subordinate to his capacity to function as the process that mediates between order and chaos. Each "properly socialized" individual therefore comes to serve as Horus (the sun king, the son of the Great Father), after painstakingly acquiring the wisdom of Osiris.

The adoption of group identity—the apprenticeship of the adolescent—disciplines the individual, and brings necessary predictability to his or her actions within the social group. Group identity, however, is a construct of the past, fashioned to deal with events characteristic of the past. Although it is reasonable to view such identity as a necessary developmental stage, it is pathological to view it as the end point of human development. The present consists in large part of new problems, and reliance on the wisdom of the dead, no matter how heroic, eventually compromises the integrity of the living. The well-trained apprentice, however, has the skills of the dead, and the dynamic intelligence of the living. This means that he can benefit from—even welcome—inevitable contact with anomaly, in its many guises. The highest level of morality therefore governs behavior in those spaces where tradition *does not* rule. The exploratory hero is at home in unexplored territory—is friend of the stranger, welcoming ear for the new idea, and cautious, disciplined social revolutionary.

4

THE APPEARANCE OF ANOMALY

Challenge to the Shared Map

———————————————○———————————————

M oral theories necessarily share common features with other theories. One of the most fundamental shared features of theories, in general, is their reliance on "extra-theoretical" presuppositions. The "extra-theoretical" presuppositions of explicit moral theorems appear to take
implicit form in image and, more fundamentally, in action. Moral behaviors and schemas of valuation arise as a consequence of behavioral interaction undertaken in the social world: every individual, motivated to regulate his emotions through action, modifies the behavior of others, operating
in the same environment. The consequence of this mutual modification, operating over time, is the
emergence of a stable pattern of behavior, "designed" to match individual and social needs, simultaneously. Eventually, this behavioral pattern comes to be coded in image, heralded in narrative,
and explicitly represented in words. In the integrated individual—or the integrated state—action,
imagination and explicit verbal thought are isomorphic: explicit and image-mediated beliefs and
actual behaviors form a coherent unit. Verbal theories of morality (explicit rules) match traditional
images of moral behavior, and action undertaken remains in concordance with both. This integrated
morality lends predictability to behavior, constitutes the basis for the stable state, and helps ensure
that emotion remains under control.

The emergence of anomaly constitutes a threat to the integrity of the moral tradition governing
behavior and evaluation. Strange things or situations can pose a challenge to the structure of a
given system of action and related beliefs; can pose such a challenge at comparatively restricted
("normal") or broader ("revolutionary") levels of organization. A prolonged drought, for example,
destructive at the social level—or the occurrence of a serious illness or disability, destructive at the
personal—can force the reconstruction of behavior and the reanalysis of the beliefs that accompany, follow, or underlie such behavior. The appearance of a stranger—or, more commonly, a group of
strangers—may produce a similar effect. The stranger acts out and holds different beliefs, using different implements and concepts. The mere existence of these anomalous beliefs, actions and tools—

233

generally the consequence of prolonged, complex and powerful social-evolutionary processes—may be sufficient to totally transform or even destroy the culture which encounters them, unprepared. Cultures may be upset internally, as well, as a consequence of the "strange idea"—or, similarly, by the actions of the revolutionary.

The capacity to abstract, to code morality in image and word, has facilitated the communication, comprehension and development of behavior and behavioral interaction. However, the capacity to abstract has also undermined the stability of moral tradition. Once a procedure has been encapsulated in image—and, particularly, in word—it becomes easier to modify, "experimentally"; but also easier to casually criticize and discard. This capacity for easy modification is very dangerous, in that the explicit and statable moral rules that characterize a given culture tend to exist for reasons that are still implicit and fundamental. The capacity to abstract, which has facilitated the communication of very complex and only partially understood ideas, is therefore also the capacity to undermine the very structure that lends predictability to action, and which constrains the *a priori* meaning of things and situations. Our capacity for abstraction is capable of disrupting our "unconscious"—that is, imagistic and procedural—social identity, upsetting our emotional stability, and undermining our integrity (that is, the isomorphism between our actions, imaginings, and explicit moral theories or codes). Such disruption leaves us vulnerable to possession by simplistic ideologies, and susceptible to cynicism, existential despair, and weakness in the face of threat.

The ever-expanding human capacity for abstraction—central to human "consciousness"—has enabled us to produce self-models sufficiently complex and extended to take into account the temporal boundaries of individual life. Myths of the "knowledge of good and evil" and the "fall from paradise" represent emergence of this representational capacity, in the guise of a "historical event." The consequence of this "event"—that is, the development of "self-consciousness"—is capacity to represent death, and to understand that the possibility of death is "part" of the unknown. This "contamination of anomaly with the possibility of death" has dramatically heightened the emotional power and motivational significance of the unknown, and led to the production of complex systems of action and belief designed to take that terrible possibility into account.

These complex systems of action and belief are religious. They are the traditional means of dealing with the shadow cast on life by knowledge of mortality. Our inability to understand our religious traditions—and our consequent conscious denigration of their perspectives—dramatically decrease the utility of what they have to offer.

We are conscious enough to destabilize our beliefs and our traditional patterns of action, but not conscious enough to understand them. If the reasons for the existence of our traditions were rendered more explicit, however, perhaps we could develop greater intrapsychic and social integrity. The capacity to develop such understanding might help us use our capacity for reason to support, rather than destroy, the moral systems that discipline and protect us.

INTRODUCTION: THE PARADIGMATIC STRUCTURE OF THE KNOWN

The "known" is a hierarchical structure, composed of "walls within walls." The individual sits at the middle of a series of concentric rings, composed of the integrated "personalities" of his ancestors,

nested (at least in the ideal) within the figure of the exploratory hero. The inner walls are dependent for their protection—for their continued existence and validity—on the integrity of the outer
walls. The farther "out" a given wall, the more "implicit" its structure—that is, the more it is incarnated in behavior and image, rather than explicit in word. Furthermore, the farther "out" the wall,
the older the "personality," the broader range of its applicability, and the greater the magnitude of
emotion it holds in check. Groups—and individuals—may share some levels of the known, but not
others. The similarities account for "shared group identity," insofar as that exists; the differences,
for the identification of the other with the forces of chaos.

Rituals designed to strengthen group identity hold chaos at bay, but threaten individual identification with the exploratory hero—an identity upon which maintenance of the group ultimately
depends. For the sake of the group, therefore, the individual must not be rendered subservient to
the group.

> The aspects of things that are most important for us are hidden
> because of their simplicity and familiarity. (One is unable to notice
> something—because it is always before one's eyes.) The real founda-
> tions of his enquiry do not strike a man at all. Unless *that* fact has at
> some time struck him.—And this means: we fail to be struck by
> what, once seen, is most striking and powerful.[375]

A moral system—a system of culture—necessarily shares features in common with other systems. The most fundamental of the shared features of systems was identified by Kurt Godel.
Godel's Incompleteness Theorem demonstrated that any internally consistent and logical
system of propositions must *necessarily* be predicated upon assumptions that cannot be
proved from within the confines of that system. The philosopher of science Thomas Kuhn,
discussing the progress of science, described similar implicit-presumption-ridden systems as
paradigmatic. Explicitly scientific paradigmatic systems, the focus of Kuhn's attention, are
concerned with the prediction and control of events whose existence can be verified, in a
particular formal manner, and offer "model problems and solutions to a community of practitioners."[376] Pre-experimental thinking—which primarily means moral thinking (thinking
about the meaning or significance of events [objects and behaviors])—also appears necessarily characterized by paradigmatic structure.

A paradigm is a complex cognitive tool, whose use presupposes acceptance of a limited
number of axioms (or *definitions of what constitutes reality, for the purposes of argument and
action*), whose interactions produce an internally consistent explanatory and predictive structure. Paradigmatic thinking might be described as thinking whose domain has been formally
limited; thinking that acts "as if" some questions have been answered in a final manner. The
"limitations of the domain" or the "answers to the questions" make up the axiomatic statements of the paradigm, which are, according to Kuhn, "explicitly" formulated—semantically
represented, according to the argument set forth here—or left "implicit"—embedded in
(episodic) fantasy or embodied behavior. The validity of the axioms must either be accepted on

faith, or (at least) demonstrated using an approach which is external to the paradigm in question (which amounts to the same thing as faith, from a "within-the-paradigm" perspective).

In some regards, a paradigm is like a game. Play is optional, but, once undertaken, must be governed by (socially verified) rules. These rules cannot be questioned, while the game is on (or if they are, *that is a different game*. Children arguing about how to play football are not playing football. They are engaging, instead, in a form of philosophy). Paradigmatic thinking allows for comprehension of an infinity of "facts," through application of a finite system of presuppositions—allows, in the final analysis, for the limited subject to formulate sufficient provisional understanding of the unlimited experiential object (including the subject).

Human culture has, by necessity, a paradigmatic structure—devoted not toward objective description of what is, but to description of the cumulative affective relevance, or meaning, of what is. The capacity to determine the motivational relevance of an object or situation is dependent, in turn, upon representation of a (hypothetically) ideal state (conceived in constrast to conceptualization of the present), and upon generation of an action sequence designed to attain that ideal. It is (stated, unstated and unstatable) articles of faith that underlie this tripartite representation, and that keep the entire process in operation. These "articles of faith" are axioms of morality, so to speak—some explicit (represented declaratively, in image and word), most still implicit—which evolved in the course of human exploration and social organization, over the course of hundreds of thousands of years. In their purely implicit states, such axioms are extremely resistant to alteration. Once made (partially) explicit, however, moral axioms rapidly become subject to endless careful and thoughtful or casual careless debate. Such debate is useful, for continuance and extension of adaptation, but also very dangerous, as it is the continued existence of unchallenged moral axioms that keeps the otherwise unbearable significance of events constrained, and the possibility for untrammeled action alive.

A paradigmatic structure provides for determinate organization of (unlimited) information, according to limited principles. The system of Euclidean geometry provides a classic example. The individual who wishes to generate a desired outcome of behavior, as a consequence of the application of Euclidean principles, is bound by necessity to accept certain axioms "on faith." These axioms follow:

1. A straight line segment can be drawn joining any two points.
2. Any straight line segment can be extended indefinitely in a straight line.
3. Given any straight line segment, a circle can be drawn having the segment as radius and one end point as center.
4. All right angles are congruent.
5. If 2 lines are drawn which intersect a 3rd in such a way that the sum of the inner angles on one side is less than two right angles, then the 2 lines inevitably must intersect each other on that side if extended far enough.[377]

It is the interaction of each of the five initial postulates—which are all that necessarily have to be remembered, or understood, for geometry to prove useful—that gives rise to the internally consistent and logical Euclidean structure we are all familiar with. What constitutes truth, from within the perspective of this structure, can be established by reference to these initial postulates. However, the postulates themselves must be accepted. Their validity cannot be demonstrated, within the confines of the system. They might be "provable" from within the confines of another system, however—although the integrity of that system will still remain dependent, by necessity, on different postulates, down to an indeterminate end. The validity of a given structure appears necessarily predicated on "unconscious" presuppositions—the presupposition that space has three dimensions, in the case of Euclidean geometry (a presupposition which is clearly questionable).

It appears, in many cases, that the assumptions of explicit semantic statements take episodic or imagistic form. The Euclidean postulates, for example, appear to be based upon "observable facts" (images of "the world of experience" as interpreted). Euclid grounded his explicit abstract (semantic) system in observable "absolutes." It can be concretely demonstrated, for example, that any two points drawn in the sand can be joined by a given line. Repeated illustration of this "fact" appears (acceptably) convincing—as does, similarly, ("empirical") demonstration that any straight line segment can be extended indefinitely in a straight line. These postulates (and the remaining three) cannot be proved from within the confines of geometry itself, but they appear true, and will be accepted as such, *as a consequence of practical example*. What this means is that belief in Euclidean presumptions is dependent upon acceptance of practical experience as sufficient certainty. The Euclidean draws a line in the sand, so to speak, and says "the questions stop here."

Similarly, it appears that what constitutes truth from the episodic perspective is predicated upon acceptance of the validity and sufficiency of specific procedural operations. How a thing is represented in episodic memory, for example—which is what a thing is, insofar as we know what it is—appears dependent upon how it was investigated and on the implicit "presuppositions" driving or limiting the behavioral strategies applied to it in the course of creative exploration. Kuhn states:

Scientists can agree that a Newton, Lavoisier, Maxwell or Einstein has produced an apparently permanent solution to a group of outstanding problems and still disagree, sometimes without being aware of it, about the particular abstract characteristics that make those solutions permanent. They can, that is, agree in their *identification* of a paradigm without agreeing on, or even attempting to produce, a full *interpretation* or *rationalization* of it. Lack of a standard interpretation or of an agreed reduction to rules will not prevent a paradigm from guiding research. Normal science can be determined in part by the direct inspection of paradigms, a process that is often aided by but does not depend upon the formulation of rules and assumptions. Indeed, the existence of a paradigm need not even imply that any full set of rules exists.

He continues, in a footnote:

Michael Polyani[378] has brilliantly developed a very similar theme, arguing that much of the scientist's success depends upon "tacit knowledge," i.e., upon knowledge that is acquired through practice and that cannot be articulated explicitly.[379]

The Euclidean draws a line connecting two points in the sand, and accepts on faith the sufficiency of that behavioral demonstration and the evident certainty of its outcome (in part, because no alternative conceptualization can presently be imagined). Euclidean geometry worked and was considered complete for centuries because it allowed for the prediction and control of all those experienceable phenomena that arose as a consequence of human activity, limited in its domain by past behavioral capacity. Two hundred years ago, we did not know how to act concretely, or think abstractly, in a manner that would produce some situation whose nature could not be described by Euclid. That is no longer the case. Many alternative, and more inclusive geometries have been generated during the course of the last century. These new systems describe the nature of "reality"—the phenomena that emerge as a consequence of ongoing behavior—more completely.

All representations of objects (or situations, or behavioral sequences) are of course conditional, because they may be altered unpredictably, or even transformed, entirely, as a consequence of further exploration (or because of some spontaneous anomaly-emergence). The (anxiety-inhibiting, goal-specifying) model of the object of experience is therefore inevitably *contingent*—dependent, for its validity, on the maintenance of those (invisible) conditions which applied and those (unidentified) contexts which were relevant when the information was originally generated. Knowledge is mutable, in consequence—as Nietzsche observed:

There are still harmless self-observers who believe that there are "immediate certainties"; for example, "I think," or as the superstition of Schopenhauer put it, "I will"; as though knowledge here got hold of its object purely and nakedly as "the thing in itself," without any falsification on the part of either the subject or the object. But that "immediate certainty," as well as "absolute knowledge" and "the thing in itself," involve a *contradictio in adjecto*, I shall repeat a hundred times; we really ought to free ourselves from the seduction of words!

Let the people suppose that knowledge means knowing things entirely; the philosopher must say to himself: When I analyze the process that is expressed in the sentence, "I think," I find a whole series of daring assertions that would be difficult, perhaps impossible, to prove; for example, that it is I who think, that there must necessarily be something that thinks, that thinking is an activity and operation on the part of a being who is thought of as a cause, that there is an "ego," and, finally, that it is already determined what is to be designated by thinking—that I *know* what thinking is. For if I had not already decided within myself what it is, by what standard could I determine whether that which is just happening is not perhaps "willing" or "feeling"? In short, the assertion "I think" assumes that I *compare* my state at the present moment with other states of myself which I know, in order to determine what it is; on account of this retrospective connection with further "knowledge," it has, at any rate, no immediate certainty for me.

In place of the "immediate certainty" in which the people may believe in the case at hand, the philosopher thus finds a series of metaphysical questions presented to him, truly searching questions of the intellect; to wit: "From where do I get the concept of thinking? Why do I believe in cause and effect? What gives me the right to speak of an ego, and even of an ego as a cause, and finally of an ego as the cause of thought?" Whoever ventures to answer these metaphysical questions at once by an appeal to a sort of *intuitive* perception, like the person who says, "I think, and know that this, at least, is true, actual, and certain"—will encounter a smile and two question marks from a philosopher nowadays. "Sir," the philosopher will perhaps give him to understand, "it is improbable that you are not mistaken; but why insist on the truth?" [380]

The "object" always remains something capable of transcending the "bounds" of its representation; it is something that inevitably retains its mysterious essence, its connection with the unknown, and its potential for the inspiration of hope and fear. The "actual" or "transcendent" object, in and of itself, insofar as such a thing can be considered, is the sum total of its explored properties, *plus that which remains unexplored—the unknown itself.*

Our understanding of a given phenomenon is always limited by the temporal, economic and technological resources that we have at our disposal. Knowledge is necessarily contingent, although it is neither less "objective," necessarily, nor less "knowledge," because of that. Our representations of objects (or situations, or behavioral sequences) are currently accepted as valid, *because they serve their purposes as tools.* If we can manipulate our models in imagination, apply the solutions so generated to the "real" world, and produce the outcome desired, we presume that our understanding is valid—and sufficient. It isn't until we do something, and produce an unexpected outcome, that our models are deemed insufficient. This means that our current representations of a given phenomena are predicated on the (implicit) presumption that sufficient exploration of that phenomena has taken place. "Sufficient exploration" is a judgment rendered as a consequence of a sequence of action attaining its desired end ("what works" is "true"). *A procedure is deemed sufficient when it attains its desired end—when it meets its goal.* The nature of that goal, archetypally, is establishment of or movement toward a paradisal state characterized by stable, dynamic relief from (unbearable) suffering, freedom from (paralyzing) anxiety, abundance of hope, and bountiful provision of primary reward—the peaceful land of "milk and honey," in mythical language. This is merely to say that knowledge serves the ends of life, rather than existing in and of itself.

Some contingent forms of knowledge—behaviors, say, and schemas of value—prove of lasting worth, producing the desired outcome across a broad range of contexts. These are "remembered"—stored in ritual and myth—and transmitted down the generations. Over the course of time, they become integrated with all other extant behaviors and schemas of value, in a hierarchy that allows for their various expression. This hierarchy, as described previously, is composed of the actions and valuations of past heroes, organized by other heroes into a stable social character, shared by all members of the same culture (as the Christian church constitutes the symbolic body of Christ). This hierarchy has been and currently is shaped by endless loops of affective feedback, as the means and goals chosen by each individual and the society at large are modified by the actions and reactions of society and the eternally ineradicable presence of the unknown itself. The resultant "hierarchy of motivation" can be most accurately characterized as a *personality*—the mythic "ancestral" figure that everyone imitates, consciously (with full participation of the semantic and episodic system, rational thought and imagination) or unconsciously (in action only, despite express "disbelief"). The hierarchically structured behavioral pattern (personality) that constitutes culture comes, with the passage of time, to be represented secondarily, isormorphically, in episodic memory, and then coded explicitly insofar as current cognitive development makes that possible. The explicit moral code is therefore predicated upon presumptions which are valid purely from the episodic perspective; in turn, these episodic representations derive their validity from procedural knowledge, designed to meet affective requirements, in the social community and in the presence of the unknown.

A moral philosophy, which is a pattern for behavior and interpretation, is therefore dependent for its existence upon a mythology, which is a collection of images of behaviors, which emerge, in turn, as a consequence of social interaction (cooperation and competition), designed to meet emotional demands. These demands take on what is essentially a universally constant and limited form, as a consequence of their innate psychobiological basis and the social expression of that basis. Hence (as implied previously) the limited "forms" of myth. Northrop Frye states, in this regard:

I should distinguish primary and secondary concern, even though there is no real boundary line between them. Secondary concerns arise from the social contract, and include patriotic and other attachments of loyalty, religious beliefs, and class-conditioned attitudes and behaviors. They develop from the ideological aspect of myth, and consequently tend to be directly expressed in ideological prose language. In the mythical stage, they often accompany a ritual. Such a ritual may be designed, for example, to impress on a boy that he is to be admitted to the society of men in a ritual for men only; that he belongs to this tribe or group and not that one, a fact which will probably determine the nature of his marriage; that these and not those are his special totems or tutelary deities.

Primary concerns may be considered in four main areas: food and drink, along with related bodily needs; sex; property (i.e. money, possessions, shelter, clothing and everything that constitutes property in the sense of what is "proper" to one's own life); liberty of movement. The general object of primary concern is expressed in the Biblical phrase "life more abundantly." In origin, primary concerns are not individual or social in reference so much as generic, anterior to the conflicting claims of the singular and the plural. But as society develops they become the claims of the individual body as distinct from those of the body politic. A famine is a social problem, but only the individual starves. So a sustained attempt to express primary concerns can develop only in societies where the sense of individuality has also developed. The axioms of primary concerns are the simplest and baldest platitudes it is possible to formulate: that life is better than death, happiness better than misery; health better than sickness, freedom better than bondage, for all people without significant exception.

What we have been calling ideologies are closely linked to secondary concerns, and in large measure consist of rationalizations of them. And the longer we look at myths, or storytelling patterns, the more clearly their links with primary concern stand out.... This rooting of poetic myth in primary concern accounts for the fact that mythical themes, as distinct from individual myths or stories, are limited in number.[381]

The (explicit) moral code is validated by reference to the (religious, mythic) narrative; the narrative is (primarily episodic) representation of behavioral tradition; the tradition emerges as a consequence of individual adaptation to the demands of natural conditions, manifest (universally) in emotion, generated in a social context. The episodic representation—which is representation of the outcome of a procedure and the procedure itself—is predicated upon belief in the sufficiency and validity of that procedure; more subtly, it has the same structure—at least insofar as it is an accurate representation of behavior—and therefore contains the (implicit) hierarchical structure of historically determined procedural knowledge in more explicit form.

Over lengthy historical periods, therefore, the "image" ever more accurately encapsulates the behavior, and stories find their compelling essential form. Frye states, with regard to the process underlying "construction" of the Old and New Testaments:

The Bible's literary unity is a by-product of something else—we might call it an unconscious by-product if we knew anything at all about the mental processes involved. The earlier part of the Old Testament, with its references to the Book of Jasher and the like, gives the effect of having distilled and fermented a rich poetic literature to extract a different kind of verbal essence, and on a smaller scale the same process can be seen in the New Testament.... The editorial work done on this earlier poetic material was not an attempt to reduce it from poetry to a kind of plain prose sense, assuming that there is such a thing. This kind of sense implies a direct appeal to credulity, to the infantilism which is so exasperating a feature of popular religious and other ideologies. What we have is rather an absorption of a poetic and mythic presentation that takes us past myth to something else. In doing so it will elude those who assume that myth means only something that did not happen.[382]

The second-order semantic codification is grounded in the episodic representation; tends, over time, to duplicate the hierarchical structure of that representation; and is predicated upon acceptance of the validity of the procedural and episodic memories. Semantic, episodic and procedural contents therefore share (in the intrapsychically integrated, "conscious" or psychologically healthy individual) identical hierarchical structure, in their respective forms of action or representation. This integrated morality lends predictability to individual and interpersonal behavior, constitutes the basis for the stable state, and helps ensure that emotion remains controlled and regulated.

Figure 47: The Paradigmatic Structure of the Known presents the "personality" of a typical Western individual—in this case, a middle-class businessman and father. His individual life is nested within an increasingly transpersonal, shared "personality," with deep, increasingly implicit historical roots. The "smaller stories," nested within the larger, are dependent for their continued utility on maintenance of the larger—as the middle-class family, for example, is dependent for its economic stability on the capitalist system, as the capitalist system is nested in humanistic Western thought, as humanism is dependent on the notion of the inherent value of the individual (on the notion of "individual rights"), and as the inherent value of the individual is dependent on his association, or ritual identification, with the exploratory communicative hero. The more encompassing "outer" levels of organization may be extant purely in behavior—that is, the individual in question may have little or no explicit imagistic or semantic knowledge of his historical roots, although he still "acts out" a

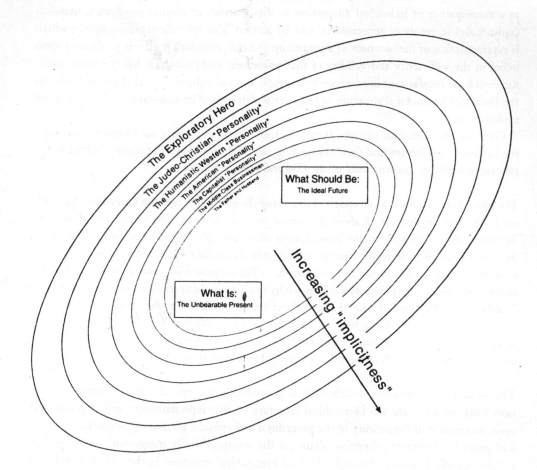

Figure 47: The Paradigmatic Structure of the Known

historically conditioned personality. It is certainly possible, as well—and is increasingly the norm—for an individual to deny explicit "belief" in the validity of the Judeo-Christian ethic or the existence of any transpersonal "exploratory hero" whatsoever. This denial, at the explicit (verbalizable) level of "consciousness" merely interferes with the integrity of the personality in question. The procedural aspect that largely constitutes Judeo-Christian belief (for example)—and even ritual identification with the hero, to some degree (the "imitation of Christ")—almost inevitably remains intact (at least in the case of the "respectable citizen"). The modern educated individual therefore "acts out" but does not "believe." It might be said that the lack of isomorphism between explicit abstract self-representation and actions undertaken in reality makes for substantial existential confusion—and for susceptibility to sudden dominance by any ideology providing a "more complete" explanation. Equally or even more troublesome is the tendency of lack of explicit "belief" to manifest itself, slowly, in

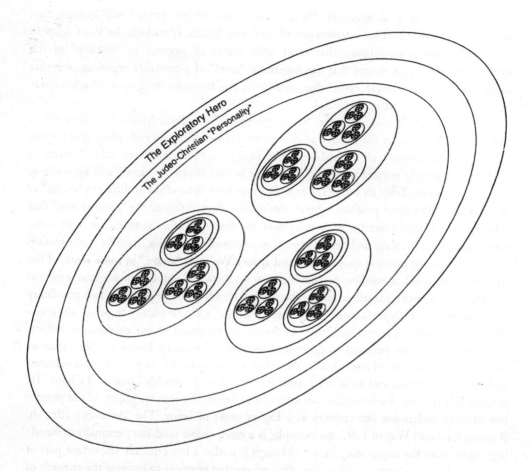

The Exploratory Hero

The Judeo-Christian "Personality"

Figure 48: The Known: Nested Groups and Individuals

alteration of imagistic representation and behavior (as ideas change actions, over time) and to "invisibly" undermine intrapsychic and social stability.

Groups and individuals may differ in their goals, values and behaviors at one level of analysis, while sharing features in common at "higher," more implicit levels. *Figure 48: The Known: Nested Groups and Individuals* portrays three such groups. This number is arbitrary: Catholic, Protestant and Greek Orthodox Christians, for example, might all be regarded as enveloped by their participation in the Judeo-Christian "personality"; although they may well fight among themselves, at the drop of a hat ("within" the confines of that personality), they are liable to eagerly join forces, to eliminate a threat, real or perceived, from Jews or Muslims. *Within* each of these three groups there are going to be differences and similarities, as well. Each community of believers is likely to have its separate sects, separated from one another by a certain historical duration (and the alterations in value structure and behavior that accompany such divergence). Finally, individuals within groups will diverge, too, according to their individual interests and idiosyncratic beliefs. (Paradoxically, it is fidelity to these *individual* characteristics that most truly unites all persons in "worship" of the exploratory hero. This means that the innermost "level" of personality organization—that aspect which is truly unique, rather than shared—is also the outer level, upon which the stability of the entire structure depends.)

The emergence of anomaly—the "re-emergence of the Great Mother"—constitutes a threat to the integrity of the moral tradition governing behavior and evaluation. It is for this reason that adjustment to anomaly—in the many "mythologically equivalent" forms it takes—is frequently resisted passively (by "failure to take it into account") and aggressively (by attempts to eradicate its source). Anomalies may have their effect at different "levels," as we have seen. The most profound threats undermine the stability of the "personalities" that encompass the largest number of people, have the deepest historical roots, are most completely grounded in image and behavior—are most broadly applicable, regardless of situation ("cover" the largest possible span of time and space). We seem "aware," in some sense, of the danger of profound anomalies, perhaps because a substantial amount of negative emotion and abstract cognitive consideration can be elicited merely through positing their possibility ("what if we *were* truly threatened by the foreign devils?"). Our tendency to personally identify with our respective countries, say—to foster and be proud of our patriotism—reflects "knowledge" that our personal integrity and security is integrally bound up, for better or worse, with the destiny of our cultures. We are therefore motivated to protect those cultures, to defend our societies and ourselves against the "return of the terrible Dragon of Chaos." [It is frequently the case, however, that our attempts to bolster the security of part of our protective identity undermine our stability at a higher order of being. The "American (British, Russian, Chinese) Way of Life," for example, is a more visible (and less personally demanding) figure than the exploratory hero—although it is also a less critically important part of our core cultural and personal identities. This means that attempts to increase the strength of the state at the cost of the individual are counterproductive, even though they may serve to heighten the sense of order and regulate emotion in the short term. Patriotism—or any similar attempt at strengthening group identity—must necessarily be bounded by supreme regard for the creative capacity of the individual.]

The individual is protected from chaos in its full manifestation by the many "walls" that surround him. All the space outside a given wall, however—despite its probable encapsulation by additional protective structures—appears relatively dangerous to anyone currently within that wall. All "outside territory" evokes fear. This "equivalence" does not mean, however, that all threats are equivalently potent—just that anything "outside" shares the capacity to frighten (or enlighten) anything "inside." Challenges posed to the "highest" levels of order are clearly the most profound, and are likely to engender the most thorough reactions. Observation of response to such threats may be complicated, however, by the problem of time frame: challenge posed to extremely "implicit" personalities may evoke reactions that extend over centuries, in the form of abstract exploration and argumentation, revision of action, and war between opposing alternative viewpoints (as in the case, for example, of the Catholic and Protestant Christians). The fact that threats posed to the "highest" levels of order are the most profound is complicated, to say it another way, by the "implicitness" of those levels, and their "invisibility." Furthermore, the structures nested within a given personality may have enough intrinsic strength to stand for a long while after the outer walls that protected them and provided them with structural integrity have been breached and destroyed. The stability of a political or social structure once nested in a damaged religious preconception might be likened to a building standing after an earthquake: superficially, it looks intact, but one more minor shake may be sufficient to bring it crashing down. The "death of God" in the modern world looks like an accomplished fact, and perhaps an event whose repercussions have not proved fatal. But the existential upheaval and philosophical uncertainty characteristic of the first three-quarters of the twentieth century demonstrate that we have not yet settled back on firm ground. Our current miraculous state of relative peace and economic tranquillity should not blind us to the fact that gaping holes remain in our spirits.

The chaos "hidden" or given form by the establishment of temporal order may remanifest itself at any time. It may do so in a number of guises, of apparent diversity. Any re-emergence of chaos, however—whatever the reason—may be regarded as the same sort of event, from the perspective of emotion, motivational significance or meaning. This is to say that all things that threatens the status quo, regardless of their "objective" features, tends to be placed into the same "natural category," as a consequence of their affective identity. The barbarian at the gates is therefore indistinguishable from the heretic within; both are equivalent to the natural disaster, to the disappearance of the hero, and to the emergent senility of the king. The "re-emergence of the Dragon of Chaos," whatever his form, constitutes the unleashing of dangerous, fear-producing (and promising) potential. The different "guises" of this potential, and the reasons for and nature of their equivalence, constitute our next topic of discussion. The nature of the response evoked by that potential provide subject matter for the remainder of the book.

PARTICULAR FORMS OF ANOMALY:
THE STRANGE, THE STRANGER, THE STRANGE IDEA
AND THE REVOLUTIONARY HERO

Anomalous events share capacity to threaten the integrity of the known, to disrupt the "familiar and explored." Such events, while differing in their specific details and manner of manifestation, tend to occupy the same natural category. Threats to the stability of cultural tradition emerge in four "mythologically inseparable" manners: through rapid natural environmental shift, "independent" of human activity; through contact with a heretofore isolated foreign culture; through application of novel (revolutionary) linguistically or episodically mediated critical skill—the inevitable consequence of increasing ability to abstract, learn and communicate; and as a consequence of revolutionary heroic activity.

The "natural" human tendency to respond to the stranger, the strange idea and the creative individual with fear and aggression can be more easily comprehended, once it is understood that these diverse phenomena share categorical identity with the "natural disaster." The problem with this "natural" response pattern, however, is that the upsetting capacity of the anomalous is simultaneously the vital source of interest, meaning and individual strength. Furthermore, the ability to upset ourselves—to undermine and revitalize our own beliefs—is an intrinsic, necessary and "divine" aspect of the human psyche (part of the seminal "Word" itself).

The Word—in its guise as painstakingly abstracted action and object—can create new worlds and destroy old; can pose an unbearable threat to seemingly stable cultures, and can redeem those that have become senescent, inflexible and paralytic.

To those who have sold their souls to the group, however, the Word is indistinguishable from the enemy.

The Strange

Transformation of "environmental" circumstances, as the consequence of purely natural causes, constitutes the single most immediately evident cause for the deterioration of cultural stability. Prolonged drought, floods, earthquakes, plagues—nature's most horrifying and arbitrary occurrences—are capable of rendering the most carefully adapted societies impotent at a single blow.

Natural disasters of this sort might merely be considered rapid transformation—situations where previously noted affectively relevant environmental relationships alter faster than adaptive movement keeps pace. This means that the insufficiency of cultural adaptation cannot easily be distinguished from natural catastrophe. A society light on its feet, so to speak, is constantly in a position to adapt to the unexpected—even the catastrophic—and to transform such change into something beneficial (consider, for example, the postwar Japanese). The relationship "natural disaster/cultural adaptation" therefore constitutes the social analogue to that obtaining between "emotion" and "cognition": affect generated, in large part, as a consequence of novelty, always emerges where something is not known (and is therefore always dependent on what is known); is always experienced in relationship to some conceptualization of the present, the future and the means to get from one to the other. What constitutes "novelty," then, is dependent on what is not novel in a particular circumstance. What constitutes "trauma" depends, likewise, on the behavioral repertoire and value schema available for use at the time of a given event or transformation. A blizzard that would incapacitate Washington for a month barely makes the residents of Montreal blink.

Mythic representations of the rapid mutation of environmental contingency (portrayed as the reappearance of the Great Mother or the Dragon of Chaos) are in consequence necessarily "contaminated" with images of the sterile, senescent or tyrannical king, whose inflexibility renders all inevitable environmental transformation deadly. When is a disaster not a disaster? When the community is prepared to respond appropriately. Conversely, any minor change in

the natural world might be regarded as terminal, catastrophic—and actually be so—when
the adaptive structure designed to fit that world has become so authoritarian that any change
whatsoever is reflexively deemed forbidden, heretical.[383] A society with this attitude—such
as the former Soviet Union—is an accident waiting to happen. An interesting example of
the consequences of such inflexibility, on the personal scale, is offered by Kuhn:

In a psychological experiment that deserves to be far better known outside the trade, Bruner and
Postman[384] asked experimental subjects to identify on short and controlled exposure a series of playing
cards. Many of the cards were normal, but some were made anomalous, e.g., a red six of spades and a
black four of hearts. Each experimental run was constituted by the display of a single card to a single
subject in a series of gradually increased exposures. After each exposure the subject was asked what he
had seen, and the run was terminated by two successive correct identifications.

Even on the shortest exposures many subjects identified most of the cards, and after a small increase
all the subjects identified them all. For the normal cards these identifications were usually correct, but
the anomalous cards were almost always identified, without apparent hesitation or puzzlement, as normal. The black four of hearts might, for example, be identified as the four of either spades or hearts.
Without any awareness of trouble, it was immediately fitted to one of the conceptual categories prepared by prior experience. One would not even like to say that the subjects had seen something different from what they identified. With a further increase of exposure to the anomalous cards, subjects did
begin to hesitate and to display awareness of anomaly. Exposed, for example, to the red six of spades,
some would say: "That's the six of spades, but there's something wrong with it—the black has a red
border." Further increase of exposure resulted in still more hesitation and confusion until finally, and
sometimes quite suddenly, most subjects would produce the correct identification without hesitation.
Moreover, after doing this with two or three of the anomalous cards, they would have little further difficulty with the others. A few subjects, however, were never able to make the requisite adjustment of
their categories. Even at forty times the average exposure required to recognize normal cards for what
they were, more than 10 per cent of the anomalous cards were not correctly identified. And the subjects who then failed often experienced acute personal distress. One of them exclaimed: "I can't make
the suit out, whatever it is. It didn't even look like a card that time. I don't know what color it is now or
whether it's a spade or a heart. I'm not even sure now what a spade looks like. My God!"[385]

Myth and literature constantly represent the "parched kingdom," the society victimized
(most frequently) by drought—which is the absence of water, concretely, and the "water of
life" or spirit, symbolically—brought on by the over-prolonged dominance of the (once
great) ruling idea. This idea, in the narrative (and frequently, in actuality), is the king, the
ancestral spirit, representative of his people, made tyrannical by age, pride, or unbearable disappointment, withering under the influence of some willfully misunderstood malevolent
advising force. The development of such an unpleasant and dangerous situations calls, of
course, for the entrance of the hero: the "lost son" of the true king, raised in secrecy by alternative parents; the rightful ruler of the kingdom, whose authority was undermined or who
was supposedly killed during vulnerable youth; the proper heir to the throne, who had been

journeying in far-off lands and was presumed dead. The hero overturns the tyrant and regains his proper place; the gods, pleased by the re-establishment of proper order, allow the rain once more to fall (or stop it from falling in dangerous excess). In a story of this type, the creative aspect of the unknown (nature) is "locked away," metaphorically, by the totalitarian opinion of the current culture. Such a state of affairs might be represented, for example, by the sleeping princess, in the kingdom brought to a standstill (or by some alternative variant of the existence of the "treasure hard to attain"[386]). Paralyzed by patriarchal despotism[387] (or, frequently, by fear of the Terrible Mother), the kingdom remains stagnant, while the princess—nature, in her benevolent guise—waits for the kiss of the hero to wake. Her awakened and revitalized beauty subsequently reanimates her people.

Rituals of the death and renewal of the king act out this transformation of cultural adaptation long before the concept of rebirth can be rendered abstractly comprehensible. Frye states:

The hypothetical ritual studied in Frazer's *Golden Bough* may be vulnerable enough in various anthropological contexts, but as a mythical structure it is as solid as the pyramids. Here a king regarded as divine is put to death at the height of his powers, for fear that his physical weakening will bring a corresponding impotence to the fertility of the land he rules.... When sacrificed, the divine king is immediately replaced by a successor, and his body is then eaten and his blood drunk in a ritual ceremony. We have to make a rather violent effort of visualization to see that there are now two bodies of the divine king, one incarnate in the successor, the other concealed in the bellies of his worshippers. The latter causes the society to become, through eating and drinking the same person, integrated into a single body, which is both their own and his.[388]

The extensive and universal corpus of dying-and-resurrecting-god myths[389] (acted in sacrificial ritual) dramatize two notions. The first is that the actual ideas/patterns of behavior governing adaptation must die and be reborn to ensure constant update of the techniques of survival. The second, more fundamental, is that the hero—the active agent of adaptation—must eternally upset the protective structure of tradition and enter into "sacrificial union" with the re-emergent unknown. Cosmological phenomena themselves "act out" (are utilized as descriptive tools for, more accurately) this eternal drama: the sun (god), born in the east, "dies" in the west, and passes into the underworld of night (into the lair of the dragon of chaos). Nightly, the sun hero battles the terrible forces of chaos, cuts himself out of the belly of the beast and is reborn triumphant in the morning.

The master of the strange in its "natural form" is the hero in his technological guise (more particularly, say, than in his role as social revolutionary). Marduk, who faced Tiamat in single combat, is a very focused representative of man's "mastery" over nature. The pattern of action signified by this god—that is, courageous and creative approach in the face of uncertainty—was regarded "unconsciously" by the Mesopotamians as necessary, as stated previously, to the "creation of ingenious things" from the "conflict with Tiamat."[390] The hero fashions defenses out of nature to use against nature. This idea, which underlies man's cultural adaptation, manifests itself "naturally" in the human psyche:

Spontaneous fantasy manifested August 10, 1997, by my daughter, Mikhaila (age five years, eight months) while playing "prince and princess" with Julian (her three-year-old brother): Dad, if we killed a dragon, we could use his skin as armor, couldn't we? Wouldn't that be a good idea?

The hero uses the positive aspect of the Great Mother as protection from her negative counterpart. In this manner, the "natural disaster" is kept at bay or, better yet, transformed from a crisis into an opportunity.

The Stranger

Arrival of the stranger, concretely presented in mythology, constitutes a threat "to the stability of the kingdom," metaphorically indistinguishable from that posed by "environmental transformation." The stable *meaning* of experiential events, constrained by the hierarchical structure of group identity, is easily disrupted by the presence of the "other," who practically poses a concrete threat to the stability of the present dominance structure, and who, more abstractly—as his actions "contain" his moral tradition—exists as the literal embodiment of challenges to the *a priori* assumptions guiding belief. The stranger does not act in the manner expected. His inherent unpredictability renders him indistinguishable from the unknown, as such, and easily identified with the force constantly working to undermine order. From a within-group perspective, so to speak, such identification is not purely arbitrary, either, as the mere existence of the (successful) stranger poses serious threat to the perceived utility of the general culture—and, therefore, to its ability to inhibit existential terror and provide determinate meaning to action.

When the members of one isolated group come into contact with the members of another, the stage is therefore set for trouble. Each culture, each group, evolved to protect its individual members from the unknown—from the abysmal forces of the Great and Terrible Mother, from unbearable affect itself. Each evolved to structure social relationships and render them predictable, to provide a goal and the means to attain it. All cultures provide their constituent individuals with particular modes of being in the face of terror and uncertainty. All cultures are stable, integrated, hierarchically arranged structures predicated upon assumptions held as absolute—but the particular natures of these assumptions differ (at least at the more comprehensible and "conscious" levels of analyses). Every culture represents an idiosyncratic paradigm, a pattern of behaving in the face of the unknown, and the paradigm cannot be shifted (its basic axioms cannot be modified), without dramatic consequences—without dissolution, metaphoric death—prior to (potential) reconstruction.

Every society provides protection from the unknown. The unknown itself is a dangerous thing, full of unpredictability and threat. Chaotic social relationships (destructured dominance hierarchies) create severe anxiety and dramatically heighten the potential for interpersonal conflict. Furthermore, the dissolution of culturally determined goals renders individual life, identified with those goals, meaningless and unrewarding in intrinsic essence. It is neither reasonable nor possible to simply abandon a particular culture, which is a pattern of general adaptation, just because someone else comes along who does things a different way, whose actions are predicated on different assumptions. It is no simple matter to rebuild social relationships in the wake of new ideas. It is no straightforward process, furthermore, to give up a goal, a central unifying and motivating idea. Identification of an individual with a group means that individual psychological stability is staked on maintenance of group welfare. If the group founders suddenly as a consequence of external circumstance or internal strife, the individual is laid bare to the world, his social context disappears, his reason for being vanishes, he is swallowed up by the unbearable unknown, and he cannot easily survive. Nietzsche states:

In an age of disintegration that mixes races indiscriminately, human beings have in their bodies the heritage of multiple origins, that is, opposite, and often not merely opposite, drives and value standards that fight each other and rarely permit each other any rest. Such human beings of late cultures and refracted lights will on the average be weaker human beings: their most profound desire is that the war they *are* should come to an end.[391]

Of course, the unstated conclusion to Nietzsche's observation is that the war typifying the person of "mixed race" (mixed *culture,* in more modern terminology) is the affectively unpleasant precursor to the state of mind characterizing the more thoroughly integrated individual, who has "won" the war. This "victor"—who has organized the currently warring diverse cultural standpoints into a hierarchy, integrated once more—will be stronger than his "unicultural" predecessor, as his behavior and values will be the consequence of the more diverse and broader ranging union of heretofore separate cultures. It is reasonable to presuppose that it was the "unconscious" consideration of the potentially positive outcome of such mixing that led Nietzsche to the revelation of the dawning future "superman."[392] It is not the mere existence of various previously separated presuppositions in a single psyche that constitutes the postcontact victory, however. This means that the simplistic promotion of "cultural diversity" as panacea is likely to produce anomie, nihilism and conservative backlash. It is the molding of these diverse beliefs into a single hierarchy that is precondition for the peaceful admixture of all. This molding can only be accomplished by war conducted between paradoxical elements, within the "postcontact" individual psyche. Such a war is so difficult—so emotionally upsetting and cognitively challenging—that murder of the anomalous "other" in the morally acceptable guise of traditional war frequently seems a comforting alternative.

Fundamental threats can be posed very easily between groups of people. Most concretely, foreign *behaviors* are threatening, unpredictable in particular, terrifying in general—because essential beliefs, challenging beliefs, are most convincingly expressed through actions:

He became to us a reproof of our thoughts;
the very sight of him is a burden to us, because his manner of life is unlike that of others, and his ways
    are strange. (Wisdom 2:14–15 RSV)

A foreign man, a stranger, is threatening because he is not firmly fixed within a social hierarchy, and may therefore behave unpredictably—with unpredictable consequences for the social hierarchy. Signals of safety and threat vary, or may vary, between members of different groups. Unpredictable means potentially dangerous. More abstractly, what the stranger *believes*, specifically, threatens the integrated structure of historically determined belief, in general. This does not present a problem, when his foreign actions or ideas do not produce fundamental conflict—do not threaten key beliefs. When basic concepts are threatened, however, the unbearable, terrible unknown once again rises up, and once-firm ground begins to give way.

The Strange Idea

Increasing ability to abstract makes previous learning, established through nonabstract means, increasingly modifiable—and increasingly vulnerable. In a way, this is the whole point of abstraction, and the very capacity to learn. Words, deceptively simple and harmless, are sufficient to create disruption and conflict, because *Homo sapiens* can verbalize his beliefs. It could be said, therefore, with sufficient rationale that a *new idea* is an *abstract stranger* (or, by the same logic, a *natural disaster*). It is for this reason that the pen is mightier than the sword.

The process of increased abstraction allows for increasing self-understanding (self-consciousness)—at least in potential—and for the prediction of the behaviors of others [which is a capacity integrally linked to the development of self-consciousness (how would I behave in a situation like that?)]. In addition, abstraction eases communication of morality (instruction in how to behave), by making it unnecessary to wait around to watch until something important actually happens. The use of drama, for example—which is the representation of behavior, *in behavior and image*—allows us to watch the interplay of issues of mortal consequence, without the actors or the observers actually suffering that consequence.

The capacity to abstract has not come without price, however. The incautious, imaginative (and resentful) can easily use their gift of socially constructed intelligence to undermine moral principles that took eons to generate and that exist for valid but invisible reasons. Such "invisible" principles can be subjected to facile criticism, by the historically ignorant, once they take imagistic, written or spoken form. The consequence of this "criticism" is the undermining of necessary faith, and the consequent dissolution of interpersonal predictability, dysregulation of emotion, and generation of anomie, aggression and ideological gullibility (as the naked psyche strives to clothe itself, once again).

The danger of such criticism can be more particularly appreciated when the effect of what might be described as *cascade* is considered. We can change our behaviors because we change how we think—although this is not as simple as is generally considered. We can change how we think, facilely, and without regard for the consequences, partly because we do not understand why we think what we think (because all the facts that govern our behavior are not at our "conscious" disposal) and because the effects of that change are often not immediately apparent. The fact that changes in tradition have unintended and often dangerous "side effects" accounts for the conservatism of most human cultures. "Cascade" means that threat to the perceived validity of any presupposition, at any level (procedural, imagistic or episodic, explicit or semantic) threatens all levels simultaneously. This means that the casual criticism

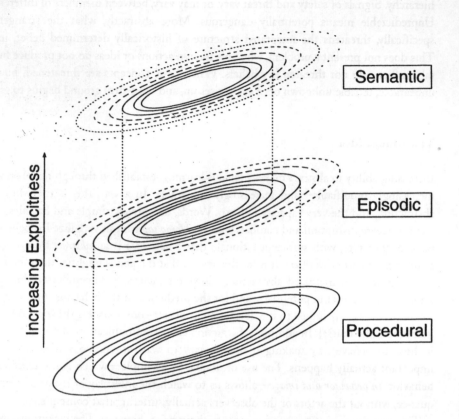

Figure 49: The Fragmentary Representation of "Procedure and Custom" in Image and Word

of a given explicit presupposition can come, over time, to undermine the unconscious imagistic and procedural personality and the emotional stability that accompanies it. Words have a power that belie their ease of use.

Figure 49: The Fragmentary Representation of "Procedure and Custom" in Image and Word provides a schematic representation of the organization of behavior and schemas of value in "memory." Customs—that is, predictable and stable patterns of behavior—emerge and are stored "procedurally" as a consequence of constant social interaction, over time, and as a result of the exchange of emotional information that characterizes that interaction. You modify me, I modify you, we both modify others, and so on, in a cycle that involves thousands of individuals, over thousands of years. Most of this information is a more or less permanent part of the social network (is part of the structure of society), but can become represented in part or whole in image, and then more explicitly, in verbal code. The imagistic representation of the morality constituting a given society is likely to be incomplete, as the complexity of the patterns emerging consequential to the totality of social interaction exceeds (current) representational capacity. The semantic representations perched above the images are likely to be even more incomplete. This means that the verbal systems utilized in abstract thinking, for example, contain only "part of the puzzle," at best; they have only partial information regarding the structure of the whole. So, while some of the rules governing behavior have become completely explicit and understood, others will remain partially implicit (and poorly understood). Some of these partially implicit rules are likely to exist for completely implicit (and therefore completely invisible) reasons. It is rules like these on the ragged edge of comprehension that are likely to attract ill-informed but nonetheless potentially devastating criticism. Abstract verbal intelligence may therefore pick holes in the "absurd mythological structure" that supports it, without understanding either that it is supported, or that the act of undermining is existentially—mortally—dangerous. It is easy to criticize the notion of the "immortal soul," for example, and the traditional forms of morality that tend to accompany such a belief, without realizing that there is much more to the idea than meets the eye.

"Cascade" means that threat to the perceived validity of any presupposition, at any level—generally verbally mediated—now becomes threat to that presupposition and to everything that rests on it. The socially mediated capacity to abstract—to reason and represent in behavior, imagination and word—means that an ill-chosen action, fantasy or thought may have devastating consequences. This is true in particular of the word. One well-chosen phrase can change *everything* ("from each according to his ability ..."). The word, in a particular context (one established by behavior and episodic representation) has a polysemous significance—it excludes more (constrains more) than it appears to and means more than it "contains," considered as an isolated or decontextualized element. It has this capacity in part because it is capable of referring to phenomena outside its domain, in order to make itself understood (this is use of metaphor). The word brings to mind events and actions, sequenced in a particular manner; it is the imaginary presentation of these events and actions that contains much of the meaning—the words merely act as cues for retrieval. The information retrieved is not necessarily yet semantic; it may still remain embodied in episodic memory and procedure. The polysemous quality of the meaningful word, which implies something for imagistic representation and for the structuring of behavior, is what makes it potent and dangerous. An entire behavioral hierarchy can be undermined by a well-chosen creative phrase, because the phrase brings with it, as integral part of an integrated whole, moral presuppositions of entirely different, and perhaps logically (or at least apparently) contrary nature.

There is an apocryphal story about a cosmologist, lecturing to a rural audience of laypeople in the late 1800s. He describes the basic structure of the solar system, laying emphasis on the fact that the earth floats unsupported in space, endlessly circling the sun. After the lecture, an old woman approaches the podium and says:

"That was a very interesting story, young man. Of course, it is completely absurd."
"Absurd, madam?" the lecturer inquired. "Whatever do you mean?"

"It is a well-known fact," replied the old woman, "that the earth rests on the back of a giant turtle."

"Is that so, ma'am. What, then, does the turtle rest on?"

"Don't play games with me, young man," responded the matron. "It's turtles all the way down."[393]

Douglas Hofstadter presented a similar idea in a fictional discussion between Achilles, the Greek hero, and a tortoise (of *Zeno's paradox* fame):

Tortoise: ... For purposes of illustration, let me suggest that you consider the simpler statement "29 is prime." Now in fact, this statement really means that 2 times 2 is not 29, and 5 times 6 is not 29, and so forth, doesn't it?

Achilles: It must, I suppose.

Tortoise: But you are perfectly happy to collect all such facts together, and attach them in a bundle to the number 29, saying merely, "29 is prime?"

Achilles: Yes ...

Tortoise: And the number of facts involved is actually infinite, isn't it? After all, such facts as "4444 times 3333 is not 29" are all part of it, aren't they?

Achilles: Strictly speaking, I suppose so. But you and I both know that you can't produce 29 by multiplying two numbers which are both bigger than 29. So in reality, saying "29 is prime" is only summarizing a FINITE number of facts about multiplication.

Tortoise: You can put it that way if you want, but think of this: the fact that two numbers which are bigger than 29 can't have a product equal to 29 involves the entire structure of the number system. In that sense, that fact in itself is a summary of an infinite number of facts. You can't get away from the fact, Achilles, that when you say "29 is prime," you are actually stating an infinite number of things.

Achilles: Maybe so, but it feels like just one fact to me.

Tortoise: That's because an infinitude of facts are contained in your prior knowledge—they are embedded implicitly in the way you visualize things. You don't see an explicit infinity because it is captured implicitly inside the images you manipulate.[394]

Jerome Bruner's comments on "triggers" are equally apropos here. He gives the following sentences as examples: Trigger: "John saw/didn't see the chimera." Presupposition: "There exists a chimera." Trigger: "John realized/didn't realize he was broke." Presupposition: "John was broke." Trigger: "John managed/didn't manage to open the door." Presupposition: "John tried to open the door." There "exists" a virtually infinite number of "presuppositions" for every "trigger." Bruner states: "Obviously you cannot press a reader (or a listener) to make endless interpretations of your obscure remarks. But you can go a surprisingly long way—provided only that you start with something approximating what Joseph Campbell called a "mythologically instructed community."[395] The transmission of what is generally regarded as spiritual wisdom is in fact able to take (to be "reduced to") narrative form precisely because the word—in the context of the story, which is description of episodic representation of events and behaviors—has this deceptively simple, yet infinitely meaningful "triggering" property:

Another parable put he forth unto them, saying, The kingdom of heaven is like to a grain of mustard seed, which a man took, and sowed in his field:

Which indeed is the least of all seeds: but when it is grown, it is the greatest among herbs, and becometh a tree, so that the birds of the air come and lodge in the branches thereof.

Another parable spake he unto them; The kingdom of heaven is like unto leaven, which a woman took, and hid in three measures of meal, till the whole was leavened.

All these things spake Jesus unto the multitude in parables; and without a parable spake he not to them:

That it might be fulfilled which was spoken by the prophet, saying, I will open my mouth in parables; I will utter things which have been kept secret from the foundation of the world. (Matthew 13:31–35)

It is not merely the story which is saturated with meaning; it is imagination, behavior and the practical consequences of imagination and behavior, as well. The individual ideas, particular fantasies and personal actions of individuals presuppose the culture from which they are derived. The word, in meaningful context, is meaningful precisely because it provides information relevant to episodic representation, *per se*, and because it has relevance—which may not be "consciously" comprehensible or declarable—for behavior. Likewise, the behavior and fantasies of self and other—in context—are predicated upon culturally determined values and beliefs, and could be said, in a manner of speaking, to contain them. It is for this reason that Jung could claim, with regard to the fantasies of a modern dreamer:

He is in fact an unconscious exponent of an autonomous psychic development, just like the medieval alchemist or the classical Neoplatonist. Hence one could say—*cum grano salis*—that history could be constructed just as easily from one's own unconscious as from the actual texts.[396]

Even the more concrete implement or tool—like the word—is not an artifact separable from the culture in which it is produced. It is failure to comprehend this fact that dooms many well-meaning "foreign aid" projects and, no less, the foreigners to which such aid is granted. Even something as simple as the shovel or hoe presupposes the existence of a culture that has granted the individual dominion over nature, so that the individual has the right to make the Great Mother subservient to the claims of man. This notion constitutes the central idea of complexly civilized patriarchal culture, and emerges into consciousness, against competing claims, with the greatest of difficulty:

An American-Indian prophet, Smohalla, of the tribe of Umatilla, refused to till the soil. "It is a sin," he said, "to wound or cut, to tear or scratch our common mother by working at agriculture." And he added: "You ask me to dig in the earth? Am I to take a knife and plunge it into the breast of my mother? But then, when I die, she will not gather me again into her bosom. You tell me to dig up and take away the stones. Must I mutilate her flesh so as to get at her bones? Then I can never again enter into her body and be born again. You ask me to cut the grass and the corn and sell them, to get rich like the white men. But how dare I crop the hair of my mother?"[397]

Every society shares a moral viewpoint, which is essentially an identity composed of unquestioned fidelity to a particular conception of "reality" (*what is* and *what should be*), and of agreement upon the nature of those behaviors that may reasonably be manifested. All the individuals in a particular nation agree, fundamentally, about the nature of the unbearable present, the ideal future, and the means to transform one into the other. Every individual plays out that conceptualization, in terms of his or her own actions, more or less successfully: more successfully, or at least more easily, when nothing unintended arises to make the act of questioning necessary; less successfully when the moral action does not produce the proper consequence. Any assumption can be challenged. The most fundamental expectation of my fantasies—whatever they might be—is that my assumptions are valid. Mismatch between what I desired and what actually occurred constitutes evidences that one or more of my assumptions are invalid (but not necessarily information about which one, or at what level). The outcome of such a mismatch is application of other (assumption-predicated) patterns of action, and associated expectations, associated with gathering of new information, through active exploration. The further down the hierarchy of assumption that mismatch occurs, the more *stressful* the occurrence, the more fear is disinhibited, the more motivation for denial, the more necessity for exploration, the more necessary reprogramming of behavioral assumption and matching sensory expectation.

A truly unexpected event sequence upsets the implicit assumptions upon which the original particular fantasy was predicated—and not only that fantasy, but *innumerable presently implicit others, equally dependent for their existence upon those violated presuppositions.* The inevitable consequence of such violation is the breakdown of expectation, and consequent generation of fear and hope, followed by exploration, the attempt to adapt to the new environment (to behave appropriately, to fulfill motivational demands under new conditions, and to map new conditions). This consequence requires the paralysis of the old model, reversion of otherwise stably maintained affects to competition and chaos, and exploration-guided reconstruction of order.

The more basic the level, the more that assumption is shared by virtually every conceivable fantasy. The more basic the level undermined, the more anxiety and depression [and other motivation—particularly (and non-evidently) *hope*] released from containment; the more behavioral adaptation cast into disrepute—the more motivation for denial, deceit, fascistic readaptation, degeneration and despair—the more wish for redemption. The undermining and reconstruction of more basic levels is, as we have seen, a revolutionary act, even in the scientific domain. The "normal" scientist works within the constraints of great models; the revolutionary changes the models. The normal scientist accepts the (current) game as valid, and tries to extend its relevant domain. The revolutionary scientist, who alters the rules of the game themselves, is playing a different game (with different and dangerous rules, from a within-the-game perspective). Kuhn states:

The transition from a paradigm in crisis to a new one from which a new tradition of normal science can emerge is far from a cumulative process, one achieved by an articulation or extension of the old paradigm. Rather it is a reconstruction of the field from new fundamentals, a reconstruction that changes some of the field's most elementary theoretical generalizations as well as many of its paradigm methods and applications. During the transition period there will be a large but never complete overlap between the problems that can be solved by the old and the new paradigm. But there will also be a decisive difference in the modes of solution. When the transition is complete, the profession will have changed its view of the field, its methods, and its goals.[398]

The normal scientist is often antithetical to his more extreme (more creative/destructive) counterpart, like the good citizen opposes the heretic, in part because alteration of the rules changes the motivational significance of previously valued action and thought—often, apparently, reducing it to zero (which means that the revolutionary can completely destroy the significance of the career, past, present and future, of the dedicated plodder), in part, because restructuring of the rules temporarily returns everything to a state of anxiety-provoking chaos. Kuhn states:

A paradigm is prerequisite to perception itself. What a man sees depends both upon what he looks at and also upon what his previous visual-conceptual experience has taught him to see. In the absence of such training there can only be, in William James's phrase, "a bloomin' buzzin' confusion."[399]

That "bloomin' buzzin' confusion"—the Great Dragon of Chaos—is not affectively neutral: in fact, its affective significance, threat and promise, is perhaps all that can be experienced of it before it has been categorized.

Sometimes new information means mere lateral adjustment of behavior—the modification of approach, within a domain where still defined by the familiar goal. Sometimes, however, the unknown emerges in a manner that demands a qualitative adjustment in adaptive strategy: the revaluation of past, present and future, and acceptance of the suffering and confusion this necessarily entails. Kuhn comments on the effect (and affect) of emergent and persistent unknown in the domain of science. The pattern he describes characterized all cognitive revolutions, including those that take place in the universe of normal morality:

When ... an anomaly comes to seem more than just another puzzle of normal science, the transition to crisis and to extraordinary science has begun. The anomaly itself now comes to be more generally recognized as such by the profession. More and more attention is devoted to it by more and more of the field's most eminent men. If it still continues to resist, as it usually does not, many of them may come to view its resolution as *the* subject matter of their discipline. For them the field will no longer look quite the same as it had earlier. Part of its different appearance results simply from the new fixation point of scientific scrutiny. An even more important source of change is the divergent nature of the numerous partial solutions that concerted attention to the problem has made available. The early attacks upon the resistant problem will have followed the paradigm rules quite closely. But with continuing resistance, more and more of the attacks upon it will have involved some minor or not so minor articulation of the paradigm, no two of them quite alike, each partially successful, but none sufficiently so to be accepted as paradigm by the group. Through this proliferation of divergent articulations (more and more frequently they will come to be described as *ad hoc* adjustments), the rules of normal science become increasingly blurred. Though there still is a paradigm, few practitioners prove to be entirely agreed about what it is. Even formerly standard solutions of solved problems are called in question.

When acute, this situation is sometimes recognized by the scientists involved. Copernicus complained that in his day astronomers were so "inconsistent in these [astronomical] investigations ... that they cannot even explain or observe the constant length of the seasonal year. With them," he continued, "it is as though an artist were to gather the hands, feet, head and other members for his images from diverse models, each part excellently drawn, but not related to a single body, and since they in no way match each other, the result would be monster rather than man."[400] Einstein, restricted by current usage to less florid language, wrote only, "It was as if the ground had been pulled out from under one, with no firm foundation to be seen anywhere, upon which one could have built."[401] And Wolfgang Pauli, in the months before Heisenberg's paper on matrix mechanics pointed the way to a new quantum theory, wrote to a friend, "At the moment physics is again terribly confused. In any case, it is too difficult for me, and I wish I had been a movie comedian or something of the sort and had never heard of physics." That testimony is particularly impressive if contrasted with Pauli's words less than five months later: "Heisenberg's type of mechanics has again given me hope and joy in life. To be sure it does not supply the solution to the riddle, but I believe it is again possible to march forward."[402 403]

Now, Kuhn drew a qualitative distinction between the normal and revolutionary modes of operation. No such qualitative differences exist (although exemplars of the two types, drawn from the "extreme poles" of the process of knowledge-production, can be easily be brought to mind). The distinction is more along the lines of "transformation of what the group wants to transform" vs. "transformation of what the group would like to remain stable"—with the revolutionary changing more than might presently be desired (for the maintenance of the extant social hierarchy, for example). The "transformation of what the group wants to transform" is a form of bounded revolution, as we have discussed previously. Optimally bounded revolutions produce positive affect. Revolutions that upset the desired bounds—which are what Kuhn's revolutionary scientist produces—evoke fear (and denial and aggression as defense mechanisms). The revolutionary produces involuntary alteration in the "articles of faith" of the normal individual. It is this capacity that makes him revolutionary and necessary—and feared and despised. It may be more generally said that processes of "discovery" which upset large-scale space-time "maps" produce disruption of affect on an equivalent scale (and that it is such large-scale disruption that we entitle revolution).

Mythologically structured social and individual "presumptions"—articles of faith—provide the environment in which a given culture-specific adaptive pattern retains its conditional validity. This pre-rational mythic environment is analogous in structure to the physical or natural environment itself—as the structure adapted to the environment rapidly becomes a constituent element of the environment itself, with the same essential characteristics. (Or, to

say it somewhat differently, everything contained outside the wall defining "presently considered space" is "environment," even though much of it is actually the consequence of historical
or even individual activity). Disruption of the "pre-rational mythic 'environment'" is just as
catastrophic as disruption of the "physical or natural environment" (the two "disruptions"
may not really be distinguishable, in the final analysis). This means essentially that to give
serious consideration to another's viewpoint means to risk exposure to indeterminate uncertainty—to risk a rise in existential anxiety, pain and depression; to experience temporally
indeterminate affective, imagistic and cognitive chaos. It is much more likely, in consequence, that a foreign viewpoint will appear evil or will come to be defined as such (especially during times rendered unstable—unbearably novel—for additional alternative reasons).
Once such definition occurs, application of aggression, designed to obliterate the source of
threat, appears morally justified, even required by duty. The alternative or foreign viewpoint
is in fact reasonably considered evil (although this consideration is dangerously one-sided),
when viewed in terms of its potential destructive capacity, from within the strict confines of
the historically determined social-psychological adaptive structure. It is only within the
domain of meta-morality (which is the morality designed to update moral rules) that the
strange may be tolerated or even welcomed.

   The group, in its external social and intrapsychic incarnations, is the current expression of
a form of acting and thinking that has been given particular content over the course of thousands of years. These particular contents, patterns of behavior and their representations, were
established initially by individuals who faced the unknown and prevailed, who were able to
do or think something that no one had been able to do or think before. In this manner, heroic individuals create new assumptions and formulate new values. The integration of these
assumptions and values into the group, through the competitive process that begins with
imitation and ends with verbal abstraction, increases the permanent behavioral and abstract
logical repertoire of the individuals that form that group. The sum total of such behavioral
patterns (and second- and third-order descriptions thereof), shared within a social group,
constitutes that group. Groups are predicated upon a collective, historically determined
structure of (abstractly represented) behavioral patterns (and consequences thereof), which
tends toward internal consistency and stability over time. Internalization of this behavioral
pattern and representations thereof protects the individuals who compose the group against
fear of their own experience. The group is the culturally determined hierarchical structure of
behavior—and abstracted conceptualization thereof—which inhibits fear of novelty, the
Terrible Mother, source of all nightmares. The group is the historical structure that humanity has erected between the individual and the unknown. The group, in its beneficial guise,
serves to protect the individuals who compose it from threat and the unknown. The social
establishment of how to behave, when presented with a given situation, inhibits the paralyzing fear that situation would otherwise instinctively induce.

   The group is also simultaneously the concrete historical expression of *Homo sapiens'*
unique heroic "thesis," as stated previously: that *the nature of experience can be altered, for the
better, by voluntary alteration of action and thought*. This central thesis is expressed in the myth

of the way. Loss of (previously extant) paradise initiates the "redemptive" activity, history; restoration of paradise—in the course or as a consequence of proper behavior—is its goal. This general pattern appears characteristic of all civilizations, every philosophy, every ideology, all religions. The general idea that change may bring improvement—upon which all voluntary change is predicated—is in itself based in the ideal upon the assumption [on the (necessary) fiction] that through historical process perfection might be attained. This myth—even in its earliest ritual incarnation—therefore provides the basis for the idea of progress itself. The group, history incarnate, is the embodiment of a specific mode of being designed to attain perfection, and contains the concrete expression of the goal of a people; it is the objective and subjective realization of the mode by which they improve their tragic condition. History not only protects people from the unknown; it provides them with rules for achieving what they desire most, and, therefore, for expressing the (essentially undeclarable) meaning of their lives.

Human moral knowledge progresses as procedural knowledge expands its domain, as episodic memory encodes, ever more accurately, the patterns that characterize that knowledge; as the semantic system comes to explicitly represent the implicit principles upon which procedural knowledge and episodic representation of that knowledge rest—and, of course, as the consequences of this second- and third-order representation alter the nature of procedure itself. Thus the democratic political theorist, for example, can finally put into words the essence of religious myth after the myth had captured in image the essence of adaptive behavior; can talk about "intrinsic right" as if that notion were something *rational*. This process of increasing abstraction and representation is equivalent to development of "higher" consciousness (especially if the ever-more enlightened words are in fact—utopian wish—transformed back down the hierarchy to the level of action).

The major advantage of increased abstraction of representation, apart from ease of communication, is increased adaptive flexibility: alterations in abstract thought can proceed "as if" a game, without immediate practical consequences, positive or negative.[404] The disadvantage of this adaptive flexibility is the emergence of ability to constantly (and inappropriately, in most cases) undermine the *a priori* presumptions of the game: to call the rules into question; to dissolve impetus for action and to disinhibit existential anxiety. A game is fun until the rules appear childish. Then the fun disappears. This might be progress, in time. Until a new game appears, however, it is merely troublesome. The process of abstract (semantic) inquiry is capable of undermining moral adaptation at each level—semantic, episodic and procedural—simultaneously. This possibility might be regarded, once again, as a (destructive/beneficial) side effect of the ability to abstract.

The evolutionary construction of an adaptive social structure, simultaneously extant in behavior and in semantic/episodic representation of that behavior, means abstraction and hierarchical organization of knowledge hard won in the physical battle for survival, and consequent capacity for immediate communication of that knowledge, in the absence of direct demonstration. Furthermore, it means potential for alteration and experimentation in the abstract (in play, episodic and semantic), prior to application in the real world. Acquisition of such ability—the capacity for abstract creative thought, and social exchange thereof—means tremendous heightening of adaptive ability, as concepts constructed purely semantically attain the capacity for alteration of episodic representation and procedure itself. Once the nature of morality is coded semantically, so that the implicit hierarchically structured presuppositions of behavior have been rendered explicit, they can be considered, debated and altered in their essential nature. Such alteration is capable of resonating down the cognitive chain to procedure itself. Likewise, alterations in procedure are (and should be) capable of producing profound effects upon episodic and semantic representation. This increased flexibility, the result of a tremendously complex and lengthy historical development, is tremendously useful for the purposes of rapid adaptation and change, but also equally promotes conflict, social and intrapsychic. Such conflict emerges as a consequence of destabilization of historical tradition.

It is the essential flexibility of the human brain, its very capacity to learn, and therefore to unlearn, that renders *Homo sapiens* so appallingly susceptible to group and intrapsychic conflict. An animal's behavioral pattern, its procedural knowledge, is set; its way of being in the unknown cannot easily be altered in its fundament. The assumptions and values by which an individual human being lives can, by contrast, be threatened with a few well-chosen and revolutionary words, whose ease of communication belies their elaborately complex evolutionary history, the depth of heroic endeavor necessary to their formulation and their extreme current potency. Sufficiently novel verbally transmitted information may disturb semantic, episodic and procedural paradigm simultaneously, although the totality of such effects may not become manifest for years—not infrequently, for generations.

Every culture maintains certain key beliefs that are centrally important to that culture, upon which all secondary beliefs are predicated. These key beliefs cannot be easily given up, because if they are, everything falls, and the unknown once again rules. Western morality and behavior, for example, are predicated on the assumption that every individual is sacred. This belief was already extant in its nascent form among the ancient Egyptians, and provides the very cornerstone of Judeo-Christian civilization. Successful challenge to this idea would invalidate the actions and goals of the Western individual; would destroy the Western dominance hierarchy, the social context for individual action. In the absence of this central assumption, the body of Western law—formalized myth, codified morality—erodes and falls. There are no individual rights, no individual value—and the foundation of the Western social (and psychological) structure dissolves. The Second World War and Cold War were fought largely to eliminate such a challenge.

For the man whose beliefs have become abstracted (and, therefore, more doubtful, more debatable), the mere *idea* of the stranger is sufficient to disrupt the stability of everyday presumption. Tolstoy, in his *Confessions*, recalls the impact of modern Western European ideas on the too-long-static medieval culture of Russia:

I remember that when I was eleven years old a high-school boy named Volodin'ka M., now long since dead, visited us one Sunday with an announcement of the latest discovery made at school. The discovery was that there is no God and that the things they were teaching us were nothing but fairy tales (this was in 1838). I remember how this news captured the interest of my older brothers; they even let me in on their discussions. I remember that we were all very excited and that we took this news to be both engaging and entirely possible.[405]

This "discovery," which was in fact the cumulative result of a very lengthy and traumatic Western European cognitive process, had the capacity to undermine the most fundamental presuppositions of Russian culture (as it had undermined those of the West):

Since ancient times, when the life of which I do know something began, people who knew the arguments concerning the vanity of life, the arguments that revealed to me its meaninglessness, lived nonetheless, bringing to life a meaning of their own. Since the time when people somehow began to live, this meaning of life has been with them, and they have led this life up to my own time. Everything that is in me and around me is the fruit of their knowledge of life. The very tools of thought by which I judge life and condemn it were created not by me but by them. I myself was born, educated and have grown up thanks to them. They dug out the iron, taught us how to cut the timber, tamed the cattle and the horses, showed us how to sow crops and live together; they brought order to our lives. They taught me how to think and to speak. I am their offspring, nursed by them, reared by them, taught by them; I think according to their thoughts, their words, and now I have proved to them that it is all meaningless![406]

This rational undermining eventually, inevitably, produced the following effects:

It happened with me as it happens with everyone who contracts a fatal internal disease. At first there were the insignificant symptoms of an ailment, which the patient ignores; then these symptoms recur more and more frequently, until they merge into one continuous duration of suffering. The suffering increases, and before he can turn around the patient discovers what he already knew: the thing he had taken for a mere indisposition is in fact the most important thing on earth to him, is in fact death.

This is exactly what happened to me. I realized that this was not an incidental ailment but something very serious, and that if the same questions should continue to recur, I would have to answer them. And I tried to answer them. The questions seemed to be such foolish, simple, childish questions. But as soon as I laid my hands on them and tried to resolve them, I was immediately convinced, first of all, that they were not childish and foolish questions but the most vital and profound questions in life, and, secondly, that no matter how much I pondered them there was no way I could resolve them. Before I could be occupied with my Samsara estate, with the education of my son, or with the writing of books, I had to know why I was doing these things. As long as I do not know the reason why, I cannot do anything. In the middle of my concern with the household, which at the time kept me quite busy, a question would suddenly come into my head: "Very well, you will have 6000 desyatins in the Samara province, as well as 300 horses; what then?" And I was completely taken aback and did not know what else to think. As soon as I started to think about the education of my children, I would ask myself, "Why?" Or I would reflect on how the people might attain prosperity, and I would suddenly ask myself, "What concern is it of mine?" Or in the middle of thinking about the fame that my works were bringing me I would say to myself, "Very well, you will be more famous than Gogol, Pushkin, Shakespeare, Moliere, more famous than all the writers in the world—so what?"

And I could find absolutely no reply.

My life came to a stop. I could breathe, eat, drink, and sleep; indeed, I could not help but breathe, eat, drink, and sleep. But there was no life in me because I had no desires whose satisfaction I would have found reasonable. If I wanted something, I knew beforehand that it did not matter whether or not I got it.

If a fairy had come and offered to fulfill my every wish, I would not have known what to wish for. If in moments of intoxication I should have not desires but the habits of old desires, in moments of sobriety I knew that it was all a delusion, that I really desired nothing. I did not even want to discover truth anymore because I had guessed what it was. The truth was that life is meaningless.

It was as though I had lived a little, wandered a little, until I came to the precipice, and I clearly saw that there was nothing ahead except ruin. And there was no stopping, no turning back, no closing my eyes so I would not see that there was nothing ahead except the deception of life and of happiness and the reality of suffering and death, of complete annihilation.

I grew sick of life; some irresistible force was leading me to somehow get rid of it. It was not that I wanted to kill myself. The force that was leading me away from life was more powerful, more absolute, more all-encompassing than any desire. With all my strength I struggled to get away from life. The thought of suicide came to me as naturally then as the thought of improving life had come to me before. This thought was such a temptation that I had to use cunning against myself in order not to go through with it too hastily. I did not want to be in a hurry only because I wanted to use all my strength to untangle my thoughts. If I could not get them untangled, I told myself, I could always go ahead with it. And there I was, a fortunate man, carrying a rope from my room where I was alone every night as I undressed, so that I would not hang myself from the beam between the closets. And I quit going hunting with a gun, so that I would not be too easily tempted to rid myself of life. I myself did not know what I wanted. I was afraid of life, I struggled to get rid of it, and yet I hoped for something from it.

And this was happening to me at a time when, from all indications, I should have been considered a completely happy man; this was when I was not yet fifty years old. I had a good, loving, and beloved wife, fine children, and a large estate that was growing and expanding without any effort on my part. More than ever before I was respected by friends and acquaintances, praised by strangers, and I could claim a certain renown without really deluding myself. Moreover, I was not physically and mentally unhealthy; on the contrary, I enjoyed a physical and mental vigor such as I had rarely encountered among others my age. Physically, I could keep up with the peasants working in the fields; mentally, I could work eight and ten hours at a stretch without suffering any aftereffects from the strain. And in such a state of affairs I came to a point where I could not live; and even though I feared death, I had to employ ruses against myself to keep from committing suicide.

I described my spiritual condition to myself in this way: my life is some kind of stupid and evil practical joke that someone is playing on me. In spite of the fact that I did not acknowledge the existence of any "Someone" who might have created me, the notion that someone brought me into the world as a stupid and evil joke seemed to be the most natural way to describe my condition.[407]

Group identity—inculcated morality and accepted interpretation—serves to constrain the motivational significance of experiential phenomena. When that identity (which is predicated upon implicit or explicitly held faith in a particular conceptualization of the way) is challenged, such constraints vanish. This "deconstruction" of symbolically patriarchal custom and belief subjects the individual to intrapsychic war of conflicting affect—the "clash of opposites," in Jungian terms; it subjugates him or her to unbearable cognitive, emotional and moral conflict. Nietzsche's comments on Hamlet, "sicklied o'er by the pale cast of thought," are relevant in this context:

Knowledge kills action; action requires the veils of illusion: that is the doctrine of Hamlet. . . . Now no comfort avails any more; longing transcends a world after death, even the gods; existence is negated along with its glittering reflection in the gods or in an immortal beyond. Conscious of the truth he has once seen, man now sees everywhere only the horror or absurdity of existence; now he understands what is symbolic in Ophelia's fate; now he understands the wisdom of the sylvan god, Silenus: he is nauseated.[408 409]

Dostoevsky's tragically comic bureaucratic-personality-disordered protagonist (the metaphoric mouse) in *Notes from Underground* reacts similarly, comparing his own (sophisticated) inability to respond courageously to an insult to that of *l'homme de la nature et de la verite*— the natural, and therefore truthful, yet comparatively unconscious (procedural) man:

Let us now look at the mouse in action. Suppose, for example, that it too has been insulted (and it will almost always be subjected to slights) and desires revenge. Perhaps even more fury will accumulate inside it than inside *l'homme de la nature et de la verite* because *l'homme de la nature et de la verite*, with his innate stupidity, considers his revenge to be no more than justice, while the mouse, with its heightened consciousness, denies that there is any justice about it. At last comes the act itself, the revenge. The wretched mouse has by this time accumulated, in addition to the original nastiness, so many other nastinesses in the shape of questions and doubts, and so many other unresolved problems in addition to the original problem, that it has involuntarily collected round itself a fatal morass, a stinking bog, consisting of its own doubts and agitations, and finally of the spittle rained on it by all the spontaneous men of action standing portentously round as judges and referees, and howling with laughter. Of course, nothing remains for it to do but shrug the whole thing off and creep shamefacedly into its hole with a smile of pretended contempt in which it doesn't even believe itself.[410]

The fictional characters of Shakespeare and Dostoevsky respond like the flesh-and-blood Tolstoy to the same historically determined set of circumstances—to the "death of god," in Nietzsche's terminology, brought about, inexorably, by continued development of abstract consciousness. The "first modern man," Hamlet, and those who follow him, in art and in life, characteristically respond like Nietzsche's "pale criminal"; like *Crime and Punishment's* Raskolnikov, they remain unable to bear the "terrible beauty"[411] of their deeds. Nietzsche states:

Of what is great one must either be silent or speak with greatness. With greatness—that means cynically and with innocence. What I relate is the history of the next two centuries. I describe what is coming, what can no longer come differently: *the advent of nihilism....* Our whole European culture is moving for some time now, with a tortured tension that is growing from decade to decade, as toward a catastrophe: restlessly, violently, headlong, like a river that wants to reach the end, that no longer reflects, that is afraid to reflect.

He that speaks here has, conversely, done nothing so far but to reflect: as a philosopher and solitary by instinct who has found his advantage in standing aside, outside. Why has the advent of nihilism become necessary? Because the values we have had hitherto thus draw their final consequence; because nihilism represents the ultimate logical conclusion of our great values and ideals—because we must experience nihilism before we can find out what value these "values" really had.

We require, at some time, new values.

Nihilism stands at the door: whence comes this uncanniest of all guests?

Point of departure: it is an error to consider "social distress" or "physiological degeneration," or corruption of all things, as the cause of nihilism. Ours is the most honest and compassionate age. Distress, whether psychic, physical, or intellectual, need not at all produce nihilism (that is, the radical rejection of value, meaning, and desirability). Such distress always permits a variety of interpretations. Rather: it is in one particular interpretation, the Christian moral one, that nihilism is rooted.

The end of Christianity—at the hands of its own morality (which cannot be replaced), which turns against the Christian God: the sense of truthfulness, highly developed by Christianity, is nauseated by the falseness and mendaciousness of all Christian interpretations of the world and of history; rebound from "God is the truth" to the fanatical faith "All is false"; an active Buddhism.

Skepticism regarding morality is what is decisive. The end of the moral interpretation of the world, which no longer has any sanction after it has tried to escape into some beyond, leads to nihilism.

"All lacks meaning." (The untenability of one interpretation of the world, upon which a tremendous amount of energy has been lavished, awakens the suspicion that all interpretations of the world are false.)[412]

That, in a nutshell, is "cascade."

Nihilism, alter ego of totalitarianism, is response to experience of the world, self and other, rendered devoid of certain meaning, and therefore allowed no meaning; is reaction to the world freed from the unconscious constraints of habit, custom and belief; is response to the re-emergence of the terrible unknown; is reaction of a spirit no longer able, as a consequence of abstract critical ability, to manifest unconscious or procedural identity with the hero—no longer able to muster belief in human possibility, in the face of exposure to the most dreadful imaginable. Phenomena remain constrained in their affective significance, at least partially, because the group (the dominance hierarchy) has reached agreement as to their meaning (their implications for situation-specific action). When that hierarchy falls—perhaps as a consequence of emergent disbelief in central presumption—nothing remains "sacred." This process becomes evidently manifest, from the empirical viewpoint, during a riot. When law and order are held temporarily in abeyance [when the inhibitory force of imposed threat is relieved (when the dominance hierarchy momentarily collapses)], those whose moral behavior remains predicated upon resentful obedience fall prey to their own disordered affect, and explode in aggression, greed, hatred and vengeful destructiveness. This explosion [implosion (?)] is "reduction to the precosmogonic continuum," from the pre-experimental or mythic viewpoint[413]—regression to the time and place prior to the division of things into known and unknown. This can be viewed either as alteration in affect, or transformation of the motivational significance of the phenomena whose apprehension motivates behavior. The objective mind would postulate the former; the mythic mind, concerned with subjective reality, the latter. This form of regression exists as precondition to creative restructuring. Semi-conscious (semi-declarative) apprehension of this affect-laden state manifest as paralyzing fear, exists (fortunately and catastrophically) as the greatest impediment to change.

The dominance hierarchy of value, extant socially and intrapsychically, employs fear (and promise) to regulate access to desired commodities—to determine the net motivational significance of particular events and processes. Any given phenomenon is capable of inducing a variety of affective or motivational states. It is the socially and individually determined outcome of competition between these intrapsychic states that determines behavioral output. The internalized consequence of the external dominance hierarchy—which is the "intrapsychic patriarchy," Freud's superego—is knowledge of the net motivational relevance of phenomena within a particular society. This implies, as stated previously, that the historically determined power structure of a given society could be inferred through analysis of the significance granted technological and cognitive artifacts by the individuals within that society. What is desired depends upon the goal toward which a given society moves. The goal is posited as valuable, initially, as a consequence of the operation of unconscious "presumptions," hypothetically preceding action. The value presupposed by the action is then coded episodically, then, perhaps, formalized semantically. Someone from a different culture values things differently; this difference is predicated upon acceptance of an alternate goal-directed schema. The nature and presence of this difference may be inferred (will, in fact, necessarily be inferred) from observation of foreign behavior, imagination and discussion—even inferred, perhaps, from exposure to cultural artifacts (which are generally granted the status of "mere" tools, which is to say, implements of the way) or from cues as subtle as voice or procedural melody.[414]

Movement from one schema to another—or from both to a hypothetical third, which unites both (which might constitute the consequence of revolutionary heroic effort)—presupposes dissolution, mutual or singular, not mere addition (a "qualitative" shift, not a "quantitative" shift). Mythically, as we have seen, this movement might be represented as descent from the precipice into the abyss, as the collapse of the idol with feet of clay, as dissolution to constituent bodily or material elements, as journey to the underworld or sea bottom, as sojourn through the valley of the shadow of death, as forty years (or forty days) in the desert, as encounter with the hydra, as incest with the mother. When such a journey is undertaken voluntarily—resources prepared adequately beforehand, faith in place—chance of success (return, reconstitution, resurrection, ascent) is substantially enhanced. When dissolution occurs accidentally—when encounter with the unknown is unintentional[415] or avoided beyond its time of inevitable occurrence—intrapsychic or social catastrophe, suicide or war, becomes certain.

The goal toward which behavior is devoted serves as one pole of the cognitive schema that determines the motivational significance of events. Members of the same culture share the same goal. This goal consists of a hypothetical desired state that exists in contrast to some conceptualization of the present and that can be attained through participation in a particular consensually accepted and traditionally determined process. This schema is analogous in structure to the normal mythological conception of the way, which includes a representation of the (troubled) present, a conception of the (desired) future, and a description of methods (moral prescriptions and injunctions) for transforming the former into the latter. Moral knowledge serves to further the way by reducing the infinite potential motivational significance of particular events to the particular and determinate. This process of reduction is social in nature—events take on established meaning that is socially determined, shared. The affective relevance of a given phenomenon—which, most fundamentally, is its significance for goal-directed behavior—is a consequence of the operation of the goal-oriented schema, which finds partial expression in establishment of a dominance hierarchy. A dominance hierarchy is a social arrangement which determines access to desired commodities. In most cases, these commodities are cues for consummatory reward—experiences that signify movement toward or increased likelihood of attaining the desired goal. Relative position in the dominance hierarchy—at least in the perfectly functioning society—is in itself determined through social judgment. That judgment reflects appreciation of the value of a particular individual. That value reflects how society views the ability of that individual to contribute to attainment of the goal. This interpretation of course implies that the postulation of a given way necessarily, inevitably, produces a hierarchy of value (since people and things will inevitably differ in their utility as means to the desired end). Every phenomenon, experienced within the confines of a particular society, is laden with dominance-hierarchy and goal-schema relevant information. The value of any particular item or experience is determined by the mythic foundation—upon which the entire society, consciously and unconsciously, rests. This value is the magic of the object.

Schismatic activity, semantic, episodic, or procedural, might be considered the within-group equivalent to arrival of an (abstracted or concrete) stranger. Cultural schisms emerge when once-predictable and familiar individuals become possessed by novel behavioral notions, images or semantic formulations, which present a challenge to presumptions deemed necessarily inviolable—such as the (most dangerous, authoritarian) presumption that all currently accepted presumptions are "true." Medieval horror of heresy and the drastic responses to such ideation defined as necessary by the Catholic guardians of proper thought is rendered comprehensible as a consequence of consideration (1) of the protective function of intact dogma and (2) of the methodological impossibility of "disproving," so to speak, alternative mythically founded narrative ideas. The Christian church fragmented chaotically (and, perhaps, creatively)—and continues to do so—with horrendous consequences, even under conditions where such fragmentation was severely punished. This is not stated to provide justification for repression of creativity, but to make the motivation for such repression understandable. Degeneration into chaos—decadence—might be considered the constant threat of innovation undertaken in the absence of comprehension and respect for tradition. Such decadence is precisely as dangerous to the stability and adaptability of the community and the individual and as purely motivated by underground wishes and desires as is totalitarianism or desire for absolute order. The continuing absence of a generally accepted methodology for peaceably sorting out the relative value or validity of evident mythologically predicated differences helps ensure that savage repression will remain the frequently utilized alternative.

Rapid development of semantic skill (and its second-order elaboration into empirical methodology) constitutes the third major threat to the continued stability of sociohistorically determined adaptive cultural systems (as well as the major factor in the complex elaboration of such systems). (The first two—just a reminder—were rapid natural environmental shift, independent of human activity, and contact with a heretofore isolated foreign culture). Literate individuals, members of cultures contained in express theologies or rational philosophies, can more easily incarnate and/or abstractly adopt or provisionally formulate different positions, with regard to the value of initial assumptions; can also verbalize the beliefs of other people, absorb them, and subject them to critical consideration or (theoretically) guileless acceptance; are fated necessarily to be able to become *many other people*, in imitation, imagination, and thought. Linguistically mediated criticism of the predicates of behavior undermines faith in the validity of historically established hierarchical patterns of adaptation. The final emergent process of the developmental chain of abstraction can be applied to undermine the stability of its foundation. The modern and verbally sophisticated individual is therefore always in danger of sawing off the branch on which he or she sits.

Language turned drama into mythic narrative, narrative into formal religion, and religion into critical philosophy, providing exponential expansion of adaptive ability—while simultaneously undermining assumption and expectation, and dividing knowledge from action. Civilized *Homo sapiens* can use words to destroy what words did not create. This ability has left modern individuals increasingly subject to their worst fears. Nietzsche states:

Our Europe of today, being the arena of an absurdly sudden attempt at a radical mixture of classes, and *hence* races, is therefore skeptical in all its heights and depths—sometimes with that mobile skepticism which leaps impatiently and lasciviously from branch to branch, sometimes dismal like a cloud overcharged with question marks—and often mortally sick of its will. Paralysis of the will: where today does one not find this cripple sitting? And often in such finery! How seductive the finery looks! This disease enjoys the most beautiful pomp-and-lie costumes; and most of what today displays itself in the showcases, for example, as "objectivity," "being scientific," "*l'art pour l'art*," "pure knowledge, free of will," is merely dressed-up skepticism and paralysis of the will: for this diagnosis of the European sickness I vouch.[416]

The intellectual developments which lead to the establishment of modern scientific methodology have heightened the danger of this partially pathological tendency. The construction of a powerful and accurate representation of the "objective" or shared world—a logical conclusion of the interpersonal exchange of sensory information, made possible by linguistic communication—challenged belief in the reality of the mythic world, which was in fact never objective. The mythic world was always affective—although it was shared socially—and contained procedural information (and abstracted representation thereof), arranged hierarchically in terms of value, embodied in nonverbal procedural and abstracted imagistic and semantic form. Representation of mythic value in verbal format allowed for simple experimentation in ethics, in imagination (and then, often tragically, in action), and for generation of naive but effective criticism regarding traditional foundations for behavior. Nietzsche states:

For this is the way in which religions are wont to die out: under the stern, intelligent eyes of an orthodox dogmatism, the mythical premises of a religion are systematized as a sum total of historical events; one begins apprehensively to defend the credibility of the myths, while at the same time one opposes any continuation of their natural vitality and growth; the feeling for myth perishes, and its place is taken by the claim of religion to historical foundations.[417]

Freud maintained, as an ideal nineteenth-century empiricist, that "there is no other source of knowledge of the universe but the intellectual manipulation of carefully verified observations—that is, what is called research—and that no knowledge can be obtained from revelation, intuition or inspiration." He said, furthermore, that "there is no appeal beyond reason"[418] (grounded directly in "observation," one would presume). This description leaves no place for the primal role of affect (or even of sensation, for that matter) in determination of wisdom—"what causes me and others pain is wrong," in the most basic and naive form—and also fails to address the issue of the source of scientific hypotheses in general (the narrative process). Furthermore, pure knowledge of the sensory world—what *is*, most fundamentally—does not include knowledge about how to adapt to or behave in that world (even though the gathering of such information has obvious implications for such adaptation). Tolstoy states:

As presented by the learned and the wise, rational knowledge denies the meaning of life, but the huge masses of people acknowledge meaning through an irrational knowledge. And this irrational knowledge is faith, the one thing that I could not accept. This involves the God who is both one and three, the creation in six days, devils, angels and everything else that I could not accept without taking leave of my senses.

My position was terrible. I knew that I could find nothing in the way of rational knowledge except a denial of life; and in faith I could find nothing except a denial of reason, and this was even more impossible than a denial of life. According to rational knowledge, it followed that life is evil, and people know it. They do not have to live, yet they have lived and they do live, just as I myself had lived, even though I had known for a long time that life is meaningless and evil. According to faith, it followed that in order to understand the meaning of life I would have to turn away from reason, the very thing for which meaning was necessary.[419]

Mythic thinking, so to speak, is also based on observation—but on observation of behavior in the world of affective experience. This means cyclical observation of action predicated upon an implicit or explicitly formulated theory of what should be, and derivation of procedural, episodic or semantic representations thereof. This is knowledge, as well—and appears, in the light of careful analysis, no more arbitrary than empirical description of the objective world.

Perhaps it was necessary for science, struggling to escape from a cognitive world dominated by religious and mythical thinking, to devalue that world in order to set up an independent existence. That existence has long been established, however—but the process of devaluation, implicit and explicit, continues (even in fields theoretically separate from the strictly empirical). Frye states:

Ever since Plato, most literary critics have connected the word "thought" with dialectical and conceptual idioms, and ignored or denied the existence of poetic and imaginative thought. This attitude continued into the twentieth century with I.A. Richards's *Science and Poetry*, with its suggestion that mythical thinking has been superseded by scientific thinking, and that consequently poets must confine themselves to pseudo-statements. The early criticism of T.S. Eliot, though considerably more cautious than this, also exhibited an array of confusions clustering around the word "thought." Since then there has been a slowly growing realization that mythological thinking cannot be superseded, because it forms the framework and context for all thinking. But the old views still persist, if in more sophisticated forms, and there are still far too many literary critics who are both ignorant and contemptuous of the mental processes that produce literature.[420]

Nietzsche states, similarly, but with somewhat more scorn:

Every age has its own divine type of naivety for whose invention other ages may envy it—and how much naivety, venerable, childlike and boundlessly clumsy naivety lies in the scholar's faith in his superiority, in the good conscience of his tolerance, in the unsuspecting simple certainly with which his instinct treats the religious man as an inferior and lower type that he has outgrown, leaving it behind, *beneath* him—him, that presumptuous little dwarf and rabble man, the assiduous and speedy head- and handiworker of "ideas," of "modern ideas!"[421]

Mythological thinking is not mere arbitrary superstition. Its denigration—cascading even through literary criticism, in recent years—is not only unwarranted but *perilous*. This is not to say that religious institutions and dogmas are not prey to the same weaknesses as all other human creations. The ideas and patterns of action that underlay and generated those institutions remain of critical importance, however—remain important for sustaining individual emotional stability, maintaining group tolerance, cohesion and flexibility, supporting capacity to adapt to the strange, and strengthening ability to resist domination by one-sided and murderous ideologies. The idea that we have superseded such thinking is a prime example of the capacity of the "semantic system" to partially represent and to thoroughly criticize. This is wrong, arrogant and dangerous.

The group promotes an integrated pattern of behavior and conception of values. This is strength, in that an integrated pattern provides one message, and therefore promotes unity and direction. It is also weakness, in that integration—stable, hierarchically organized structure—is inflexible, and therefore brittle. This means the group, and those who identify with it, cannot easily develop new modes of perception or change direction when such change or development becomes necessary. Under stable environmental and social conditions, this is an advantage, as what worked in the past will continue to work in the present. However, in times of transition, of rapid environmental transformation, of multicultural contact, of technological or ideological advance, stability is not necessarily sufficient. The Russian neuropsychologist Sokolov stated, as cited previously,[422] "One way to improve the quality of extrapolation [judgment of match between intent and outcome] is to secure additional information; another method is to change the principles by which such information is handled, so that the process of regulation will prove more effective." This fundamental idea is embodied in mythology in the figure of the revolutionary hero. He is the fourth manner in which threat to the stability of cultural tradition may be presented and, simultaneously, is solution to the ever-recurring problem of such threat.

The Revolutionary Hero

The revolutionary hero reorders the protective structure of society, when the emergence of an anomaly makes such reordering necessary. He is therefore the agent of change, upon whose actions all stability is predicated. This capacity—which should make him a welcome figure in every community—is exceedingly threatening to those completely encapsulated by the status quo, and who are unable or unwilling to see where the present state of adaptation is incomplete and where residual danger lies. The archetypal revolutionary hero therefore faces the anger and rejection of his peers, as well as the terrors of the absolutely unknown. He is nonetheless the "best friend" of the state.

Analysis of the archaic ecstatic practice of *shamanism*—prevalent throughout "the immense area comprising Central and North Asia"[423]—sheds further light on the nature of the actions and typical experiences of the revolutionary hero. Europeans who made initial contact with these tribal healers frequently deemed them insane. The reverse was in fact true: the genuine shaman was the most sane man of the tribe (that is, the man whose extent of adaptation was greatest). Furthermore, he served as primordial "unified ancestor" of the lately differentiated or specialized creative agent: explorer, mystic, artist, scientist and physician. The Asian shaman was master of religious life, embodiment and keeper of the sacred doctrine, dominant authority and creator of culture.

The widespread practices and viewpoints of shamanism constitute a cohesive philosophy, so to speak, embedded "unconsciously" in behavior and image. This ritual philosophy comprises a set of observations about the nature of radical personality transformation, and a set of practices designed to bring such alteration about. Shamanism is devoted to furtherance of the possibility of qualitative improvements in "consciousness" or general adaptive ability; it has captured the essence of such possibility in image, to minimize the accompanying terror. Shamanism is prototypical of those religious practices designed to modify human behavior and interpretation—to induce and regulate the processes of spiritual reconfiguration. These practices are not merely cultural in nature. They originate in the observation of spontaneous psychological transmutation, a psychobiologically grounded human capacity. Shamanic rituals are therefore not merely anachronistic, without modern relevance, except as curiosity dictates—but prime exemplars of a process we must come to understand.

The shaman is *not* simply an archaic figure, an interesting anomaly from the dead past—he is the embodiment, in cultures we do not comprehend, of those people we admire most in the past. The phenomenon of the "creative illness," described in detail by Henri Ellenberger, in his massive study of the history of the unconscious, is alive and well in our own culture. Ellenberger described its characteristic elements:

A creative illness succeeds a period of intense preoccupation with an idea and search for a certain truth. It is a polymorphous condition that can take the shape of depression, neurosis, psychosomatic ailments, or even psychosis. Whatever the symptoms, they are felt as painful, if not agonizing, by the subject, with alternating periods of alleviation and worsening. Throughout the illness the subject never loses the thread of his dominating preoccupation. It is often compatible with normal, professional activity and family life. But even if he keeps to his social activities, he is almost entirely absorbed with himself. He suffers from feelings of utter isolation, even when he has a mentor who guides him through the ordeal (like the shaman apprentice with his master). The termination is often rapid and marked by a phase of exhilaration. The subject emerges from his ordeal with a permanent transformation in his personality and the conviction that he has discovered a great truth or a new spiritual world.[424]

Many of the nineteenth- and twentieth-century figures recognized unquestionably as "great"—Nietzsche, Darwin, Dostoyevsky, Tolstoy, Freud, Jung, Piaget—were additionally characterized by lengthy periods of profound psychological unrest and uncertainty. Their "psychopathology," a term ridiculous in this context, was generated as a consequence of the revolutionary nature of their personal experience (their action, fantasy and thought). It is no great leap of comparative psychology to see their role in our society as analogous to that of the archaic religious leader and healer.

For the average "tribal" individual, socially imposed initiation signifies the death of childhood and reintegration on the level of social maturity. For the future shaman, voluntarily undertaken initiation signifies the disintegration of socially determined adult personality and

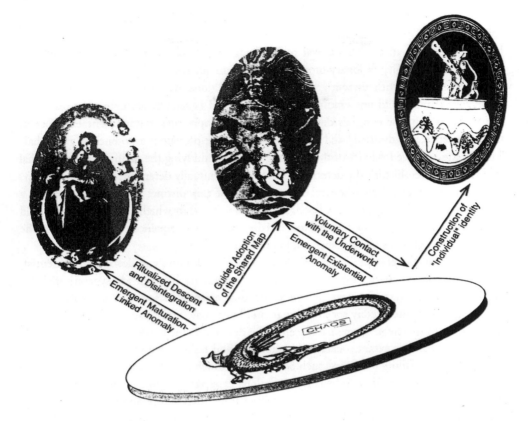

Figure 50: The "Dual Death" of the Revolutionary Hero

reintegration at the level of unique individuality. This process is illustrated in *Figure 50: The "Dual Death" of the Revolutionary Hero*. Those who undergo a second initiation suffer more deeply and profoundly from life than their peers; they are, in Jung's phrase, the most "complex and differentiated minds of their age."[425] These creative individuals detect emergent anomaly, and begin the process of adaptation to it, long before the average person notices any change whatsoever in circumstance. In his ecstasy the shaman lives the potential future life of his society. This dangerous individual can play a healing role in his community because he has suffered more through experience than his peers. If someone in the community (or the community itself) becomes ill, breaks down—begins the journey, so speak, to the land of the dead, the terrible unknown—the shaman is there to serve as guide, to provide rationale for current experience, to reunite the suffering individual with his community or to renew the community—to restabilize the paradigmatic context of expectation and desire within which individual and social experience remains tolerable. The truly creative individual has "been there and done that," and can therefore serve as a guide to others voluntarily beginning—or roughly thrown into—similar voyages.

The archaic shamanic initiate was commonly someone uniquely marked by fate, by the "will of the gods"—by particular heredity, "magical" (novel) occurrence in early childhood or later in life (birth in a caul; survival of lightning strike), or by intrapsychic idiosyncrasy (epileptic susceptibility, visionary proclivity).[426] His unique personality or experiential history, in combination with presently extant social conditions, doomed him to experience so anomalous that it could not simultaneously be accepted as actually occurring—as real—*and* as possible within the confines determined by ruling social presumption. The existence of this experience, if "admitted" and "processed," therefore presented a potentially fatal challenge to the perceived validity of the axioms currently underlying the maintenance of normal "sanity"—the sociohistorically determined stability of mutually determined behavioral adaptation and experiential significance. The existence of this distinct experience served as a gateway to the unknown, or as a floodgate, a portal, through which the unexpected could pour, with inevitably destructive and potentially creative consequences. The shaman is the individual who chooses to meet such a flood head on.

The shaman, the ecstatic in general—equally, the revolutionary philosopher or scientist, true to himself—is characterized by stubborn adherence to his own idiosyncratic field of experience, in which occurrences emerge, procedural, episodic or semantic in structure, that are foreign to the predictably socialized man and his prosaic moral expectation. The experiential range of the creative agent transcends the domain of the current adaptive sufficiency of his culture, as it is extant socially and embodied and represented intrapsychically. Rather than ignoring or failing to process such occurrences (which exist in contradiction to or completely outside his conditional, socially determined expectations), and acting as though they do not exist, the creative individual (voluntarily) admits their reality, and submits himself to the dissolution of his current (moral) worldview and pattern of action. This dissolution of personality, equivalent in episodic representation to death, temporarily "renovelizes" experience; furthermore, it provides the precondition for more inclusive resurrection of order, personal and social.

The future shaman is in fact tormented by the incomplete or self-contradictory state of his cultural structure, as it is intrapsychically represented; is undergoing a breakdown induced by some aspect of personal experience, some existential anomaly, that cannot be easily integrated into that structure. This breakdown re-exposes him to the unknown—previously covered, so to speak, by his culture. His comportment during the period of incubation preceding his emergence as shaman is generally marked by commission of acts considered characteristic, in modern and archaic culture alike, of serious mental breakdown. He behaves idiosyncratically, seeking solitude, flying into fits of rage, losing consciousness, living in the mountains or woods alone, and suffering from visions and periods of absentmindedness. His peers explain his odd behavior as possession. This experience of dissolution and re-exposure to chaos accompanies intrapsychic subjugation to the operation of innate, involuntary [episodic, limbic, right-hemisphere-governed (?)] mechanisms responsible for the deconstruction and renewal of conditional knowledge. This operation manifests itself subjectively in structured mythic experience—in spontaneous personal experience, which adheres to the pattern associated with ritualized social initiation, and which may also have served, original-ly, as its source.

The soul of the shaman is "carried away by spirits," habituants of the episodic realm, and returned to "the place of the gods." This place exists outside of time and space itself, on the same plane of pleromatic reality as the prehistoric and postapocalyptic Paradise. Entry into this domain is preceded by complete psychic disintegration, accompanied by horrifying visions of torture, dismemberment and death. The shamanic initiate descends into the matriarchal hell that preceded and coexists with creation, passing through clashing rocks, or gates in the shape of jaws; he is reduced to a skeleton, while his disembodied head observes the procedure; he has his internal organs removed or restructured; his bones are broken, his eyes gouged out. He is devoured by a serpent or a giantess; is boiled, roasted or otherwise reduced to his essential and fundamental structure—to his very bones. Eliade states:

The total crisis of the future shaman, sometimes leading to complete disintegration of the personality and to madness, can be valuated not only as an initiatory death but also as a symbolic return to the pre-cosmogonic Chaos, to the amorphous and indescribable state that precedes any cosmogony. Now, as we know, for archaic and traditional cultures, a symbolic return to Chaos is equivalent to preparing a new Creation. It follows that we may interpret the psychic Chaos of the future shaman as a sign that the profane man is being "dissolved" and a new personality being prepared for birth.[427]

This disintegration is the removal of experience—objects and processes—from their socially determined state of provisional paradigm-governed significance, and their return to the affectively numinous unknown, infinitely threatening and promising. Exposure to conse-quently renovelized experience constitutes the affective and motivational core of the ecstatic experience, the basis for the religious experience (and the experience of meaning) as such— prior to its entrapment and canalization in dogma. Dissolution is experienced in imaginal or episodic representation, as death—an accurate conceptualization, *death of socialized personali-ty*: dissolution of the presently constituted intrapsychic representation and procedural embodiment of action patterns historically constructed and currently deemed morally acceptable. The justifiable terror consideration of the consequences of such decomposition induces constitutes a major impediment to the pursuit of redemptive change, a formidable barrier to intrapsychic integration.

The shamanic "process of transformation" appears as the means by which cognitive sys-tems are updated, when necessary; the affect that is released, during the process, is necessari-ly part of the experience. Every major "step forward" therefore has some of the aspect of the revolutionary "descent into madness"; change shades gradually from the normal to the radi-cal. The structure of this process formulates itself easily into imagistic representation—even among children far too young to develop any "explicitly statable" knowledge about such occurrences.

The following dream was described by my daughter, Mikhaila (then three years, nine months old), about my son, Julian (one year, eleven months old) on October 5, 1995. Julian was in a time of toilet training and rapid speech development, and was having some trouble controlling his emotions. Mikhaila liked to call him "baby." We had several discussions about the fact that he really wasn't a baby anymore. She told me this story, while I was at the computer, so I was able to get it verbatim:

> Mikhaila: Julian's eyes falled out
> and then
> he falled into pieces
> Dad: (what sort of pieces?)
> Mikhaila: Julian pieces
> and the bones falled out too
> then
> a hole got him
> and there was water in it
> and when he came out he was big
> Mom: (Julian isn't a baby anymore?)
> Mikhaila: No he's a big boy
> and a bug with legs got him out
> 'cause bugs can swim
> and the hole was in the park
> and it moved into the back yard
> and he falled in it
> a tree burned
> and left the hole.

It was the partial "dissolution" of Julian's previous infantile personality that was causing his emotional distress. Mikhaila, upset by his trouble (and curious about the disappearance of her "baby") was trying to understand what her brother was going through. Her dream represented his transformation as a "death" and rebirth: First his eyes fell out, then he fell into pieces, then his bones came out. Everything went into a "hole," which originally inhabited the nearby park. (The park by our house was forty wooded acres; the children and I had gone there at night several times. They found it spooky, but exciting. For them, it was the nearest manifestation of the unknown, outside explored and familiar territory—prime locale for metaphoric application as source of the "hole," in which transformation takes place.) The hole was full of water, whose symbolism we have partially discussed (as the rejuvenating/destroying "water of life"). The "bug with legs" that could "swim" was, I think, a theriomorphized representation of the very archaic intrapsychic systems that guide or underlie the transformation of more sophisticated cortical or personality "contents." The notion that a "tree" burned and left the hole is very complex. A tree, at minimum, is a sophisticated structure that emerges from basic material (from the "ground"). It is also commonly used as a metaphoric representative of the essence of the individual human—even of the nervous system itself[428]—as we shall see. The tree in this case was therefore also representative of Julian, but in a more impersonal way. It stood for, among other things, the personality that was currently undergoing transformation.

Adaptive ability remains necessarily limited to the domain encompassed by a single set of principles—a single pattern of action, a single mode of apprehension—in the absence of capacity to reconfigure present conceptualizations of morality (morality: description of unbearable present, ideal future and means of transformation). Such limitation—which is the inability to play games with the rules of the games—means dangerous restriction of behavioral and representational flexibility, and increased susceptibility to the dangers posed by inevitable "environmental" shift (that is, by inevitable re-emergence of the dragon of the unknown). Biologically determined capacity for such dissolution—and for its satisfactory resolution—provides the necessary precondition for the existence of human capacity for qualitative alteration in adaptation. Resolution of crisis—symbolic rebirth—follows, attendant upon initiatory dissolution, dismemberment and death. Eliade states:

The initiatory operations proper always include the renewal of the organs and viscera, the cleaning of the bones, and the insertion of magical substances—quartz crystals, or pearl shell, or "spirit snakes." Quartz is connected with the "sky world and with the rainbow"; pearl shell is similarly connected with the "rainbow serpent," that is, in sum, still with the sky. This sky symbolism goes along with ecstatic ascents to Heaven; for in many regions the candidate is believed to visit the sky, whether by his own power (for example, by climbing a rope) or carried by a snake. In the sky he converses with the Supernatural Beings and mythical Heroes. Other initiations involve a descent to the realm of the dead; for example, the future medicine man goes to sleep by the burying ground, or enters a cave, or is transported underground or to the bottom of a lake. Among some tribes, the initiation also includes the novice's being "roasted" in or at a fire. Finally, the candidate is resuscitated by the same Supernatural Beings who had killed him, and he is now "a man of Power." During and after his initiation he meets with spirits, Heroes of the mythical Times, and souls of the dead—and in a certain sense they all instruct him in the secrets of the medicine man's profession. Naturally, the training proper is concluded under the direction of the older masters. In short, the candidate becomes a medicine man through a ritual of initiatory death, followed by a resurrection to a new and superhuman condition.[429]

The shaman travels up and down the *axis mundi*, the central pole of the world, the tree of life connecting the lower, chthonic reptilian and the upper, celestial avian worlds with the central domain of man. This is the "constituent elements of experience" conceived in an alternative but familiar arrangement, as heaven above (father above), underworld/matter/earth below (mother below)—conceived in the configuration arranged originally by the cosmos-creating hero. The shaman's success at completing the journey "from earth to the domain of the gods" allows him to serve the role of *psychopomp*, intermediary between man and god; to aid the members of his community in adjusting to what remains outside of conditional adaptation, when such adaptation fails. The shaman therefore serves his society as active intermediary with the unknown; as the conduit, so to speak, through which the unknown speaks to man; as the agent through which the information which compels adaptive change flows. It is important to note that the shaman's journey into "unknown lands" must be bounded by return to the community for the voyage to be of value. Otherwise, the prototypal ecstatic experience—central to the shamanic vocation (and to creative thought and action in general)—is mere insanity; will be regarded socially and experienced intrapsychically as such. Resolution is psychological reconstruction, reincorporation, rebirth "on a higher level"—with redemptive personal experience intact, but reintegrated in the corpus of current sociocultural myth and history.

The ineradicable anomaly that comprises an eternal aspect of existence periodically undermines the stability of a subset of unfortunate but gifted individuals. Those who maintain their heads during the "journey into the underworld" return, contaminated by that underworld, from the perspective of their compatriots, but rife with possibility for reordering the world. Such recovery is *in essence* the transformation of assumption and value—individual, then cultural. History is an invaluable storehouse of the creative experience and wisdom of the past. Past wisdom is not always sufficient to render present potentiality habitable. If the structure of experience itself was static and finite, like the past, all things would have been conquered long ago, and the lives of the ancestors and their children would differ little in kind. But the structure of experience is dynamic and infinite in possibility. The nature of experience itself varies with time. New challenges and dangers appear out of the future, into the present, where none existed before. History, as description of the past, is incomplete, as well as static. It must therefore exist in constant conflict with new experiences. The spirit underlying the transmutation of culture resolves unbearable intrapsychic conflict with shattering revelation, first to the individual, then to society at large. The creative individual "dies"—metaphysically and, too often, literally—for those who follow him, instead of sharing the common destiny of his peers. Those who bear the initial burden for the forward movement of history are capable of transforming personal idiosyncrasy and revelation into collective reality, without breaking down under the weight of isolation and fear. Such creativity is feared and hated and desired and worshiped by every human individual and by human society in general. Creative individuals destroy old values and threaten with chaos, but they also bear light and the promise of better things. It is in this manner that the "sacrifice of the revolutionary savior" redeems and rekindles the cosmos.

The revolutionary hero is the individual who decides voluntarily, courageously, to face some aspect of the unknown and threatening. He may also be the only person who is presently capable of perceiving that social adaptation is incompletely or improperly structured in a particular way; only he understands that there still remain unconquered evil spirits, dangerous unknowns and threatening possibilities. In taking creative action, he (re)encounters chaos, generates new myth-predicated behavioral strategies, and extends the boundaries (or transforms the paradigmatic structure) of cultural competence. The well-adapted man identifies with what has been, conserves past wisdom, and is therefore protected from the unknown. The hero, by contrast, author and editor of history, masters the known, exceeds its bounds, and then subjects it to restructuring—exposing chaos once more to view in the process—or pushes back unknown frontiers, establishing defined territory where nothing but fear and hope existed before. The hero overcomes nature, the Great Mother, entering into creative union with her; reorganizing culture, the Great Father, in consequence. Such reintegration and resurrection is in essence the metamorphosis of individual, and then cultural, moral presumption. Cumulative socially mediated transmission of the past consequences of such creation and intrapsychic reorganization constitutes group identity, culture itself—the canon of assumptions and values that underlie behavior, the eternal shield against the terrible unknown.

The hero is the first person to have his "internal structure" (that is, his hierarchy of values and his behaviors) reorganized as a consequence of contact with an emergent anomaly. His "descent into the underworld" and subsequent reorganization make him a savior—but his contact with the dragon of chaos also contaminates him with the forces that disrupt tradition and stability. The reigning status-quo stability may be only apparent—that is, the culture in its present form may already be doomed by as-of-yet not fully manifested change. The hero detects the dragon, or at least admits to its presence, before anyone else, and leads the charge. His return to the kingdom of threatened order may hardly be accompanied by praise, however, since the information he now carries (or perhaps is) will appear disruptive and destructive long before it proves redemptive. It is very easy to view the hero as the most profound danger to the state, in consequence—and this would in fact be true if the absolute stasis of the state did not constitute a more fundamental danger. *Figure 51: The Crucified Redeemer as Dragon of Chaos and Transformation*[430] presents the savior as serpent, in keeping with his "contamination" by the unknown.[431]

Figure 52: The Socially Destructive and Redemptive "Journey" of the Revolutionary Hero schematically presents the "way of the savior." The individual troubled by anomalous and anxiety-provoking experience is suffering equally from the disintegration, rigidity or senility of the society within. The decision to "mine" such experience for significance—and to destabilize the socially constructed intrapsychic hierarchy of behavior and values, is in consequence equivalent, mythologically speaking, to the "descent to the underworld." If this descent is successful—that is, if the exploring individual does not retreat to his previous personality structure, and wall himself in, and if he does not fall prey to hopelessness, anxiety and despair—then he may "return" to the community, treasure in hand, with information whose incorporation would benefit society. It is very likely, however, that he will be viewed with fear and even hatred, as a consequence of his "contamination with the unknown," particularly if those left behind are unconscious of the threat that motivated his original journey. His contamination is nothing to be taken lightly, besides. If the exploratory figure has in fact derived a new mode of adaptation or representation, necessary for the continued success and survival of the group, substantial social change is inevitable. This process of change will throw those completely identified with the group into the realm of chaos, against their will. Such an involuntary descent into the underworld is a very dangerous undertaking, as we have seen, particularly in the absence of identification with the hero. This means that it is primarily those persons who have sold their soul to the group who cannot distinguish

Figure 51: The Crucified Redeemer as Dragon of Chaos and Transformation

between the hero and the dragon of chaos (between the hero and the environmental disaster, the death of the king, the dangerous stranger or the heretical idea).

The more tyrannical the attitude, the more those who hold it hate and fear the hero, victim and beneficiary of the creative illness:

Let us lie in wait for the righteous man, because he is inconvenient to us and opposes our actions; he reproaches us for sins against the law, and accuses us of sins against our training.

He professes to have knowledge of God, and calls himself a child of the Lord.

Let us see if his words are true, and let us test what will happen at the end of his life; for if the righteous man is God's son, he will help him, and will deliver him from the hand of his adversaries.

Let us test him with insult and torture, that we may find out how gentle he is, and make trial of his forbearance.

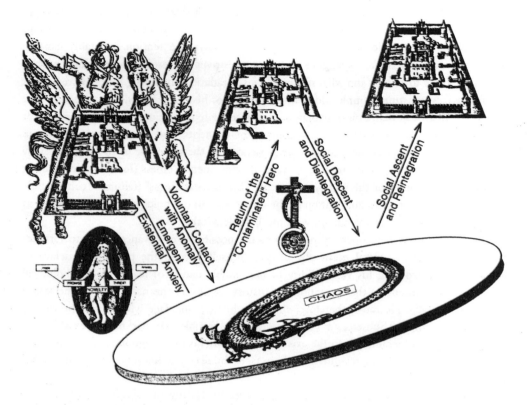

Figure 52: The Socially Destructive and Redemptive "Journey" of the Revolutionary Hero

Let us condemn him to a shameful death, for, according to what he says, he will be protected.
(Wisdom 2:12–13, 16–20 RSV)

The tyrannical attitude maintains society in homogeneity and rigid predictability, but dooms it to eventual collapse. This arrogant traditionalism, masquerading as moral virtue, is merely unexpressed fear of leaving the beaten path, of forging the new trail—the entirely comprehensible but nonetheless unforgivable shrinking from destiny, as a consequence of lack of faith in personal ability and precisely equivalent fear of the unknown. The inevitable result of such failure is *restriction of meaning*—by definition, as meaning exists on the border between the known and the unknown. Repression of personal experience, which is failure to update action and representation in the face of an anomalous occurrence, means damming up the river of life; means existence on the barren plain, in the paralyzed kingdom, in the eternal drought. It is *personal* experience—anathema to the fascist; eternally superseding group categorization and the interpretations of the dead—that is novel and endlessly refreshing.

The security of predictable society provides an antidote to fear, but a too-rigid society ensures its own eventual destruction. The future brings with it the unknown; inflexibility and unwillingness to change therefore bring the certainty of extinction. Adaptive behavior is created and/or transformed by those driven to resolve the tension inevitably existing between dynamic personal experience and society—driven to resolve the tension between what they know to be true and what history claims. Readaptation, during times of crisis, does not necessarily constitute simple addition to the body of historical knowledge, although that is heroic endeavor as well. Full readaptation may necessitate revolutionary measures, partial or complete reincarnation—dissolution to constituent elements, and systemic reorganization. Such reorganization alters the meaning of experience, and therefore, the mythology of history and being. If resolution is not reached in time of crisis, mental illness (for the individual) or cultural degeneration (for the society) threatens. This "mental illness" (failure of culture, failure of heroism) is return to domination by the unknown—in mythological terms, expressed as involuntary incest (destructive union) with the Terrible Mother.

The revolutionary hero opens himself up to the possibility of advancement—to furtherance of his culture's central myth—by placing himself beyond the protective enclave of history and by exposing his vulnerability to the terrible nature of reality. In psychological terms, the hero discovers the limitations of history; discovers the nakedness of the father (Genesis 9:20-25). He must, therefore, challenge history, and face what it had previously protected him from. Subsequent contact with the Terrible Mother means exposure to absolute mortal vulnerability—to the existence and consequence of ignorance, insanity, cruelty, disease and death. The revolutionary hero faces the reality of his vulnerability and fights a battle with terror.

The constant transcendence of the future serves to destroy the absolute sufficiency of all previous historically determined systems, and ensures that the path defined by the revolutionary hero remains the one constant route to redemption. The "revolutionary hero" is embodiment and narrative representation of the action of consciousness itself. This mythically masculine principle emerges from its identity with chaos and culture, and stands as an independently divine phenomenon, equivalent in potential strength to the destructive, generative, protective and tyrannical forces that make up human experience. The hero is the individual who has found the "third solution" to his existential problems, the alternative to decadence and authoritarianism. When faced with a paradox whose solution is impossible in terms of the historical canon (that established axiomatically predicated hierarchy of values and assumptions) he takes inspired action and transcends his culturally determined limitations. Instead of denying the existence of the problem—and, therefore, tormenting those who cannot help but posit it—the revolutionary hero accepts the apparently impossible task of solution, and of reuniting the warring opposites. He admits the possibility of successful solution not because the problem can be minimized, but because he believes human nature can expand to meet it. Such belief—faith—provides the precondition for courage. His act of voluntary transcendence re-exposes him to the brute force of the unknown (and to the anger of the social group) but enables creative action. The hero's ability to risk standing alone—neither rejecting his culture because he is ignorant of its value, nor running away from it in panic because of fear—offers him the possibility of attaining true stature, although not necessarily acclaim or popularity.

The true absolute in the individual, which can meet the absolute unknown, is the heroic aspect, which cannot be made finally subject to tyranny and is not ruled by the past. This is the spirit that created civilization, which must not be bound, within the individual, by abject subjugation to what has already been. The man who stands outside of culture necessarily places himself against nature and the world. This seems a hopeless position. But man knows little of his true potential, and in that ignorance lies his hope:

This is the stone which was set at nought of you builders, which is become the head of the corner.
    (Acts 4:11)

The Rise of Self-Reference, and the Permanent Contamination of Anomaly with Death

The appearance of anomaly can be less or more upsetting. Small "manifestations of the unknown" disrupt relatively small tracts of "explored territory." Larger manifestations may disrupt all things previously taken for granted, even things invisible.

Upsetting manifestations of the unknown may occur as a consequence of "outside forces," geological, meteorological, even cosmological. Similarly, social transformations may upset the stable and familiar. Wars, revolutions and migrations make the conditional nature of everything taken for granted evident once more.

Internal transformations are just as likely to introduce instability. The process of maturation, in and of itself, is sufficient to disrupt the previously stable and well-adapted personality and the little society of the family. Crises in adaptation may be brought about in childhood at the onset of schooling and first independent contact with the unmediated social world. The hormonal changes and new social demands of youth may likewise transform the happy and reasonable child into the depressed and hostile adolescent.

Some internal transformations are also natural and social events. The ever-expanding human capacity for abstract thinking, for example, appears to be a consequence of biological and social forces, working synergistically. The human brain has evolved exceptionally quickly, from the phylogenetic perspective. The language-mediated interpersonal interaction characteristic of ever-larger human societies has provided that rapidly developing biological capacity with information whose sophistication and breadth is increasing exponentially. This means that the human mind increasingly manifests the capacity to upset itself, to produce revelations, so to speak, that knock gaping holes in the previously sufficient adaptive and protective social and intrapsychic structures.

The human capacity for abstraction has enabled us as a species and as individuals to produce self-models that include the temporal boundaries of existence. We have become able to imagine our own deaths, and the deaths of those we love, and to make a link between mortal fragility and every risk we encounter. Emergence of such capacity—which reoccurs with the maturation of every new human being—introduces the most intractable anomaly imaginable into the developmental course of every life.

Myth represents the ever-recurring appearance of this representational ability—this emergent "self-consciousness," the heritable sin of Adam—as incorporation of the "forbidden fruit," development of knowledge of good and evil, and consequent expulsion from paradise. This appearance is an event of "cosmic significance," driving the separation of heaven and earth, making human experience something "eternally fallen," something ever in need of redemption.

The unknown has become permanently contaminated with death, for *Homo sapiens*. This contamination has tremendously heightened our general motivation—our fear and curiosity—as we are able to perceive the potential that lurks behind every anomalous event. Our cultural creations—our great societies and the beliefs that accompany them—can be profitably viewed as driven by our knowledge of mortality, and by the energy (the heightened alertness and penetrating consciousness) such knowledge inspires.

Our great transpersonal cognitive power, however, has not yet rescued us from the valley of the shadow of death.

> What man is found such an idiot as to suppose that God planted trees in Paradise, in Eden, like a husbandman, and planted therein the tree of Life, perceptible to the eyes and sense, which gave life to the eater thereof; and another tree which gave to the eater thereof a knowledge of good and evil? I believe that every man must hold these things for images, under which the hidden sense lies concealed.[432]

The meta-mythology of the Way portrays the manner in which specific ideas about the present, the future, and the mode of transforming one into the other are initially constructed, and then reconstructed in their entirety, when such transformation becomes necessary. This meta-myth provides the deep structure linking other classes of myths, including those describing the current or pre-existence stable state, those that portray the emergence of something unexpected into that state, those that represent the dissolution of paradise, in consequence, and those that describe the regeneration of stability. This cyclic pattern is essentially characteristic of the development of consciousness, of the capacity to act and represent—which is regarded from the mythic perspective as akin to the creation of the world.

The "previous place of stability," destroyed as a consequence of emergent anomaly, may be apprehended either as "the paradise that once reigned," from the perspective of the chaos engendered by its collapse, or as "the rigid and tyrannical past," from the perspective of revitalized and renewed order. Myths of paradise and the fall typically describe the first dynamic elements of the way from the perspective of "the chaos presently reigning"—that is, from the position of the uncertainty and fear that characterizes profane and worldly life. From this standpoint, human life is existence in the "valley of the shadow of death," contaminated by the unbearable and unreturnable gift of the knowledge of good and evil. Myths of redemption—that is, of the ascent from chaos, of the return to paradise, or of the "flight" to heaven—are tales "designed" to describe the process of remediation for the "prehistoric" fall. Such myths lay out a morality whose incorporation or incarnation constitutes cure for the spiritual paralysis engendered by emergent knowledge of death.

The idea of primeval paradise, then paradise lost—of the origin of experience, the rise of (self)consciousness, then permanent, heritable fall, descent from grace—appears as a constant predicate of human culture, distributed throughout the world. Even the most technologically primitive of people, whose styles of existence were often mistaken for paradisal by the Europeans who first encountered them, generally considered themselves fallen from an earlier condition of perfection. For them, like us, the noble savage was an ancestral Adamic figure, who could communicate directly with God:

When Heaven had been abruptly *separated* from the earth, that is, when it had become *remote*, as in our days; when the tree or liana connecting Earth to Heaven had been cut; or the mountain which used to touch the sky had been flattened out—then the paradisiac stage was over, and man entered into his present condition. In effect, all [myths of paradise] show us primordial man enjoying a beatitude, a spontaneity and freedom, which he has unfortunately lost in consequence of the *fall*—that is, of what followed upon the mythical event that caused the *rupture* between Heaven and Earth.[433]

The idea of paradise encompasses somewhat more than the "previous place of stability." It is actually all previous places of stability, concatenated into a single representation. Every previous place of stability becomes in this manner *order*, as such, balanced perfectly with potential—becomes existence without suffering, in Eden or Paradise, in the "walled garden of delight" ("Eden, signifies in Hebrew 'delight, a place of delight' ... our own English word Paradise, which is from the Persian, *pairi*—"around," *daeza*—"a wall," means properly a walled enclosure. Apparently, then, Eden is a walled garden of delight."[434]). Paradise is the place where the perfect harmony of order and chaos eliminates suffering, while bringing forth the necessities and pleasures of life without work or effort. Chaos and order are integrated, perfectly, in the paradisal state.

Paradise therefore also partakes of the state of the "cosmos" before its division into the ever-warring constituent elements of experience. This *uroboric* condition or state, conceptualized as a mode of being that is free from or beyond opposition, is also by necessity that place or state of being where suffering—a consequence of limitation and opposition—does not exist. This form of symbolic representation seems somewhat paradoxical, as it is the "dragon of chaos" that generates dread anxiety, when it manifests itself unexpectedly. However, context determines salience—determines *meaning*—in mythology as elsewhere. The conditions of existence—that is, the balance obtained by the forces of order, chaos and consciousness—not infrequently appear as intolerable, in and of themselves (in the state of anxiety and pain characterized by severe grief or depression, for example). From this perspective, the state of nonbeing (equivalent to identity with precosmogonic chaos) is the absence of all possibility of suffering. In the state of ideation characterizing suicide, for example, the Great Mother beckons. A student of mine, who had undergone a relatively severe crisis of identity, told me the following story:

I took a trip to the ocean. There were cliffs behind the beach. I was standing on one of the cliffs, looking out over the water. I was in a depressed state of mind. I looked out to the horizon. I could see the figure of a beautiful woman in the clouds. She gestured for me to come forward. I almost went over the edge, before I came out of my fantasy.

My wife told me a very similar tale. When she was in late adolescence, feeling somewhat unsettled, she took a camping trip on the sides of a deep river bank near her hometown. She stayed overnight on a bluff overlooking a steep drop. In the morning, the fog came off the river and filled the valley. She walked to the edge.

I saw the clouds below me. They looked like big, soft pillows. I imagined diving in, where it was warm and comfortable. But part of me knew better.

The state of non-existence—the state before the opening of Pandora's box—can under many conditions appear a state worth (rē)attaining.

The common metaphor of Paradise as geographic place serves to concretize a complex state of affairs, whose intrinsic nature would otherwise remain entirely beyond grasp. It brings down to earth the *a priori* conditions of the spirit and renders them initially comprehensible, at least in the symbolic sense. Paradise as place or state is perfected interpersonal interaction—the harmony of the lion and the lamb—as well as spiritual harmony (is the "internal kingdom" and the "external kingdom" simultaneously united as the "kingdom of God"). Paradise is also the world before it became profane—before innocence was lost.

Myths of "the paradise of childhood" use the circumstances applying at the dawn of each individual life—prior to separation of mother and child—as metaphor for the "place of beginnings." The symbiotic mother-child relationship is a union of elements that will in time become separate. The intimate union of two individuals at the beginning of a life comprises a state that is one thing, and more than one thing, simultaneously. This concrete example of a unity that is at the same time a plurality can be used in abstraction, to represent the hypothetical pretemporal state itself, where everything that would be more than one thing still "existed" in inseparable identity. This unity—the unviolated original state—tends to take on the affective evaluation of perfection (since it is the place where there is no conflict, no "separation of opposites").

Widespread iconic representations of the Holy Virgin Mother and Child, for example—Christian and non-Christian—might be regarded as crystallized fantasies about the affective nature of the origin. In the ideal mother-infant union, every desire remains absolutely bounded by love. The state of early childhood, more generally, symbolizes freedom from conflict; symbolizes honest, innocent, idyllic human existence, immersion in love, life before the necessary corruption of social contact, life preceding exposure to the harshly punitive conditions of physical existence. Childhood represents (perhaps, *is*) existence prior to the discovery of mortality. This lack of contamination by knowledge of death lends childhood experience an ideal quality, which easily comes to serve the mythic imagination as model for the state of being that transcends the existential anxiety of adulthood. The child, father to the man, represents the past of man; additionally, represents human potential, and man's eternal hope for the future. The Hasidim believe, for example, that "the Zaddik [the perfect, righteous man] finds that which has been lost since birth, and restores it to man."[435] In the Christian tradition, likewise, it is held that "except ye be converted, and become as little children, ye shall not enter into the Kingdom of Heaven" (Matthew 3:3). Maturation means expansion of ability, differentiation of self and world, transformation of possibility into actuality, but loss of potential as well, as anything developed develops in one direction, and not in any of the innumerable alternatives. Growth therefore also means decline, as each step toward adulthood is one step closer to death.

The initial paradisal state is typically disrupted, in mythological representation, by some fateful act undertaken by man—by some act that places him in opposition to his heavenly source. Such opposition is painful, and is often portrayed as a dreadful mistake or sin. It is nonetheless the case that the origin of experience and history—that is, the origin of being itself—appears inextricably bound up with such opposition, with differentiation from the origin. The initial paradisal state, although characterized by absolute totality, nonetheless seems paradoxically flawed; it suffers from an indeterminate form of non-existence—lacks reality itself:

There was something formless, yet complete, that existed before Heaven and Earth;
Without sound, without substance,
dependent on nothing, unchanging, all-pervading, unfailing.
One may think of it as the Mother of all things under Heaven.[436]

Such non-existence appears as an inevitable consequence of the absence of limitation, or of opposition. This absence deprives whatever constitutes the origin of a point of reference, distinguishable from itself—and, therefore, deprives it of existence. As a place (as the "previous state of innocent being"), paradise retains a patina of carefree existence. This is diminished by the comparative *unreality* of that existence. Things have not yet fallen apart in the Garden of Eden—have not yet separated (completely) into their constituent elements. Two things that cannot be discerned from one another are not two things, however, and one thing with no discernible features whatsoever may not even be.

Paradise is the world, before it has become realized. In such a state, nothing suffers, and nothing dies, because there is no defined one to suffer—no one aware of either the nature of subjective being, or the meaning of such being, once it has become "detached" from the whole. The "primordial ancestor," simultaneously male and female, dwells in this unrealized place, prior to division into husband and wife;[437] exists, *unselfconsciously*, even after that division:

And they were both naked, the man and his wife, and were not ashamed. (Genesis 2:25)

To "know" nakedness and to be shamed by it is to understand exposure, weakness and vulnerability. To be exposed before a crowd and the world is to have the essential frailty of individual being dramatically and incontrovertibly demonstrated. To be unaware of nakedness—to lack "self-consciousness"—is to be much less troubled, but also to be much less. The "paradisal" world of the child is much less—much less *manifest*, that is—than the world of the adult. The child has fewer responsibilities, and fewer defined concerns, than the adult. This lends childhood a glamour that mature existence lacks, at least from a certain adult perspective. But it is also the case that the child has a terrible vulnerability that the adult has transcended. The child does not explicitly perceive his vulnerability, and therefore does not suffer, until that vulnerability tragically manifests itself. The adult, by contrast, knows he can be hurt, and suffers constantly for that knowledge. His "heightened consciousness"—self-consciousness, really—means that he can take steps to ensure his healthy survival, however (even though he must in consequence worry for the future). The world of the child is circumscribed, incompletely realized, but nonetheless vulnerable. The paradisal world is incomplete, yet threatened, in the same manner.

It is primordial separation of light from darkness—engendered by *Logos*, the Word, equivalent to the process of consciousness—that initiates human experience and historical activity (which is reality itself, for all intents and purposes). This initial division provides the prototypic structure, and the fundamental precondition, for the elaboration and description of more differentiated attracting and repulsing pairs of opposites:

In the beginning God created the heaven and the earth.
And the earth was without form, and void; and darkness was upon the face of the deep. And the Spirit
    of God moved upon the face of the waters.
And God said, Let there be light; and there was light.
And God saw the light, that it was good; and God divided the light from the darkness. (Genesis 1:1–4)

Light and darkness constitute mythic totality; order and chaos, in paradoxical union, provide primordial elements of the entire experiential universe. Light is illumination, inspiration; darkness, ignorance and degeneration. Light is the newly risen sun, the eternal victor of the endless cyclical battle with the serpent of the night; it is the savior, the mythic hero, the deliverer of humanity. Light is gold, the king of metals, pure and incorruptible, a symbol for civilized value itself. Light is Apollo, the sun king, god of enlightenment, clarity and focus; spirit, opposed to black matter; bright "masculinity," opposed to the dark and unconscious "feminine." Light is Marduk, the Babylonian hero, god of the morning and spring day, who struggles against Tiamat, monstrous goddess of death and the night. Light is Horus, who fights against evil and redeems the father. Light is Christ, who transcends the past, and extends to all individuals identity with the divine *Logos*. To exist in the light means to be born, to live, to be redeemed. To depart from the light means to choose the path of evil—spiritual death—or to perish bodily altogether.

Myth equates the origin of the universe of experience with the partition of light from darkness because of the analogical or metaphorical identity between that separation and the mysterious differentiation of conscious experience from unconscious nonawareness. Awareness and daytime experience are inextricably united, like oblivion and the night. Darkness places severe, uncontrollable external transpersonal limitations upon waking human awareness, by eliminating or dramatically restricting visually dependent temporal and spatial sensory extension. The blackness of the night brings with it the re-emergence of the unknown, and the eternal human sense of subjugation to those terrors still incomprehensibly embedded in experience:

> When sacred Night sweeps heavenward, she takes
> the glad, the winsome day, and folding it,
> rolls up its golden carpet that had been
> spread over an abysmal pit.
> Gone vision-like is the external world,
> and man, a homeless orphan, has to face,
> in utter helplessness, naked, alone,
> the blackness of immeasurable space.
> Upon himself he has to lean; with mind
> abolished, thought unfathered, in the dim
> depths of his soul he sinks, for nothing comes
> from outside to support or limit him.
> All life and brightness seem an ancient dream—
> while in the substance of the night,
> unravelled, alien, he now perceives
> a fateful something that is his by right.[438]

External "cosmic" forces veil the day with the night. Similarly, and as a consequence of equally uncontrollable and impersonal "internal" forces, consciousness vanishes, into sleep, in the night: [439]

The central metaphor underlying "beginning" is not really birth at all. It is rather the moment of waking from sleep, when one world disappears and another comes into being. This is still contained within a cycle: we know that at the end of the day we shall return to the world of sleep, but in the meantime there is a sense of self-transcendence, of a consciousness getting 'up' from an unreal into a real, or at least more real, world. This sense of awakening into a greater degree of reality is expressed by Heraclitus ... as a passing from a world where everyone has his own "logos" into a world where there is a common "logos." Genesis presents the Creation as a sudden coming into being of a world through articulate speech (another aspect of logos), conscious perception, light and stability. Something like this metaphor of awakening may be the real reason for the emphasis on "days," and such recurring phrases as "And the evening and the morning were the first day," even before the day as we know it was established with the creating of the sun.[440]

The temporary nocturnal state of nonexistence appears similar to the more permanent situation theoretically prevailing prior (?) to the dawn of awareness as such, where there was no subject, no object, and no experience at all—but where the *possibility* of such things somehow lay dormant.

There is no suffering in the Garden of Eden. In such a state, however, things do not really exist. In consequence, myth appears to have equated the establishment of the *opposition necessary to being* with the appearance and evolution of the limited subject, who serves creation as the mirror of God. In the mythic world, the very existence of experience—past, present and future—appears dependent upon experience of the spatially and temporally limited observer. Restricted in their manifestation in this manner—that is, manifest in the domain of individual experience—things attain a brief, differentiated existence, before they crash into their opposites and vanish forever. An ancient midrash states, in this vein, that "God and man are in a sense twins."[441] The modern physicist John Wheeler states, analogously:

In every elementary quantum process the act of observation, or the act of registration, or the act of observer-participancy, or whatever we choose to call it, plays an essential part in giving "tangible reality" to that which we say is happening. [Paradoxically:] The universe exists "out there" independent of acts of registration, but the universe does not exist out there independent of acts of registration.[442]

From the standard perspective, objective things exist, in and of themselves. But this viewpoint eliminates the necessity of the observer, who gives to all things a necessary vantage point, reducing indefinable virtuality to extant actuality. Myth makes no such mistake, equating the very presence of being and becoming with the emergence of consciousness and self-consciousness. [443] It is this equation that allows the mythic imagination to place man at the center of the universe, and to draw an analogy between the principle that makes order out of chaos, and the individual himself.

The mythic world—the world, as it is experienced—might in fact be considered an emergent property of first-order self-reference; might be regarded as the interaction between the universe as subject and the universe as object. Myth equates the origin with the dawning of light, with the emergence of consciousness: equates the universe with the world of experience; assumes that the subjective is a precondition of the real. This idea seems exceedingly foreign to modern sensibility, which is predicated upon the historically novel proposition that the objective material in and of itself constitutes the real, and that subjective experience, which in fact provides source material for the concept of the object, is merely an epiphenomenal appendage. However, it is the case that self-referential systems (like that consisting of *being* as subject and object, simultaneously) are characterized by the emergence of unexpected and qualitatively unique properties. The world as subject (that is, the individual) is an exceedingly complex phenomenon—more complex, by far, than anything else (excepting other subjects). The world as object is hardly less mysterious. It is reasonable to regard the interaction of the two as something even more remarkable. We think: *matter first, then subject*—and presume that *matter*, as we understand it, is that which exists in the absence of our understanding. But the "primal matter" of mythology (a more comprehensive "substance" than the matter of the modern world) is much more than mere substance: it is the source of everything, objective and subjective (*is matter and spirit, united in essence*). From this perspective, consciousness is fundamental to the world of experience—as fundamental as "things" themselves. The matter of mythology therefore seems more than "superstition, that must be transcended"—seems more than the dead stuff of the modern viewpoint.

Furthermore, the world of experience appears generated by the actions of consciousness—by dawning awareness—in more than one "stage." The "purely conscious awareness" which hypothetically exists prior to the generation of active representations of the self—that is, which accompanied the mere division of "object" and "subject"—still retains essential unity and associated "paradisal" elements. Adam and Eve exist as independent beings prior to their "fall," but still commune with the animals and walk with God. Sheltered in an eternally productive garden, blissfully ignorant of their essential nakedness and vulnerability, they exist without anxious care or toil. It is the emergence of *second-order* self-reference—awareness of the self; self-consciousness—which finally disrupts this static state of perfection, and irreversibly alters the nature of experience. (The development of consciousness—the apprehension of the system by "itself"—adds one form of self-reference to the universal structure. *Self*-consciousness—the apprehension of the subject by himself—appears to have added another.) The modern mind would consider nothing fundamental altered by such internal transformation (as it considers consciousness epiphenomenal to reality). The mythological mind adopts another stance, entirely, presuming as it does that consciousness is allied with the very creator of things. From this viewpoint, cognitive transformations alter the structure of existence—change the very relationship between heaven and earth, creator and created; permanently restructure the cosmos itself. The modern materialist would consider such a theory arrogant and presumptious, to say the least. Nonetheless, the great societies of East and West are predicated precisely upon such a viewpoint—upon myths of origin and fall, characterized by uncanny structural parallel:

The father of Prince Gautama, the Buddha, savior of the Orient, determined to protect his son from desperate knowledge and tragic awareness, built for him an enclosed pavilion, a walled garden of earthly delights. Only the healthy, the young, and the happy were allowed access to this earthly paradise. All signs of decay and degeneration were thus kept hidden from the prince. Immersed in the immediate pleasures of the senses, in physical love, in dance, and music, in beauty, and pleasure, Gautama grew to maturity, protected absolutely from the limitations of mortal being. However, he grew curious, despite his father's most particular attention and will, and resolved to leave his seductive prison.

Preparations were made, to gild his chosen route, to cover the adventurer's path with flowers, and to display for his admiration and preoccupation the fairest women of the kingdom. The prince set out, with full retinue, in the shielded comfort of a chaperoned chariot, and delighted in the panorama previously prepared for him. The gods, however, decided to disrupt these most carefully laid plans, and sent an aged man to hobble, in full view, alongside the road. The prince's fascinated gaze fell upon the ancient interloper. Compelled by curiosity, he asked his attendant:

"What is that creature stumbling, shabby, bent and broken, beside my retinue?"

and the attendant answered:

"That is a man, like other men, who was born an infant, became a child, a youth, a husband, a father, a father of fathers. He has become old, subject to destruction of his beauty, his will, and the possibilities of life."

"Like other men, you say?" hesitantly inquired the prince. "That means ... this will happen to me?"

and the attendant answered:

"Inevitably, with the passage of time."

The world collapsed in upon Gautama, and he asked to be returned to the safety of home. In time, his anxiety lessened, his curiosity grew, and he ventured outside again. This time the gods sent a sick man into view.

"This creature," he asked his attendant, "shaking and palsied, horribly afflicted, unbearable to behold, a source of pity and contempt: what is he?"

and the attendant answered:

"That is a man, like other men, who was born whole, but who became ill and sick, unable to cope, a burden to himself and others, suffering and incurable."

"Like other men, you say?" inquired the prince. "This could happen to me?"

and the attendant answered:

"No man is exempt from the ravages of disease."

Once again the world collapsed, and Gautama returned to his home. But the delights of his previous life were ashes in his mouth, and he ventured forth a third time. The gods, in their mercy, sent him a dead man, in funeral procession.

"This creature," he asked his attendant, "laying so still, appearing so fearsome, surrounded by grief and by sorrow, lost and forlorn: what is he?"

and the attendant answered:

"That is a man, like other men, born of woman, beloved and hated, who once was like you, and now is the earth."

"Like other men, you say?" inquired the Prince. "Then ... this could happen to me?"

"This is your end," said the attendant, "and the end of all men."

The world collapsed, a final time, and Gautama asked to be returned home. But the attendant had orders from the prince's father, and took him instead to a festival of women, occurring nearby in a grove in the woods. The prince was met by a beautiful assemblage, who offered themselves freely to him, without restraint, in song, and in dance, in play, in the spirit of sensual love. But Gautama could think only of death, and the inevitable decomposition of beauty, and took no pleasure in the display.

The myth of the Buddha is the story of individual development, considered in the ideal. The

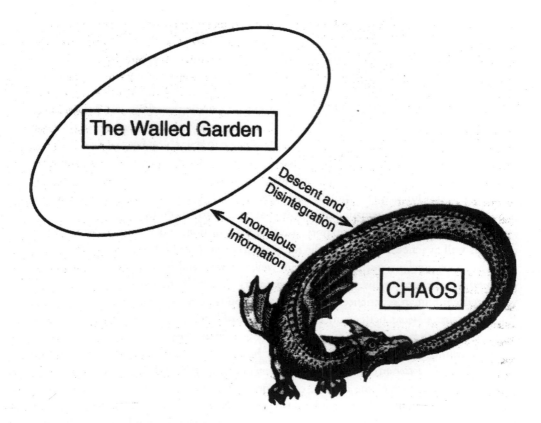

Figure 53: The (Voluntary) Descent of the Buddha

story opens with Gautama's father, shielding his child from the dangers of the world, much as any child in a healthy family is shielded. As the young prince matures, however, and becomes increasingly curious, he starts to wonder about the "world beyond." Children who develop within a safe and secure family grow into individuals who can no longer be contained by that family. It is the "good parent" who "fails," necessarily, by fostering a child who rapidly becomes so independent that parenting no longer suffices. Each foray out into the world produces an increase in knowledge, and a commensurate decrease in the ability of the childhood family constellation and personality to "map the world"—to provide acceptable patterns of action and representation, for existence as a true individual. The future Buddha's encounter with his intrinsic mortal limitations destroyed his childhood paradise, tragically—but propelled him out into the world as an independent being. This story can be portrayed, in the familiar manner, as in *Figure 53: The (Voluntary) Descent of the Buddha*. The story of Gautama's maturation details the consequential contamination of existence with unbearable anxiety; describes the association, in potential, even of beauty and the most fundamental and necessary of biological pleasures with the inevitability of decay and of death, the ultimate punishment. The Buddha's struggle with and eventual victory over his emergent tragic self-consciousness comprises the rest of the great tale: first, Gautama incorporated the knowledge of his ancestors; then, he transcended and restructured that knowledge.

After leaving the "walled garden" of his childhood, Gautama became a master of tradition, in his attempt to make sense of the world of experience as it now presented itself to him. He developed extensive knowledge of various philosophies, including Samkhya and Yoga, leaving each in turn as insufficient, and then took asceticism—worldly renunciation—to an extreme: "reduced almost to the state of a skeleton, he finally came to resemble a heap of dust."[444] That approach, too, proved insufficient. Finally, having tasted everything life had to offer, and having developed the discipline of a dedicated adherent, he prepared himself for his final battle. He entered a vast forest (the spiritual home of the unknown), placed himself at the foot of a pipal tree, and resolved to remain immobile in that place until he attained awakening.

Gautama experienced a true initiatory ordeal in that position, undergoing all the terrors of death (as well as renewed assault by the temptations of profane life). The discipline he had acquired in his previous journeys served him well, however, and he was able to remain single-mindedly devoted to his task—to the discovery of a truth that would serve life, that would redeem human experience. His final temptation is perhaps the most interesting. The Buddha attains *nirvana*, perfection, as a consequence of his ordeal, and is offered the option of remaining in that state by the God of Death. The offer is rejected: Buddha returns to the world, accepting his mortal condition, so that he can disseminate the knowledge he has acquired. *It is this latter action that truly marks him as a revolutionary hero.* Acquisition of wisdom—the consequence of the creative endeavor—is insufficient. The circle of redemptive action is not closed until information hard won on the battleground of the individual psyche has been integrated into the larger community. There can be no salvation for one in the presence of the continued suffering of all. It is Buddha's return from the heaven that is in his grasp that makes him truly great.

The story of the Buddha is perhaps the greatest "literary" production of the East. It is of great interest to note, therefore, that its theme also informs the most fundamental levels of Western sensibility. The Judeo-Christian tale of redemption is predicated upon representation of the individual subject, marred with Original Sin, fallen from grace, conscious of life and the borders of life, irretrievably blessed and cursed with knowledge of good and evil. The ability to develop such knowledge appears in the stories of Genesis as a "heritable characteristic of the race," as the precondition for generation of knowledge of the objective boundaries of subjective existence, as the fundamental precondition of tragic self-awareness:

And they were both naked, the man and his wife, and were not ashamed.
Now the serpent was more subtil than any beast of the field which the Lord God had made. And he
    said unto the woman, Yea, hath God said, Ye shall not eat of every tree of the garden?
And the woman said unto the serpent, We may eat of the fruit of the trees of the garden:
But of the fruit of the tree which is in the midst of the garden, God hath said, Ye shall not eat of it,
    neither shall ye touch it, lest ye die.

And the serpent said unto the woman, Ye shall not surely die:

For God doth know that in the day you eat thereof, then your eyes shall be opened, and ye shall be as
    gods, knowing good and evil.

And when the woman saw that the tree *was* good for food, and that it *was* pleasant to the eyes, and a
    tree to be desired to make *one* wise, she took of the fruit thereof, and did eat, and gave also unto
    her husband with her; and he did eat. (Genesis 3:1–6)

Myths dream ideas long before ideas take on recognizable, familiar and verbally comprehensible form. The myth, like the dream, may be regarded as the birthplace of conscious abstract knowledge, as the matrix from which formed ideas spring. Every concept, no matter how new or modern it appears, emerges from ground prepared by centuries of previous intellectual activity. Myth "prepares the ground" for explicit understanding by using what is presently comprehended—what has been partially explored, what has been adapted to in action—to represent that which remains unknown. Objects of experience which have been investigated can therefore come to serve as symbols of representation for description of the subject of experience, comparatively difficult to comprehend. It is in this manner that the *self*, which is essentially incomprehensible, unknown, gathers metaphoric representations.

Things that are in themselves complex and mysterious in their attributes serve this metaphoric function most usefully, since their potential for symbolic application is virtually infinite in scope. The tree and the serpent, for example—complex objects of apprehension—can be understood in part through direct and active observation, and can therefore provide productive grist for the metaphorical mill. Tree and serpent, coupled and singly, have an extensive, pervasive, and detailed history as representational agents. They serve similar functions in a multitude of myths describing the loss of paradise, and must therefore serve as apt representatives of some process or structure playing a central part in that loss. It appears likely—despite the initial strangeness of the presumption—that this structure is the nervous system,[445] as such (rather than any individual nervous system), as it manifests itself in intrapsychic representation.[446]

The tree is the *axis mundi*, world-tree, grounded immovably in the maternal (or, not infrequently, "material") world of chaos, with branches reaching to the sky (to heaven, to the realm of the ancestral spirits). According to the adepts of Hatha Yoga:

The feet, firmly placed on the ground, correspond to the roots of the tree, its foundation and source of nourishment. This might indicate that in daily life you stand firmly on the ground to meet life's demands. Your head is in space, or *heaven* [accented in the original]. The word "heaven" in this instance means in contact with life's energy, with a wisdom beyond the intellect.... The spine is like the trunk of the tree, along which are located the various Cakras. The top of the head is the crowning blossom of this flowering tree, the thousand-petalled Lotus of the Sharasrara Cakra.[447]

The *axis mundi* stands at the "center of the cosmos," uniting three separate but intertwined "eternal" realms. The *lower kingdom* is the domain of the unknown, subterranean, oceanic, hellish—land of reptilian power, blind force, and eternal darkness. The ancient Scandinavians believed, for example—in keeping with this general conceptualization—that a great serpent lived underneath *Yggdrasil*, the world-tree, and gnawed at its roots, trying forever to destroy it. (Yggdrasil was constantly revivified, however, by the springs of "magical water" that also lay underneath it). The great serpent is the dragon of chaos, in his destructive aspect—the source of all things (including the "world-tree"), as well as the power that reduces created objects to the conditions of their origin. (The "magical water" is the positive aspect of the unknown, with its procreative and rejuvenating power.) The dyad of tree and serpent is an exceedingly widespread motif of mythology and a common literary theme. Frye's comments on Melville's *Moby Dick* are relevant here. Moby Dick is a great white whale who lives in the depths of the sea. Ahab is the captain of a whaling boat, passionately and unreasonably dedicated to finally conquering that leviathan:

In *Moby Dick*, Ahab's quest for the whale may be mad or "monomaniacal," as it is frequently called, or even evil so far as he sacrifices his crew and ship to it, but evil or revenge are not the point of the quest. The whale itself may be only a "dumb brute," as the mate says, and even if it were malignantly determined to kill Ahab, such an attitude, in a whale hunted to the death, would certainly be understandable if it were there. What obsesses Ahab is in a dimension of reality much further down than any whale, in an amoral and alienating world that nothing normal in the human psyche can directly confront.

The professed quest is to kill Moby Dick, but as the portents of disaster pile up it becomes clear that a will to identify with (not adjust to) what Conrad calls the destructive element is what is really driving Ahab. Ahab has, Melville says, become a "Prometheus," with a vulture feeding on him. The axis image appears in the maelstrom or descending spiral ("vortex") of the last few pages, and perhaps in a remark by one of Ahab's crew: "The skewer seems loosening out of the middle of the world." But the descent is not purely demonic, or simply destructive: like other creative descents, it is partly a quest for wisdom, however fatal the attaining of such wisdom may be. A relation reminiscent of Lear and the fool develops at the end between Ahab and the little black cabin boy Pip, who has been left so long to swim in the sea that he has gone insane. Of him it is said that he has been "carried down alive to wondrous depths, where strange shapes of the unwarped primal world glided to and fro ... and the miser-merman, Wisdom, revealed his hoarded heaps."

Moby Dick is as profound a treatment as modern literature affords of the leviathan symbolism of the Bible, the titanic-demonic force that raises Egypt and Babylon to greatness and then hurls them into nothingness; that is both an enemy of God outside the creation and, as notably in Job, a creature within it of whom God is rather proud. The leviathan is revealed to Job as the ultimate mystery of God's ways, the "king over all the children of pride" (Job 41:34), of whom Satan himself is merely an instrument. What this power looks like depends on how it is approached. Approached by Conrad's Kurtz through his Antichrist psychosis, it is an unimaginable horror: but it may also be a source of energy that man can put to his own use. There are naturally considerable risks to doing so: risks that Rimbaud spoke of in his celebrated *lettre du voyant* as a "dereglement de tous les sens." The phrase indicates the close connection between the titanic and the demonic that Verlaine expressed in his phrase *poete maudit*, the attitude of poets who feel, like Ahab, that the right worship of the powers they invoke is defiance.[448]

Above the lower kingdom is earth, the *middle kingdom*, mundane conscious existence, domain of man, trapped uncomfortably between the titanic and the heavenly—trapped in the realm where "spirit and matter" or "heaven and hell" or "order and chaos" eternally interact and transform. The *upper kingdom*, finally, is heaven, the intrapsychic ideal, abstract symbolic construction and utopian state, creation of generations of autonomous fantasy, following its own rules, governed by its own denizens, with its own non-individual transcendent existence. The fact that the *axis mundi* unites earth and heaven means that it may serve ritual purpose as a bridge between the profane individual domain and the "realm of the gods":

The symbolism of the ascension into heaven by means of a tree is ... clearly illustrated by the ceremony of initiation of the Buriat shamans. The candidate climbs up a post in the middle of the yourt, reaches the summit and goes out by the smoke-hole. But we know that this opening, made to let out the smoke, is likened to the "hole" made by the Pole Star in the vault of Heaven. (Among other peoples, the tent-pole is called the *Pillar of the Sky* and is compared to the Pole-Star, which is also the hub of the celestial pavilion, and is named, elsewhere, the *Nail of the Sky*.) Thus, the ritual post set up in the middle of the yourt is an image of the Cosmic Tree which is found at the *Center of the World*, with the Pole Star shining directly above it. By ascending it, the candidate enters into Heaven; that is why, as soon as he comes out of the smoke-hole of the tent he gives a loud cry, invoking the help of the gods: up there, he finds himself in their presence.[449]

Figure 54: The World-Tree as Bridge Between "Heaven" and "Hell"[450] offers a visual interpretation of the cosmic tree, connecting "heaven, earth and hell." The cosmic tree—Yggdrasil, in this representation—is grounded in the domain of the dragon of chaos (the "serpent" who "gnaws at its roots"), passes through "earth," and reaches up into "heaven," the realm of the ancestor/gods. It was unconscious apprehension of this tripartite structure that led Freud, for example, to his model of the psyche: id (the "natural" world of dark instinctive "drive"), ego (the world of the individual), and superego (the gods of tradition). It is Freud's inclusion of all the elements of the world-tree (negative and positive) that has given his mythology its remarkable strength, influence and power.

Figure 55: The World-Tree and the Constituent Elements of Experience offers another interpretation and explanation of this tree, relating its "place" in the cosmos to the "constituent elements of experience." This diagram suffers somewhat from its precise symbolic equation of the tree and the "archetypal son." Christ and Satan, for example—Christian exemplars of the ambivalent son—may also be viewed as *products* of the tree (as well as particular incarnations or forms of the tree, or as phenomena otherwise inextricably associated with the tree). The world-tree as "forbidden tree of knowledge of good and evil" is, for example, the cross upon which Christ, the archetypal individual, crucified, suspended and tormented, manifests for all eternity his identity with God; the tree upon which Odin, Norse savior, is likewise suspended:

I ween that I hung
on the windy tree

Figure 54: The World-Tree as Bridge Between "Heaven" and "Hell"

> Hung there for nights full nine
> With the spear I was wounded
> and offered I was
> To Odin, myself to myself
> On the tree that none
> may never know
> What root beneath it runs.[451]

The tree is to Christ, therefore, as Christ is to the individual ("I am the vine, ye are the branches: He that abideth in me, and I in him, the same bringeth forth much fruit: for without me ye can do nothing." John 15:5). Satan, by contrast, is something that *lurks* in the forbidden tree. The (devastating) wisdom he promises—the knowledge of the gods—is that

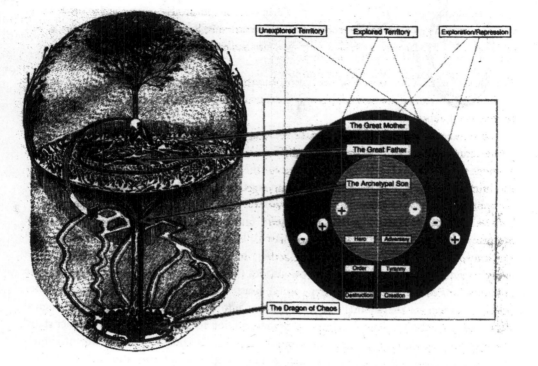

Figure 55: The World-Tree and the Constituent Elements of Experience

tree's "first fruit." This makes the world-tree the source of the revelation that destroys—the source of the anomalous "idea," for example, that disrupts the static past and plunges it into chaos—as well as the eventual source of the revelation that redeems.

In the book of Genesis, the fruits of the tree of knowledge are ingested in mythic action by the free (though sorely tempted) act of the individual. The myth uses a particular act, the incorporation of food, as metaphor for the assimilation of knowledge and ability. Erich Neumann states:

Wherever liquor, fruit, herbs, etc., appear as the vehicles of life and immortality, including the "water" and "bread" of life, the sacrament of the Host, and every form of food cult down to the present day, we have [an] ancient mode of human expression before us. The materialization of psychic contents, by which contents that we would call "psychic"—like life, immortality, and death—take on material form in myth and ritual and appear as water, bread, fruit, etc., is a characteristic of the primitive mind.

Conscious realization is "acted out" in the elementary scheme of nutritive assimilation, and the ritual act of concrete eating is the first form of assimilation known to man....

The assimilation and ingestion of the "content," the eaten food, produces an inner change. Transformation of the body cells through food intake is the most elementary of animal changes experienced by man. How a weary, enfeebled, and famished man can turn into an alert, strong, and satisfied being, or a man perishing of thirst can be refreshed or even transformed by an intoxicating drink: this is, and must remain, a fundamental experience so long as man shall exist.[452]

The act of defiant incorporation, initiating alienation from paradise and God, is instigated by the serpent, an ancient and dangerous creature of base matter, who can shed his skin, and be renewed, reborn.

The snake serves mythology in a dual role, as agent and symbol of transformation, and as prime representative of fundamental, undifferentiated uroboric power. The Edenic serpent provides the individual with the knowledge of the gods, without their compensatory power and immortality. His "enlightenment" of man engenders an unparalleled catastrophe—a catastrophe sufficiently complete to engender not only the "final division of heaven and earth," but, on that earth, a more or less permanent (and unfortunate) association between the promise of knowledge and the appearance of evil. The Edenic serpent occupies the same categorical space in the Christian psyche as Lucifer, "bringer of light," spirit of "unbridled rationality"—in large part because the anomalous idea (the "product of rationality") has the same potential for destruction as any other natural disaster. This identification is somewhat one-sided, however, as the anomaly-inspired descent into chaos is only half the mythological story, and can also be viewed as a necessary precondition for emergence into a "higher state" of consciousness (even for the incarnation of Christ, the "second fruit of the tree of knowledge"). The medieval alchemists tended to adopt a gnostic interpretation of the Edenic story, for this reason, as Jung states:

Hence we get the parallel of the dragon's head with Christ, corresponding to the Gnostic view that the son of the highest divinity took on the form of the serpent in paradise in order to teach our first parents the faculty of discrimination, so that they should see that the work of the demiurge [the god who created the world in the first place] was imperfect.[453]

The Edenic serpent is, above all, the unknown (power) still lurking "inside" the nervous system, inside the "world-tree." It is the innate capacity of the mind, its ability to generate revelatory thought, its capacity to disrupt the stable cosmos and to extend the domain of consciousness. It was "unconscious" (imagistic) apprehension of this idea that led medieval alchemy to treat the serpent as the "arcane substance" that transformed itself inside the tree, and to regard the serpent as the tree's "life."[454]

It is curiosity that kills the cat but, equally, curiosity that guides the discovery. The forbidden or unknown object exists, shrouded in mystery, "outside" the mundane, familiar and explored world. The command "you can't explore that" inevitably contaminates the forbidden object or situation with mystery: what could possibly be so dangerous (powerful, interesting) that it must be treated as if it was not there? To explicitly forbid something contaminates it with the "dragon of chaos"—places a serpent inside it, so to speak. To explicitly forbid something virtually ensures that it will attract attention, at least (as the unknown inevitably compels approach, as well as fear). The serpent/dragon–chaos/forbidden object connection can therefore be profitably viewed from a more "physiological" perspective, as well. The snake is regarded as the regulator of conscious intensity, by the adepts of Kundalini Yoga. This snake is a creature of the spine, a storehouse of intrapsychic energy, whose activation leads to ecstasy and enlightenment. The goal of Kundalini Yoga is to "awaken" this serpent, and to thereby reach enlightenment.

The snake shares obvious—and subtle—features with the spine. First is the shape; second is shared evolutionary history. The human nervous system is composed in part of structures as phylogenetically ancient as the reptile, in whose recesses lurk tremendous excitatory power. The deep structures of the brainstem—the "head" of the spinal snake—perform activities upon which maintenance of consciousness absolutely depends.[455] An individual, lost in sleep (in "unconsciousness") can be brought instantaneously awake and alert by stimulated operation of these structures, in a situation (for example) where something unexpected and potentially dangerous occurs. A sleeping mother can be brought instantly awake and motivated for exploration by the unexpected cry of her baby, for example. The process of contrast between desire and current status (between ideal future and present) does not disappear even in sleep. *The unknown brings wakefulness to the sleeping.* Threat—more generally, the appearance of the unknown—propels active exploration, designed to expand adaptive competence (or, terrified cessation of activity) and produces dramatic heightening of interest and consciousness. This means that consciousness as a phenomena depends in large part on activation of the ancient circuitry designed for response to the unknown. As the human brain evolved, much more "territory for activation" developed; nonetheless, alertness still depends on very archaic nervous system substructures. Knowledge of this dependence echoes through myth and literature. Goethe's Mephistopheles is therefore able to state, for example:

> Follow the adage of my cousin Snake
> From dreams of god-like knowledge you will wake
> To fear, in which your very soul shall quake.[456]

The most "conscious" animal is the most *motivated* animal. The most motivated animal lives in apprehension of the ever-present possibility of the greatest possible threat (that of its own demise) and in eternal desire for rectification of this threat—in *hope*, in consideration of the possibilities of the dangerous unknown for generation of "redemptive" information. It is clear apprehension of the mortal danger and infinite possibility lurking everywhere that has boosted human consciousness far beyond that of its nearest kin, in a process that extended over eons. We can see the unknown in everything, as a consequence of our elaborate cognitive systems: worse (better)—we can see mortal danger in everything unknown. This makes us anxious, certainly—but also (if we don't run away) *awake*. The "serpent" of the "external unknown" works in concert, therefore, with the "serpent" of the internal unknown: apprehension of the mystery which transcends the current realm of adaptation (that is, the permanent mystery of mortal limitation) produces *permanent consciousness*, at least in principle. It is for this reason that the Buddha is the "awakened one." Our expanding brain, "designed" to produce adaptation, instead sees risk and opportunity everywhere. The circuitry "designed" to explore anomaly, and then to cease its actions, once exploration has produced its desired consequences, is instead always operating—as it can never reach its ever-receding goal. And so we are forever unsettled, unhappy, unsatisfied, terrified, hopeful—and awake.

Individual incorporation of socially predicated knowledge—expanded exponentially in scope during the course of centuries of human cultural endeavor, culminating in development of an elaborated self-model—*produced within the individual clear apprehension of mortality as a defining feature of human existence*. This act of self-definition inextricably associated every aspect of human experience with threat—eternally contaminated all human experience with the intimation of mortality, with the hint of death, with the absolutely inexplicable unknown. This act of self-definition drove us to consider the world we had built as forever insufficient, forever lacking in security; drove us to regard the unknown "place of death," in addition, as simultaneous eternal source of new redemptive information. This contamination rendered every object, every facet of experience, permanently mysterious, and sufficiently motivating to maintain heightened consciousness, as an interminable, awful and beneficial feature of human existence.

The myth of the Fall, Christian or Buddhist, describes the development of self-consciousness as voluntary, though pre-arranged, in a sense, by the gods, whose power remains outside human control. Lucifer, in the guise of the serpent, offers Eve the apple, with truly irresistible promise of expanded knowledge. Fate arranges the future Buddha's introduction to old age, sickness, and death—but Gautama chose, voluntarily, to leave the confines of the paradise his father endeavored to render perfect. It is the expansive exploratory tendency of man, his innate curiosity, that is simultaneously saving grace and mortal error. For this reason, the stories of Genesis and of the Buddha are predicated on the implicit assumption that contact with the unbearable, in the course of maturation, is predetermined, inevitable—and *desired*, catastrophic but *desired*. Voltaire tells a story, of the Good Brahmin—an admirable, tragic figure—which clarifies the role of voluntarism (and pride) in the reach for heightened human awareness:

"I wish I had never been born!"

"Why so?" said I.

"Because," he replied, "I have been studying these forty years, and I find that it has been so much time lost.... I believe that I am composed of matter, but I have never been able to satisfy myself what it is that produces thought. I am even ignorant whether my understanding is a simple faculty like that of walking or digesting, or if I think with my head in the same manner as I take hold of a thing with my hands.... I talk a great deal, and when I have done speaking I remain confounded and ashamed of what I have said."

The same day I had a conversation with an old woman, his neighbour. I asked her if she had ever been unhappy for not understanding how her soul was made? She did not even comprehend my question. She had not, for the briefest moment in her life, had a thought about these subjects with which the good Brahmin had so tormented himself. She believed in the bottom of her heart in the metamorphoses of Vishnu, and provided she could get get some of the sacred water of the Ganges in which to make her ablutions, she thought herself the happiest of women. Struck with the happiness of this poor creature, I returned to my philosopher, whom I thus addressed:

"Are you not ashamed to be thus miserable when, not fifty yards from you, there is an old automaton who thinks of nothing and lives contented?"

"You are right," he replied. "I have said to myself a thousand times that I should be happy if I were but as ignorant as my old neighbour; and yet it is a happiness which I do not desire."

This reply of the Brahmin made a greater impression on me than anything that had passed.[457]

The shame the Brahmin feels at his own words is a consequence of his realization of their insufficiency, of his self-comprehended inability to address the problems of life in some manner that he regards as final and complete. His shame and unhappiness is, paradoxically, a consequence of the activities of the same process that enables him to seek redress—a process that is troublesome in the extreme, but so valuable that it will not be abandoned, once attained. The questing spirit undermines its own stability, but will not give up that destabilizing capacity for return to the "unconscious" source. This is, I suppose, part of the "pride" of man, that serves as predestiny for the fall—but is, as well, part of another "unconscious" apprehension: something that destroys, initially, may still save, in its later development, and the process that undermines may also be the same process that rebuilds, out of wreckage, something stronger.

The birth of tragedy and the evolution of shame might be considered inevitable consequences of voluntarism itself, of the heroic exploratory tendency, diabolically predetermined in its unfolding, leading inexorably to development of unbearable but potentially redemptive (self)consciousness. Extension of objective knowledge to the self means permanent establishment of a conceptual connection between individual existence and certain mortality. Development of this connection means existence in endless conflict, as all human activity henceforth takes place in the valley of the shadow of death. Fate compels all members of the human race to comprehend their isolation, their individuality, their abject subjugation to the harsh conditions of mortal existence.[458] Recognition of the naked self, exposed indignantly[459] to the ravages of time and the world, unbearable and highly motivating, condemns man and woman to toil and suffer for life and for death:

And the eyes of both of them were opened, and they knew that they were naked, and they sewed fig leaves together, and made themselves aprons. (Genesis 3:7)

Acquisition of this unbearable knowledge renders unquestioning acceptance of biological necessity impossible, and destroys all possibility of simple adherence to the paradisal way. Adam and Eve immediately cover themselves up—erect a *protective barrier*, symbolic of culture itself, between their vulnerable bodies and the terrible world of experience. This emergent fear of vulnerability—a direct consequence of the development of self-consciousness (no, an intrinsic aspect of self-consciousness)—permanently undermines capacity for faith in blind instinctual action:

> And they heard the voice of the Lord God walking in the garden in the cool of the day: and Adam and
>   his wife hid themselves from the presence of the Lord God amongst the trees of the garden.
> And the Lord God called unto Adam, and said unto him, Where *art* thou?
> And he said, I heard thy voice in the garden, and I was afraid, because I *was* naked; and I hid myself.
>   (Genesis 3:8–10)

Paradise is the place where heaven, earth and nature still touch—the place where man, who lives in harmony with the animals, is still unrebellious and still "walks with God." The conscious (?), but unselfconscious animal lives within the undisputed sway of natural processes. It cannot develop referential perspective regarding its own perceptions, impulses, and behaviors, because it lacks access to the experience of others. The self-aware individual human being, by contrast, lives *in history*, in an experiential field whose every aspect has been shaped and modified by experience communicated from extant personage and ancestral figure. This socially predicated historical construct appears to provide the basis for sophisticated self-consciousness, which is (innate) capacity for self-reference, provided with content through action of culture. Construction of self-consciousness requires elaboration of a self-model; extension of the notion of the independent other to the self; internalization of a socially determined conceptual representation of the self. The capacity for such objective description arose as a consequence of the communication of disembodied or abstracted thought from person to person, through processes ranging in complexity from concrete imitation to generalized philosophical discourse.

The ability to communicate skill and representation makes it possible for the individual to internalize and formulate a complex self-representation, to conceive of him or herself in terms of the experience of others—that is, in terms of the experience of specific others, offering (and embodying) their defining opinion, and the general other, historical humanity. This process apparently occurs (occurred) as each person becomes subject not only to those unique experiences that constitute his or her own being, but to the experience of every other individual, transmitted imitatively, dramatically, linguistically. This spatially and temporally summed wealth of culturally predicated experience, whose cumulative breadth and depth far exceeds the productive capability of a single individual lifetime, must acquire tremendous intrapsychic power once transmitted and cortically represented, must become capable of fundamentally altering—restricting and extending—innate personal experience. One inevitable consequence of this shared perception is self-definition, development of individual self-awareness, under the pressure of immediate transient experience, individual past and historical opinion regarding the nature of one's own experience and human experience in general. The expansion of detailed communication allows the individual to become at least partly aware of his or her own "objective" nature.

Individual intrapsychic representation of cumulative historically predicated human experience makes the one into the many, so to speak; makes the individual into the embodiment of group experience, to date. Development of moral sense, and moral choice, constitutes an emergent property of such incorporation of knowledge. Knowledge of morality, of good and evil, presupposes the presence of alternative possibilities for action in a given situation—means capacity for conceptualization of alternative ideals, toward which behavior can be devoted. The animal, guided purely by its individual, biologically determined perceptual and motivational structures, essentially unaltered in function as a consequence of summed and stored communication, develops no possibility for self-criticism, has no platform upon which to stand, to criticize from—has no basis for comparison, no expanded repertoire of adaptive behavior, no capacity for fantasizing about what could be and no cultural experience to flesh out that capacity. Animal perception and action—animal experience—has not been made subject to historically predicated self-conscious analysis.

The animal, in its natural and constant environment, remains beyond (or before) good and evil, gripped by its biologically determined destiny, which is the will of God, from the mythic perspective. The human being, by contrast, with a head full of alternative opinions (the abstracted residue of individual ancestral choice), can use the internalized or freely offered opinion of the group to criticize spontaneous manifestations of subjective perception and motivation—to judge, alter, or inhibit pure subjectivity itself. This ability allows the human being tremendous interpretive and behavioral possibility, freedom, but lays subjective experience, untarnished instinct, bare to insult. This separation of man from immersion in the natural way constitutes a remarkable achievement, with eternally unsettling consequences.

The birth of tragedy and the evolution of shame might be considered emergent properties of self-consciousness. The idea of redemption, which compensates for self-conscious existential anxiety, might be considered another, higher-order emergent property. The tradition of the "fall from paradise" is predicated on the idea that the appearance of self-consciousness dramatically altered the structure of reality. The explicitly religious accept the notion that man and God have been tragically estranged—that human actions have shattered the divine order. This idea is so central to our worldview that it crops up everywhere—in the ease with which we can all be made to feel guilty, in the (often explicitly areligious) view that human existence is "foreign" to the "natural order" (that human activity is detrimental to the environment, that the planet would somehow be "better off" with no people on it), that our species is somehow innately disturbed or even insane. Our constantly emerging self-reference (our constantly developing self-consciousness) has turned the world of experience into a tragic play:

Unto the woman he said, I will greatly multiply thy sorrow and thy conception; in sorrow thou shalt bring forth children;[460] and thy desire shall be to thy husband, and he shall rule over thee.
And unto Adam he said, Because thou has hearkened unto the voice of thy wife, and hast eaten of the tree, of which I commanded thee, saying, Thou shalt not eat of it: cursed is the ground for thy sake; in sorrow shalt thou eat of it for all the days of thy life;
Thorns also and thistles shall it bring forth for thee; and thou shalt eat the herb of the field;
In the sweat of thy face shalt thou eat bread, till thou return unto the ground; for dust thou art, and unto dust shalt thou return." (Genesis 3:16–19)

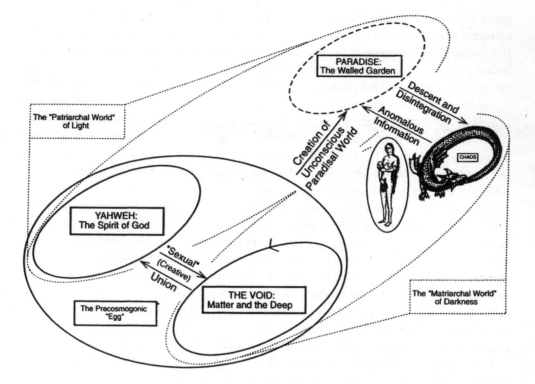

Figure 56: Genesis and Descent

Foreknowledge of destiny seals fate, and paradise is forever lost to man:

And the Lord God said, Behold, the man is become as one of us, to know good and evil: and now, lest
he put forth his hand, and take also of the tree of life, and eat, and live forever:
Therefore the Lord God sent him forth from the garden of Eden, to till the ground from whence he
was taken.
So he drove out the man: and he placed at the east of the Garden of Eden Cherubims, and a flaming
sword which turned every way, to keep the way of the tree of life. (Genesis 3:22–24)

Why hide from God? Because knowledge of vulnerability makes us shrink from our own
potential. To live, fully, is to risk—to risk everything, to risk death. Why hide from God?
How, under such conditions, could we not hide? Survival has become terror and endless
toil—necessitating discipline, compelled by burdensome wisdom, rife with intrapsychic conflict, motivated by anxiety—instead of spontaneous natural activity. We remain eternally
hung on the cross of our own vulnerability. The creation and fall of man is portrayed
schematically in *Figure 56: Genesis and Descent*. [461]

5

THE HOSTILE BROTHERS

Archetypes of Response to the Unknown

───────────────○───────────────

The "contamination of anomaly with the threat of death," attendant on the development of self-consciousness, amplifies the valence of the unknown to a virtually unbearable point. This unbearable amplification has motivated the development of two transpersonal patterns of behavior and schemas of representation, constituting the individual as such, embodied in mythology as the "hostile brothers." One of these "hostile brothers" or "eternal sons of God" is the mythological hero. He faces the unknown with the presumption of its benevolence—with the (unprovable) attitude that confrontation with the unknown will bring renewal and redemption. He enter, voluntarily, into creative "union with the Great Mother," builds or regenerates society, and brings peace to a warring world.

The other "son of God" is the eternal adversary. This "spirit of unbridled rationality," horrified by his limited apprehension of the conditions of existence, shrinks from contact with everything he does not understand. This shrinking weakens his personality, no longer nourished by the "water of life," and makes him rigid and authoritarian, as he clings desperately to the familiar, "rational," and stable. Every deceitful retreat increases his fear; every new "protective law" increases his frustration, boredom and contempt for life. His weakness, in combination with his neurotic suffering, engenders resentment and hatred for existence itself.

The personality of this adversary comes in two forms, so to speak—although these two forms are inseparably linked. The fascist sacrifices his soul, which would enable him to confront change on his own, to the group, which promises to protect him from everything unknown. The decadent, by contrast, refuses to join the social world, and clings rigidly to his own ideas—merely because he is too undisciplined to serve as an apprentice. The fascist wants to crush everything different, and then everything; the decadent immolates himself, and builds the fascist from his ashes. The bloody excesses of the twentieth century, manifest most evidently in the culture of the concentration camp, stand as testimony to the desires of the adversary and as monument to his power.

307

The pitfalls of fascism and decadence may be avoided through identification with the hero, the true individual. The hero organizes the demands of social being and the responsibilities of his own soul into a coherent, hierarchically arranged unit. He stands on the border between order and chaos, and serves the group as creator and agent of renewal. The hero's voluntary contact with the unknown transforms it into something benevolent—into the eternal source, in fact, of strength and ability. Development of such strength—attendant upon faith in the conditions of experience—enables him to stand outside the group, when necessary, and to use it as a tool, rather than as armor. The hero rejects identification with the group as the ideal of life, preferring to follow the dictates of his conscience and his heart. His identification with meaning—and his refusal to sacrifice meaning for security—renders existence acceptable, despite its tragedy.

INTRODUCTION: THE HERO AND THE ADVERSARY

The culture bequeathed to us by our forebears degenerates of its own accord, as the flux of the present invalidates the static presuppositions of the past. We may speed this process of degeneration by our "sins"—by voluntary refusal to attend to our errors, when they manifest themselves, and by failure to adjust behaviors and attitudes, in consequence. Through such refusal and failure, we transform the irritations of the present into the catastrophes of the future, and invite a wrathful God to drown us beneath the waves.

> As has been well known since the compilations made by R. Andree, H. Usener, and J.G. Frazer, the deluge myth is almost universally disseminated; it is documented in all the continents (although very rarely in Africa) and on various cultural levels. A certain number of variants seem to be the result of dissemination, first from Mesopotamia and then from India. It is equally possible that one or several diluvial catastrophes gave rise to fabulous narratives. But it would be risky to explain so widespread a myth by phenomena of which no geological traces have been found. The majority of the flood myths seem in some sense to form part of the cosmic rhythm: the old world, peopled by a fallen humanity, is submerged under the waters, and some time later a new world emerges from the aquatic "chaos."

> In a large number of variants, the flood is the result of the sins (or ritual faults) of human beings: sometimes it results simply from the wish of a divine being to put an end to mankind.... the chief causes lie *at once in the sins of men and the decrepitude of the world.* By the mere fact that it exists—that is, that it *lives* and *produces*—the cosmos gradually deteriorates and ends by falling into decay. This is the reason why it has to be recreated. In other words, the flood *realizes,* on the macrocosmic scale, what is symbolically effected during the New Year festival: the "end of the world" and the end of a sinful humanity in order to make a new creation possible.[462]

No discussion of the architecture of belief can possibly be considered complete, in the absence of reference to *evil*. Evil is no longer a popular word, so to speak—the term is generally considered old-fashioned, not applicable in a society that has theoretically dispensed with its religious preoccupations. Acts once defined as evil are now merely considered the consequence of unjust familial, social or economic structures (although this view is not as widespread as it once was). Alternatively, the commission of incomprehensible acts of cruelty and destruction are viewed as symptomatic of some physiological weakness or disease. Seldom are acts of evil considered voluntary or purposeful—committed by someone possessed by an aesthetic that makes art of terror and pain.

In the Egyptian cosmology, Osiris, the king, mythic image of the known, the Great Father, has an eternal evil twin and opposite, Seth, who eventually leads Osiris to his death. Four thousand years later, the moral of this great story has not yet been grasped: failure to understand the nature of evil leads to its eventual victory. At the end of this, the most cruel and bloodthirsty of centuries, we are in danger not only of failing to understand evil, but of denying its very existence. Invisibility, however, is what the devil craves most.

I have spent a substantial amount of time, up to this point, describing the nature of culture, and the manner in which it is generated. Culture, the Great Father, protects us from the terrors of the unknown; defines around us a sacred space into which nothing unbearably foreign is allowed. Culture is generated by the process whose essential features have been captured in the pervasive and recurrent myths of the hero. This hero is the individual who voluntarily faces the dragon of the unknown, cuts it up, and creates the world from its pieces; the individual who overcomes the too-long-senescent tyrant and frees the virgin mother from his grasp. Such myths present a world starkly ambivalent in its nature: the "natural" world is endlessly creative, and equally destructive; the inextricably associated social "environment" is simultaneously tyrannical and protective. So far in our discussion, however, the hero has stood alone. This means that our story is far from over. The essential ambivalence characterizing the "constituent elements of experience" extends to the individual as well, who is capable of thoughts and actions as dark and destructive as anything in society or nature.

Mythology envelops the transpersonal capacity for evil characterizing the individual, as such, in the guise of a personality, duplicating its encapsulation of chaos and order. The dark side of the individual is the absolute adversary of the hero; the personality who shrinks from contact with the unknown, or who denies that it exists, instead of actively approaching and exploring; the personality whose "advice" speeds the decline of society rather than its renewal. The image of this personality—like the phenomenon itself—has developed and become elaborated in complexity and sophistication over the course of centuries: proper understanding of its nature is terrifying, in a salutary sort of way. This informative terror is the "purpose" of encapsulation in narrative, in transpersonal memory: the image of the Christian devil, for example, is the best "bad example" available. His implicit or explicit imitation leads to disaster; the stories that portray his central features exist as object lessons in the consequences of resentment, hatred, totalitarian arrogance and jealousy.

Evil, like *good*, is not something static: it does not merely mean breaking the rules, for example, and is not simply aggression, anger, force, pain, disappointment, anxiety or horror. Life is of course endlessly complicated by the fact that what is bad in one circumstance is positively necessary in the next. I noted previously[463] that the answer to the question "what is the good?" must in fact be sought in the meta-domain, so to speak: the more fundamental mystery—given the context-dependent nature of "the good"—is "how are answers to the question 'what is the good?' endlessly and appropriately generated?" The "good," then, becomes the set of circumstances that allow the process of moral construction to flourish or becomes the process of moral construction itself. The problem "what then is evil?" must be addressed similarly.

Evil is rejection of and sworn opposition to the process of creative exploration. Evil is proud repudiation of the unknown, and willful failure to understand, transcend and transform the social world. Evil is, in addition—and in consequence—hatred of the virtuous and courageous, precisely on account of their virtue and courage. Evil is the desire to disseminate darkness, for the love of darkness, where there could be light. The spirit of evil underlies all actions that speed along the decrepitude of the world, that foster God's desire to inundate and destroy everything that exists.

Great evils are easily identifiable, at least in retrospect, and are usually the result (at least in interpretation) of the act of another. We build endless memorials to the Holocaust, for example, and swear never to forget. But what is it that we are remembering? What is the lesson we are supposed to have learned? We don't know how the Holocaust came about—don't know what it is that the people involved did, or failed to do, step by step, that made them behave in such an appalling manner; don't know what or who made German society take such a terrible turn. How could Hitler fail to believe that he was correct, when everyone around him bowed to his orders? Would it not take character of exceptional magnitude to resist the temptation of absolute power, freely offered, democratically granted—even insisted upon? How would it be possible for anyone to remain properly humble under such conditions? Most of us have personal frailties that remain constrained by our social environments. Our neurotic tendencies are checked by the people around us, who care for us, who complain and protest when we lose our self-control and take things, in our weakness, one step too far. If everyone around thinks you are the savior, who is left to point out your defects and keep you conscious of them? This is not an apology for Hitler, merely recognition that he was all too human. And what does that statement mean? Hitler was human; Stalin, too; Idi Amin, too. What does that say about being human?

Our tyrannical tendencies and moral decadences generally find their expression limited by our narrow domains of personal power. We cannot doom millions to death, at a whim, because we do not have the resources to do so. We satisfy ourselves, in the absence of such power, with riding roughshod over those near to us—and congratulate ourselves on our moral virtue. We use aggression and strength to bend dependent others to our will—or, in the absence of strength, use sickness and weakness to harness the force of empathy, and deceive our way to dominance, underground. Granted the opportunity, how many of us would *not* be Hitlers? Assuming we had the ambition, dedication and power of organization—which is highly unlikely. Paucity of skill, however, does not constitute moral virtue.

Many kings are tyrants, or moral decadents, because they are *people*—and many people are tyrants, or moral decadents. We cannot say "never again" as a consequence of the memory of the Holocaust, because we do not understand the Holocaust, and it is impossible to remember what has not been understood. We do not understand the Holocaust because we do not comprehend ourselves. Human beings, very much like ourselves, produced the moral catastrophes of the Second World War (and of Stalin's Soviet Union, and of Pol Pot's Cambodia …). "Never forget" means "know thyself": recognize and understand that evil twin, that mortal enemy, who is part and parcel of every individual.

The heroic tendency—the archetypal savior—is an eternal spirit, which is to say, a central and permanent aspect of human being. The same is true, precisely, of the "adversarial" tendency: the capacity for endless denial, and the desire to make everything suffer for the outrage of its existence, is an ineradicable intrapsychic element of the individual. The great dramatists and religious thinkers of the world have been able to grasp this fact, at least implicitly, and to transmit it in story and image; modern analytic thinkers and existential theorists have attempted to abstract these ideas upward into "higher consciousness," and to present them in logical and purely semantic form. Sufficient material has been gathered to present a compelling portrait of evil.

THE ADVERSARY: EMERGENCE, DEVELOPMENT AND REPRESENTATION

The figure of Satan is arguably the most well-developed representation of evil extant in religious and mythological thought. Although it is tempting to identify this "personage" with particular personality attributes, such as aggression—or with the differences of the stranger—it is more realistic to view him as the embodiment of a personal and social process. The devil is the spirit who underlies the development of totalitarianism; the spirit who is characterized by rigid ideological belief (by the "predominance of the rational mind"), by reliance on the lie as a mode of adaptation (by refusal to admit to the existence of error, or to appreciate the necessity of deviance), and by the inevitable development of hatred for the self and world. Each of these characteristics is intrinsically and causally related to the others; they are linked inextricably together, and can be aptly conceptualized as a transpersonal and eternal personality.

The devil is willful rejection of the process that makes life bearable, out of spite for the tragic conditions of existence. This rejection is intellectually arrogant, because the "conditions" are interpreted—which is to say: development of self-consciousness tainted everything with death, but self-consciousness is contained within a global understanding that is still exceptionally limited in its scope. The present, as currently interpreted, is indeed the unbearable present: but that interpretation may change, if the possibility for change is not disallowed, as a consequence of absolutist belief, conceit and resentment.

The devil works to eliminate the world, as something whose weakness and vulnerability makes it contemptible. He has produced dreadful suffering in the twentieth century—not least among cultures who have dispensed with his image. We are fortunate to have survived without an irreversible tragedy. Our luck to date should not blind us to the dangers of continued ignorance, or to the

necessity of bringing our poorly comprehended selves under control. Every technological advance
we make increases our power; every increase in power makes our internal integration and expanded
self-consciousness that much more necessary.

The noble soul has reverence for itself.[464]

As I looked deeper into the problem of evil over the last fourteen years, I found myself
increasingly and involuntarily fascinated with the myth of Satan, and curious about the position of this story in Western thought. The idea of the devil has exerted a powerful influence
on the development of Christianity and Christian culture—and, therefore, on Western and
world society—although there is little direct reference to Satan in the Old and New Testaments (surprisingly little: no description of hell, to speak of, limited and oblique reference to
the rebellion of the angels and the war in heaven preceding the establishment of hell; nothing of any consequence regarding the terrible afterlife that theoretically awaits all sinners).

It is my understanding that traditional and literary representations of Satan, the ruling
angel of hell, constitute *true mythology*. These ideas surround the established central writings
and ideas of Christianity like fog surrounds a mountain. They have been transmitted to us,
in part, as religious doctrine; in part, as oral tradition; in part, because of the efforts of Dante
and Milton. I had a very limited religious education, in my youth, and all I ever knew about
the devil was "hearsay"—bits and pieces I picked up while reading other material (such as
Joyce's *Portrait of the Artist as a Young Man*, which contains a terrifying Jesuit sermon on the
wages of sin). All I knew was the outline of the story codified by Milton: Satan, the highest
angel in God's heavenly hierarchy, desired to become like the Most High, and fostered a
rebellion in heaven. He was defeated, and cast, unrepentant, into Hell, where he rules eternally over the spirits of the sinful dead. I had no real idea what this story meant, although it
was obvious to me that these characters and events could have never "really" existed.

I learned later that the association made between the serpent in Eden and the devil was
essentially speculative. Indeed, certain Gnostics had even posited that the deity who brought
Adam and Eve to the light of self-consciousness was a higher spirit than the unconscious
demiurge who had created everything in the beginning. This idea was predicated upon
Gnostic "recognition" that a fall from paradise—from a previous "plane of stability"—often
comprised the necessary precondition for movement to a "higher place." Traditional
medieval Christians developed a similar idea. For them, Original Sin was a "fortunate error"
that made necessary the incarnation of Christ. This meant that the Christian fall, although
tragic in and of itself, could be regarded as positively beneficial, since it brought about the
redemptive incarnation of God (which was the most stupendous event in history, from the
Christian perspective). Adoption of this broader viewpoint allowed even the Edenic serpent,
who propelled mankind into chaos, to be interpreted as a "tool of God"—as a tool of the
beneficial God who is endlessly working to bring about the perfection of the world, despite

the troublesome existence of free choice and demonic temptation. (The name *Lucifer* means "bringer of light," after all.) I also knew, more or less "unconsciously," that the devil has been long associated with the power and arrogance of rational thought (in Goethe's *Faust*, for example). This association has enabled the dogmatic forces of the church to adopt an antiscientific stance, frequently—science ⇒ rationality ⇒ devil—and to justify unfortunate church opposition to emergent truth. However, a mythological idea is not invalidated *as an idea* in consequence of its misapplication. The capacity for rational thought *is* a dangerous force, without doubt, because it is a powerful force—and the conditions under which thinking plays a purely destructive role are still not well comprehended.

This plethora of vaguely related ideas and stories kept entering and re-entering my mind, most frequently in combination with remembrance of a telling symbolic historic act: the transformation of the Notre Dame Cathedral into the "Temple of Reason," in the midst of the terrors of the French Revolution. It is no easy matter to come to a clear understanding of such notions, to grasp their nature logically or emotionally, or even to determine how they could possibly be related. After all, we tend to regard the development of "clear understanding" as equivalent to the construction of a "proper set" and assume that the reality of a thing can be clearly defined. Ideas about evil, however, do not form a "proper set." They form a "natural category," containing diverse material—just like ideas about the "known" or the "unknown." To complicate things further, evil, like good, is not something static (although it may align itself with all that is stubbornly static). It is rather a *dynamic process*, a spirit that partakes of the motivational or affective states of pride, resentment, jealousy and hatred but cannot be identified unerringly with the presence of any or all. The morality of an aggressive act, for example, depends on the nature of the context in which it is manifested, just as the meaning of a given word is defined by the sentence, the paragraph—even the book or culture—in which it appears. Evil is a living *complex*. Its nature can be most clearly comprehended through examination of the "personality" it has "adopted" in mythology, literature and fantasy, elaborated in the lengthy course of historical development. This personality consists of those "meta"-attributes of evil that have remained stable over time despite dramatic shifts in the particulars of human existence and human morality.

The image of the devil is the form that the idea of evil has taken, for better or worse, at least in the West. We have not yet developed an explicit model of evil that would allow us to forget, transcend or otherwise dispense with this mythological representation. We rationalize our lack of such understanding by presuming that the very notion of evil is archaic. This is a truly ridiculous presumption, in this century of indescribable horror. In our ignorance and complacency, we deride ancient stories about the nature of evil, equating them half-consciously with childish things best put away. This is an exceedingly arrogant position. There is no evidence whatsoever that we understand the nature of evil any better than our forebears, despite our psychology, even though our expanded technological power has made us much more dangerous when we are possessed. Our ancestors were at least constantly concerned with the problem of evil. Acceptance of the harsh Christian dogma of Original Sin, for example (despite its pessimism and apparent inequity) at least meant *recognition of evil;*

such dogma at least promoted some consideration of the tendency toward evil as an intrinsic, heritable aspect of human nature. From the perspective informed by belief in Original Sin, individual actions and motivations must always be carefully scrutinized and considered, even when apparently benevolent, lest the ever-present adversarial tendencies "accidentally" gain the upper hand. The dogma of Original Sin forces every individual to regard himself as the (potential) immediate source of evil and to locate the terrible underworld of mythology and its denizens *in intrapsychic space*. It is no wonder that this idea has become unpopular: nonetheless, evil exists somewhere. It remains difficult not to see hypocrisy in the souls of those who wish to localize it somewhere else.

Once I understood these things in a provisional sort of way, the ancient ideas began to sort themselves out. I learned from Eliade how to make sense out of the notion of a "heavenly hierarchy." The monotheism of Judaism and Christianity has its roots in older, more polytheistic thinking. The many gods of archaic conceptualization became the single ruler of more modern religious thinking as a consequence of spiritual competition—so to speak. This competition is the battle of ideas with implication for action—fought in abstraction, image and in the course of genuine earthly combat—portrayed in mythology as *spiritual war*, played out in heaven (which is the place where transpersonal ideas exist). The deity who came to prevail over all is One God, with a complex set of attributes, "surrounded" by a panoply of angels and divine "echoes" of previous gods (who represent those transpersonal and eternal psychological processes rendered subordinate in the course of the spiritual phylogenesis of man).

Christian mythology portrays Satan as the "highest angel" in God's "heavenly kingdom." This fact renders his association with reason more comprehensible. Reason may well be considered the "highest angel"—which is to say, the most developed and remarkable psychological or spiritual faculty, characteristic of all men (and therefore, something transpersonal and eternal). *Figure 57: The Devil as Aerial Spirit and Ungodly Intellect* portrays Eugene Delacroix's imagistic interpretation, an illustration for Part One of *Faust*.[465] Reason, the most exceptional of spirits, suffers from the greatest of temptations: reason's own capacity for self-recognition and self-admiration means endless capacity for pride, which is the act of presuming omniscience. It is reason's remarkable ability and its own recognition of that ability that leads it to believe it possesses absolute knowledge and can therefore replace, or do without, God:

> He trusted to have equalled the Most High,
> If he opposed, and with ambitious aim
> Against the throne and monarchy of God,
> Raised impious war in Heaven.[466]

It is reason's belief in its own omniscience—manifest in procedure and image, if not in word—that "unconsciously" underlies totalitarianism in its many destructive guises. Frye notes:

Figure 57: The Devil as Aerial Spirit and Ungodly Intellect

A demonic fall, as Milton presents it, involves defiance of and rivalry with God rather than simple disobedience, and hence the demonic society is a sustained and systematic parody of the divine one, associated with devils or fallen angels because it seems far beyond normal human capacities in its powers. We read of ascending and descending angels on Jacob's and Plato's ladders, and similarly there seem to be demonic reinforcements in heathen life that account for the almost superhuman grandeur of the heathen empires, especially just before they fall.

Two particularly notable passages in the Old Testament prophets linked to this theme are the denunciation of Babylon in Isaiah 14 and of Tyre in Ezekiel 28. Babylon is associated with Lucifer the morning star, who said to himself: "I will be like the Most High"; Tyre is identified with a "Covering Cherub," a splendid creature living in the garden of Eden "till the day that iniquity was found in thee." In the New Testament (Luke 10:18) Jesus speaks of Satan as falling from heaven, hence Satan's traditional identification with Isaiah's Lucifer and his growth in legend into the great adversary of God, once the prince of the angels, and, before being displaced, the firstborn son of God. The superhuman demonic force behind the heathen kingdoms is called in Christianity the Antichrist, the earthly ruler demanding divine honors.[467]

It is not that easy to understand why the act of presuming omniscience is reasonably construed as precisely opposite to the act of creative exploration (as the adversary is opposite to the hero). What "knowing everything" means, however—at least in practice—is that the unknown no longer exists, and that further exploration has therefore been rendered superfluous (even treacherous). This means that absolute identification with the "known" necessarily comes to replace all opportunity for *identification with the process that comes to know*. The presumption of absolute knowledge, which is the cardinal sin of the rational spirit, is therefore *prima facie* equivalent to rejection of the hero—to rejection of Christ, of the Word of God, of the (divine) *process* that mediates between order and chaos. The arrogance of the totalitarian stance is ineradicably opposed to the "humility" of creative exploration. [*Humility*—it is only constant admission of error and capacity for error (admission of "sinful and ignorant nature") that allows for recognition of the unknown, and then for update of knowledge and adaptation in behavior. Such humility is, somewhat paradoxically, courageous—as admission of error and possibility for error constitutes the necessary precondition for confrontation with the unknown. This makes *genuine cowardice* the "underground" motivation for the totalitarian presumption: the true authoritarian wants everything unpredictable to vanish. The authoritarian protects himself from knowledge of this cowardice by a show of patriotic advocacy, often at apparent cost to himself.]

In the fifth book of *Paradise Lost*, which Milton constructed from Biblical and mythological allusions, Lucifer is "passed over" by God in honor of the "second son," Christ.[468] This "shift in the dominance hierarchy of heaven" seems to me to indicate that reason (which, in consequence of its self-recognition as the "highest angel" of God, believes itself capable of single-handedly engendering redemption) must remain subordinate to the processes of the exploratory hero. Reason can serve health only when it plays a *secondary* role. The option of ruling in hell, rather than serving in heaven, nonetheless appears as an attractive alternative to the rational mind, under a wide variety of circumstances.

The devil is the spirit who eternally states, "all that I know is all that there is to be known"; the spirit who falls in love with his own beautiful productions and, in consequence, can no longer see beyond them. The devil is the desire to be right, above all, to be right once and for all and finally, rather than to constantly admit to insufficiency and ignorance, and to therefore partake in the process of creation itself. The devil is the spirit which endlessly denies, because it is afraid, in the final analysis, afraid and weak.

It is lack of discrimination between the existence of the adversary *as process* with the existence of anomaly *as constituent element of experience* that has led to some of the worst excesses of Christianity (and not just Christianity). It has constantly been the case that "proper-thinking" people confuse the existence of threats to their security and moral integrity with evil. This means that the proper-thinking confuse the being of the genius and the stranger, who offer up experience that exists in contrast with established belief, with the process of rejection of such experience. This lack of discrimination is both comprehensible and motivated: comprehensible, because the strange/stranger/strange idea/revolutionary hero upsets the applecart and produces affective dysregulation (which is the state most devoutly desired by the devil); motivated, because categorizing anomaly with evil allows for its "justified" repression. The heroic act of updating current morality, however—through the promotion of uncomfortable contact with the unknown—creates chaos only in the service of higher order. To repress that process and cling "patriotically" to tradition is to ensure that tradition will collapse precipitously—and far more dangerously—at some point in the not-too-distant future.

The fact of my lascivious or aggressive fantasy—to take a specific example of things generally regarded in a dim light—is not evil, if I am a devout Christian: evil is the act of denying that such fantasy exists (or, perhaps, the act of realizing that fantasy, without consideration of its proper place). The fantasy itself merely constitutes information (unacceptable information, to be sure, from the current, merely provisional standpoint: but information with the capacity to transform, if admitted). The existence of the Muslim, and the Muslim viewpoint, likewise is not evil, if I am a devout Christian. Evil is instead my presumption of personal omniscience—my certainty that I understand my Christian belief well enough to presume its necessary opposition to the stranger and his ideas; my certainty that identification with a static "comprehended" moral structure is sufficient to guarantee my integrity—and my consequent ignorant and self-righteous persecution of the Muslim. The devil is not the uncomfortable fact but the act of shrinking from that fact. The weaknesses, stupidities, laxities and ignorances that ineradicably constitute the individual are not evil in and of themselves. These "insufficiencies" are a necessary consequence of the limitations that make experience possible. It is the act of denying that stupidity exists, once it has manifested itself, that is evil, because stupidity cannot then be *overcome*. Such denial brings spiritual progress to a halt. Consciousness of ignorance and cupidity manifests itself in shame, anxiety and pain—in the guise of the visitor whose arrival is most feared—and such consciousness may consequently come to be considered the embodiment of evil itself. But it is the bearer of bad news who brings us closer to the light, if the significance of the news is allowed to manifest itself.

Elaine Pagels has recently written a book, *The Origin of Satan*,[469] in which she describes how the idea of the devil as the eternal enemy of Christ enabled those who profess Christianity to persecute those who do not. The presuppositions of the persecutor are, for example: "the devil is the enemy, the Jew is not a Christian—the Jew is an enemy, the Jew is the devil." Pagels presents the not unreasonable and justifiably popular hypothesis that the invention of Satan was *motivated* by desire to transform the act of persecuting others into a moral virtue. It appears, however, that the historical "developmental path" of the "idea of the adversary" is somewhat more complex. Transpersonal notions of the breadth of the "image of

the Devil" cannot emerge as a consequence of conscious motivation, because their development requires many centuries of transgenerational work (which cannot be easily "organized"). The image of the devil, although endlessly applied to rationalize the subjugation of
others (as all great ideas can be subverted) emerged as a consequence of endless geniune
attempts to encapsulate the "personality" of evil. The logic that associates the *other* with the
devil works only for those who think that religion means belief—that is, identification with a
set of static and often unreasonable "facts"—and not action, meta-imitation, or the incarnation of the creative process in behavior. The existence of the anomalous fact, properly considered—the uncomfortable fact, embodied in the stranger or rendered abstract in the form
of differing philosophy—is a *call to religious action, and not an evil.*

It has taken mankind thousands of years of work to develop dawning awareness of the
nature of evil—to produce a detailed dramatic representation of the process that makes up
the core of human maladaptation and voluntarily produced misery. It seems premature to
throw away the fruit of that labor or to presume that it is something other than what it
appears before we understand what it signifies. Consciousness of evil emerged first as ritual
enactment, then as dynamic image, expressed in myth. This representation covers a broad
spatial and temporal territory, whose examination helps flesh out understanding of the personality of the adversary. The most thoroughly developed archaic personification of evil can
perhaps be found in the ideas of Zoroastrianism, which flourished in relatively explicit form
from 1000–600 B.C. (and which undoubtedly depended for its form on much more ancient,
less explicit "ideas"). The Zoroastrians developed a number of ideas which were later incorporated into Christianity, including "the myth of the savior; the elaboration of an optimistic
eschatology, proclaiming the final triumph of Good and universal salvation; [and] the doctrine of the resurrection of bodies."[470]

Zarathustra, the mythic founder of Zoroastrianism, was a follower of *Ahura Mazdā* (the
central deity in this essentially monotheistic religion). Ahura ("sky") Mazdā was surrounded
by a pantheon of divine *entities*—the *Amesha Spentas*, analogous to angels—who were very
evidently psychological in their nature (at least from the modern perspective).[471] These "spirits" include *Asha* (justice), *Vohu Manah* (good thought), *Ármaiti* (devotion), *Xshathra*
(power), *Haurvatāt* (integrity) and *Ameretāt* (immortality). Ahura Mazdā was also the father
of twin "brothers," *spirits*—*Spenta Mainyu* (the beneficient spirit) and *Angra Mainyu* (the
destroying spirit). Eliade states:

In the beginning, it is stated in a famous *gāthā* (*Yasna* 30, authored by Zarathustra), these two spirits
chose, one of them good and life, the other evil and death. Spenta Mainyu declares, at the "beginning
of existence," to the Destroying Spirit: "Neither our thoughts nor our doctrines, nor our mental
powers; neither our choices, nor our words, nor our acts; neither our consciences nor our souls are in
agreement." This shows that the two spirits—the one holy, the other wicked—differ rather by *choice*
than by *nature.*

Zarathustra's theology is not dualistic in the strict sense of the term, since Ahura Mazdā is not confronted by an anti-God; in the beginning, the opposition breaks out between the two Spirits. On the

other hand, the unity between Ahura Mazdā and the Holy [Good] Spirit is several times implied (see *Yasna* 43.3, etc.). In short, Good and Evil, the holy one and the destroying demon, proceed from Ahura Mazdā; but since Angra Mainyu freely chose his mode of being and his malificient vocation, the Wise Lord cannot be considered responsible for the appearance of Evil. On the other hand, Ahura Mazdā, in his omniscience, knew from the beginning what choice the Destroying Spirit would make and nevertheless did not prevent it; this may mean either that God transcends all kinds of contradictions or that the existence of evil constitutes the preliminary condition for human freedom.[472]

The mythic "hostile brothers"—Spenta Mainyu and Angra Mainyu, Osiris and Seth, *Gilgamesh* and *Enkidu*, Cain and Abel, Christ and Satan—are representative of two eternal individual tendencies, twin "sons of god," heroic and adversarial. The former tendency, the archetypal savior, is the everlasting spirit of creation and transformation, characterized eternally by the capacity to admit to the unknown and, therefore, to progress toward "the kingdom of heaven." The eternal adversary, by contrast, is incarnation in practice, imagination and philosophy of the spirit of denial, eternal rejection of the "redeeming unknown," and the adoption of rigid self-identification. Myths of the "hostile brothers"—like those of the Zoroastrians—tend to emphasize the role of free choice in determination of essential mode of being. Christ, for example (and Gautama Buddha) are tempted constantly and potently toward evil, but choose to reject it. Angra Mainyu and Satan accept evil, by contrast, and revel in it (despite evidence that it produces their own suffering). The choice of these spirits cannot be reduced to some more essential aspect, such as the particular conditions of existence (which are identical, anyway, for both "beings") or the vagaries of intrinsic nature. It is *voluntary willingness to do what is known to be wrong, despite the capacity to understand and avoid such action*, that most particularly characterizes evil—the evil of spirit and man. So Milton's God can comment, on the degeneration of Satan and mankind:

> So will fall
> Hee and his faithless progeny: whose fault?
> Whose but his own? ingrate, he had of mee
> All he could have; I made him just and right,
> sufficient to have stood, though free to fall.[473]

Refusal of the good is, I think, most effectively and frequently *justified* by reference to the terrible affective consequences of (self)consciousness. This means that comprehension of the vulnerability and mortality of man, and the suffering associated with that vulnerability—apprehension of the ultimate cruelty and pointlessness of life—may be used as rationale for evil. Life *is* terrible, and appears, at some moments, *ultimately* terrible: unfair, irrational, painful and meaningless. Interpreted in such a light, existence itself may well appear as something reasonably eradicated. Goethe's Mephistopheles, "prince of lies," defines his philosophy, in consequence, in the following terms (in Part One of *Faust*):

> The spirit I, that endlessly denies.
> And rightly, too; for all that comes to birth
> Is fit for overthrow, as nothing worth;
> Wherefore the world were better sterilized;
> Thus all that's here as Evil recognized
> Is gain to me, and downfall, ruin, sin
> The very element I prosper in.[474]

He repeats this credo, in slightly elaborated form, in Part Two:

> Gone, to sheer Nothing, past with null made one!
> What matters our creative endless toil,
> When, at a snatch, oblivion ends the coil?
> "It is by-gone"—How shall this riddle run?
> As good as if things never had begun,
> Yet circle back, existence to possess:
> I'd rather have Eternal Emptiness.[475]

Spiritual reality plays itself out endlessly in profane reality (as man remains eternally subject to the "dictates of the gods"). Individual persons therefore "unconsciously" embody mythological themes. Such embodiment becomes particularly evident in the case of great individuals, where the play of "divine forces" becomes virtually tangible. We analyzed sections of Leo Tolstoy's autobiography, previously[476]—using his self-reported personal experience as universal exemplar for the catastrophic primary affective consequences of revolutionary anomaly. Tolstoy's secondary ideological response to such anomaly is equally archetypal. The "news" from Western Europe—the revelation of the "death of God"—cascaded through the great author's implicit and explicit culturally determined beliefs and action schemas, propelling him headlong, over a very lengthy period of time, into emotional turmoil and existential chaos. Identification with the spirit of denial lurked as a profound temptation in the midst of that chaos.

Tolstoy begins the relevant section of his confession with an allegory, derived "from a tale of the East." A traveler, chased by a wild beast, jumps down an old well. He grabs the branch of a vine that happens to be growing there, and clings to it. At the bottom of the well lurks an ancient dragon, mouth gaping. Above the well is the terrible beast—so there is no turning back. The traveler's arms grow weak, clinging to the vine, but he still holds on. Then he sees two mice—one black, one white—gnawing at either side of the very branch that supports him. Soon they will chew their way through, and send him plummeting into the dragon's gullet. The traveler sees some drops of honey, on the leaves of the vine. He stretches out his tongue, tastes the honey, and is comforted. For Tolstoy, however, the pleasures of life had lost their analgesic sweetness:

I could not be deceived. All is vanity. Happy is he who has never been born; death is better than life; we must rid ourselves of life.

Having failed to find an explanation in knowledge, I began to look for it in life, hoping to find it in the people around me. And so I began to observe people like myself to see how they lived and to determine what sort of relation they had with the question that had led me to despair.

And this is what I found among people whose circumstances were precisely the same as mine with respect to education and way of life.

I found that for the people of my class there were four means of escaping the terrible situation in which we all find ourselves.

The first means of escape is that of ignorance. It consists of failing to realize and to understand that life is evil and meaningless. For the most part, people in this category are women, or they are very young or very stupid men; they still have not understood the problem of life that presented itself to Schopenhauer, Solomon, and the Buddha. They see neither the dragon that awaits them nor the mice gnawing away at the branch they cling to; they simply lick the drops of honey. But they lick these drops of honey only for the time being; something will turn their attention toward the dragon and the mice, and there will be an end to their licking. There was nothing for me to learn from them, since we cannot cease to know what we know.

The second escape is that of epicureanism. Fully aware of the hopelessness of life, it consists of enjoying for the present the blessings that we do have without looking at the dragon or the mice; it lies in licking the honey as best we can, especially in those places where there is the most honey on the bush. Solomon describes this escape in the following manner:

"And I commended mirth, for there is nothing better for man under the sun than to eat, drink, and be merry; this will be his mainstay in his toil through the days of his life that God has given him under the sun.

So go and eat your bread with joy and drink your wine in the gladness of your heart. . . . Enjoy life with a *woman* you love through all the days of your life of vanity, through all your vain days; for this is your fate in life and in the labors by which you toil under the sun. . . . Do whatever you can do by the strength of your hand, for there is no work in the grave where you are going, no reflection, no knowledge, no wisdom."

Most people of our class pursue this second means of escape. The situation in which they find themselves is such that it affords them more of the good things in life than the bad; their moral stupidity enables them to forget that all the advantages of their position are accidental, that not everyone can have a thousand women and palaces, as Solomon did; they forget that for every man with a thousand wives there are a thousand men without wives, that for every palace there are a thousand men who built it by the sweat of their brows, and that the same chance that has made them a Solomon today might well make them Solomon's slave tomorrow. The dullness of the imagination of these people enables them to forget what left the Buddha with no peace: the inevitability of sickness, old age, and death, which if not today then tomorrow will destroy all these pleasures. The fact that some of these people maintain that their dullness of thought and imagination is positive philosophy does not, in my opinion, distinguish them from those who lick the honey without seeing the problem. I could not imitate these people, since I did not lack imagination and could not pretend that I did. Like every man who truly lives, I could not turn my eyes away from the mice and the dragon once I had seen them.

The third means of escape is through strength and energy. It consists of destroying life once one has realized that life is evil and meaningless. Only unusually strong and logically consistent people act in this manner. Having realized all the stupidity of the joke that is being played on us and seeing that the blessings of the dead are greater than those of the living and that it is better not to exist, they act and put an end to this stupid joke; and they use any means of doing it: a rope around the neck, water, a knife in the heart, a train. There are more and more people of our class who are acting in this way. For the most part, the people who perform these acts are in the very prime of life, when the strength of the soul is at its peak and when the habits that undermine human reason have not yet taken over. I saw that this was the most worthy means of escape, and I wanted to take it.

The fourth means of escape is that of weakness. It consists of continuing to drag out a life that is evil and meaningless, knowing beforehand that nothing can come of it. The people in this category know that death is better than life, but they do not have the strength to act rationally and quickly put an end to the delusion by killing themselves; instead they seem to be waiting for something to happen. This is the escape of weakness, for if I know what is better and have it within my reach, then why not surrender myself to it? I myself belonged in this category.

Thus the people of my class save themselves from a terrible contradiction in these four ways. No matter how much I strained my intellectual faculties, I could see no escape other than these four.[477]

Tolstoy's "intellectual faculties"—his rationality—could see no way out of the dilemma posed by his incorporation of an indigestible idea. Furthermore, logic clearly dictated that existence characterized only by inevitable and pointless suffering should be brought to an abrupt end, as an "evil joke." It was Tolstoy's clear apprehension of the endless conflict between the individual and the conditions of existence that destroyed his ability to work and undermined his desire to live. He was unable to see (at least at that point in his journey) that man was fashioned to confront chaos constantly—to eternally work toward transforming it into real being—rather than to master it finally, once and for all (and to therefore render everything intolerably static!).

The fact of mortal vulnerability—that defining characteristic of the individual, and the "reason" for his emergent disgust with life—may be rendered even more "unjust" and "intolerable" by the specific manifestations of such vulnerability. Some are poorer than others, some weaker, some unsightlier—all less able, in some regard (and some apparently less able in all regards). Recognition of the seemingly arbitrary distribution of skill and advantage adds additional rationally "justifiable" grounds for the development of a philosophy based on resentment and antipathy—sometimes, "on behalf" of an entire class, other times, sheerly for the purposes of a specific individual. Under such circumstances, the desire for revenge on life itself may become paramount above all else, particularly for the "unfairly oppressed." Shakespeare's crippled Richard the Third speaks for all revolutionaries, all rebels, so motivated:

Then, since the heavens have shaped my body so,
Let hell make crooked my mind to answer it.
I have no brother, I am like no brother:
And this word "love" which greybeards call divine,
Be resident in men like one another
And not in me: I am myself alone.[478]

Evil is voluntary rejection of the process that makes life tolerable, justified by observation of life's terrible difficulty. This rejection is presumptious, premature, because it is based on acceptance of a provisional judgment as final: "everything is insufficient, and is therefore without worth, and nothing whatsoever can be done to rectify the situation." Judgment of this sort precludes all hope of cure. Lack of belief in hope and meaning (which appear more than willing to vanish, in the face of rationale critique) seldom means commensurate "lack of belief in anxiety and despair" (even though recognition of the pointlessness of everything should also undermine one's faith in suffering). Suffering cannot be disbelieved away, however: rejection of the process that constantly renews the positive aspect of the "constituent elements of experience" merely ensures that their negative counterparts gain the upper hand. Such additional torture—added to that already considered sufficient to bring about hatred for life—is sure to produce a character motivated to perform acts worse than mere suicide. The development of the adversary therefore follows a predictable path, from pride ("Pride and worse Ambition threw me down"[479]), through envy, to revenge[480]—to the ultimate construction of a character possessed by infinite hatred and envy:

To do aught good never will be our task,
But ever to do ill our sole delight,
As being contrary to his high will
Whom we resist. If then his Providence
Out of our evil seeks to bring forth good,
Our labor must be to pervert that end,
And out of good still to find means of evil.[481]

Tolstoy's nihilism—disgust with the individual and human society, combined with the desire for the eradication of existence—is one logical "evil" consequence of heightened self-consciousness. It is not, however, the only consequence, and may not even be the most subtle. Far more efficient—far more hidden from the perpetrator himself, and from his closest observers—is heightened identification with tradition and custom. This is envelopment in the guise of patriotism, to facilitate the turning of state power toward destruction. Nietzsche described such "loyalty" in the following manner:

Definition of morality: Morality—the idiosyncrasy of decadents, with the ulterior motive of revenging oneself against life—successfully. I attach value to this definition.[482]

This description of initial motivated decision and consequent dissolution seems to me to characterize the processes and bifurcated final state of moral (and, therefore, psychological) degeneration more accurately and potently than any purely "scientific" theory of psychopathology generated to date. Of course, we are at present unable to take our rationally reduced selves seriously enough to presume a relationship between evil as a "cosmic force" and our petty transgressions and self-betrayals. We believe that in reducing the scope and importance of our errors, we are properly humble; in truth, we are merely unwilling to bear the weight of our true responsibility.

The Adversary in Action: Voluntary Degradation of the Map of Meaning

> Who alone has reason to *lie himself out* of actuality?
> He who *suffers* from it.[483]

Tragic encounter with the forces of the unknown is inevitable, in the course of normal development, given continued expansion of conscious awareness. Even socialized identification with the cultural canon cannot provide final protection. Unshielded personal contact with tragedy is inextricably linked with emergence of self-consciousness, which has as its mythic consequence (its virtual equivalent) heightened awareness of human limitation. This awareness is manifested in shame, and has been expressed mythologically as shame of nakedness, which is knowledge of essential vulnerability and weakness before the world.

The intrinsic nature of human experience ensures that potent motivation for deceitful adaptation is always present. It is the encounter with what is *truly horrible and terrifying,* after all, that inspires fear and engenders avoidance. The human tendency to flee into false havens of security can therefore be viewed with sympathy and understanding. Maturation is a frightening process. Transformation from the paradisal matriarchal world of childhood to the fallen masculine, social world is fraught with peril. The same might be said about the dangers of post-apprenticeship individuality. It is not easy to become an adolescent after being a child. It might be said that this transition is in itself a heroic act. It therefore happens, upon occasion, that those who have abandoned heroism as a style of adaptation do not take even this first step. The relative advantages that accompany increased freedom may seem frightening and of dubious value, given the comparative responsibility and lack of security that are part and parcel of maturity.

As maturation takes place, the "environment" transforms. As the developing individual masters his powers, his behavioral capacity expands. He can do more things and, in consequence, experience more things. The ability to bring heretofore unknown and, therefore, frightening phenomena into being constantly increases, and the boundaries of the experiential domain of the individual eventually extend beyond the area shielded by the parents. The capability to endlessly further apprehension is central to the adaptive capacity of the individual; this capability, however, comes at an immense price, which is knowledge of finitude and death. Potent motivation therefore exists to resist such development, when it duly emerges; to fight desperately for maintenance of childhood ignorance, or to hide in the commands of others. Individuality—which is the ability to establish a realm of experience that is unique to the self; the capacity for the creation of purely subjective experience—also means acceptance of vulnerability and mortality. The creative capacity is divine *Logos*, which in the course of its development necessitates recognition of the inevitability of failure and death. That is in part the meaning of the symbol of the Christian crucifixion, which paradoxically melds mortality with divinity; which portrays the "mortal god," infinitely creative, responsible and vulnerable.

Individual existence means limited existence—limited in space and time. The existence of the limits makes experience possible; the fact of them makes experience unbearable. We have been granted the capacity for constant transcendence, as an antidote, but frequently reject that capacity, because using it means voluntarily exposing ourselves to the unknown. We run away because we are afraid of the unknown, at bottom; such fear also makes us cling to our protective social identities, which shield us from what we do not understand. So, while running away, we necessarily become slave to convention and habit, and deny the troublesome best within our selves. Why run away? It is fear—fear of the unknown, and its twin, fear of rejection by the protective social world, which leads to pathological subjugation of unique individual personality, to rejection of the totality of personal being (which, when manifested, has truly redemptive capability). The Great Father hates innovation and will kill to prevent it; the Great Mother, source of all new knowledge, has a face that paralyzes when encountered. How can we not run away, when confronted by such powers? But running away means that everything worthwhile ages, then dies.

When a child is born, he is protected from the vagaries of existence by the benevolence of circumstance, through maternal presence; the infant is prepared, *a priori*, instinctively, to respond to such protection, and to form a relationship—a bond with the mother. The helpless baby is at the mother's mercy, but is sheltered, as well, from the terrible world. Culture intercedes, in the form of proscriptions on behavior, when mortality nonetheless threatens, but adherence to such requirements means increased responsibility, separation from the good mother, and sacrifice of the primary dependent relationship. Culture molds the maturing personality, offering knowledge, but at the same time limitation, as the social world mangles individuality, interest and meaning.

Spirit is offered up to the group to maintain the group's benevolent nature, ensuring its continued protection and its grant of knowledge derived from history. It is *necessary* to identify with the group, in the course of normal development—that identification fosters maturity, and separation from blind maternal solicitude—but ultimately the group is tyrannical, and demands obedience at the cost of unique being. This is not to say, naively, that the group is intrinsically evil, that the root of human suffering is buried in the ground of the social world. Society is more purely expansion of power, which may be directed according to individual choice. The past contains within it the behavioral wisdom of generations, established in pain and fear, and offers the possibility of immense extension of individual power and ability. Culture and civilization offers each individual the opportunity to stand on the shoulders of giants. Adoption of group identity should constitute apprenticeship, not capitulation; should constitute a developmental stage in disciplined maturation, requiring temporary subjugation and immolation of immature individuality, prior to its later re-emergence, in controllable form, under voluntary direction.

Group membership, social being, represents a necessary advance over childish dependence, but the spirit of the group requires its pound of flesh. Absolute identification with the group means rejection of individual difference: means rejection of "deviation," even "weakness," from the group viewpoint; means repression of individuality, sacrifice of the mythic *fool*; means abandonment of the simple and insufficient "younger brother." The group, of course, merely feels that it is doing its duty by insisting upon such sacrifice; it believes, with sufficient justification, that it is merely protecting its structure. However, the group is not capable of making final judgments regarding what is necessary—what is good and what is evil—because it is incomplete by its very nature: it is a static structure, composed of the past. Individual difference, even weakness—anathema to the absolutist—is strength, from a more inclusive viewpoint; is that force capable of transcending inevitable group limitation and extending the reach of all.

Absolutists, rejecting the necessity of all change, necessarily deny to themselves and others even their own strength, because true heroism, regardless of its source, has the capacity to upset the *status quo*. Through such denial the absolutist hopes to find protection from his individual vulnerability. In truth, however, he has suppressed and pathologized the sole element within himself that could actually provide such protection; he has undermined his ability to utilize the sole process capable of actually providing security and freedom:

A traveler who had seen many countries and peoples and several continents was asked what human traits he had found everywhere; and he answered: men are inclined to laziness. Some will feel that he might have said with greater justice: they are all timorous. They hide behind customs and opinions. At bottom, every human being knows very well that he is in this world just once, as something unique, and that no accident, however strange, will throw together a second time into a unity such a curious and diffuse plurality: he knows it, but hides it like a bad conscience—why? From fear of his neighbor who insists on convention and veils himself with it.

But what is it that compels the individual human being to fear his neighbor, to think and act herd-fashion, and not to be glad of himself? A sense of shame, perhaps, in a few rare cases. In the vast majority it is the desire for comfort, inertia—in short, that inclination to laziness of which the traveler spoke. He is right: men are even lazier than they are timorous, and what they fear most is the troubles with which any unconditional honesty and nudity would burden them.

Only artists hate this slovenly life in borrowed manners and loosely fitting opinions and unveil the secret, everybody's bad conscience, the principle that every human being is a unique wonder; they dare to show us the human being as he is, down to the last muscle, himself and himself alone—even more, that in this rigorous consistency of his uniqueness he is beautiful and worth contemplating, as novel and incredible as every work of nature, and by no means dull.

When a great thinker despises men, it is their laziness that he despises: for it is on account of this that they have the appearance of factory products and seem indifferent and unworthy of companionship or instruction. The human being who does not wish to belong to the mass must merely cease being comfortable with himself; let him follow his conscience, which shouts at him: "Be yourself! What you are at present doing, opining, and desiring, that is not really you."[484]

Denial of unique individuality turns the wise traditions of the past into the blind ruts of the present. Application of the letter of the law when the spirit of the law is necessary makes a mockery of culture. Following in the footsteps of others seems safe, and requires no thought—but it is useless to follow a well-trodden trail when the terrain itself has changed. The individual who fails to modify his habits and presumptions as a consequence of change is deluding himself—is denying the world—is trying to replace reality itself with his own feeble wish. By pretending things are other than they are, he undermines his own stability, destabilizes his future, and transforms the past from shelter to prison.

The individual embodiment of collective past wisdom is turned into the personification of inflexible stupidity by means of the lie. The lie is straightforward, voluntary rejection of what is *currently known* to be true. Nobody knows what is *finally* true, by definition, but honest people make the best possible use of their experience. The moral theories of the truthful, however incomplete from some hypothetical transcendent perspective, account for what they have seen and for who they are, insofar as that has been determined in the course of diligent effort. It is not necessary to define truth, to have seen and heard everything—that would make truth itself something impossible. It is necessary only to have represented and adapted to what *has* been seen and heard—to have represented and adapted to those phenomena characterizing the natural and social worlds, as encountered, and the self, as manifested. This is to say, merely, that the truth of children and adults differs, because their experience—their reality—differs. The truthful child does not think like an adult: he thinks like a child, with his eyes open. The adult, however, who still uses the morality of the child—despite his adult capacities—he is lying, and he knows it.

The lie is willful adherence to a previously functional schema of action and interpretation—a moral paradigm—despite new experience that cannot be comprehended in terms of that schema; despite new desire, which cannot find fulfillment within that framework. The lie is willful rejection of information apprehended as anomalous on terms defined and valued by the individual doing the rejecting. That is to say: the liar chooses his own game, sets his own rules and then cheats. This cheating is failure to grow, to mature; it is rejection of the process of consciousness itself.

The lie is therefore not so much a sin of commission, in most cases, as a sin of omission (although it may take the former condition as well). The lie is a matter of voluntary failure to explore and to update. The appearance of an anomalous occurrence in the ongoing stream of experience indicates only that the present goal-directed schema within which behavior is being undertaken and evaluated is characterized *by the presence of a flaw*. The place of the flaw, the reasons for its existence, the meaning of the flaw (its potential for altering interpretation and behavior)—all that is *hypothetical*, at the first stage of anomaly emergence and analysis. The unknown has to be "mined" for precise significance, before it can be said to have been experienced, let alone comprehended; has to be transformed, laboriously, from pure affect into revision of presumption and action (into "psyche" or "personality"). "Not doing" is therefore the simplest and most common lie: the individual can just "not act," "not investigate," and the pitfalls of error will remain unmanifest, at least temporarily. This rejection of the process of creative exploration means lack of effortful update of procedural and declarative memory, adaptation to the present as if it still were the past, refusal to *think*. The rectification of error is, after all, not inevitable; it is neither effortless nor automatic. Mediation of order and chaos requires courage and work.

Adoption of identity with the heroes of the past is necessary, but rife with pathological potential. It becomes certain corruption when the identified individual is a liar, who has voluntarily rendered himself incapable of personal heroism. Adoption of group identity and position means access to the power embodied in the past—means access to the collective strength and technical ability of the culture. This power is terribly dangerous in cowardly and deceitful hands. The liar cannot see any value in weakness or deviance in himself or others—only the potential for chaos—and he cannot see any value in chaos or uncertainty. He has no sympathy or patience for or appreciation of his own weaknesses—or his own strengths—and can therefore have none for the weakness or strength of others. The liar can only pretend to embody what is best of the past, because he cannot support or tolerate the presence of necessary deviance in the present. This means that the liar is a tyrant, because he cannot stand being a fool.

The liar cannot tolerate anomaly, because it provokes anxiety—and the liar does not believe that he can or should withstand anxiety. This means that he is motivated to first *avoid* and then to *actively suppress* any behavioral pattern or experience of world that does not fit comfortably into his culturally determined system of affect-regulating moral presuppositions. *Avoidance* means that anomalous experience is kept "unconscious," so to speak—which means incompletely realized. The implications of the dangerous thought remain unconsidered; the presence of the threatening fantasy remains unadmitted; the existence of the unacceptable personal action remains unrecognized. *Active suppression* does not mean intrapsychic "repression," in the classic sense, but aggressive action undertaken in the world to forcibly eliminate evidence of error. This may mean treachery, spiritual cruelty, or the outright application of power: may mean application of whatever maneuver is presumed necessary, to destroy all indication of insufficiency. The bearer of bad news therefore inevitably suffers at the hand of the deceitful individual, who would rather kill the source of potential wisdom than benefit from its message.

The lie is easy, and rewarding, as it allows for the avoidance of anxiety, at least in the short term. In the long run, however, the lie has terrible consequences. The "avoidance or suppression" of novel or unexpected experience, which is the abstract equivalent of running away, transforms it perforce into determinate threat (is the categorical equivalent of labeling *as* threat). The domain of unprocessed novelty—defined *prima facie* by inaction and avoidance as "threat too intolerable to face"—expands inevitably with time, when the past is held as absolute. More and more experience is therefore rendered intolerable, inexplicable and chaotic, as the cumulative effects of using the lie as a mode of adaptation inexorably manifest themselves. The lie transforms culture into tyranny, change into danger, while sickening and restricting the development and flexibility of adaptive ability itself. Reliance on the lie ensures, as fear grows, heightened, pathologized identification with the past (manifested as fascism: personal and political intolerance), or decadent degeneration (manifested as nihilism: personal and social deterioration).

Identification with the spirit of denial eventually makes life unbearable, as everything new—and, therefore, everything defining hope—comes to be axiomatically regarded as punishment and threat; makes life unbearable, as the realm of acceptable action shrinks inexorably. The attendant and unavoidable suffering experienced generates the desire for—and motivates actions predicated on the attainment of—the end of all experience, as compensation and revenge for sterility, absence of meaning, anxiety, hatred and pain:

The Marabout draws a large circle in the dirt, which represents the world. He places a scorpion, symbolic of man, inside the circle. The scorpion, believing it has achieved freedom, starts to run around the circle—but never attempts to go outside. After the scorpion has raced several times around the inside edge of the circle, the Marabout lowers his stick and divides the circle in half. The scorpion stops for a few seconds, then begins to run faster and faster, apparently looking for a way out, but never finding it. Strangely enough, the scorpion does not dare to cross over the line. After a few minutes, the Marabout divides the half circle. The scorpion becomes frantic. Soon the Marabout makes a space no bigger than the scorpion's body. This is "the moment of truth." The scorpion, dazed and bewildered, finds itself unable to move one way or another. Raising its venomous tail, the scorpion turns rapidly 'round and 'round in a veritable frenzy. Whirling, whirling, whirling until all of its spirit and energy are spent. In utter hopelessness the scorpion stops, lowers the poisonous point of its tail, and stings itself to death. Its torment is ended.[485]

The individual who lives by the lie continually shrinks his domain of competence, his "explored and familiar territory." Eventually, in consequence, he has nowhere left to turn—except to himself. But his own personality has, in the meantime, become shrunken and inept, as a consequence of underdevelopment—as a consequence of repeated failure to

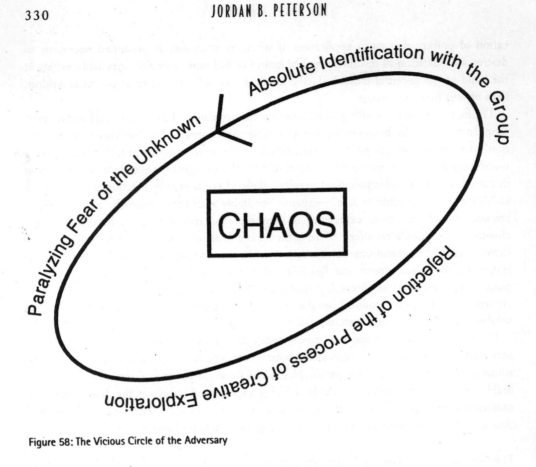

Figure 58: The Vicious Circle of the Adversary

participate in the process that turns "precosmogonic matter" into "spirit" and "world." Nothing remains but weakness, resentment, hatred and fear. Thus the chaos that is rejected out of desire for too much security attains its inevitable victory. The "vicious circle" created by the liar spirals down inevitably to the "underworld." This process is schematically represented in *Figure 58: The Vicious Circle of the Adversary*.

The "patriarchal" system, the known, is the concrete consequence of past adaptation, the hierarchically integrated and represented residue of the heroic past. Such adaptation is necessarily incomplete, in that the full scope of "natural" phenomena always exceeds the capacity for interpretation. Absolutist application of the past, motivated by fear of the unknown, transforms the past perforce into tyranny, which does not tolerate inevitable individual or deviant experience. This process of "absolute ancestral deification" is the consequence of security-seeking, made necessary by the abandonment of individual heroism as a potential mode of adaptation. Such abandonment occurs as a consequence of premature and arrogant self-definition—definition that makes of evident human vulnerability final and sufficient evidence for the unbearable cruelty of God and the uselessness of man.

The constant search for security, rather than the embodiment of freedom, is wish for rule by law's letter, rather than law's spirit. The resultant forcible suppression of deviance is based upon desire to support the pretence that the unknown does not exist. This suppression has as its consequence the elimination of creative transformation from the individual and social spheres. The individual who denies his individual identification with the heroic will come to identify with and serve the tyrannical force of the past—and to suffer the consequences. This principle is aptly illustrated by the mythic story of Judas. Judas sacrifices Christ, the hero, to the authorities of tradition—for all the best reasons—and is then driven to destroy himself in despair:

> Then Judas, which had betrayed him, when he saw that he was condemned, repented himself, and brought again the thirty pieces of silver to the chief priests and elders,
> Saying, I have sinned in that I have betrayed the innocent blood. And they said, What is that to us? see thou to that. And he cast down the pieces of silver in the temple, and departed, and went and hanged himself. (Matthew 27:3–5)

Sacrifice of the hero to the great and terrible father means abandonment of identification with the process that makes cosmos out of chaos. Rejection of the process whereby the endlessly negative and terrifying is transformed into the acceptable and beneficial means, by definition, the end of all hope:

> Wherefore I say unto you, All manner of sin and blasphemy shall be forgiven unto men: but the blasphemy against the Holy Ghost shall not be forgiven unto men.
> And whosoever speaketh a word against the Son of man, it shall be forgiven him: but whosover speaketh against the Holy Ghost, it shall not be forgiven him, neither in this world, neither in the world to come. (Matthew 12:31–32)

The individual lies to convince himself, and others, that he embodies the greatness of the past. He pretends to be upright and courageous, instead of acting morally and bravely. Truly courageous actions might turn the group against him, and it is only identity with that group that keeps his head above water. The lie means denial of self, means the abandonment of mythic identity with God, means certain involuntary "revolutionary" collapse, in time. The lie means conscious refusal to modify and reconfigure historically predicated behavior and belief to incorporate novelty and alleviate threat.

Endless failure to voluntarily update means the generation of a morass around the individual, where the "water of life" once existed: what could be life-giving liquid becomes a deadly swamp, composed of past errors, unresolved traumas and current difficulties. This is the domain characterized by Freud as the "unconscious," into which "repressed memories" are cast. But unprocessed information is not precisely memory. That which has not yet been explored is not yet memorable, not yet even "real." The consequence of untaken action is more accurately "potential from which 'spirit' and 'world' could be constructed"—much of it implicit in the world as it currently exists (instead of "stored in memory"). (Implicit, that is, in the form of unencountered but latent "trouble"—in the form of the unanswered letter, the unpaid debt or the unresolved dispute.)

This self-generated swamp grows increasingly impenetrable, as time passes; becomes increasingly "uninhabitable," as the consequences of long-term avoidance propagate (as the monsters of the bog sprout new and hungry heads). This "accumulation of precosmogonic potential" is tantamount to reanimation of the dragon of chaos (is precisely equivalent to the reawakening of Tiamat, who eternally sleeps under the secure and familiar world). The more restricted, fear-bound, faithless and repressive the particular mode of adaptation—that is, the more extreme the lie—the more horrendous, dangerous, intolerable and powerful the associated dragon. It is in this manner that attitude comes to define the world. Every attempt to wish any aspect of experience out of existence transforms it into an enemy. Every facet of being hidden from the light leads a corrupt and sun-starved existence, underground. Experience—absolute reality itself, in the final analysis—cannot be denied without consequence. Reality cannot be fantasized out of existence. The enforcement of a wish merely ensures that the information contained in the denied experience can neither be removed from the domain of threat nor utilized for adaptive purposes.

It is possible that we are in fact *adapted* to the world—that we are adapted to the world as it actually exists, rather than to the world as we wish it might be. It is possible that our experience contains information precisely sufficient to ensure our happy survival. This means that every task left undone—every emergent "territory" left unexplored—comprises latent information from which competent personality could yet be extracted. If experience is valid as source of world and spirit then those elements of experience that have been avoided or suppressed or devalued may yet contain within them what is absolutely essential to continued successful existence. Voluntary movement toward "the good" would therefore mean reintegration of cast-off "material"—voluntary incorporation of that which appears, at present, indigestible. The alternative to this "voluntary pursuit of the inedible" is eventual psychological catastrophe, at the social or individual level, engendered through involuntary contact with the "hostile forces" of rejected being. From the mythological perspective, this psychological catastrophe is accidental reunion with the Terrible Mother, on territory of her choosing. This "Oedipal incest" culminates in certain suffering, on the part of the unwilling "hero": culminates in suicide, dismemberment, castration—ends in the final sacrifice of "masculine" consciousness, and in the victory of the underworld.

The identity of the individual with his culture protects him from the terrible unknown, and allows him to function as an acceptable member of society. This slavish function strengthens the group. But the group states that certain ways of thinking and acting are all that are acceptable, and these particular ways do not exhaust the unknown and necessary capabilities of the human being. The rigid, grinning social mask is the individual's pretence that he is "the same person" as everyone else (that is, the same dead person)—that he is not a natural disaster, not a stranger, not strange—that he is not deviant, weak, cowardly, inferior and vengeful. The true individual, however—the honest fool—stands outside the protective enclave of acceptance, unredeemed—the personification of weakness, inferiority, vengefulness, cowardice, difference. He cannot make the cut, and because he cannot make the cut, he is the target of the tyranny of the group (and of his own judgment, insofar as he is that group). But man as a fool, weak, ignorant and vulnerable, is what the group is not: a true individual, truly existing, truly experiencing, truly suffering (if it could only be admitted). Consciousness of intrinsic personal limitation and apprehension of its consequences brings clear definition of the nature of subjective experience, when allowed to surface, and fosters attempts to adapt to that experience. It is for this reason that only the unredeemed—the outcast, the sick, the blind, and the lame—can be "saved." Apprehension of the true nature of subjective experience—of individual reality, outside the delusionary constraints of the group—is of sufficient power to demoralize, absolutely. The eternal consequence of self-consciousness is therefore the expulsion from Eden, in its maternal and patriarchal forms. But such a fall is a step on the way to the "true paradise"—is a step toward adoption of identity with the hero, who is not protected from the vagaries of existence, but who can actively transform the terrible unknown into the sustenant and productive world. Acceptance (at least recognition) of the mortal limitation characterizing human experience therefore constitutes the precondition for proper adaptation. The lie, which denies individual experience, is denial of the fool—but the fool is the truth.

Acceptance of mortal weakness is the paradoxical humility that serves as precondition for true heroism. The heroic attitude is predicated on the belief that something new and valuable still exists, to be encountered and assimilated, regardless of the power and stability of the current position. This belief is further based upon faith in human potential—upon faith that the individual spirit will respond to challenge and flourish. Such belief must be posited—voluntarily, freely—prior to participation in any heroic endeavor. This is the necessary leap that makes courageous and creative action possible; that makes religion something real. Humility means, therefore: *I am not yet what I could be.* This an adage both cautious and hopeful.

The adversarial position, deceit, is predicated on the belief that the knowledge of the present comprises all necessary knowledge—is predicated on the belief that the unknown has finally been conquered. This belief is equivalent to denial of vulnerability, equivalent to the adoption of omniscience—"what I do is all there is to do, what I know is all there is to know." Inextricably associated with the adoption of such a stance is denial, implicit or explicit, of the existence, the possibility, and the necessity of the heroic—as everything worthwhile has already been done, as all problems have been solved, as paradise has already been spread before us. This is a *terrible* position, as the axiom of faith "we are redeemed" makes human suffering itself (which can never be eradicated as a consequence of ideological identification) something *heretical*—something that can exist only as an insult to the guardians of traditional order. The authoritarian is thereby necessarily stripped of his empathy, even for himself: in the "perfect world," presently extant, nothing imperfect may be allowed to exist. So the adversary backs himself into a position where he cannot admit even to his own misery (let alone the misery of others). A more hopeless position cannot possibly be imagined.

Acceptance of insufficiency paradoxically catalyzes identification with the hero and opens up the possibility of participation in the process of creation and renewal. Rejection of insufficiency produces, by contrast, identification with the adversary, whose eternal dwelling place is hell. This hell is something whose nature can be rendered explicitly comprehensible, despite its mythological character; is something that has familiar and defined features; is something that can be understood, first and foremost, as a consequence of the "imbalance of the constituent elements of reality." Adoption of the deceitful or adversarial mode of adaptation produces an accelerated search for security and increased likelihood of aggression, in those cases where identification with the cultural canon is deemed possible—or degeneration of personality and decadent breakdown (where the costs of cultural identity are regarded as "too high," where no such identity is waiting to offer protection, or where even fascistic behavior appears as something too positive to manifest in the all-too-unbearable world).

Denial of the heroic promotes *fascism*, absolute identification with the cultural canon. Everything that is known, is known within a particular historically determined framework, predicated upon mythologically expressed assumptions. Denial or avoidance of the unknown therefore concomitantly necessitates deification of a particular previously established viewpoint. The way that things are, under such circumstances, must be the way they forever remain. Questioning the wisdom of the past necessarily exposes the anxiety-provoking unknown once again to view. This exposure of the unknown can be regarded as beneficial under those circumstances where positive adaptation to the unknown is viewed as possible, but only as destructive where lack of faith in the heroic rules. All that lives, however, grows. When conservatism destroys the capacity for individual creativity—when it becomes tyranny—then it works against life, not for it. The "spirit within" has withdrawn from the group afraid to develop. An absolutely conservative society cannot survive, because the future transcends the limitations of the past, and the absolute conservative wants to limit what could be to what has already been. If history was complete, and perfect, if the individual had fully exploited his highest potential, then the human race would be run, for all would be explored, all known, all accomplished. But this pinnacle of attainment has not yet been reached—and perhaps never will be. Those who pretend the contrary rapidly come to actively oppose the very process that delivers what they claim to have already obtained.

Denial of the heroic promotes *decadence*, equally—absolute rejection of the order of tradition; absolute rejection of order itself. This pattern of apprehension and behavior seems far removed from that of the fascist, but the decadent is just as arrogant as his evidently more rigid peer. He has merely identified himself absolutely with *no thing*, rather than with *one thing*. He is rigidly convinced of the belief that nothing matters— convinced that nothing is of value, despite the opinions of (clearly deluded, weak and despicable) others; convinced that nothing is worth the effort. The decadent functions in this manner like an anti-Midas—everything he touches turns to ashes.

Under normal circumstances, the individual who reaches adolescence identifies with the tribe—with the collective historically determined structure designed to deal with threat. The normal individual solves his problem of adaptation to the unknown by joining a group. A group, by definition, is composed of those who have adopted a central structure of value, and

who therefore behave, in the presence of other group members, identically—and if not identically, at least predictably.

The fascist adapts to the group with a vengeance. He builds stronger and stronger walls
around himself and those who are "like him," in an ever more futile attempt to keep the
threatening unknown at bay. He does this because his worldview is incomplete. He does not
believe in the heroic aspect of the individual, does not see the negative aspect of the social
world, and cannot visualize the beneficial aspect of chaos. He is frightened enough to develop the discipline of a slave, so that he can maintain his protected position in the group, but
he is not frightened enough to transcend his slavish condition. He therefore remains twisted
and bent. The decadent, by contrast, sees nothing but the tyranny of the state. Since the
adversarial aspect of the individual remains conveniently hidden from his view, he cannot
perceive that his "rebellion" is nothing but avoidance of discipline. He views chaos as a beneficial home, seeing the source of human evil in social regulation, because he cannot imagine
the Terrible Mother as soul-devouring force. So he abandons his father in the belly of the
beast, unredeemed, and has no tools to rely on when he finally faces a true challenge.

The decadent looks to subvert the process of maturation—looks for a "way out" of group
affiliation. Group membership requires adoption of at least adolescent responsibility, and
this burden may seem too much to bear, as a consequence of prolonged immaturity of outlook. The decadent therefore acts "as if" the paradigmatic structure of the group has been
rendered "insufficient" as a consequence of environmental, cultural or intellectual change,
and refuses to be the fool who risks belief. The proper response to "the illness of the father,"
is, of course, "the journey to the land of living water." The decadent makes his intellectual
superiority to the "superstitions of the past" an article of faith, instead, and shirks his responsibility. (That is to say—it is the desire to shirk that responsibility [and the "heroic sacrifice"
it entails], that constitutes motivation for belief in "intellectual superiority.") The "suffering
rebel" stance that such adoption allows, as a secondary consequence, also serves admirably as
mask for cowardice.

The fascist and the decadent regard each other as opposites, as mortal enemies. They are
in actuality two sides of the same bent coin:

Today is Christmas day, and I have just come home from Julie's. While I was there it struck me, as I sat
on their couch between the two girls, just how foolish and idiotic I have been in this my only life. I hope
you will have patience while I unburden myself on you, because I need desperately to confess my sins to
someone, and I know that if I were sitting in a little cubicle talking to an unseen clergyman I wouldn't
do a proper job of it. You fit the definition of a religious man as someone who gives careful consideration
to the demonic and irrational in humanity, so I think you will find my confession interesting.

Imagine if you can a grown man who harbors in his heart the most vicious resentment for his fellow
man, his neighbor, who is guilty of nothing more than embodying a superior consciousness of what it
means to be a man. When I think of all the black, scathing thoughts I have directed at those who I
could not look in the eye, it is almost unbearable. All of my lofty disdain for the "common" man, who,
so I thought, was guilty of the sin of unconsciousness, was, I now realize, founded on nothing more

than jealousy and spite. I hated, I absolutely loathed anyone who had wrestled with their fear of leaving the maternal confines of a childish mentality and won their battle, only because I had not done so. I equated independence and success with egotism and selfishness, and it was my fondest hope, my highest ambition, to witness and participate in the destruction of everything that successful, independent people had built for themselves. This I considered a duty. In fact there was a decidedly fanatical element in my urge to cleanse the world of what I perceived to be selfishness.

Think what would have happened if I had been in a position to realize my fine feelings! The memory makes me fear that any moment the earth will crack open and swallow me up, because if there is any justice it would. I, who had not the faintest inkling of a capacity for moral judgment, traipsing around passing judgment on anyone who dared cross my path. It makes me wonder that I have even one friend in this world. But of course I had friends before. Anyone with enough self-contempt that they could forgive me mine.

It is fortunate for humanity that there are few saviors of the caliber of myself. Did you know that I used to identify with Christ? I considered myself entirely, immaculately free of aggression and every other form of antisocial feeling. But what about the hatred I have just now confessed, you ask? That didn't count. Those feelings were based on sound common sense, you see. After all, there are sons of bitches in the world, and one needs to be ready for them. (Do I smell ozone? They say you get a tingling sensation just before the lightning bolt strikes.)

That is a very apt phrase, son of a bitch. There is a passage in Jung's *Phenomenology of the Self* which runs: "Often a mother appears beside him who apparently shows not the slightest concern that her little son should become a man, but who, with tireless and self-immolating effort, neglects nothing that might hinder him from growing up and marrying. You now behold the secret conspiracy between mother and son, and how each helps the other to betray life." This insight would be useful for me as an excuse, being a perfectly accurate description of my situation, were it not for the fact that I am almost daily presented with a residual bit of undiluted evil in myself. For example, when I am faced with a frustrating situation I do not ask myself what I am going to do about it. I ask myself who is responsible for it—and I am always ready to conclude that if the other person were to act properly then the problem would not exist. What is evil about that, you ask? Obviously if I am determined to overlook my own part in the failure to resolve my own frustrations, if I am determined to find a scapegoat for my problems, then I am just a stone's throw away from the mentality that was responsible for Hitler's final solution, or for the Spanish inquisition, or for Lenin's cultural cleansing.

What was it you told me when I complained about the flaws in capitalism, about the fact that so many people take advantage of the capitalist system? Something like "the fact that people go on consolidating their financial position *ad nauseum* is another problem, but it is no reason to conclude that there is anything virtuous in refusing to even try to consolidate one's position in the first place." But it is so much easier to crown one's cowardice and laziness with the accolade of virtue. Just ask Lenin's henchmen, who swaggered around the countryside robbing every farmer who had managed any success whatever, and called themselves friends of the common people and patted each other on the back for their moral uprightness! I wonder if I have changed so much that I would not join them when put to the test. The idea that morality stems from a lack of personal interests is thoroughly ingrained in my mind. "Good people are those who don't want anything for themselves" is the way I think. But I never ask myself why such a person should put any effort into disciplining himself, or take any pains to keep his motives clear in his own mind, because there is nothing of value to him in this world.

In his essay *Relations Between the Ego and the Unconscious* Jung says that in an unconscious state the individual is torn by the conflict of opposites, and that achieving consciousness involves resolving that conflict on a higher level. (I understand that this particular state of adult unconsciousness is different from the original state of child-like unconsciousness, in which there is no long-term conflict.) Just last week I was stuck in that dead-end again. I was sitting and thinking about what course my life should be taking, and at every imagined scenario of a fulfilling or meaningful activity I was met by a counterpoint coming from somewhere in my head, showing me how this or that aspect of my scenario was wrong because it would result in this or that problem, to the point where it was unacceptable to consider any career at all because just by being alive I would contribute to the destruction of the planet. And as badly as I wanted to refute this echo of wrong to my every imagined right as an irrational chimera, the fact is, so I told myself, that we see daily in the newspapers how the activities of humanity, which are also the activities of individual men and women, are causing incalculable harm.

It is of course due to my being influenced by yourself that I do not remain stuck in that particular bog too long these days. If our industrialism is causing problems, I now answer myself, then I should hope that there are people out there working to solve those problems, or perhaps I should try to do something about them myself, but by sitting idly by I do not solve a thing. Of course what is most daunting, and also most snively, about being stuck in that bog is the fact that the rational mind wants to be absolutely sure about the successful outcome of its life plan, and obviously there is another part of the mind that knows that such certainty is impossible, so one is then faced with the need to accept on faith that things will turn out for the better with some luck and perseverance. And being a fine upstanding modern mouse with an enlightened rational mind, I have no use for faith and other such religious sounding claptrap and nonsense. Faith is obviously irrational, and I'll not have any irrationality influencing my behavior.

Previously my solution to this problem was to allow chance to make my career choices for me, letting my own interests influence my decisions as little as possible, and I then believed that I had somehow avoided personal responsibility for the state of the modern world, because I was not really responsible for the state of my life, and that I had escaped from the possibility that my plans wouldn't work out because I had no plans. It was on this rock-solid foundation that I looked out at the world, and saw around me people who were stupid enough to add their own selves into the equation.

To put this kind of faith in oneself, to believe that there exists inside of oneself a motive force, call it an interest, which will respond to life and carry one through uncertainty and adversity is an irrational attitude without equal, and it is with this irrational approach to life that the conflict of the opposites is resolved, it seems to me. But the problem now is this: in order to have this faith in one's irrational nature one needs proof that personal interests and passions are capable of sustaining one through the uncertainties and adversities of life that the rational mind foresees so clearly, and the only way to get that proof is to risk oneself and see the result. It is a very exceptional person who can take such an undertaking on their own. Most of us need guidance and support from others, from believers, so to speak. Strange, isn't it, that religious terms should become useful for this discussion?

As I wrote that last paragraph I was suddenly reminded of your idea that the devil as he is represented

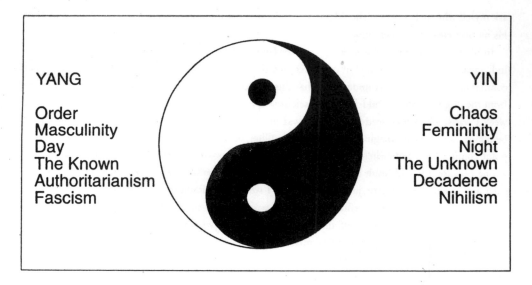

Figure 59: The Constituent Elements of Existence, Reprise

in Milton's *Paradise Lost* is a metaphor for the rational intellect, placed in the position of the highest psychic authority. "Better to rule in hell than to serve in heaven." Hell, then, is a condition in which the rational mind, with its acute consciousness of the many perils of life, holds sway over the individual and effectively prevents him from engaging in life, which results in the morally degenerate state of weakness that I described in the first pages of this letter. And heaven, I presume, would be a condition in which the rational mind subordinates itself to faith in … in God. But what is God?

You have a chapter in the manuscript of your book titled *The Divinity Of Interest*. Your ideas are starting to make sense to me now—at least I think they are. Faith in God means faith in that which kindles one's interest, and leads one away from the parental sphere out into the world. To deny those interests is to deny God, to fall from heaven and land squarely in hell, where one's passions burn eternally in frustration. What was it God said when he cast Adam out of Eden? Something about working in the dust to the end of his days, with the spectre of death always looming in the future. I can certainly relate to that. One of the most vivid impressions I get from recalling all those years I spent moving from one job to the next is the pointlessness of my daily life back then, and the glaring knowledge that the end was drawing near. But when I'm doing something that has meaning for me, something that interests me, as I am right now, death seems far away, and work seems quite agreeable, even joyous.[486]

This "theory of the genesis of social psychopathology"—this theory of a *direct* relationship obtaining between personal choice and fascistic or decadent personality and social movement—finds its precise echo in Taoist philosophy, and can be more thoroughly comprehended through application of that perspective. The traditional Taoist believes that profane human experience consists of the differentiated parts of an essentially uncategorizable background—the Tao, which may also be interpreted as "meaning" or as "the way."[487] Tao manifests itself as the eternal flux of being. The "natural categories" of Yin and Yang, represented in *Figure 59: The Constituent Elements of Existence, Reprise*, constitute the most fundamental "divisions" of Tao, the basic maternal and patriarchal constituent elements of experience. Much of ancient Chinese philosophy (cosmology, medicine, political theory, religious thinking) is predicated on the idea that pathology is caused by a relative excess of one primordial "substance" or the other. The goal of the Chinese sage—physician, spiritual leader or social administrator—is to establish or re-establish harmony between the fundamental "feminine" and "masculine" principles, and to diagnose and cure the faulty action or irresponsible inactions that led to their original discord. The schematic representation of Yin and Yang, portrayed in Figure 59, utilizes the image of a circle to represent totality; the paisleys that make up that circle are opposed but balanced. The image is rendered additionally sophisticated by the presence of the white circle in the black paisley, and vice versa. Too much chaos breeds desire for order. Yin may therefore serve as mother to Yang. Conversely, too much order breeds the desire for novelty, as antidote to stultifying predictability. In this manner, Yang serves as father to Yin.

The fascist, who will not face the reality and necessity of the unknown, hides his vulnerable face in a "pathological excess of order." The decadent, who refuses to see that existence is not possible without order, hides his immaturity from himself and others in a "pathological excess of chaos." The fascist is willing to sacrifice painful freedom for order, and to pretend that his unredeemed misery is meaningless, so that he does not have to do anything for himself. The decadent believes that freedom can be attained without discipline and responsibility, because he is ignorant of the terrible nature of "the undifferentiated ground of reality" and is unwilling to bear the burden of order. When he starts to suffer, as he certainly will, he will not allow the reality of his suffering to prove to him *that some things are real*, because acceptance of that proof would force him to *believe and to act* (would force him as well toward painful realization of the counterproductive and wasteful stupidity of his previous position).

The fascistic mode of adaptation is, above all, a method for the direct control of the unknown and unpredictable. Modern human beings, like the ancients, identify the stranger implicitly with the dragon of chaos. The stranger acts unpredictably, and thinks unpredictable things—things which might have dramatic and unsettling effect, if they were allowed full expression. Extreme conservatism allows for restriction of uncertainty, for evasion of the unknown. It performs this function by ensuring that each member of the group acts, imagines and thinks precisely like every other member (generally, precisely like the leader—a dark parody of the hero). In times of heightened uncertainty, periods of increased unemployment or unsettled political structure, the call for return to the "glorious past" therefore always arises. The fascist, dominated by his fear, believes that the world should only be order, because disorder is too frightening to consider. This makes the cosmos he creates—when granted the opportunity—a place of endless sterility and machinelike organization. This increased conformity allows at least for the temporary alleviation and restriction of anxiety, but damages the capacity of the group (his group, that is) to respond flexibly to inevitable change. It is as if, to use a biological metaphor, the fascist strives to force all the genetic diversity out of his "species." No diversity means no variance in response to new challenges; means one solution (likely the wrong one) to every problem. The suppression of deviance, of the unknown, therefore merely ensures its irrepressible emergence in negative guise, at some undetermined point in the future (as problems ignored do not go away, but get worse, as they follow their own peculiar developmental path). The order the fascist imposes, therefore, bears within it the seeds of its own destruction.

The fascist tends to be cruel as well as rigid and will pursue his cruelty even at the cost of his own stability. The Nazi persecution of the Jews, for example, continued at an increasing rate (and with increasingly harsh procedures) as it became more and more evidently a burden to the war effort. Nazi hatred therefore became a force so potent, as the Third Reich developed, that it overcame Nazi patriotism, motivated by mortal terror of the unknown. Underneath the fascist's professed patriotism and cowardly love of order is an even more profound phenomenon: hatred for the tragic conditions of existence, and for the vulnerable life that makes those conditions evident.

For they reasoned unsoundly, saying to themselves, "Short and sorrowful is our life, and there is no remedy when a man comes to his end, and no one has been known to return from Hades

Because we were born by mere chance, and hereafter we shall be as though we had never been; because the breath in our nostrils is smoke, and reason is a spark kindled by the beating of our hearts.

When it is extinguished, the body will turn to ashes, and the spirit will dissolve like empty air. Our name will be forgotten in time and no one will remember our works; our life will pass away like the traces of a cloud, and be scattered like mist that is chased by the rays of the sun and overcome by its heat.

For our allotted time is the passing of a shadow, and there is no return from our death, because it is sealed up and no one turns back.

Come, therefore, let us enjoy the good things that exist, and make use of the creation to the full as in youth.

Let us take our fill of costly wine and perfumes, and let no flower of spring pass by us.

Let us crown ourselves with rosebuds before they wither.

Let none of us fail to share in our revelry, everywhere let us leave signs of enjoyment, because this is our portion, and this our lot.

Let us oppress the righteous poor man; let us not spare the widow nor regard the gray hairs of the aged.

But let our might be our law of right, for what is weak proves itself to be useless." (Wisdom 2:1–11, RSV)

Fascist cruelty is motivated by the affective consequences of pathologically increased order. When the "water of life" dries up, nothing is left of existence but its inevitable pains and frustrations, compounded by terrible boredom. Furthermore, anomaly inevitably accumulates, as order is imposed in an increasingly strict manner. This adds increased apprehension of chaos to pain, frustration and stultification. Individuals "subjected" to a surfeit of such emotions "have every reason" to be vengeful, aggressive and cruel; have placed themselves in a state where the emergence of such motivation is certain.

The unknown appears only when there has been error. The fascist says, "I know everything there is to know." He cannot, therefore, make an error. But error is the mother of all things. The inability to admit to imperfection, therefore, means withdrawal from every informative situation. This means death of continued adaptation—and certain future re-emergence of the unknown, in negative guise. If you do not change, in the face of constant slow transformation, then the discrepancies and unresolved errors pile up, and accumulate. The more stubborn (read: arrogant) you are, the longer the period of time during which this accretion occurs. Sooner or later so much unknown surrounds you that it is no longer avoidable. At that point the dragon of the underworld emerges, and swallows you whole. Then you live in the belly of the beast, in darkness, in the night, in the kingdom of the dead. Hatred comes easily, in such an environment.

The decadent says, "there is no such thing as *to know*"—and never attempts to accomplish anything. Like his authoritarian counterpart, he makes himself "immune from error," since mistakes are always made with regard to some valued, fixed and desired end. The decadent says, "look, here is something new, something inexplicable; that is evidence, is it not, that everything that I have been told is wrong. History is unreliable; rules are arbitrary; accomplishment is illusory. Why do anything, under such circumstances?" But he is living on borrowed time—feeding, like a parasite, on the uncomprehended body of the past. If he works sufficiently hard, and saws off the branch on which he is sitting, then he will fall, too, into the jaws of the thing he ignored.

The habitual act of avoidance—of rejection—weakens the personality, in a direct causal manner. The strength of a personality might be defined, in part, as its breadth of explored territory, its capacity to act appropriately in the greatest number of circumstances. Such strength is evidently dependent upon prior learning—at least upon learning how to act—and knowledge of how to act is generated and renewed as a consequence of constant, voluntary, exploratory behavior. If everything new and different is rejected out of hand, the personality cannot adjust itself to changing circumstances. Circumstances change inevitably, however, as a consequence of simple maturation; as a consequence of entropy itself. It is of little use to be entirely prepared for the past; furthermore, it is only possible to remain prepared for the future, by facing the present. Anomaly is, therefore, spiritual "food," in the most literal sense: the unknown is the raw material out of which the personality is manufactured, in the course of exploratory activity. The act of rejecting anomaly transforms the personality into something starved, something senile, and something increasingly terrified of change, as each failure to face the truth undermines capacity to face truth in the future. The man who comes to adopt an inappropriate attitude toward the unknown severs his connection with the source of all knowledge, undermining his personality, perhaps irreparably. This dissolution of strength is self-perpetuating: every weakness increases the likelihood of further weakness:

For he that hath, to him shall be given: and he that hath not, from him shall be taken even that which he hath. (Mark 4:25)

The individual who turns away from indications of his own insufficiency increases the probability that he will seek to repress and destroy all information that indicates threat to current belief. Avoided, suppressed or otherwise undeveloped elements of personality are not accessible for use in conscious adaptation—will in fact offer resistance to such adaptation, as a consequence of their "resentment" at being twisted and ignored. Failure to utilize full human potential severely undermines strength of individual character. Dissociation of action, imagination and ideation weakens personality. Weakness of personality means inability to bear the weight of the conscious world. The result of the hypocritical suppression of individual differences, in the service of the social unit and its intrapsychic representative, is frailty in the face of the unknown:

And if a house be divided against itself, that house cannot stand. (Mark 3:25)

The Adversary In Action: A Twentieth-Century Allegory

Jung once said that "any internal state of contradiction, unrecognized, will be played out in the world as fate." This statement carries with it the apparent stamp of mysticism. How could the world play out a psychological condition (or the refusal to recognize a psychological condition)? Well, the purpose of abstraction is to represent experience, and to manipulate the representations, to further successful adaptation. If we both want the same toy, we can argue about our respective rights to it; if the argument fails, or if we refuse to engage in it, we can fight. If we are suffering from moral uncertainty, at the philosophical level, and cannot settle the internal war, then our behavior reflects our inner disquiet, and we act out our contradictions in behavior, much to our general discredit. Thus the means of settling a dispute cascade, with each failure, down the chain of abstraction: from the word, to the image, to the deed—and those who will not let their outdated identities and beliefs die, when they must, kill themselves instead. Aleksandr Solzhenitsyn describes how "order and predictability" were thereby established in the Soviet Union, during Stalin's extensive reign of terror:

A. B———v has told how executions were carried out at Adak—a camp on the Pechora River. They would take the opposition members "with their things" out of the camp compound on a prisoner transport at night. And outside the compound stood the small house of the Third Section. The condemned men were taken into a room one at a time, and there the camp guards sprang on them. Their mouths were stuffed with something soft and their arms were bound with cords behind their backs. Then they were led out into the courtyard, where harnessed carts were waiting. The bound prisoners were piled on the carts, from five to seven at a time, and driven off to the "Gorka"—the camp cemetery. On arrival they were tipped into big pits that had already been prepared and *buried alive*. Not out of brutality, no.

It had been ascertained that when dragging and lifting them, it was much easier to cope with living people than with corpses.

The work went on for many nights at Adak.

And that is how the moral-political unity of our Party was achieved.[488]

The invention, establishment and perfection of the concentration camp, the efficient genocidal machine, might be regarded as the crowning achievement of human technological and cultural endeavor, motivated by resentment and loathing for life. Invented by the English, rendered efficient by the Germans, applied on a massive scale by the Soviets and the Chinese, revivified by the Balkan conflict—perfection of the factory whose sole product is death has required truly multinational enterprise. Such enterprise constitutes, perhaps, the prime accomplishment of the cooperative bureaucratization of hatred, cowardice and deceit. Tens of millions of innocent people have been dehumanized, enslaved and sacrificed in these efficient disassembly lines, in the course of the last century, to help their oppressors maintain pathological stability and consistency of moral presumption, enforced through terror, motivated by adherence to the lie.

The very name has an uncanny aspect: horrifying, ironical, allegorical. *Camp*—that is summer sun and holiday, satirical comedy and masquerade, military rule, obedience and efficiency: *death* camp—the very devil's idea of a joke, of camp; black humor and vacation paradise; the dystopian state induced in reality by diligent pursuit of fantastic ideal, ideological purity, statist heaven on earth. *Concentration* camp—that is concentration of people in arbitrary association, restriction of movement and thought to a particular area; concentration of the processes of human life, distillation, reduction to essence, forcing attention to, concentration on, the central values underlying human endeavor.

The concentration camp has spawned its own literature, remembrance of survival under conditions as harsh as imagination can construct—human imagination, capable of positing the existence and describing the nature of an eternally torturous hell, with walls seven miles thick, lit by a fire which consumes and renews flesh, simultaneously, so it can once again be burned away.[489] This camp literature has a strange affective and descriptive consistency—a consequence of the constant re-emergence of patterned, innate modes of adaptive action and thought, arising naturally in response to experience of overwhelming anomaly and extreme threat.

Camp life is still human existence, analogous to normal life in all its facets, but made starker, less ambiguous, clarified, laid bare:

"Pardon me, you ... love life? You, you! You who exclaim and sing over and over and dance it too: 'I love you, life! Oh, I love you, life!' Do you? Well, go on, love it! Camp life—love that too! It too, is life!

> There where there is no struggle with fate,
> there you will resurrect your soul ...

You haven't understood a thing. When you get there, you'll collapse."[490]

The extreme nature of camp conditions appears merely to augment tendencies for behavior always present, under normal conditions; appears merely to exaggerate the expression of possibilities innately characteristic of the human soul.

Camp incarceration, in the typical case, begins with the fall, with arrest: unexpected, unjust, arbitrary, implacable and terrifying. The prisoner-to-be starts his involuntary descent into the underworld with his historically determined defenses intact, firmly embedded in his cultural context, entrenched in his persona—identified with his job, his social status, his view of the present, his hopes for the future. The initial intrusion of fate into this self-deceptive security occurs at night. Arrest takes place without warning, in the early hours of the morning, when people are easily frightened, dazed and less likely to offer resistance, more willing to cooperate, in their fear and naive hope—afraid for the security of their nervously gathered family, standing helpless in their household, at the mercy of state authority, in its most contemptible and repressive incarnation:

"That's all there is to it! You are arrested!"

And you'll find nothing better to respond with than a lamblike bleat: "Me? What for?"

That's what arrest is: it's a blinding flash and a blow which shifts the present instantly into the past and the impossible into omnipotent actuality.

That's all. And neither for the first hour nor for the first day will you be able to grasp anything else.

Except that in your desperation the fake circus moon will blink at you: "It's a mistake! They'll set things right!"

And everything which is by now comprised in the traditional, even literary, image of an arrest will pile up and take shape, not in your own disordered memory, but in what your family and your neighbours in your apartment remember: The sharp nighttime ring or the rude knock at the door. The insolent entrance of the unwiped jackboots of the unsleeping State Security operatives. The frightened and cowed civilian witness at their backs. . . .

The traditional image of arrest is also trembling hands packing for the victim—a change of underwear, a piece of soap, something to eat; and no one knows what is needed, what is permitted, what clothes are best to wear; and the Security agents keep interrupting and hurrying you:

"You don't need anything. They'll feed you there. It's warm there." (It's all lies. They keep hurrying you to frighten you.) . . .

The kind of night arrest described is, in fact, a favourite, because it has important advantages. Everyone living in the apartment is thrown into a state of terror by the first knock at the door. The arrested person is torn from the warmth of his bed. He is in a daze, half-asleep, helpless, and his judgment is befogged. In a night arrest the State Security men have a superiority in numbers; there are many of them, armed, against one person who hasn't even finished buttoning his trousers.[491]

Arrest means instantaneous depersonalization, isolation from family, friends and social position. This forcibly induced shift of context removes, by design, all concrete reminders of group identity, all hallmarks of social hierarchy, destroys all previous ideals, undermines all goal-directed activity—exposes essential human vulnerability and subjects it to ruthless exploitation. The arrested individual is brutally stripped of every reminder of previous identity, his predictable environment, his conditional hope—left bereft even of his clothes and hair. He is treated with utmost contempt and derision, regardless of his previous social status. This complete destruction of social context, of social identity, heightens the newly arrested individual's sense of self-consciousness, of nakedness and vulnerability. This leaves him unbearably anxious, tremendously uncertain, miserably subject to a new and uncertain world—or underworld:

We waited in a shed which seemed to be the anteroom to the disinfecting chamber. SS men appeared and spread out blankets into which we had to throw all our possessions, all our watches and jewellry. There were still naive prisoners among us who asked, to the amusement of the more seasoned ones who were there as helpers, if they could not keep a wedding ring, a medal, or a good-luck piece. No one could yet grasp the fact that everything would be taken away.

I tried to take one of the old prisoners into my confidence. Approaching him furtively, I pointed to the roll of paper in the inner pocket of my coat and said, "Look, this is the manuscript of a scientific book. I know what you will say; that I should be glad to escape with my life, that that should be all I can expect of fate. But I cannot help myself. I must keep this manuscript at all costs; it contains my life's work. Do you understand that?"

Yes, he was beginning to understand. A grin spread slowly over his face, first piteous, then more amused, mocking, insulting, until he bellowed one word at me in answer to my question, a word that was ever present in the vocabulary of the camp inmates: "Shit!" At that moment I saw the plain truth and did what marked the culminating point of the first phase of my psychological reaction: I struck out the whole of my former life.[492]

The arrested individual has no specific sociohistorically determined intrapsychic structure to protect himself from the terrible world of incarceration and slavery; no model of desire and expectation to inhibit his mortal terror, to guide his activity and to channel his hope. He has been forcibly ejected from paradise, made intolerably aware of his own essential limitations, his own nakedness, has been sentenced to endless labor and subjugation. In consequence, he has been rendered vulnerable to his worst fears, his most chaotic psychological states, and his most severe depressions:

Here is how it was with many others, not just with me. Our initial, first prison sky consisted of black swirling storm clouds and black pillars of volcanic eruptions—this was the heaven of Pompeii, the heaven of the Day of Judgment, because it was not just anyone who had been arrested, but I—the center of this world.

Our last prison sky was infinitely high, infinitely clear, even paler than sky-blue.

We all (except religious believers) began from one point: we tried to tear our hair from our head, but our hair had been clipped close! ... How could we? How could we not have seen those who informed against us?! How could we not have seen our enemies? (And how we hated them! How could we avenge ourselves on them?) And what recklessness! What blindness! How many errors! How can they be corrected? They must be corrected all the more swiftly! We must write ... We must speak out ... We must communicate....

But—there is nothing we can do. And nothing is going to save us! At the appropriate time we will sign Form 206. At the appropriate time the tribunal will read us our sentence, in our presence, or we will learn it in absentia from the OSO.

Then there begins the period of transit prisons. Interspersed with our thoughts about our future camp, we now love to recall our past: How well we used to live! (Even if we lived badly.) But how many unused opportunities there were! How many flowers we left uncrumpled! ... When will we now make up for it? If I only manage to survive—oh, how differently, how wisely, I am going to live! The day of our future *release*? It shines like a rising sun!

And the conclusion is: Survive to reach it! Survive! At any price!

This is simply a turn of phrase, a sort of habit of speech: "at any price."

But then the words swell up with their full meaning, and an awesome vow takes shape: to survive *at any price*.

And whoever takes that vow, whoever does not blink before its crimson burst—allows his own misfortune to overshadow both the entire common misfortune and the whole world.

This is the great fork of camp life. From this point the roads go to the right and to the left. One of them will rise, and the other will descend. If you go to the right you lose your life, and if you go to the left—you lose your conscience.[493]

Work at a killing pace characterizes concentration camp life—work under conditions made deathly harsh, for the sake of the sheer aesthetic quality of the misery; senseless labor—mere parody of productive work[494]—accompanied by constant, consciously arranged privation:

The most ghastly moment of the twenty-four hours of camp life was the awakening, when, at a still nocturnal hour, the three shrill blows of a whistle tore us pitilessly from our exhausted sleep and from the longings in our dreams. We then began the tussle with our wet shoes, into which we could scarcely force our feet, which were sore and swollen with edema. And there were the usual moans and groans about petty troubles, such as the snapping of wires which replaced shoelaces. One morning I heard someone, whom I knew to be brave and dignified, cry like a child because he finally had to go to the snowy marching grounds in his bare feet, as his shoes were too shrunken for him to wear. In those ghastly moments, I found a little bit of comfort: a small piece of bread which I drew out of my pocket and munched with absorbed delight.[495]

In cold lower than 60 degrees below zero [!], workdays were written off: in other words, on such days the records showed that the workers had not gone out to work; but they chased them out anyway, and whatever they squeezed out of them on those days was added to the other days, thereby raising the percentages. (And the servile Medical Section wrote off those who froze to death on such cold days on some other basis. And the ones who were left who could no longer walk and were straining every sinew to crawl along on all fours on the way back to camp, the convoy simply shot, so that they wouldn't escape before they could come back to get them).[496]

No one can immerse himself in description of the consciously perpetrated horrors of the twentieth century without recognizing, first, that such evil was perp-trated in large part by the well-socialized and obedient. No one can come to such recognition—which is self-comprehension, as well—and remain unimpressed by the power and profundity of literary and mythic representations of the power of evil: that living force, that eternally active, transcendent personality, intrapsychically incarnate; that permanent aspect of man—every man—dedicated solely, vengefully to destruction, dissolution, suffering and death:

> O Rose, thou art sick!
> The invisible worm
> That flies in the night
> In the howling storm,
> Hath found out thy bed
> Of crimson joy
> And his dark secret love
> Does thy life destroy.[497]

The Rwandan massacres, the killing fields in Cambodia, the tens of millions dead (by Solzhenitsyn's estimate) as a consequence of internal repression in the Soviet Union, the untold legions butchered during China's Cultural Revolution [the Great Leap Forward (!), another black joke, accompanied upon occasion, in the particular, by devouring of the victim], the planned humiliation and rape of hundreds of Muslim women in Yugoslavia, the holocaust of the Nazis, the carnage perpetrated by the Japanese in mainland China—such events are not attributable to human kinship with the animal, the innocent animal, or even by the desire to protect territory, interpersonal and intrapsychic, but by a deep-rooted spiritual sickness, endemic to mankind, the consequence of unbearable self-consciousness, apprehension of destiny in suffering and limitation, and pathological refusal to face the consequences thereof.

Man is not merely innately aggressive, a poorly socialized, and therefore uncontrolled predator; at best, such theory can account for his criminal aggression; it is in fact slavish adherence to the forces of socialization—to the very principle of domestication itself—that enables him to participate in production of the most truly efficient and organized of human evils. It was the discipline of the Germans, not their criminality, that made the Nazis fearsome. It was the loyalty, patriotism and commitment of the Soviet and Chinese communists that enabled the mass persecution and destructive-labor camp elimination of their countrymen. Neither is man a victim of society, innocent lamb perverted by social forces beyond individual control. Man created society in his own image; it enables him as much as it corrupts him. Man chooses evil, for the sake of the evil. Man exults in agony, delights in pain, worships destruction and pathology. Man can torture his brother and dance on his grave. Man despises life, his own weak life, and the vulnerability of others, and constantly works to lay waste, to undermine, to destroy, to torment, to abuse and devour:

From the side we looked like two reddish stones in the field.

Somewhere young men of our age were studying at the Sorbonne or at Oxford, playing tennis during their ample hours of relaxation, arguing about the problems of the world in student cafes. They were already being published and were exhibiting their paintings. They were twisting and turning to find ways of distorting the insufficiently original world around them in some new way. They railed against the classics for exhausting all the subjects and themes. They railed at their own governments and their own reactionaries who did not want to comprehend and adopt the advanced experience of the Soviet Union. They recorded interviews through the microphones of radio reporters, listening all the time to their own voices and coquettishly elucidating what they *wished to say* in their last or their first book. They judged everything in the world with self-assurance, but particularly the prosperity and justice of our country. Only at some point in their old age, in the course of compiling encyclopedias, would they notice with astonishment that they could not find any worthy Russian names for our letters—for all the letters of our alphabet.

The rain drummed on the back of our heads, and the chill crept up our wet backs.

We looked about us. The half-loaded cars had been overturned. Everyone had left. There was no one in the entire clay pit, nor in the entire field beyond the compound. Out in the gray curtain of rain lay the hidden village, and even the roosters had hidden in a dry place.

We, too, picked up our spades, so that no one would steal them—they were registered in our names. And dragging them behind us like heavy wheelbarrows, we went around Matronina's plant beneath the shed where the empty galleries wound all around the Hoffman kilns that fired the bricks. There were drafts here and it was cold, but it was also dry. We pushed ourselves down into the dust beneath the brick archway and sat there.

Not far away from us a big heap of coal was piled. Two zeks[498] were digging into it, eagerly seeking something there. When they found it, they tried it in their teeth, then put it in their sack. Then they sat themselves down and each ate a similar black-grey lump.

"What are you eating there, fellows?"

"It's 'sea clay.' The doctor doesn't forbid it. It doesn't do any good, but it doesn't do any harm either. And if you add a kilo of it a day to your rations, it's as if you had really eaten. Go on, look for some; there's a lot of it among the coal."

And so it was that right up to nightfall the clay pit did not fulfill its work norms. Matronina gave orders that we should be left out all night. But ... the electricity went out everywhere, and the work compound had no lights, so they called everyone in to the gatehouse. They ordered us to link arms, and with a beefed-up convoy, to the barking of dogs and curses, they took us to the camp compound. Everything was black. We moved along without seeing where it was wet and where the earth was firm, kneading it all up in succession, losing our footing and jerking one another.

And in the camp compound it was dark. Only a hellish glow came from beneath the burners for "individual cooking." And in the mess hall two kerosene lamps burned next to the serving window. And you could not read the slogan, nor see the double portion of nettle gruel in the bowl, and you sucked it down with your lips by feel.

And tomorrow would be the same and every day: six cars of red clay—three scoops of black gruel. In prison, too, we seemed to have grown weak, but here it went much faster. There was already a ringing in the head. That pleasant weakness, in which it is easier to give in than to fight back, kept coming closer.

And in the barracks—total darkness. We lay there dressed in everything wet on everything bare, and it seemed it was warmer not to take anything off—like a poultice.

Open eyes looked at the black ceiling, at the black heavens.

Good Lord! Good Lord! Beneath the shells and the bombs I begged you to preserve my life. And now I beg you, please send me death![499]

It is reassuring to presume that the individuals who constructed, organized and ran the concentration camps of Germany and the Soviet Union were in some profound manner different from the people that we know, and love, and are. But there is no reason to make this presumption, except for convenience and naive peace of mind.[500] The image of the concentration camp guard, much as the inmate, defines the modern individual. Hell is a bottomless pit, and why? Because nothing is ever so bad that we cannot make it worse:

Fire, fire! The branches crackle and the night wind of late autumn blows the flame of the bonfire back and forth. The compound is dark; I am alone at the bonfire, and I can bring it still some more carpenters' shavings. The compound here is a privileged one, so privileged that is almost as if I were out in freedom—this is an Island of Paradise; this is the Marfino "sharashka"—a scientific institute staffed with prisoners—in its most privileged period. No one is overseeing me, calling me to a cell, chasing me away from the bonfire. I am wrapped in a padded jacket and even then it is chilly in the penetrating wind.

But *she*—who has already been standing in the wind for hours, her arms straight down, her head drooping, weeping, then growing numb and still. And then again she begs piteously: "Citizen Chief! Forgive me! Please forgive me! I won't do it again."

The wind carries her moan to me, just as if she were moaning next to my ear. The citizen chief at the gatehouse fires up his stove and does not answer.

This was the gatehouse of the camp next door to us, from which workers came into our compound to lay water pipes and to repair the old ramshackle seminary building. Across from me, beyond the artfully intertwined, many-stranded barbed-wire barricade and two steps away from the gatehouse, beneath a bright lantern, stood the punished girl, head hanging, the wind tugging at her gray work skirt, her feet growing numb from the cold, a thin scarf over her head. It had been warm during the day, when they had been digging a ditch on our territory. And another girl, slipping down into a ravine, had crawled her way to the Vladykino Highway and escaped. The guard had bungled. And Moscow city buses ran right along the highway. When they caught on, it was too late to catch her. They raised the alarm. A mean, dark major arrived and shouted that if they failed to catch the girl, the entire camp would be deprived of visits and parcels for a whole month, because of her escape. And the women brigadiers went into a rage, and they were all shouting, one of them in particular, who kept viciously rolling her eyes: "Oh, I hope they catch her, the bitch! I hope they take scissors and—clip, clip, clip—take off all her hair in front of the line-up!" (This wasn't something she had thought up herself. This was the way they punished women in the Gulag.) But the girl who was now standing outside the gatehouse in the cold had sighed and said instead: "At least she can have a good time out in freedom for all of us!" The jailer overheard what she said, and now she was being punished; everyone else had been taken off to the camp, but she had been set outside there to stand "at attention" in front of the gate-house. This had been at 6 PM, and it was now 11 PM. She tried to shift from one foot to another, but the guard stuck out his head and shouted: "Stand at attention, whore, or else it will be worse for you!" And now she was not moving, only weeping: "Forgive me, Citizen Chief! Let me into the camp, I won't do it any more!"

But even in the camp no one was about to say to her: *All right, idiot! Come on in!*

The reason they were keeping her out there so long was that the next day was Sunday, and she would not be needed for work.

Such a straw-blond, naive, uneducated slip of a girl! She had been imprisoned for some spool of thread. What a dangerous thought you expressed there, little sister! They want to teach you a lesson for the rest of your life!

Fire, fire! We fought the war—and we looked into the bonfires to see what kind of Victory it would be. The wind wafted a glowing husk from the bonfire.

To that flame and you, girl, I promise: the whole wide world will read about you.[501]

Who would admit, even to himself: "given the choice, I would be the citizen chief, in prefer-ence to the punished girl?" And without this admission there is no reason to change, and no reason to combat the evil within:

> for whence
> But from the author of all ill could spring
> So deep a malice, to confound the race
> Of mankind in one root, and Earth with Hell
> To mingle and involve, done all to spite
> The great Creator?[502]

Faced with the horror of camp life—"It too, is life"—many become corrupted:

Let us admit the truth: At that great fork in the camp road, at that great divider of souls, it was not the majority of the prisoners that turned to the right.[503]

This corruption was not caused by camp conditions, appalling as they were:

Bread is not issued in equal pieces, but thrown onto a pile—go grab! Knock down your neighbours, and tear it out of their hands! The quantity of bread issued is such that one or two people have to die for each that survives. The bread is hung up on a pine tree—go fell it. The bread is deposited in a coal mine—go down and mine it. Can you think about your own grief, about the past and the future, about humanity and God? Your mind is absorbed in vain calculations which for the present moment cut you off from the heavens—and tomorrow are worth nothing. You *hate* labor—it is your principal enemy. You hate your companions—rivals in life and death. You are reduced to a frazzle by intense *envy* and alarm lest somewhere behind your back others are right now dividing up that bread which could be yours, that somewhere on the other side of the wall a tiny potato is being ladled out of the pot which could have ended up in your own bowl.[504]

Such conditions merely provided the precondition for the emergence of the consequences of decisions already taken, prior to incarceration, in the large part—decisions to choose security instead of maintenance of conscience; to embrace safety instead of soul:

> In looking at the people, we might see
> that in the space twixt birth and death,
> one third follow life, and one third death,
> and those who merely pass from birth to death,
> are also one third of those we see.[505]

Greed and fear in the everyday world culminate in the same blind inability Solzhenitsyn recognized in the camps—the same incapacity to think about grief, about the past and future, about man and God—but with far less evident justification. Fear of mortality, in normal life, is most generally dealt with in the same manner as in the camp situation: through absolute identification with the system and consequent rejection of the self; through acceptance of ideological promise, offer of material security and guarantee of (unearned) intrapsychic stability:

My friend Panin and I are lying on the middle shelf of a Stolypin compartment and have set ourselves up comfortably, tucked our salt herring in our pockets so we don't need water and can go to sleep. But at some station or other they shove into our compartment ... a Marxist scholar! We can even tell this from his goatee and spectacles. He doesn't hide the fact: he is a former Professor of the Communist Academy. We hang head down in the square cutout—and from his very first words we see that he is: impenetrable. But we have been serving time for a long while, and have a long time left to serve, and we value a merry joke. We must climb down to have a bit of fun! There is ample space left in the compartment, and so we exchange places with someone and crowd in:

"Hello."

"Hello."

"You're not too crowded?"

"No, it's all right."

"Have you been in the jug a long time?"

"Long enough."

"Are you past the halfway mark?"

"Just."

"Look over there: how poverty-stricken our villages are—straw thatch, crooked huts."

"An inheritance from the Tsarist regime."

"Well, but we've already had thirty Soviet years."

"That's an insignificant period historically."

"It's terrible that the collective farmers are starving."

"But have you looked in *all* their ovens?"

"Just ask any collective farmer in our compartment."

"Everyone in jail is embittered and prejudiced."

"But I've seen collective farms myself."

"That means they were uncharacteristic."

(The goatee had never been in any of them—that way it was simpler.)

"Just ask the old folks: under the Tsar they were well fed, well clothed, and they used to have so many holidays."

"I'm not even going to ask. It's a subjective trait of human memory to praise everything in the past. The cow that died is the one that gave twice the milk. (Sometimes he even cited proverbs!) And our people don't like holidays. They like to work."

"But why is there a shortage of bread in many cities?"

"When?"

"Right before the war, for example."

"Not true! Before the war, in fact, everything had been worked out."

"Listen, at that time in all the cities on the Volga there were queues of thousands of people ..."

"Some local failure in supply. But more likely your memory is failing you."

"But there's a shortage now!"

"Old wives' tales. We have from seven to eight billion poods of grain."

"And the grain itself is rotten."

"Not at all. We have been successful in developing new varieties of grain." ...

And so forth. He is imperturbable. He speaks in a language which requires no effort of the mind. And arguing with him is like walking through a desert.

It's about people like that that they say: "He made the rounds of all the smithies and came home unshod."

And when they write in their obituaries: "perished tragically during the period of the cult," this should be corrected to read: "perished comically."

But if his fate had worked out differently, we would never have learned what a dry, insignificant little man he was. We would have respectfully read his name in the newspaper. He would have become a people's commissar or even ventured to represent all Russia abroad.

To argue with him was useless. It was much more interesting to play with him ... no, not at chess, but at the game of "comrades." There really is such a game. It is a very simple game. Just play up to him a couple of times or so, use some of his own pet words and phrases. He will like it. For he has grown accustomed to find that all around him ... are enemies. He has become weary of snarling and doesn't like to tell his stories because all those stories will be twisted around and thrown right back in his face. But if he takes you for one of his own, he will quite humanly disclose to you what he has seen at the station: People are passing by, talking, laughing, life goes on. The Party is providing leadership, people are being moved from job to job. Yet you and I are languishing here in prison, there are a handful of us, and we must *write* and write petitions, begging a review of our cases, begging for a pardon....

Or else he will tell you something interesting: In the Communist Academy they decided to *devour* one comrade; they decided he wasn't quite genuine, *not one of our own*, but somehow they couldn't manage it: there were no errors in his essays, and his biography was clean. Then all of a sudden, going through the archives, what a find! They ran across an old brochure written by this comrade which Vladimir Ilyich Lenin himself had held in his hands and in the margin of which he had written in his own handwriting the notation: "As an economist he is shit." "Well now, you understand," our companion smiled confidentially, "that after *that* it was no trouble at all to make short work of that muddle-head and impostor. He was expelled from the Academy and deprived of his scholarly rank."

The railroad cars go clicking along. Everyone is already asleep, some lying down, some sitting up. Sometimes a convoy guard passes along the corridor, yawning.

And one more unrecorded episode from Lenin's biography is lost from *view*.[506]

Human life is bounded on all fronts by vulnerability, is eternally subject to fear for lack of resources and security of person. Moral knowledge, knowledge of good and evil, is the capability to choose pattern of adaptation—authoritarian, decadent or creative—in the face of mortal limitation, explicit and implicit, in camp, and outside:

Those people became corrupted in camp who had already been corrupted out in freedom or who were ready for it. Because people are corrupted in freedom too, sometimes even more effectively than in camp.

The convoy officer who ordered that Moiseyevaite be tied to a post in order to be mocked—had he not been corrupted more profoundly than the camp inmates who spat on her?

And for that matter did every one of the brigade members spit on her? Perhaps only two from each brigade did. In fact, that is probably what happened.

Tatyana Falike writes: "Observation of people convinced me that no man could become a scoundrel in camp if he had not been one before."

If a person went swiftly bad in camp, what it might mean was that he had not just gone bad, but that that inner foulness which had not previously been needed had disclosed itself.

Voichenko has his opinion: "In camp, existence did not determine consciousness, but just the opposite: consciousness and steadfast faith in the human essence decided whether you became an animal or remained a human being."

A drastic, sweeping declaration! . . . But he was not the only one who thought so. The artist Ivashev-Musatov passionately argued exactly the same thing.[507]

This corruption—this turning to the left—led some to dissolve, to decay; to fall into the sickness unto death, despair; to embrace death, with final hope—necessary faith—irrevocably, understandably crushed and shattered:

The prisoner who had lost faith in the future—his future—was doomed. With his loss of belief in the future, he also lost his spiritual hold; he let himself decline and became subject to mental and physical decay. Usually, this happened quite suddenly, in the form of a crisis, the symptoms of which were familiar to the experienced camp inmate. We all feared this moment—not for ourselves, which would have been pointless, but for our friends. Usually it began with the prisoner refusing one morning to get dressed and wash to to go out on the parade grounds. No entreaties, no blows, no threats had any effect. He just lay there, hardly moving. If this crisis was brought about by an illness, he refused to be taken to the sick-bay or to do anything to help himself. He simply gave up. There he remained, lying in his own excreta, and nothing bothered him anymore.

I once had a dramatic demonstration of the close link between the loss of faith in the future and this dangerous giving up. F———, my senior block warden, a fairly well-known composer and librettist, confided in me one day: "I would like to tell you something, Doctor. I have had a strange dream. A voice told me that I could wish for something, that I should only say what I wanted to know, and all my questions would be answered. What do you think I asked? That I would like to know when the war would be over for me. You know what I mean, Doctor—for me! I wanted to know, when we, when our camp, would be liberated and our sufferings come to an end."

"And when did you have this dream?" I asked.

"In February, 1945," he answered. It was then the beginning of March.

"What did your dream voice answer?"

Furtively he whispered to me, "March thirtieth."

When F——— told me about this dream, he was still full of hope and convinced that the voice of his dream would be right. But as the promised day drew nearer, the war news which reached our camp made it appear very unlikely that we would be free on the promised date. On March twenty-ninth, F——— suddenly became ill and ran a high temperature. On March thirtieth, the day his prophecy had told him that the war would be over for him, he became delirious and lost consciousness. On March thirty-first, he was dead. To all appearances, he had died of typhus.[508]

Other camp victims, by contrast—but equally comprehensibly—chose to identify with the camp authorities, when allowed such honor, and therefore arrayed themselves against those who shared their fate, those who were also starved and deprived, frightened and worked to exhaustion. The persecution of others presents no difficulty—is, in fact, inevitable—in the wake of bondage and persecution of the self. Frankl states:

The process of selecting Capos [trustees] was a negative one; only the most brutal of the prisoners were chosen for this job (although there were some happy exceptions). But apart from the selection of Capos which was undertaken by the SS, there was a sort of self-selecting process going on the whole time among all of the prisoners.

On the average, only those prisoners could keep alive who, after years of trekking from camp to camp, had lost all scruples in their fight for existence; they were prepared to use every means, honest and otherwise, even brutal force, theft, and betrayal of their friends, in order to save themselves....[509]

Many of the Capos fared better in the camp than they had in their entire lives. Often they were harder on the prisoners than were the guards, and beat them more cruelly than the SS men did.[510]

Likewise, Solzhenitsyn:

You—had fallen. You—were punished. You—had been uprooted from life—but you want to avoid the very bottom of the pile? You want to hover over someone else, rifle in hand? Over your brother? Here! Take it! And if he runs—shoot him! We will even call you *comrade*. And we will give you a Red Army man's ration.

And ... he grows proud. And ... he tightens his grip on his gun stock. And ... he shoots. And ... he is even more severe than the free guards. (How is one to understand this: Was it really a purblind faith in social initiative? Or was it just an icy, contemptuous calculation based on the lowest human feelings?)[511]

Most camp inmates were normal, well-adapted members of society before their imprisonment. These normal people had identified with the structure and successes of that society, with its definitions of present and ideal future, with its means to its ends. Unjust incarceration meant loss of status, heightened fear of morality; demonstrated as nothing else could the evidently pathological operations of the state, constructed in theory precisely to protect against such deprivation and anxiety. Unjust incarceration presented anomaly sufficient in its fundamental import to undermine faith in previous identification, to demonstrate the incomplete, or even corrupt, nature of previous status, to foster anxiety, depression and desire, often realized, for dissolution and death. How could such a threat be countered?

Conscious, rationalized denial of evident injustice made social identification possible once again—but at the cost of substantial intrapsychic damage, dissociation. The lie involved meant sacrifice of more personal experience, more individual possibility, more divine meaning to the group. The inevitable result of such sacrifice—the sin against the Holy Spirit—is fanatical adherence to the letter of the law—

> Farewell happy Fields
> Where Joy for ever dwells: Hail horrors, hail
> Infernal world, and thou profoundest Hell
> Receive thy new possessor—one who brings
> A mind not to be changed by place or time.[512]

is fidelity to the lie—

> Farewell remorse! All good to me is lost
> Evil, be thou my Good: by thee at least
> Divided empire with Heaven's King I hold,
> By thee, and more than half perhaps will reign;
> As Man erelong, and this new World, shall know.
> Thus while he spake, each passion dimmed his face,
> Thrice changed with pale ire, envy and despair,
> Which marred his borrowed visage, and betrayed

> Him counterfeit, if any eye beheld,
> For Heavenly minds from such distempers foul
> Are ever clear. Whereof he soon aware
> Each perturbation smoothed with outward calm,
> Artificer of fraud, and was the first
> That practised falsehood under saintly show,
> Deep malice to conceal, couched with revenge.[513]

allegiance to cruelty and deceit—

> Oh why should wrath be mute and fury dumb?
> I am no baby, I, that with base prayers
> I should repent the evils I have done.
> Ten thousand worse than ever yet I did
> Would I perform if I might have my will.
> If one good deed in all my life I did,
> I do repent it from my very soul.[514]

and hatred of the good:

> The more I see
> Pleasures about me, so much more I feel
> Torment within me, as from the hateful siege
> Of contraries; all good to me becomes
> Bane, and in Heaven much worse would be my state,
> But neither here seek I, no, nor in Heaven,
> To dwell, unless by mastering Heaven's Supreme,
> Nor hope to be myself less miserable
> By what I seek, but others to make such
> As I, though thereby worse to me redound:
> For only in destroying I find ease
> To my relentless thoughts.[515]

Human beings are emotionally attached to those whom with they identify; sympathy for the victim of injustice means inability to perpetrate such injustice. Identification with tyranny, on the other hand, means temporary effortless surcease from painful (intra- and extrapsychic) moral conflict. Such identification merely requires denial of the injustice committed to one's own person, and the subsequent falsification of individual experience. This falsification cuts the empathic bonds, connecting prisoner to prisoner, connecting man to man, connecting man to himself:

I shall despair. There is no creature loves me;
And if I die, no soul will pity me:
Nay, wherefore should they, since that I myself
Find in myself no pity to myself.[516]

The victim who finds personal security in identity with his persecutor has become that
persecutor. He has eliminated the possibility of further adaptation, integration and growth,
and has voluntarily forfeited possibility of redemption. Solzhenitsyn describes the reactions
and actions of staunch Communist Party members, imprisoned and devoured by the system
they supported and produced:

To say that things were *painful* for them is to say almost nothing. They were incapable of assimilat-
ing such a blow, such a downfall, and from their *own people*, too, from their own dear Party, and, from
all appearances, for nothing at all. After all, they had been guilty of nothing as far as the Party
was concerned—nothing at all. It was painful for them to such a degree that it was considered
taboo among them, uncomradely, to ask: "What were you imprisoned for?" The only squeamish generation of prisoners! The rest of us, in 1945, with tongues hanging out, used to recount our arrests,
couldn't wait to tell the story to every chance newcomer we met and to the whole cell—as if it were
an anecdote.

Here's the sort of people they were. Olga Sliozberg's husband had already been arrested, and they
had come to carry out a search and arrest her too. The search lasted four hours—and she spent those
four hours sorting out the minutes of the congress of Stakhanovites of the bristle and brush industry, of
which she had been the secretary until the previous day. The incomplete state of the minutes troubled
her more than her children, whom she was now leaving forever! Even the interrogator conducting the
search could not resist telling her: "Come on now, say farewell to your children!"

Here's the sort of people they were. A letter from her fifteen-year old daughter came to Yelizaveta
Tsetkova in the Kazan Prison for long-term prisoners: "Mama! Tell me, write to me—are you guilty or
not? I hope you weren't guilty, because then I won't join the Komsomol [a Soviet youth organization],
and I won't forgive them because of you. But if you are guilty—I won't write you any more and will
hate you." And the mother was stricken by remorse in her damp gravelike cell with its dim little lamp:
How could her daughter live without the Komsomol? How could she be permitted to hate Soviet
power? Better that she should hate me. And she wrote: "I am guilty. . . . Enter the Komsomol!"

How could it be anything but hard! It was more than the human heart could bear: to fall beneath
the beloved axe—then to have to justify its wisdom. But that is the price a man pays for entrusting his
God-given soul to human dogma. Even today any orthodox Communist will affirm that Tsetkova
acted correctly. Even today they cannot be convinced that this is precisely the "perversion of small
forces," that the mother perverted her daughter and harmed her soul.

Here's the sort of people they were: Y.T. gave sincere testimony against her husband—anything to
aid the Party!

Oh, how one could pity them if at least now they had come to comprehend their former wretchedness! This whole chapter could have been written quite differently if today at least they had forsaken their earlier views!

But it happened the way Mariya Danielyan had dreamed it would: "If I leave here someday, I am going to live as if nothing had taken place."

Loyalty? And in our view it is just plain pigheadedness. These devotees to the theory of development construed loyalty to that development to mean renunciation of any personal development whatsoever! As Nikolai Adamovich Vilenchuk said, after serving seventeen years: "We believed in the Party—and we were *not mistaken*!" Is this loyalty or pigheadedness?

No, it was not for show and not out of hypocrisy that they argued in the cells in defense of all the government's actions. They needed ideological arguments in order to hold on to a sense of their own rightness—otherwise, insanity was not far off.[517]

"The evidence is intolerable—so much the worse for the evidence!" The hero, the savior, is metaphorical or narrative description of the pattern by which the existence of anomalous information is accepted, mined for significance, and incorporated into the body of cultural adaptation. The devil, incarnation of evil, is embodiment, in procedure, episode and word, of the tendency that denies, rather than accepts; embodiment of the process that consciously inhibits life and its development, and brings to a halt the spirit's revolutionary process of adaptation:

Just tell us one thing: Who laid the bricks, who laid the bricks in the wall? Was it you, you hard-heads?[518]

Ideology confines human potential to a narrow and defined realm. Adaptation undertaken within that realm necessarily remains insufficient, destined to produce misery—as it is only relationship with the transcendent that allows life to retain its savor. Ideology says "it must be thus," but human behavior constantly exceeds its realm of representation; such capacity for exception must therefore be denied, lest faith in ideology vanish, and intolerable chaos reappear. The ideologue says: anomaly means dissolution, dissolution means terror—that which frightens is evil: anomaly is evil. It is not the existence of anomalous information that constitutes evil, however—such information rejuvenates, when properly consumed. Evil is the process by which the significance of the anomaly is denied; the process by which meaning itself—truth itself—is rejected. This rejection means, necessarily, life rendered unbearable, hellish:

> For now the thought
> Both of lost happiness and lasting pain
> Torments him: round he throws his baleful eyes,
> That witnessed huge affliction and dismay,
> Mixed with obdurate pride and steadfast hate;

At once, as far as angel's ken, he views
The dismal situation waste and wild;
A dungeon horrible, on all sides round,
As one great furnace flamed, yet from these flames
No light, but rather darkness visible
Served only to discover sights of woe,
Regions of sorrow, doleful shades, where peace
And rest can never dwell, hope never comes
That comes to all, but torture without end
Still urges, and a fiery deluge, fed
With ever-burning sulphur unconsumed.
Such place Eternal Justice had prepared
For those rebellious, here their prison ordained
In utter darkness, and their portion set,
As far removed from God and light of Heaven
As from the center thrice to the utmost pole.[519]

The fact, regardless of content, is not evil; it is mere (terrible) actuality. *It is the attitude to the fact that has a moral or immoral nature.* There are no evil facts—although there are facts about evil; it is denial of the unacceptable fact that constitutes evil—at least insofar as human control extends. The suppression of unbearable fact transforms the conservative tendency to preserve into the authoritarian tendency to crush; transforms the liberal wish to transform into the decadent desire to subvert. Confusing evil with the unbearable fact, rather than with the tendency to deny the fact, is like equating the good with the static product of heroism, rather than with the dynamic act of heroism itself. Confusion of evil with the fact—the act of blaming the messenger—merely provides rationale for the act of denial, justification for savage repression, and mask of morality for decadence and authoritarianism.

Denial of (anomalous) experience eliminates possibility of growth; culminates in establishment of a personality whose weakness in the face of inevitable tragic circumstance and suffering produces desire for annihilation of life. Repression of fact—of truth—ensures deterioration of personality; assures transformation of subjective experience into endless meaningless sterility and misery. Acceptance, by contrast—in the spirit of ignorant humility, courage disguised—provides the necessary precondition for change.

Myth offers an imitative schema for the generation of such acceptance—for development of the ability to adapt—through encouragement of identification with the hero, whose form is constantly represented in ethical behavior, portrayed in ritual and described in narrative. The story of the hero is symbolic depiction of the man who chooses the third way, when confronted by facts whose significance undermines personal or social stability—who risks intrapsychic dissolution as voluntarily chosen alternative to adoption of tyranny or acceptance of decadence. Failure to adopt such identification ensures constant restriction of action and imagination; generates hatred, cruelty, disgust for the weak; assures adoption of the lie as prime, and perhaps most common, adaptive strategy. Such adoption inevitably transforms human experience into hell on earth:

> Him the Almighty Power
> Hurled headlong flaming from the ethereal Sky
> With hideous ruin and combustion down
> To bottomless perdition, there to dwell
> In adamantine chains and penal fire.[520]

The definitions of moral and immoral accepted by the members of a given society remain dependent upon the conceptualization of the way accepted by that society. From within the confines of a particular conceptualization, certain behaviors, productions of the imagination and ideas are attributed status of good and status of evil, in accordance with their perceived utility, with regard to a particular goal. Any act or idea that interferes with current individual desire becomes the fool, or worse—the enemy. This means that if the individual or the group desires anything more than to live in the light, so to speak, then truth and wisdom necessarily become foreign, abhorrent. What may be regarded as useful and necessary from a higher order of morality may look positively useless and counterproductive from a lower—and will come to be treated in that manner. So the individual (or the attribute) that is serving such a higher function may well appear contaminated by the dragon of chaos to those who have not yet seen or who will not admit to the necessity and desirability of that function. Such devaluation of the "revolutionary best" dooms the individual and those he can affect to weakness and misery. Restricted and narrow goals produce warped and stunted personalities, who have cast the best within them into the domain of anomaly—defining their true talents and "deviances" as (threatening and frustrating) impediments to their too-narrow ambitions. The stunted personality will experience life as a burden, a responsibility too heavy to bear, and turn to resentment and hatred as "justifiable" responses.

Reconceptualization of the way, by contrast, means revaluation of behavior, episode and semantic proposition, so that a new order might exist; means uncomfortable return to chaos, however, prior to the reconstruction of that more inclusive order. Re-evaluation of the goal, of the ideal undertaken voluntarily, as a consequence of exposure to anomalous information, may bring suppressed material, action potential, imagination and thought, back to light. The shift toward conceptualization of the process of heroism as goal—to valuation of truth, courage and love—allows for reincorporation and subsequent development of hitherto repressed, stunted and pathologized possibilities:

The great epochs of our lives come when we gain the courage to rechristen our evil as what is best in us.[521]

This is not to say, stupidly, that all motivational states—or all facts, or all behavioral possibilities—are equally "beneficial," in all situations. It is rather that our acts of self-definition (many of which have an "arbitrary" cultural grounding) determine what we are willing to accept, at any given place and moment, as "good" and as "evil." The self-sacrificing domestic martyr, for example, who accepts her husband's every authoritarian demand—she has defined her capacity for violence as ethically unsuitable, and will regard it as something forbidden and evil. This makes aggression something contaminated by the dragon of chaos, from the martyr's perspective. She will therefore remain a miserable doormat, unless she learns to bite—which means, until she drops her present all-too-restrictive "easygoing and affectionate" stance (which is not doing her husband any good, either, by reinforcing his weak, fascistic tendencies; nor improving the society of which her marriage is a part. Her inability to be angry—which is, in reality, her inability to regard herself as possessed of intrinsic worth—removes necessary limitations to her husband's inappropriate and socially dangerous expansion of power). It is thus the "unsuitable desire" that is often precisely what is necessary to lift us beyond our present uncomfortable stasis. This does not mean simpleminded application of a poorly designed strategy of "previously repressed" motivated behavior. It means, instead, true integration of what has not yet been expressed—or even admitted—into the structure of harmonious intrapsychic and social relations:

Ye have heard that it hath been said, Thou shalt love thy neighbour, and hate thine enemy.
But I say unto you, Love your enemies, bless them that curse you, do good to them that hate you, and
    pray for them which despitefully use you, and persecute you;
That ye may be the children of your Father which is in heaven: for he maketh his sun to rise on the
    evil and on the good, and sendeth rain on the just and on the unjust.
For if ye love them which love you, what reward have ye? do not even the publicans the same?
And if ye salute your brethren only, what do ye more [than others]? do not even the publicans so?
Be ye therefore perfect, even as your Father which is in heaven is perfect. (Matthew 5:43–48)

The adoption of a particular (socially determined) conceptualization of the way allows for provisional establishment of the meaning of experiences—objects, situations and processes. The nature of the goal toward which action and ideation is devoted, in the ideal, determines what behaviors and products of imagination and abstract thought come to be regarded as acceptable, and are therefore developed, and which are forbidden, and are therefore left repressed and stunted. If the individual or social ideal remains undeveloped, immature in conceptualization, or twisted in the course of development, then aspects of behavior and cognition necessary for redemption—for deliverance from the unbearable weight of tragic self-consciousness—will be suppressed, with intrapsychic and social pathology as the inevitable result. If the individual strives primarily for material security, or social acceptance, rather than for the mythic love of God and fellow man, then respect for truth will suffer, and complete adaptation will become impossible.

And when he was gone forth into the way, there came one running, and kneeled to him, and asked
    him, Good Master, what shall I do that I may inherit eternal life?
And Jesus said unto him, Why callest thou me good? there is none good but one, that is, God.

Thow knowest the commmandments, Do not commit adultery, Do not kill, Do not steal, Do not bear
    false witness, Defraud not, Honour thy father and mother.
And he answered and said unto him, Master, all these have I observed from my youth.
Then Jesus beholding him loved him, and said unto him, One thing thou lackest: go thy way, sell
    whatsoever thou hast, and give to the poor, and thou shalt have treasure in heaven: and come,
    take up the cross, and follow me.
And he was sad at that saying, and went away grieved: for he had great possessions.
And Jesus looked round about, and saith unto his disciples, How hardly shall they that have riches
    enter into the kingdom of God!
And the disciples were astonished at his words. But Jesus answereth again, and saith unto them,
    Children, how hard is it for them that trust in riches to enter into the kingdom of God!
It is easier for a camel to go through the eye of a needle than for a rich man to enter into the kingdom
    of God.
And they were astonished out of measure, saying among themselves, Who then can be saved? (Mark
    10:17–26)

The highest value toward which effort is devoted determines what will become elevated, and what subjugated, in the course of individual and social existence. If security or power is valued above all else, then all will become subject to the philosophy of expedience. In the long term, adoption of such a policy leads to development of rigid, weak personality (or social environment) or intrapsychic dissociation and social chaos:

Jesus said, "A man had received visitors. And when he had prepared the dinner, he sent his servant to invite the guest. He went to the first one and said to him, 'My master invites you.' He said, 'I have some claims against some merchants. They are coming to me this evening. I must go and give them my orders. I ask to be excused from the dinner.' He went to another and said to him, 'My master has invited you.' He said to him, 'I have just bought a house and am required for the day. I shall not have any spare time.' He sent to another and said to him, 'My master invites you.' He said to him, 'My friend is going to get married, and I am to prepare the banquet. I shall not be able to come. I ask to be excused from the dinner.' He went to another and said to him, 'I have just bought a farm, and I am on my way to collect the rent. I shall not be able to come. I ask to be excused.' The servant returned and said to his master, 'Those whom you invited to the dinner have asked to be excused.' The master said to his servants, 'Go outside to the streets and bring back those whom you happen to meet, so that they may dine.' Businessmen and merchants [will] not enter the places of my father."[522]

A man who has put his faith in what he owns, rather than what he stands for, will be unable to sacrifice what he owns, for what he is. He will necessarily choose—when the re-emergence of uncertainty forces choice—*what he has gathered around him* instead of *what he could be*. This decision will weaken his nature and make him unable to cope with the tragic weight of his consciousness; it will turn him toward the lie and make him an active agent in production of his own and his society's insufficiency.

If the goal toward which behavior is devoted remains pathologically restricted—if the highest ideal remains, for example, sensual pleasure, social acceptance, power or material security—then aspects of behavior and ideation which exist in conflict to these goals will become pathologically twisted in orientation, because forced to serve a contemptible master; will be defined as evil, left undeveloped in consequence, and remain unavailable for use in more potentially inclusive redemptive activity; will become subjugated, suppressed, and stagnant. This lack of development and associated sickening will decrease adaptive flexibility, in the face of the true challenges of life; will ensure that life's challenges appear devastating. The myth of the way, which describes human experience itself, encompasses loss of paradise, emergence of tragedy and then, redemption—and it is the fool, the true individual, who is in desperate need of redemption. To identify with the group is to deny the fool—

> Then shall they also answer him, saying, Lord, when saw I Thee an hungred, or athirst, or a stranger, or naked, or sick, or in prison, and did not minister unto thee?
> Then shall He answer them, saying, Verily I say unto you, Inasmuch as ye did *it* not to one of the least of these, ye did *it* not to Me. (Matthew 25: 44–45)

and to therefore lose all hope.

We have a model of ourselves, in imagination and semantic representation, which is in ill accord with the actuality of ourselves, in procedure, imaginative capacity and potential for thought. This lack of isomorphism, this willful lack of attention to truth, means that behavior and potential thereof exist, which is anomalous with regard to the ideal of behavior in representation—"I couldn't do that"; that episodic capacity and content exist, which is anomalous to the ideal of such capability and content in representation—"I couldn't imagine that"; and that semantic ability and content exists, which is anomalous with regard to the ideal of semantic ability and content, in representation—"I couldn't think that." Pathological use of this model (the replacement of actuality, with an insufficient idea, conceptualized in fantasy) restricts adaptation to the unexpected—to change itself. This means existence in boundless and ever-expanding misery:

> Which way I fly is Hell; myself am Hell;
> And in the lowest deep a lower deep
> Still threatening to devour me opens wide,
> To which the Hell I suffer seems a Heaven.[523]

The devil, traditional representation of evil, refuses recognition of imperfection, refuses to admit "I was in error, in my action, in my representation"; accepts, as a consequence of unbending pride, eternal misery—refuses *metanoia*, confession and reconciliation; remains forever the spirit that refuses and rejects:

> O, then at last relent; is there no place
> Left for repentance, none for pardon left?
> None left but by submission; and that word

Disdain forbids me, and my dread of shame
Among the Spirits beneath, whom I seduced
With other promises and other vaunts
Than to submit, boasting I could subdue
The Omnipotent. Ay me! they little know
How dearly I abide that boast so vain,
Under what torments inwardly I groan;
While they adore me on the throne of Hell,
With diadem and sceptre high advanced,
The lower still I fall, only supreme
In misery—such joy ambition finds!
But say I could repent, and could obtain,
By act of grace, my former state; how soon
Should height recall high thoughts, how soon unsay
What feigned submission swore; ease would recant
Vows made in pain, as violent and void—
For never can true reconcilement grow
Where wounds of deadly hate have pierced so deep:
Which would but lead me to a worse relapse,
And heavier fall: so should I purchase dear
Short intermission bought with double smart.
This knows my Punisher; therefore as far
From granting he, as I from begging, peace:
All hope excluded thus.[524]

Such refusal—the inability to say, "I was wrong, I am sorry, I should change," means the death of hope, existence in the abyss. Rejection of fact means alienation from God, from meaning, from truth—and life without meaning is suffering without recourse, worthy of nothing but destruction, in accordance with self-definition. Frye states:

The way of life is described as beginning in *metanoia*, a word translated "repentance" by the [Authorized Version], which suggests a moralized inhibition of the "stop doing everything you want to do" variety. What the word primarily means, however, is change of outlook or spiritual metamorphosis, an enlarged vision of the dimensions of human life. Such a vision, among other things, detaches one from one's primary community and attaches him to another. When John the Baptist says "Bring forth fruits worthy of *metanoia*" (Matthew 3:8) he is addressing Jews, and goes on to say that their primary social identity (descent from Abraham) is of no spiritual importance. . . .

The dialectic of *metanoia* and sin splits the world into the kingdom of genuine identity, presented as Jesus' "home," and a hell, a conception found in the Old Testament only in the form of death or the grave. Hell is that, but it is also the world of anguish and torment that man goes on making for himself all through history.[525]

The act of *metanoia* is adaptation itself: admission of error, founded on faith in ability to tolerate such admission and its consequences; consequential dissolution, subjugation to the "hateful siege of contraries, and—*Deo concedente*—restoration of intrapsychic and interpersonal integrity:

As soon as you have renounced that aim of "surviving at any price," and gone where the calm and simple people go—then imprisonment begins to transform your former character in an astonishing way. To transform it in a direction most unexpected to you.

And it would seem that in this situation feelings of malice, the disturbance of being oppressed, aimless hate, irritability, and nervousness ought to multiply. But you yourself do not notice how, with the impalpable flow of time, slavery nurtures in you the shoots of contradictory feelings.

Once upon a time you were sharply intolerant. You were constantly in a rush. And you were constantly short of time. And now you have time with interest. You are surfeited with it, with its months and years, behind you and ahead of you—and a beneficial calming fluid pours through your blood vessels—patience.

You are ascending. . . .

Formerly you never forgave anyone. You judged people without mercy. And you praised people with equal lack of moderation. And now an understanding mildness has become the basis of your uncategorical judgments. You have come to realize your own weakness—and you can therefore understand the weakness of others. And be astonished at another's strength. And wish to possess it yourself.

The stones rustle beneath our feet. We are ascending. . . .

With the years, armor-plated restraint covers your heart and all your skin. You do not hasten to question and you do not hasten to answer. Your tongue has lost its flexible capacity for easy oscillation. Your eyes do not flash with gladness over good tidings, nor do they darken with grief.

For you still have to verify whether that's how it is going to be. And you also have to work out—what is gladness, and what is grief.

And now the rule of your life is this: do not rejoice when you have found, do not weep when you have lost.

Your soul, which formerly was dry, now ripens from suffering. And even if you haven't come to love your neighbours in the Christian sense, you are at least learning to love those close to you.

Those close to you in spirit who surround you in slavery. And how many of us come to realize: It is particularly in slavery that for the first time we have learned to recognize genuine friendship!

And also those close to you in blood, who surrounded you in your former life, who loved you—while you played the tyrant over them. . . .

Here is a rewarding and inexhaustible direction for your thoughts: Reconsider all your previous life. Remember everything you did that was bad and shameful and take thought—can't you possibly correct it now?

Yes, you have been imprisoned for nothing. You have nothing to repent of before the state and its laws.

But . . . before your own conscience? But . . . in relation to other individuals?[526]

Refusal of metanoia means inevitable intermixture of Earth and the underworld; conscientious acceptance, by contrast, produces a characteristic transformation of personality, of
action, imagination and thought. Frankl states:

We who lived in concentration camps can remember the men who walked through the huts comforting others, giving away their last piece of bread. They may have been few in number, they they offer
sufficient proof that everything can be taken from a man but one thing: the last of the human free-
doms—to choose one's attitude in any given set of circumstances, to choose one's own way.

And there were always choices to make. Every day, every hour, offered the opportunity to make a
decision, a decision which determined whether you would or would not submit to those powers which
threatened to rob you of your very self, your inner freedom; which determined whether or not you
would become the plaything of circumstance, renouncing freedom and dignity to become moulded
into the form of the typical inmate.[527]

Solzhenitsyn echoes these sentiments, almost precisely:

And how can one explain that certain unstable people found faith right there in the camp, that they
were strengthened by it, and that they survived uncorrupted?

And many more, scattered about and unnoticed, came to their alloted turning point and made no
mistake in their choice. Those who managed to see that things were not only bad for them, but even
worse, even harder, for their neighbours.

And all those who, under the threat of a penalty zone and a new term of imprisonment, refused to
become stoolies?

How, in general, can one explain Grigory Ivanovich Grigoryev, a soil scientist? A scientist who volunteered for the People's Volunteer Corps in 1941—and the rest of the story is a familiar one. Taken
prisoner near Vyazma, he spent his whole captivity in a German camp. And the subsequent story is
also familiar. When he returned, he was arrested by us and given a tenner. I came to know him in winter, engaged in general work in Ekibastuz. His forthrightness gleamed from his big quiet eyes, some
sort of unwavering forthrightness. This man was never able to bow in spirit. And he didn't bow in
camp, either, even though he worked only two of his ten years in his own field of specialization, and
didn't receive food parcels from home for nearly the whole term. He was subjected on all sides to the
camp philosophy, to the camp corruption of soul, but he was incapable of adopting it. In the Kemerovo
camps (Antibess) the security chief kept trying to recruit him as a stoolie. Grigoryev replied to him
quite honestly and candidly: "I find it quite *repulsive* to talk to you. You will find many willing without
me." "You bastard, you'll crawl on all fours." "I would be better hanging myself from the first branch."
And so he was sent off to a penalty situation. He stood it for about half a year. And he made *mistakes*
which were even more unforgivable: When he was sent on an agricultural work party, he refused (as a
soil scientist) to accept the post of brigadier offered him. He hoed and scythed with enthusiasm. And
even more stupidly: in Ekibastuz at the stone quarry he refused to be a work checker—only because he
would have had to pad the work sheets for the sloggers, for which, later on, when they caught up with

it, the eternally drunk free foreman would have to pay the penalty (but would he?). And so he went to break rocks! His honesty was so monstrously unnatural that when he went out to process potatoes with the vegetable storeroom brigade, he did not steal any, though everyone else did. When he was in a good post, in the privileged repair-shop brigade at the pumping-station equipment, he left simply because he refused to wash the socks of the free bachelor construction supervisor, Treivish. (His fellow brigade members tried to persuade him: Come on now, isn't it all the same, the kind of work you do? But no, it turned out it was not at all the same to him!) How many times did he select the worst and hardest lot, just so as not to have to offend against his conscience—and he didn't, not in the least, and I am a witness. And even more: because of the astounding influence on his body of his bright and spotless human spirit (though no one today believes in any such influence, no one understands it) the organism of Grigory Ivanovich, who was no longer young (close to fifty), grew stronger in camp; his earlier rheumatism of the joints disappeared completely, and he became particularly healthy after the typhus from which he recovered: in winter, he went out in cotton sacks, making holes in them for his head and his arms—and he did not catch cold![528]

The process of voluntary engagement in the "revaluation of good and evil," consequent to recognition of personal insufficiency and suffering, is equivalent to adoption of identification with Horus (who, as the process that renews, exists as something superordinate to "the morality of the past"). This means that the capacity to reassess morality means identification with the figure that "generates and renews the world"—with the figure that mediates between order and chaos. It is "within the domain of that figure" that room for all aspects of the personality actually exist—as the demands placed on the individual who wishes to identify with the savior are so high, so to speak, that every aspect of personality must become manifested, "redeemed," and integrated into a functioning hierarchy. The revaluation of good and evil therefore allows for the creative reintegration of those aspects of personality—and their secondary representations in imagination and idea—previously suppressed and stunted by immature moral ideation, including that represented by group affiliation (posited as the highest level of ethical attainment).

The act of turning away from something anomalous is the process of labeling that anomalous thing as "too terrifying to be encountered or considered," in its most fundamental form. To avoid something is also to define it—and, in a more general sense, to define oneself. To avoid is to say "that is too terrible," and that means—"too terrible for me." The impossibility of a task is necessarily determined in relationship to the abilities of the one faced with it. The act of turning away therefore means willful opposition to the process of adaptation, since nothing new can happen when everything new is avoided or suppressed. The act of facing an anomaly, by contrast, is the process of labeling that event as tolerable—and, simultaneously, the definition of oneself as the agent able to so tolerate. Adoption of such a stance means the possibility of further growth, since it is in contact with anomaly that new information is generated. This "faith in oneself and the benevolence of the world" manifests itself as the courage to risk everything in the pursuit of meaning. If the nature of the goal is shifted from desire for predictability to development of personality capable of facing chaos voluntarily, then the unknown, which can never be permanently banished, will no longer be associated with fear, and safety, paradoxically, will be permanently extablished:

It was granted me to carry away from my prison years on my bent back, which nearly broke beneath its load, this essential experience: *how* a human being becomes evil and *how* good. In the intoxication of youthful successes I had felt myself to be infallible, and I was therefore cruel. In the surfeit of power I was a murderer, and an oppressor. In my most evil moments, I was convinced that I was doing good, and I was well supplied with systematic arguments. And it was only when I lay there on rotting prison straw that I sensed within myself the first stirrings of good. Gradually it was disclosed to me that the line separating good and evil passes not through states, nor between classes, nor between political parties either—but right through every human heart—and through all human hearts. This line shifts. Inside us, it oscillates with the years. And even within hearts overwhelmed by evil, one small bridgehead of good is retained. And even in the best of all hearts, there remains ... an uprooted small corner of evil.

Since then I have come to understand the truth of all the religions of the world: They struggle with the *evil inside a human being* (inside every human being). It is impossible to expel evil from the world in its entirety, but it is possible to constrict it within each person.

And since that time I have come to understand the falsehood of all the revolutions in history: They destroy only *those carriers* of evil contemporary with them (and also fail, out of haste, to discriminate the carriers of good as well). And they then take to themselves as their heritage the actual evil itself, magnified still more.

The Nuremburg Trials have to be regarded as one of the special achievements of the twentieth century: they killed the very idea of evil, though they killed very few of the people who had been infected by it. (Of course, Stalin deserves no credit here. He would have preferred to explain less and shoot more.) And if by the twenty-first century humanity has not yet blown itself up and has not suffocated itself—perhaps it is this direction that will triumph?

Yes, and if it does not triumph—then all humanity's history will have turned out to be an empty exercise in marking time, without the tiniest mite of meaning! Whither and to what end will we otherwise be moving? To beat the enemy over the head with a club—even cavemen knew that.

"Know thyself!" There is nothing that so aids and assists the awakening of omniscience within us as insistent thoughts about one's own trangressions, errors, mistakes. After the difficult cycles of such ponderings over many years, whenever I mentioned the heartlessness of our highest-ranking bureaucrats, the cruelty of our executioners, I remember myself in my captain's shoulder boards and the forward march of my battery through East Prussia, enshrouded in fire, and I say: "So, were *we* any better?"[529]

HEROIC ADAPTATION: VOLUNTARY RECONSTRUCTION OF THE MAP OF MEANING

The group provides the protective structure—conditional meaning and behavioral pattern—that enables the individual to cast off the dependence of childhood, to make the transition from the

maternal to the social, patriarchal world. The group is not the individual, however. Psychological
development that ceases with group identification—held up as the highest attainable good by every
ideologue—severely constricts individual and social potential, and dooms the group, inevitably, to
sudden and catastrophic dissolution. Failure to transcend group identification is, in the final analysis, as pathological as failure to leave childhood.

Movement from the group to the individual—like that from childhood to group—follows the
archetypal transformative pattern of the heroic (paradise, breach, fall, redemption; stability, incorporation, dissolution, reconstruction). Such transformation must be undertaken voluntarily, through
conscious exposure to the unknown—although it may be catalyzed by sufficiently unique or traumatic experience. Failure to initiate and/or successfully complete the process of secondary maturation heightens risk for intrapsychic and social decadence, and consequent experiential chaos,
depression and anxiety (including suicidal ideation), or enhances tendency toward fanaticism, and
consequent intrapsychic and group aggression.

The Bible, considered as a single story, presents this "process of maturation" in mythological
terms. The Old Testament offers group identity, codified by Moses, as antidote for the fallen state of
man. This antidote, while useful, is incomplete—even Moses himself, a true ancestral hero, fails to
reach the promised land. The New Testament, by contrast, offers identification with the hero as the
means by which the "fallen state" and the problems of group identity might both be "permanently"
transcended. The New Testament has been traditionally read as a description of a historical event,
which redeemed mankind, once and for all: it might more reasonably be considered the description
of a process that, if enacted, could bring about the establishment of peace on earth.

The problem is, however, that this process cannot yet really be said to be consciously—that is,
explicitly—understood. Furthermore, if actually undertaken, it is extremely frightening, particularly
in the initial stages. In consequence, the "imitation of Christ"—or the central culture-hero of other
religious systems—tends to take the form of ritualistic worship, separated from other "nonreligious"
aspects of life. Voluntary participation in the heroic process, by contrast—which means courageous
confrontation with the unknown—makes "worship" a matter of true identification. This means that
the true "believer" rises above dogmatic adherence to realize the soul of the hero—to "incarnate
that soul"—in every aspect of his day-to-day life.

This is easy to say, but very difficult to understand—and to do. It is no easy matter to translate
the transpersonal myth of the hero into a template for action and representation, in the unique
conditions that make up an individual life. It appears equally troublesome—even hubristic—to presume that the individual might be a force worthy of identification with the hero. Nonetheless, we
are more than we seem—and are more trouble than we imagine, when undisciplined and unrealized.
The "banality of evil"—Hannah Arendt's famous phrase,[530] applied to the oft-unprepossessing Nazi
"personality," is more accurately "the evil of banality." Our petty weaknesses accumulate, and multiply, and become the great evils of state. As our technological power expands, the danger we pose
increases—and the consequences of our voluntary stupidity multiply. It is increasingly necessary
that we set ourselves—not others—right, and that we learn explicitly what that means.

The nature of the process of identification with the hero can be understood in great detail as a
consequence of the analysis of alchemy, which Jung made his life's work. Alchemy—considered

most generally as the precursor of modern chemistry—was in fact a twenty-centuries'-long endeav-or to understand the "transformations of matter." The alchemical matter, however, was not the matter of modern science—logically enough, as the ancient alchemists practiced in the absence of the presumptions and tools of modern science. It was a substance more like Tao—"that which pro-duced or constituted the flux of being"; something more like "information" in the modern sense (if information may be considered latent in unexplored places); something more like the unknown as such (something like the matrix of being). Investigation of this instrinsically compelling "matter"—this unknown—produced a series of internal transformations in the alchemical psyche, making it ever-more akin to the Philosopher's stone: something that could turn "base matter" into spiritual gold; something that had, in addition, the eternal, durable and indestructible nature of stone. As the alchemical endeavor progressed, through the Christian era, the "stone" became increasingly assimilated to Christ—the cornerstone "rejected by the builders," the agent of voluntary transfor-mation—whose actions eternally transform the "fallen world" into paradise.

The late-stage alchemists "posited" that a personality that had completely assimilated the "spirit of the unknown" was equivalent to Christ. Jung translated their image-laden mythological lan-guage into something more comprehensible—but not yet understood. The terrible central message of this mode of thought is this: do not lie, particularly to yourself, or you will undermine the process that gives you the strength to bear the tragic world. In your weakness—the consequence of your lie—you will become cruel, arrogant and vengeful. You will then serve as an "unconscious" emissary of the agent of destruction, and work to bring about the end of life.

The Creative Illness and the Hero

"I.N.R.I."—"Igni Natura Renovatur Integra"[531]

The "third mode" of adaptation—alternative to decadence and fascism—is heroic. Heroism is comparatively rare, because it requires voluntary sacrifice of group-fostered certainty, and indefinite acceptance of consequent psychological chaos, attendant upon (re)exposure to the unknown. This is nonetheless the creative path, leading to new discovery or reconfiguration, comprising the living element of culture. The creative actor adopts the role of hero, and places himself beyond (even in opposition to) the protective enclave of history. In conse-quence, he suffers re-exposure to the terrible unknown. Such re-exposure engenders mortal terror, but allows for union with possibility—allows for inspiration, reconstruction and advancement. It is the disintegration and disinhibition of meaning (preceding its reintegra-tion)—occurrences necessarily attendant upon the heroic process—that produce the phe-nomena linking genius, in the popular imagination, with insanity. The genius and the lunatic are separated, however, by their relative position with regard to the unknown: the genius is the fortunate hero who faces the unexpected consequences of his insufficiently adaptive behavior voluntarily, on ground that he has chosen. The unfortunate madman, by contrast, has run away from something carnivorous, something that thrives on neglect and grows larger—something that will finally devour him. The genius dissolves, is flooded with indeterminate meaning, and is then reconstituted—then dissolves, floods and reconfigures the social world. The psychotic dissolves, and drowns in the flood.

It is capacity to voluntarily face the unknown, and to reconfigure accordingly the propositions that guided past adaptation, that constitutes the eternal spirit of man, the world-creating Word. The existence and nature of that spirit has been granted due recognition in Western (and Eastern) philosophy and religion since time immemorial. The eminent theologist Reinhold Niebuhr states:

In both Plato and Aristotle "mind" is sharply distinguished from the body. It is the unifying and ordering principle, the organ of *logos*, which brings harmony into the life of the soul, as *logos* is the creative and forming principle of the world. Greek metaphysical presuppositions are naturally determinative for the doctrine of man; and since Parmenides Greek philosophy had assumed an identity between being and reason on the one hand and on the other presupposed that reason works upon some formless or unformed stuff which is never completely tractable. In the thought of Aristotle matter is "a remnant, the non-existent in itself unknowable and alien to reason, that remains after the process of clarifying the thing into form and conception. This non-existent neither is nor is not; it is not yet, that is to say it attains reality only insofar as it becomes the vehicle of some conceptual determination.[532" 533]

The notion of the spirit's intrinsic kinship with the creator was abstractly elaborated, in much more detail, in the eventual course of development of Judeo-Christian thought. From this viewpoint, man is understood most profoundly in terms of his relationship to God—as made in the "image of God"—rather than in light of his cognitive abilities or his place in nature. The essence of this "spirit identified with God" is the eternal capacity to create and to transform. Niebuhr observes that

The human spirit has the special capacity of standing continually outside itself in terms of indefinite regression.... The self knows the world, insofar as it knows the world, because it stands outside both itself and the world, which means that it cannot understand itself except as it is understood from beyond itself and the world.[534]

This capacity for infinite transcendence, which is the ability to abstract, and then represent the abstraction, and then abstract from the representation, and so on, without end, does not come without a cost, as we have previously observed. We can tumble down the deck of cards, as easily as erect it; furthermore, our capacity for evil is integrally linked with our ability to overcome boundaries.

Abstract thinking in general, and abstract moral thinking in particular, is play: the game, "what if?" Games are played by first establishing, then identifying, then altering, basic presuppositions. Before any game can be played, the rules have to be established; before the game can be altered, the rules have to be made manifest. A game (at least in its final stages) is played by constructing an image of "the world" in imagination, in accordance with certain presuppositions—which are the rules (the "environment") of the game—and then by acting in that imaginary world. This game construction, playing and modifying is a form of practice, for real-world activity. As games increase in complexity, in fact, it becomes increasingly difficult to distinguish them from real-world activity.

The game itself, at its first stages, is played at the procedural level; the rules remain implicit. Once a representation of the game has been established, then the game can be shared; later, the rules themselves can be altered. Piaget discusses the formulation of the rules of children's games:

From the point of view of the practice or application of rules four successive stages can be distinguished.

A first stage of a purely *motor* and *individual* character, during which the child handles the marbles at the dictation of his desires and motor habits. This leads to the formation of more or less ritualized schemas, but since play is still purely individual, one can only talk of motor rules and not of truly collective rules.

The second may be called *egocentric* for the following reasons. This stage begins at the moment when the child receives from outside the example of codified rules, that is to say, some time between the ages of 2 and 5. But though the child imitates this example, he continues to play either by himself without bothering to find play-fellows, or with others, but without trying to win, and therefore without attempting to unify the different ways of playing. In other words, children of this stage, even when they are playing together, play each one "on his own" (everyone can win at once) and without regard for any codification of rules. This dual character, combining imitation of others with a purely individual use of the examples received, we have designated by the term Egocentrism.

A third stage appears between 7 and 8, which we shall call the stage of incipient *cooperation*. Each player now tries to win, and all, therefore, begin to concern themselves with the question of mutual control and of unification of the rules. But while a certain agreement may be reached in the course of one game, ideas about the rules in general are still rather vague. In other words, children of 7–8, who belong to the same class at school and are therefore constantly playing with each other, give, when they are questioned separately, disparate and often entirely contradictory accounts of the rules observed in playing marbles.

Finally, between the years of 11 and 12, appears a fourth stage, which is that of the *codification of rules*. Not only is every detail of procedure in the game fixed, but the actual code of rules to be observed is known to the whole society. There is remarkable concordance in the information given by children of 10–12 belonging to the same class at school, when they are questioned on the rules of the game and their possible variations....

If, now, we turn to the consciousness of rules we shall find a progression that is even more elusive in detail, but no less clearly marked if taken on a big scale. We may express this by saying that the progression runs through three stages, of which the second begins during the egocentric stage and ends towards the middle of the stage of cooperation (9–10), and of which the third covers the remainder of this co-operating stage and the whole of the stage marked by the codification of rules.

During the first stage rules are not yet coercive in character, either because they are purely motor, or else (at the beginning of the egocentric stage) because they are received, as it were, unconsciously, and as interesting examples rather than as obligatory realities.

During the second stage (apogee of egocentric and first half of cooperating stage) rules are regarded as sacred and untouchable, emanating from adults and lasting forever. Every suggested alteration strikes the child as a transgression.

Finally, during the third stage, a rule is looked upon as a law due to mutual consent, which you must respect if you want to be loyal but which it is permissible to alter on the condition of enlisting general opinion on your side.

The correlation between the three stages in the development of the consciousness of rules and the four stages relating to their practical observance is of course only a statistical correlation and therefore very crude. But broadly speaking the relation seems to us indisputable. The collective rule is at first something external to the individual and consequently sacred to him; then, as he gradually makes it his own, it comes to that extent to be felt as the free product of mutual agreement and an autonomous conscience. And with regard to practical use, it is only natural that a mystical respect for laws should be accompanied by a rudimentary knowledge and application of their contents, while a rational and well-founded respect is accompanied by an effective application of each rule in detail.[535]

The "second-stage" child, who accepts the presuppositions of his cultural subtradition as "sacred and untouchable," thinks in a manner similar to the classic, partially hypothetical, pre-experimental or "primitive" man, who worships the past, in representation, as absolute truth. The child and the primitive are both concerned primarily with *how to behave*—how to organize behavior, *contra* nature, in the social community, to simultaneously and continuously meet ends deemed desirable. It is only much later, after these most fundamental of issues have been resolved, that the means of resolution themselves can be questioned. This act of higher-order conceptualization means emergence of ability to play games, with the rules of games—and belief in the justifiability of such activity (this rebuff to traditional order). This more abstract ability allows for answer to the meta-problem of morality, posed (much) earlier: not "how to behave?" but "how can (or is or was) how to behave be determined?"[536] Paradoxically, perhaps, the answer to this meta-problem also provides the final answer to the (apparently) less abstract question "how to behave?" or "what is the good?"

Some examples from Western religious tradition may aid in comprehending (1) the nature of the distinction between morality's central problem ("what is the good?") and meta-problem ("how are answers to the question 'what is the good?' determined?"); (2) the structure of their attendant resolutions; and (3) the manner in which the meta-problem and its solution follow in the course of historical development from the problem and its solution—attended by (cyclical) development of increasingly sophisticated and powerful (self)consciousness. Let us begin with the problem of (self)consciousness, which appears in part as enhanced ability of the declarative memory system to accurately encode the nature of human behavior, self and other. This encoding first takes the form of the narrative, or myth, which, as stated previously, is semantic use of episodic representation of procedural wisdom.

Semantic analysis of narrative—criticism[537]—allows for derivation of abstracted moral principles. First-order *pure semantic codification* of the morality implicit in behavior, and then in episodic/semantic (narrative or mythic) representation, appears to take the form of the list. A list of laws—moral rules—straightforwardly and simply defines what constitutes acceptable behavior and what does not. An explicit list serves as an admirable guide for the adolescent, emerging from the maternal world. Such a list might be regarded as the most basic form of explicit moral philosophy.

The list emerges into the narrative of Judeo-Christian consciousness as a consequence of the actions of the figure of Moses, who serves as lawgiver for the Jewish people. Moses has many of the attributes of the (typical) mythical hero, and constitutes a figure analogous to that of the supernatural ancestor of the primitive. He is characterized by endangered birth, for example, and by dual parentage (one set humble, one exalted or divine):

And the king of Egypt spake to the Hebrew midwives, of which the name of the one was Shiprah, and the name of the other Puah:

And he said, When ye do the office of a midwife to the Hebrew women, and see them upon the stools; if it be a son, then ye shall kill him[538]: but if it be a daughter, then she shall live.

But the midwives feared God, and did not as the king of Egypt comanded them, but saved the men children alive.

And the king of Egypt called for the midwives, and said unto them, Why have ye done this thing, and saved the men children alive?

And the midwives said unto Pharaoh, Because the Hebrew women are not as the Egyptian women: for they are lively, and are delivered ere the midwives come in unto them.

Therefore God dealt well with the midwives: and the people mutiplied, and waxed very mighty.

And it came to pass, because the midwives feared God, that he made them houses.

And Pharaoh charged all his people, saying, Every son that is born ye shall cast into the river, and every daughter ye shall save alive.

And there went a man of the house of Levi, and took to wife a daughter of Levi.

And the woman conceived, and bare a son: and when she saw him that he was a goodly child, she hid him three months.

And when she could not longer hide him, she took for him an ark of bulrushes, and daubed it with slime and with pitch, and put the child therein; and she laid it in the flags by the river's brink.

And his sister stood afar off, to wit what would be done to him.

And the daughter of Pharaoh came down to wash herself at the river; and her maidens walked along by the river's side; and when she saw the ark among the flags, she sent her maid to fetch it.

And when she had opened it, she saw the child: and, behold, the babe wept. And she had compassion on him, and said, This is one of the Hebrews' children.

Then said his sister to Pharaoh's daughter, Shall I go and call to thee a nurse of the Hebrew women, that she may nurse the child for thee?

And Pharaoh's daughter said to her, Go. And the maid went and called the child's mother. And Pharaoh's daughter said unto her, Take this child away, and nurse it for me, and I will give thee wages. And the woman took the child, and nursed it.

And the child grew, and she brought him unto Pharaoh's daughter, and he became her son. And she
called his name Moses: and she said, Because I drew him out of the water. (Exodus 1:15–22;
2:1–10)

After Moses reaches adulthood, he rejects his secondary Egyptian heritage and rejoins the
Hebrews—in time becoming their leader and leading them away from Egypt and Egyptian
servitude (from slavery under tyranny). He takes them on a heroic journey, from the unbearable and fallen present condition, through the (purgatorial) desert—where they act in accordance with procedures he establishes—to (earthly) paradise itself, the Promised Land of
milk and honey. Frye comments:

The Biblical pattern for the purgatorial vision is the Exodus narrative, which is in three major parts.
First is the sojourn in Egypt, the "furnace of iron," a world visited by plagues, where the Egyptian
desire to exterminate the Hebrews goes into reverse with the slaughter of the Egyptian firstborn sons.
This episode concludes with the crossing of the Red Sea, the separation of Israel from Egypt, and the
drowning of the Egyptian host. The second episode is the wandering in the wilderness, a labyrinthine
period of lost direction, where one generation has to die off before a new one can enter the Promised
Land (Psalm 95:11). This is one of several features indicating that we are in a world transcending history, and that it is in the more poetic language of the prophets that the true or symbolic meaning of
Egypt, wilderness and Promised Land emerges more clearly.

The third stage is the entry into the Promised Land, where Moses, personifying the older generation, dies just outside it. In Christian typology ... this means that the law, which Moses symbolizes,
cannot redeem mankind: only his successor Joshua, who bears the same name as Jesus, can invade and
conquer Canaan.[539,540]

Moses is a revolutionary; he teaches his people a new mode of being. This means he reval-
ues their goals as well as their means. This process of revolutionary readaptation is necessarily
preceded by a period of intense suffering, as the affects released by the new situation fight
among themselves, so to speak, until subdued. The biblical story portrays this process dramatically, presenting it in terms of the lengthy and trying desert sojourn. The provision of "heavenly" food[541] during this time is a hint to the meaning, so to speak, of the story—the interregnum
of pain and confusion that precedes the re-establishment or improvement of stability can be
tolerated only by those fed on "spiritual bread"; can be tolerated only by those who have incorporated sufficient meaning and have therefore developed wisdom, patience and faith.

In the course of the exodus, Moses begins to serve as judge for his people. He is spontaneously chosen by them, perhaps on the basis of perceived strength or integration of character, as mediator between conflicting claims of value. In this role, he is forced to determine
what was right, or what should be—and what was wrong, or comparatively wrong:

And it came to pass on the morrow, that Moses sat to judge the people: and the people stood by Moses from the morning unto the evening.

And when Moses' father in law saw all that he did to the people, he said, What is this thing that thou doest to the people? why sittest thou thyself alone, and all the people stand by thee from morning unto even?

And Moses said unto his father in law, Because the people come unto me to inquire of God:

When they have a matter, they come unto me; and I judge between one and another, and I do make them know the statutes of God, and his laws. (Exodus 18: 13–16)

Adoption of this responsibility entails voluntary acceptance of tremendous intrapsychic strain—strain associated with necessity of constant, demanding moral judgment (establishment of hierarchical order, resultant of intrapsychic quasi-Darwinian struggle of abstracted values)—and, when the ability is there, consequent generation of compensatory adaptive activity. In the mythic case of Moses, such activity took the form of *translation*—translation of moral principles from procedure, and narrative representation thereof, into an abstract semantic code. This act of translation constituted a tremendous leap forward, a qualitative shift in human cognition—regardless of whether it was actually undertaken by Moses, or by any number of individuals, over hundreds of years (a flash in time, nonetheless, from the evolutionary perspective), and conflated into a "single event" by the socially mediated process of mythological memory. The emergence of moral knowledge in explicit semantic form (as opposed to its implicit representation in narrative) appears represented in mythology as "brought about" by revelation. This revelation is reception of knowledge "from a higher source"—in this case, from the episodic to the semantic memory systems (from the mysterious domain of imagination to the concrete word).

Generation (*disinhibition*, more accurately) of overwhelming affect will necessarily characterize transitions in cognitive ability of this magnitude; generation akin to that accompanying the "insight" phenomena typical of creative or psychotherapeutic endeavor. This might be considered a consequence of initial temporary establishment of intrapsychic integration—establishment of isormorphism or concordance between the procedural, episodic and semantic memory systems—and recognition of the manifold possibilities (the hitherto unforeseeen redemptive opportunities) thereby released. The "first discovery" of a new categorization system means immediate apprehension of the broad "potential utility" of those things newly comprehended (means understanding of their *refreshed promise*). This "first discovery"—*Eureka!*—is the positive aspect of the voluntary renovelization of experience. This process is dramatically represented in Exodus by the transformation of Moses' appearance as a consequence of his protracted encounter with God:

And he was there with the Lord forty days and forty nights; he did neither eat bread, nor drink water. And he wrote upon the tables the words of the covenant, the ten commandments. And it came to pass, when Moses came down from mount Sinai with the two tables of testimony in Moses' hand, when he came down from the mount, that Moses wist not that the skin of his face shone while he talked with him.

And when Aaron and all the children of Israel saw Moses, behold, the skin of his face shone, and they were afraid to come nigh him.

And Moses called unto them; and Aaron and all the rulers of the congregation returned unto him: and
  Moses talked with them.

And afterward all the children of Israel came nigh: and he gave them in commandment all that the
  Lord had spoken with him in Mount Sinai. (Exodus 34:28–32)

This "shining face" is mythic (semantic/episodic) equation of the individual with solar
power; symbolic of illumination, enlightenment, momentary transfiguration into eternal
representative of the gods. This "great leap forward" places Moses temporarily into the company of God.

 Moses transforms what had previously been custom, embedded in behavior, represented in
myth, into an explicit semantic code. The Decalogue is the single most fundamental subset
of the "new" code:

Thou shalt have no other gods before me.

Thou shalt not make unto thee any graven image, or any likeness of anything that is in heaven above,
  or that is in the earth beneath, or that is in the water under the earth:

Thou shalt not bow down thyself to them nor serve them: for I the Lord thy God am a jealous God,
  visiting the iniquity of the fathers upon the children unto the third and fourth generation of
  them that hate me.

And shewing mercy unto thousands of them that love me, and keep my commandments.

Thou shalt not take the name of the Lord thy God in vain; for the Lord will not hold him guiltless
  that taketh his name in vain.

Remember the sabbath day, to keep it holy.

Six days shalt thou labor, and do all thy work:

But the seventh day is the sabbath of the Lord thy God: In it thou shalt not do any work, thou, nor
  thy son, nor thy daughter, thy manservant, nor thy maidservant, nor thy castle, nor thy stranger
  that is within thy gates:

For in six days the Lord made heaven and earth, the sea, and all that in them is, and rested the seventh
day: wherefore the Lord blessed the sabbath day, and hallowed it.[542]

Honour thy father and thy mother.

Thou shalt not kill.

Thou shalt not commit adultery.

Thou shalt not steal.

Thou shalt not bear false witness against thy neighbour.

Thou shalt not covet thy neighbour's house, thou shalt not covet thy neighbour's wife, nor his manservant, nor his maidservant, nor his ox, nor his ass, nor any thing that is thy neighbour's. (Exodus
  20:3–17)

Codification of tradition is necessarily dependent upon existence of tradition—established
adaptive behavior, and secondary representation thereof. The knowledge embodied in such
tradition exists as a consequence of evolutionary pressure, so to speak, operating primarily (in

the literal sense) at the level of interpersonal action, and is only secondarily, lately, translated up the hierarchy of consciousness to representation. This means that evolution proceeds at least as often (and, generally, more profoundly) from *behavior to representation* (episodic and semantic)—from adaptive action and mythic portrayal thereof (all true art included) to abstract verbal knowledge—as from *representation to behavior*.[543,544,545]

Translation of tradition into law makes verbally abstract what had previously been, at best, encoded in image—makes the morality of the culture and the moral individual "conscious" for the first time. This act of transformation, culminating in a qualitative shift of cognitive sophistication in the intrapsychic activity of a mythologized culture hero, constitutes the consequence of centuries of abstract adaptive endeavor. The actions of the hero are identified with heavenly attributes "by the story," to signify the import (and intrapsychic locale and source[546]) of the revolutionary occurrence. Nietzsche states:

> Great men, like great epochs, are explosive material in whom tremendous energy has been accumulated; their prerequisite has always been, historically and physiologically, that a protracted assembling, accumulating, economizing and preserving has preceded them—that there has been no explosion for a long time. If the tension in the mass has grown too great the merest accidental stimulus suffices to call the "genius," the "deed," the great destiny, into the world.[547]

The moral presumptions of a society emerge first in procedural form, as a consequence of individual exploratory activity, which is the process that generates novel behavioral patterns. These behavioral patterns are then hierarchically structured as a consequence of quasi-Darwinian competition, in accordance with the constraints noted previously (appeal to imagination, self-sustenance, etc.). The episodic memory systems map procedure, and outcome thereof, and thereby come to contain similar paradigmatic structure—imagistically, and then more purely semantically. Over time, the unknown, nature, thereby comes to be represented mythically as the affectively bivalent Great Mother, simultaneously creative and destructive. The known, culture, becomes the Great Father, tyrant and wise king, authoritarian and protective personality, adapted to the unknown. The knower, man, becomes the hostile mythic brothers, sons of convention, hero and antihero, Christ and Satan—eternal generator and destroyer of history and tradition. Semantic cognition, feeding on narrative—the bridge between the episode and the pure verbal abstraction—derives "rules" from behavior. Application of the rules alters the environment, including procedural and episodic representations thereof. Thus the cycle continues.

Culture protects the individual against the consequences of his or her vulnerability (at least in its positive aspect); but the price paid for absolute security is freedom and individuality, and therefore, creativity. Sacrifice of individual creativity, by choice, eventually deprives life of pleasure, of meaning—but not of anxiety or pain—and therefore renders life unbearable. Civilized or historical *Homo sapiens* grows up within a structured canon of principles implicitly and explicitly posited and held as absolute by the majority of individuals within his or her civilization. In return for this legacy, which is in fact the sum total of the efforts of mankind over thousands of years, the individual is molded and shaped, and can therefore survive independently; but all this shaping is not purely beneficial. It is an unhappy fact that the intrinsically rewarding, implicitly interesting activities associated with individual heroism often come to pose a threat to the established structure of the group.

The Great Father, positive aspect of history, protects man from the Terrible Mother. He is civilized order, education and wisdom embodied and represented, the abstracted and integrated personification of all those heroes who have come before and left their mark on the (cultural) behavior of the species. He is ritual model for emulation—Good King, Wise Judge, Man of Courage, of Action, of Art, of Thought. Insofar as he represents particular, specific patterns of action, however, he is the enemy of possibility, of life in the present itself, of the hero—and is therefore, necessarily, captor of the spirit, embodiment of the Tyrant, the Bureaucrat. This is history as the Terrible Father, dead weight of the past, crushing mass of narrow, bigoted, ill-informed opinion and popular prejudice. This is the force that oppresses the Good Mother, the creative aspect of life itself. The Terrible Father opposes anything new, anything that threatens his integral structure and absolute dominance. Identification of the well-adapted man with his culture means that as history becomes established, in counterposition to the force that nature represents, the creative hero has to battle public opinion [composed, when ideological, of contemporary slogans (*sluagh-ghairms*: battle cries of the dead)] as well as the forces of the "natural" unknown. The hero is an enemy of the historically determined structure of values and assumptions, because he may have to reorder that structure and not merely add to or maintain it, to deal with what still remains unknown. In that process of reordering, unfortunately, he risks exposing himself and all those well-adapted men who identify with and maintain that culture to the terrible forces of the unknown—to mortal anxiety and dread, to fear of the void, to terror of insanity, physical destruction, and annihilation.

These ideas are dramatically presented in Dostoevsky's *The Grand Inquisitor*. Ivan, tormented atheist, tells a story he has invented to his religiously minded younger brother, Alyosha, a novice at the local monastery. Christ returns to earth, to Seville, during the time of the Spanish Inquisition:

He came unobserved and moved about silently but, strangely enough, those who saw Him recognized Him at once. This might, perhaps, be the best part of my poem if I could explain what made them recognize Him.... People are drawn to Him by an irresistible force, they gather around Him, follow Him, and soon there is a crowd. He walks among them in silence, a gentle smile of infinite compassion on His lips. The sun of love burns in His heart; light, understanding, and spiritual power flow from His eyes and set people's hearts vibrating with love for Him. He holds His hands out to them, blesses them, and just from touching Him, or even His clothes, comes a healing power. An old man who has been blind from childhood suddenly cries out to Him: "Cure me, oh Lord, so that I may see You too!" And it is as if scales had fallen from his eyes, and the blind man sees Him. People weep and kiss the ground on which He walks. Children scatter flowers in His path and cry out to Him, "Hosannah!" "It is He, He Himself!" people keep saying. "Who else could it be!" He stops on the steps of the cathedral of Seville at a moment when a small white coffin is carried into the church by weeping bearers. In it lies a girl of seven, the only daughter of a prominent man. She lies there amidst flowers. "He will raise your child from the dead!" people shout to the weeping mother. The priest, who has come out of the cathedral to meet the procession, looks perplexed and frowns. But now the mother of the dead child throws herself at His feet, wailing, "If it is truly You, give me back my child!" and she stretches out her hands to Him. The procession stops. They put the coffin down at His feet. He looks down with compassion, His lips form the words "*Talitha cumi*"—arise, maiden—and the maiden arises. The little girl sits up in her coffin, opens her little eyes, looks around in surprise, and smiles. She holds the white roses that had been placed in her hand when they had laid her in the coffin. There is confusion among the people, shouting and weeping.

Just at that moment, the Cardinal, the Grand Inquisitor himself, crosses the cathedral square. He is a man of almost ninety, tall and erect. His face is drawn, his eyes are sunken, but they still glow as though a spark smoldered in them. Oh, now he is not wearing his magnificent cardinal's robes in which he paraded before the crowds the day before, when they were burning the enemies of the Roman Church; no, today he is wearing just the coarse cassock of an ordinary monk. He is followed by his grisly assistants, his slaves, his "holy guard." He sees the crowd gathered, stops, and watches from a distance. He sees everything: the placing of the coffin at His feet and the girl rising from it. His face darkens. He knits his thick white brows; his eyes flash with an ominous fire. He points his finger and orders his guards to seize Him.

The Grand Inquisitor's power is so great and the people are so submissive and tremblingly obedient to him that they immediately open up a passage for the guards. A death-like silence descends upon the square and in that silence the guards lay hands on Him and lead Him away.

Then everyone in the crowd, to a man, prostrates himself before the Grand Inquisitor. The old man blesses them in silence and passes on.

The guards take their prisoner to an old building of the Holy Inquisition and lock Him up there in a dark, narrow, vaulted prison cell. The day declines and is replaced by the stifling, black Southern night of Seville. The air is fragrant with laurel and lemon.

Suddenly, in the complete darkness, the iron gate of the cell opens and there stands the Grand Inquisitor himself, holding a light in his hand. The old man enters the cell alone and, when he is inside, the door closes behind him. He stops and for a long time—one or even two minutes—he looks at Him. At last he sets the light down on the table and says: "You? Is it really You?" Receiving no answer, he continues in great haste:

"You need not answer me. Say nothing. I know only too well what You could tell me now. Besides, You have no right to add anything to what You said before. Why did You come here, to interfere and make things difficult for us? For You came to interfere—You know it. But shall I tell You what will happen tomorrow? Well, I do not know who You really are, nor do I want to know whether You are really He or just a likeness of Him, but no later than tomorrow I shall pronounce You the wickedest of all heretics and sentence You to be burned at the stake, and the very people who today were kissing Your feet will tomorrow, at a sign of my hand, hasten to Your stake to rake the coals. Don't You know it? Oh yes, I suppose You do," he added, deeply immersed in thought, his eyes fixed for a moment on his prisoner.[548]

Despite his tyrannical actions, the Inquisitor feels compelled to justify his actions before Christ:

"Your great prophet had a vision and told us in an allegory that he had seen all those who were in the first resurrection and that there were twelve thousand of them from each tribe. But if there were so many, they must have been gods rather than men. They bore Your cross, they endured years and years of hunger in a barren wilderness, living on roots and locusts—and of course, You can point proudly at these children of freedom, at their freely given love, and at their magnificent suffering for Your sake. Remember, though, there were only a few thousand of them, and even these were gods rather than men. But what about the rest? Why should the rest of mankind, the weak ones, suffer because they are unable to stand what the strong ones can? Why is it the fault of a weak soul if it cannot live up to such terrifying gifts? Can it really be true that You came only for the chosen few? If that is so, it is a mystery that we cannot understand; and if it is a mystery, we have the right to preach to man that what matters is not freedom of choice or love, but a mystery that he must worship blindly, even at the expense of his conscience. And that is exactly what we have done. We have corrected your Work and have now founded it on *miracle*, *mystery* and *authority*. And men rejoice at being led like cattle again, with the terrible gift of freedom that brought them so much suffering removed from them. Tell me, were we right in preaching and acting as we did? Was it not our love for men that made us resign ourselves to the idea of their impotence and lovingly try to lighten the burden of their responsibility, even allowing their weak nature to sin, but with our permission? Why have You come to interfere with our work?"[549]

The old priest explains what historical role the institution of the church has played, and why—and provides rationale for the necessity of the impending recrucifixion:

"And we, who have taken their sins upon us to give them happiness, will stand up and say to You: 'Judge us if You can and if You dare!' Know that I am not afraid of You; know that I, too, lived in the wilderness, fed upon roots and locusts, that I, too, blessed the freedom which You bestowed upon men, and that I, too, was prepared to take my place among the strong chosen ones, aspiring to be counted among them. But I came to my senses and refused to serve a mad cause. I turned away and joined those who were endeavoring to *correct Your work*. I left the proud and turned to the meek, for the happiness of the meek. What I have told You will happen and our kingdom will come. I repeat, tomorrow You will see obedient herds, at the first sign from me, hurry to heap coals on the fire beneath the stake at which I shall have You burned, because, by coming here, You have made our task more difficult. For if anyone has ever deserved our fire, it is You, and I shall have You burned tomorrow. *Dixi!*"[550]

The story takes an unexpected twist just prior to its conclusion—a twist that illustrates Dostoyevsky's genius, and his capacity to leap beyond the ideologically and easily obvious. Ivan says:

The Grand Inquisitor falls silent and waits for some time for the prisoner to answer. The prisoner's silence has weighed on him. He has watched Him; He listened to him intently, looking gently into his eyes, and apparently unwilling to speak. The old man longs for Him to say something, however painful and terrifying. But instead, He suddenly goes over to the old man and kisses him gently on his old, bloodless lips. And that is his only answer. The old man is startled and shudders. The corners of his lips seem to quiver slightly. He walks to the door, opens it, and says to Him, "Go now, and do not come back ... ever. You must never, never, come again!" And he lets the prisoner out into the dark streets of the city. The prisoner leaves.[551]

As William James said: "The community stagnates without the impulse of the individual. The impulse dies away without the sympathy of the community."[552]

The myths of a culture are its central stories. These stories provide a dramatic record of the historically predicated transformation of human intent, and appear to exist as the episodic/semantic embodiment of history's cumulative effect on action. The mythical narratives that accompany retention of historically determined behavior constitute nonempirical episodic representation of that behavior and its method of establishment. Myth is purpose, coded in episodic memory. Mythic truth is information, derived from past experience—derived from past observation of behavior—relevant from the perspective of fundamental motivation and affect. Myth simultaneously provides a record of historical essential, in terms of behavior, and programs those historical essentials. Narrative provides semantic description of action in image, back-translatable into imaginary episodic events, capable of eliciting imitative behavior. Mythic narrative offers dramatic presentation of morality, which is the study of *what should be.* Such narrative concerns itself with the meaning of the past, with the implications of past existence for current and future activity. This meaning constitutes the ground for the organization of behavior.

Mythic drama, which plays out the exploits of exceptional individuals, appears devoted toward explication of a generally applicable pattern of adaptation. This archetypal model serves to aid in the *generation of all situation-specific individual behaviors.* Myth evolves toward declarable description of a procedural schema capable of underlying construction of all complex culturally determined hierarchies of specific behavior. This schematic pattern matches the innate, instinctual, neuropsychologically predicated individual potential for creative exploratory behavior—indeed, has been constructed in the course of historical observation of that potential in action. The expression of this potential throughout history provides for the creation of specific environmentally appropriate social contexts, procedural and episodic, which promote development of the innate capacities of the individual, protect from danger, offer hope, and inhibit existential fear.

A ring of ancestral spirits, invisible and unknown, surrounds the modern individual, and protects him or her magically from darkness and chaos. When this ring is broken—when the principles these spirits represent become subject to critical evaluation, to the onslaught of other forms of heroism, to other ideologies, or to the weight of individual experience—knowledge itself loses context, and the known reverts to the unknown. This does not mean that the Terrible Mother herself sleeps under human consciousness; it means rather *that the reasons for her "existence" thousands of years ago are still sufficient reasons today.* It is not a question of racial memory, transmitted by Lamarckian means, but of the proclivity to experience similarly under similar conditions. These conditions arise eternally when the protective veil of culture is pierced.

History protects man against overwhelming material and spiritual onslaught. It performs this function by providing a framework of meaning for those enmeshed within it. History, conceived of in this manner, comprises those *a priori* assumptions all cultures base themselves on, which guide the action of individuals, enraptured by "the spirit of the times." This framework of meaning is by necessity predicated upon various articles of faith and can be described in its totality as a myth (although it also *precedes* myth). The "highest levels" of myth provide man with the capacity to attribute meaning to or to discover meaning within the tragedy of each individual human life, forever blessed and cursed by society, forever threatened and redeemed by the unknown. To live, at this mythic level—rather than to hide—means the possibility of reaching and perhaps exceeding the highest stage of consciousness yet attained or conceptualized by a particular culture. This mythic life is symbolically represented by the *savior*—the individual who embodies the essential aspects of the mythological drama. In the Western tradition, for better or worse, like it or not, that individual is Christ. Frye states:

The significance of the life of Jesus is often thought of as a legal significance, consisting in a life of perfect morality, or total conformity to a code of right action. But if we think of his significance as prophetic rather than legal, his real significance is that of being the one figure in history whom no organized human society could possibly put up with. The society that rejected him represented all societies: those responsible for his death were not the Romans or the Jews or whoever happened to be around at the time, but the whole of mankind down to ourselves and doubtless far beyond. "It is expedient that one man die for the people," said Caiphas (John 18:14), and there has never been a human society that has not agreed with him.

What primarily distinguishes Christianity (and Judaism) from most Oriental religions, it seems to me, is this revolutionary and prophetic element of confrontation with society. This element gives meaning and shape to history by presenting it with a dialectical meaning. From this point of view, the root of evil in human life cannot be adequately described as ignorance, or the cure for it correctly described as enlightenment. The record of human cruelty and folly is too hideous for anything but the sense of a corrupted will to come near a diagnosis. Hence Jesus was not simply the compassionate Jesus as Buddha was the compassionate Buddha. His work, though it includes teaching the ways of enlightenment, does not stop there, but goes through a martyrdom and a descent into death. Two implications here are of especial importance for our present purpose. One, a specifically historical situation is latent in any "enlightenment": man has to fight his way out of history and not simply awaken from it. Two, the ability to absorb a complete individual is, so far, beyond the capacity of any society, including those that call themselves Christian.[555]

Myth has come to encapsulate and express the essential nature of the exploratory, creative, communicative psyche, as manifested in behavior, as a consequence of observation and rerepresentation of that behavior, in the temporally summed, historically determined manner beginning with imitation and ending with verbal abstraction. To what end are all behaviors (and representations of those behaviors) archetypally subjugated? Toward establishment of a state—a *spiritual kingdom*—that allows the behavioral processes that transform and establish morality to flourish. Historical cultures, after all—at least those expressly open to change— organize behavior such that the self and the other are treated, in the ideal (implicit or explicit) *with the respect due to the mediator of order and chaos*. Moral action toward other and self therefore constitutes an "as if" statement, from the perspective of the semantic system: the moral individual treats himself and others "as if" recognizing, respecting and paying homage to the ultimate source of creative adaptation (the ultimate source of "the world"). Such behavior "unconsciously" presupposes identity between the individual and the savior—the archetypal redeemer, the culture-bearer, the divine hero. This is organization of incorporated behavior in accordance with recognition of the source of incorporated behavior. Establishment of such organization, however, poses threat to morality predicated strictly upon adherence to tradition.

Heroic behavior compels imitation—a hero, by definition, serves as a model for emulation. The behavior of the culture-bearer, the archetypal hero, constitutes embodiment of an elaborate procedural code. This code is the end result of an evolutionary process, consisting of the establishment of creative behaviors, in the course of heroic endeavor, their subsequent communication in imitation and its abstract forms, and their integration, over time, into a consistent pattern of behavior, whose nature and expression constitutes the cultural character. This cultural character is the central "personality" of the healthy individual, embodied in procedure, secondarily represented in episodic and semantic memory. Ideally, this character tends toward harmonious balance between tradition and adaptation, and the needs of self and other. It is the constant attempt to accurately represent such character that constitutes the "aim" of the stories of humanity.

As history progresses, becomes more "conscious" and differentiated—or, more accurately, as the presuppositions underlying adaptive social behavior become more and more accurately abstractly formulated (more declarative)—society moves from conceptualization of the consequences or productions of heroism as the ideal toward which behavior is to be devoted toward conceptualization of the act of heroism itself as such an ideal. This is movement from product to process. This transformation in conceptualization is presented in dramatic form in the Western tradition in the New Testament's description of the passion of Christ, which portrays the process and consequences of revolutionary restructuring of the axioms of Western morality.

Christ has long been considered implicitly "contained" in the Old Testament. Frye comments:

For Paul, Christ was mainly the concealed hero of the Old Testament story and the post-Easter Christ of the resurrection. The Gospels present Christ in a form that fits this pre-Gospel conception of him: not in a biographical form but as a discontinuous sequence of appearances in which Jesus comments on the Old Testament as a series of past events, laws and images coming permanently alive in the Messianic context, and body, which he supplies.[554]

What this means, at the most fundamental level of analysis, is that the pattern of action, imagination and thought that Christ represents is necessarily "there" in any narrative or mythology, *sufficiently compelling to embed itself in memory*. The reasons for this implicit existence are clear, in a sense: Christ embodies the hero, grounded in tradition, who is narrative depiction of the basis for successful individual and social adaptation. As the Word "made flesh" (John 1:14) there "in the beginning" (John 1:1), he represents, simultaneously, the power that divides order from chaos, and tradition rendered spiritual, abstract, declarative, semantic. His manner of being is that which moves morality itself from rule of law to rule of spirit—which means *process*. *Spirit* is *process*, simultaneously opposed to and responsible for generating static being. Frye states:

We are told in the New Testament itself that the mysteries of faith have to be "spiritually discerned" (I Corinthians 2:14). This is in a passage where Paul is contrasting the letter, which he says "killeth," with the spirit that "giveth life."[555]

This idea is represented schematically in *Figure 60: The Emergence of Christ from Group Identity and Chaos*, which also portrays the Christian "story of man."

For Christ, "God is not the God of the dead, but of the living" (Matthew 22:32). Christ pushes morality beyond strict reliance on codified tradition—the explicit Law of Moses—not because such tradition was unnecessary, but because it was (and is) necessarily and eternally insufficient. He states:

For I say unto you, That except your righteousness exceed the righteousness of the scribes and the Pharisees, ye shall in no case enter into the kingdom of heaven. (Matthew 5:20)

but also

Think not that I am come to destroy the law, or the prophets: I am not come to destroy, but to fulfil. (Matthew 5: 17)

This means that identification with tradition is insufficient; that tradition cannot thereby be regarded as useless, but more in the light of a developmental precursor; and, finally, that the process that regenerates tradition is somehow implicitly contained and promoted in tradition itself.

The role of Christ, who redeems culture from enslavement to the law, is prefigured at the end of Exodus, in the sequence including and continuing after the death of Moses (as discussed previously). Christ, in fact, appears as a second Moses, who offers a spiritual (intrapsychic) kingdom as the final version of the land promised to the Israelites by God.[556]

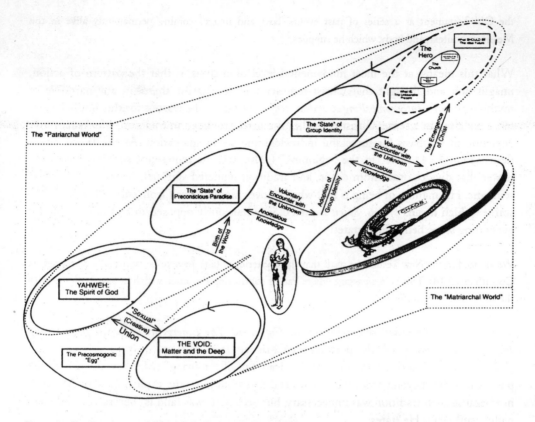

Figure 60: The Emergence of Christ from Group Identity and Chaos

He is apparently granted the authority to make such an offer with the explicit collaboration of Moses—granted such power, like Moses, on high (as befits a "solar deity"):

And after six days Jesus taketh Peter, James, and John his brother, and bringeth them up into an high mountain apart,

And was transfigured before them: and his face did shine as the sun, and his raiment was white as the light.

And, behold, there appeared unto them Moses and Elias talking with him.

Then answered Peter, and said unto Jesus, Lord, it is good for us to be here: if thou wilt, let us make here three tabernacles; one for thee, and one for Moses, and one for Elias.

While he yet spake, behold, a bright cloud overshadowed them: and behold a voice out of the cloud, which said, This is my beloved Son, in whom I am well pleased; hear ye him.

And when the disciples heard it, they fell on their face, and were sore afraid. (Matthew 17:1–9)

Like Moses, as well, Christ delivers his most famous address (which Frye interprets as a long commentary on the Decalogue) on a mountaintop. Frye notes that the Law of Moses is based upon prohibition, description of what is forbidden: "Thou shalt not." By contrast, Christ's message is more in the manner of exhortation, description of active good: "Thou shalt...."[557] This transformation is predicated upon development of heightened moral consciousness. In the beginning, for a soul steeped in sin, so to speak, it is an easier matter to identify what is evidently morally suspect, what should obviously be brought under personal control. Once a certain clarity of spirit is attained, as a consequence of conscientious, disciplined adherence to tradition, it becomes possible to determine what the good is, *what should be done*—rather than merely *what should not*. This contrast also serves as analogue to the relationship between adolescent and adult morality: strict group identity appropriately socializes the no longer properly dependent child, and allows him or her to make the leap from infancy to adulthood. The capacity to act in a disciplined manner— to follow the rules—is a necessary precondition to adult flexibility, but should not be confused with truly adult morality, which is the capacity to produce new sets of rules, with updated adaptive utility. This is also not to say, idiotically, that "Jewish" morality is adolescent, and "Christian" adult. Examples of prophetic "antagonism to tyrannical order" abound in the Old Testament, as we have seen.[558] The contrast is rather between the dogmatic and rigid and the creative and responsible *within creeds, and not between the adherents of different creeds.*

Development of the list of law—the moral wisdom of the past, cast in stone—makes the extant procedural and episodic cultural structure explicitly "conscious" for the first time. The simplicity of the list makes it easily memorable, and accessible as a "shared point of reference." The benefits of its abstraction—communicability and potential for rapid generalization—make it a potent force for the establishment and continuation of order. The list is nonetheless characterized by the presence of profound intrinsic structural limitations. It is of insufficient complexity to truly represent the nature of procedural morality (which is hierarchically organized, in a mutable and context-dependent fashion). It is unable to address suffering produced by conflict of duty—to define acceptable behavior when the situation compels conflicting behavioral response (when one listed moral prerequisite conflicts with another). The establishment of fixed law also limits capacity for judgment and choice, restricting adaptive flexibility, often dangerously, when "environmental alteration" makes such flexibility necessary:

Beware that no one lead you astray, saying, "Lo here!" or "Lo there!" For the Son of Man is within you.
    Follow after him.
Those who seek him will find him.
Go then and preach the gospel of the kingdom.
Do not lay down any rules beyond what I appointed for you, and do not give a law like the lawgiver
    lest you be constrained by it. [559] [560]

As a consequence of its intrinsic limitations, the list, which states what must not be, must give way, once fulfilled, to establishment of a more abstract form of moral order, predicated upon more flexible principles—which suggests what should be.

Descriptions of Christ's attempts to transcend the dangerous yet necessary limitations upon behavior imposed by adherence to the letter of the law take the form of *narrative about paradox*. It might be said that Christ presents (or is presented with) a series of Gordian knots—moral dilemmas—which emerge as an inevitable consequence of the structure of the list of laws. He plays a deadly serious game with the temporal representatives of then-traditional order, represented in the New Testament in the form of "Pharisees and scribes," teasing them into koanlike conundrums, which emerge as a consequence of their own beliefs:

And it came to pass on the second sabbath after the first, that he went through the corn fields; and his
disciples did eat, rubbing them in their hands.
And certain of the Pharisees said unto them, Why do ye that which is not lawful to do on the sabbath
days?[561]
And Jesus answering them said, Have ye not read so much as this, what David did, when himself was
an hungred, and they which were with him;
How he went into the house of God, and did take and eat the shewbread, and gave also to them that
were with him; which it is not lawful to eat but for the priests alone?[562]
And he said unto them, That the Son of man is Lord also of the sabbath.
And it came to pass also on another sabbath, that he entered into the synagogue and taught: and there
was a man whose right hand was withered.
And the scribes and Pharisees watched him, whether he would heal on the sabbath day; that they
might find an accusation against him.
But he knew their thoughts, and said to the man which had the withered hand, Rise up, and stand
forth in the midst. And he arose and stood forth.
Then said Jesus unto them, I will ask you one thing: Is it lawful on the sabbath days to do good, or to
do evil? to save life, or to destroy it?
And looking round upon them all, he said unto the man, Stretch forth thy hand. And he did so: and
his hand was restored whole as the other.
And they were filled with madness; and communed one with another what they might do to Jesus.
(Luke 6:1–11)

Similarly:

And it came to pass, as he went into the house of one of the chief Pharisees to eat bread on the sab-
bath day, that they watched him.
And, behold, there was a certain man before him which had the dropsy.
And Jesus answering spake unto the lawyers and Pharisees, saying, Is it lawful to heal on the sabbath
day?
And they held their peace. And he took him, and healed him, and let him go;

And answered them, saying, Which of you shall have an ass or an ox fallen into a pit, and will not straightway pull him out on the sabbath day?

And they could not answer him these things. (Luke 14: 1–6)

Also:

And as Jesus passed by, he saw a man which was blind from his birth.

And his disciples asked him, saying, Master, who did sin, this man, or his parents, that he was born blind?

Jesus answered, Neither hath this man sinned, nor his parents: but that the works of God should be made manifest in him.

I must work the works of him that sent me, while it is day: the night cometh, when no man can work.

As long as I am in the world, I am the light of the world.

When he had thus spoken, he spat on the ground, and made clay of the spittle, and he annointed the eyes of the blind man with the clay.

And said unto him, Go, wash in the pool of Siloam (which is by interpretation Sent). He went his way therefore, and washed, and came seeing.

The neighbours therefore, and they which had seen him that he was blind, said, Is not this he that sat and begged?

Some said, This is he: others said, He is like him: but he said, I am he.

Therefore they said unto him, How were thine eyes opened? He answered and said, A man that is called Jesus made clay, and annointed mine eyes, and said unto me, Go to the pool of Siloam, and wash: and I went and washed, and I received sight.

Then said they unto him, Where is he? He said, I know not.

They brought to the Pharisees him that aforetime was blind.

And it was the sabbath day when Jesus made the clay, and opened his eyes.

Then again the Pharisees also asked him how he had received his sight. He said unto them, He put clay upon mine eyes, and I washed, and do see.

Therefore said some of the Pharisees, This man is not of God, because he keepeth not the Sabbath day. Others said, How can a man that is a sinner do such miracles? And there was a division among them.

They say unto the blind man again, What sayest thou of him, that he hath opened thine eyes? He said, He is a prophet.

But the Jews did not believe concerning him, that he had been blind, and received his sight, until they called the parents of him that had received his sight.

And they asked them, saying, Is this your son, who ye say was born blind? how then doth he now see?

His parents answered them and said, We know that this is our son, and that he was born blind.

But by what means he now seeth, we know not; or who hath opened his eyes, we know not: he is of age; ask him: he shall speak for himself.

These words spake his parents, because they feared the Jews: for the Jews had agreed already, that if any man did confess that he was Christ, he should be put out of the synagogue.

Therefore said his parents, He is of age; ask him.

Then again called they the man that was blind, and said unto him, Give God the praise: we know that this man is a sinner.

He answered and said, Whether he is a sinner or no, I know not: one thing I know, that, whereas I was blind, now I see.

Then said they to him again, What did he to thee? how opened he thine eyes?

He answered them, I have told you already, and ye did not hear: wherefore would ye hear it again? will ye also be his disciples?

Then they reviled him, and said, Thou art his disciple; but we are Moses' disciples.

We know that God spake unto Moses: as for this fellow, we know not from whence he is. (John 9:1–29)

Also:

Then came together unto him the Pharisees, and certain of the scribes, which came from Jerusalem,

And when they saw some of his disciples eat bread with defiled, that is to say, with unwashen, hands, they found fault.

For the Pharisees, and all the Jews, except they wash their hands oft, eat not, holding the tradition of the elders.

And when they come from the market, except they wash, they eat not. And many other things there be, which they have received to hold, as the washing of cups, and pots, brasen vessels, and of tables.

Then the Pharisees and scribes asked him, Why walk not thy disciples according to the tradition of the elders, but eat bread with unwashen hands?

He answered and said unto them, Well hath Esias prophecied of you hypocrites, as it is written, This people honoureth me with their lips, but their heart is far from me.

Howbeit in vain do they worship me, teaching for doctrines the commandments of men. (Mark 7: 5–7)

Piaget—in what might be considered a veritable commentary on these stories—differentiates "morality of constraint" from "morality of cooperation,"[563] describing the former as a "system of rules"[564] that affective life makes use of to control behavior:[565]

Since he takes rules literally and things of good only in terms of obedience, the child will at first evaluate acts not in accordance with the motive that has prompted them but in terms of their exact conformity with established rules.[566]

Piaget associates morality of constraint with an earlier level of cognitive development—a level that nonetheless serves as a necessary precondition for further development. Piaget states: "For very young children, a rule is a sacred reality because it is traditional; for the older ones it depends upon mutual agreement."[567] Joseph Rychlak comments:

Younger children also are much harsher in assigning punishment to those who break the rules. They seem to want to emphasize the punishment for its own sake, whereas older children use punishment more to show the transgressor that a bond has been broken between people when a wrongdoing takes place. Piaget's value system considers rule by cooperation a more satisfactory equilibration in human relations than rule by authority. In order for a rule to work without authority pressures, there must be feelings of mutual respect among the persons who subscribe to the rule.[568] This necessarily brings affectivity into considerations of morality. Authoritarian constraint rules through feelings of anxiety and fear, but when mutual respect exists among people, a morality of cooperation can occur.[569]

Morality of tradition is not predicated upon the same presumptions as morality of coopera-tion. Rigid traditionalists assume that the answer to the question "what is the good?" can be—has been—answered permanently, and concretely, with the list of laws. Such a list is always insufficient, however, for the purposes of complete adaptation. Lao-Tzu can therefore say, with sufficient justification:

> The man who is truly wise and kind
> leaves nothing to be done,
> but he who only acts
> according to his nation's law
> leaves many things undone.[570]

Adherents of tradition rely on the attribution of superhuman value to ancestral figures and, equally to their current temporal and spiritual representatives. Those who embrace the morality of cooperation, by contrast, value the notion of "mutual respect"—which means simultaneous appreciation of equality and mutual value among individuals within (and, much more radically, between) social groups.

The behavior of any given social group—and, therefore, the value attributed to phenome-na that constitute the shared territory of the group—emerges as a consequence of the neces-sity of maintaining balance between opportunity for expression of individual desire and restriction of inter-individual conflict. Such balance, vital to maintenance of group stability, is established *long before* the "rules" governing such establishment can be modeled in episodic or semantic memory, from the phylo- and onto-genetic perspectives. Even the simplest social animals erect a dominance hierarchy, and behave "as if" according to principle. It is an error to presuppose, however, that simpler animals can abstractly represent either their behaviors—which is to say, form an imagistic model of them in imagination—or understand the "principles" that govern them. Likewise, socialized children, in complex human societies, embody the morality of their culture in their behavior long before they are able to abstractly represent or semantically describe rationale for that morality, and *before they can consciously (episodically or semantically) remember learning how to behave*. The same can be said for adults: the existence of morality—that intrinsic aspect of social behavior—long precedes rep-resentation of morality and rational description of grounds for its existence. Morality, at its most fundamental level, is an emergent property of social interaction, embodied in individual behavior, implicit in the value attributed to objects and situations, grounded (unconsciously) in procedural knowledge.

Two questions arise naturally from such a discussion: "is it possible to abstract out from observation of social interaction the 'rules' or patterns that characterize such interactions?" and, if so, "what might those 'rules' be?" Primeval group culture determines the nature of social interaction among the members of the group, and brings general expectation, predictability, to encounters between differentially desirable, powerful and dangerous individuals in that group. The mere fact of a stable hierarchy implies the existence of a complex procedural morality (and an implicit system of values). The behavior of social animals, within a hierarchy, constitutes *de facto* recognition of complex moral "principles," which might be regarded as the inevitable emergent properties of constant social interaction. It is very dangerous for the entire group if any of its members engage, routinely, in strenuous physical competition. Exhaustion—or outright elimination—of the power of the constituent members of the group means increased risk of attack from outside. Physical competition among social animals, necessary for dominance establishment, therefore generally has a ritualistic nature, and ends well before serious injury or death. Social animals have developed signals of submission, for example, which indicate their willingness to end the struggle for power. These signals are usually respected by the victor. The most powerful member of a given social group may come to dominate that group—at least in some circumstances—but the domination takes limited form. Even the most dominant of animals must act "as if"—*as if their expression of power is constrained by recognition of the necessity for maintenance of the group and the individuals who constitute and sustain it*. This constraint, partially manifested in social affection, provides the precondition for the emergence of complex abstract morality, which originates in innate and socialized procedural knowledge, which is "unconscious" in essence—that is, nonrepresentational, or undeclarable. It is not too difficult to recognize in this constraint the moral imperative "treat the weak as if they too are valuable"—as the Old Testament prophets insisted—or even—"love thy neighbour (even thy enemy) as thyself." Consider Christ's first sermon:

And he came to Nazareth, where he had been brought up: and, as his custom was, he went into the
    synagogue on the sabbath day, and stood up for to read.
And there was delivered unto him the book of the prophet Esasias. And when he had opened the
    book, he found the place where it was written,
The Spirit of the Lord is upon me, because he hath annointed me to preach the gospel to the poor; he
    hath sent me to heal the brokenhearted, to preach deliverance to the captives, and recovering of
    sight to the blind, to set at liberty them that are bruised,
To preach the acceptable year of the Lord.
And he closed the book, and he gave it again to the minister, and sat down. And the eyes of all them
    that were in the synagogue were fastened on him.
And he said unto them, This day is the scripture fulfilled in your ears.

And all bare him witness, and wondered at the gracious words which proceeded out of his mouth,
And they said, Is not this Joseph's son?

And he said unto them, Ye will surely say unto me this proverb, Physician, heal thyself: whatsoever we
have heard done in Capernaum, do also here in thy country.

And he said, Verily I say unto you, No prophet is accepted in his own country.

But I tell you of a truth, many widows were in Israel in the days of Elias, when the heaven was shut up
three years and six months, when great famine was throughout all the land;

But unto none of them was Elias sent, save unto Sarepta, a city of Sidon, unto a woman that was a
widow.

And many lepers were in Israel in the time of Eliseus the prophet; and none of them was cleansed,
saving Namaan the Syrian.

And all they in the synagogue, when they heard these things, were filled with wrath,

And rose up, and thrust him out of the city, and led him unto the brow of the hill whereon their city
was built, that they might cast him down headlong.

But he passing through the midst of them went his way,

And came down to Capernaum, a city of Galilee, and taught them on the sabbath days.

And they were astonished at his doctrine: for his word was with power. (Luke 4: 14–32)

also

Then Jesus went hence, and departed into the coasts of Tyre and Sidon.

And, behold, a woman of Canaan came out of the same coasts, and cried unto him, saying, Have
mercy on me, O Lord, thou Son of David; my daughter is grievously vexed with a devil.

But he answered her not a word. And his disciples came and besought him, saying, Send her away; for
she crieth after us.

But he answered and said, I am not sent but unto the lost sheep of the house of Israel.

Then came she and worshipped him, saying, Lord, help me.

But he answered and said, It is not meet to take the children's bread, and to cast it to dogs.

And she said, Truth, Lord: yet the dogs eat of the crumbs which fall from their masters' table.

Then Jesus answered and said unto here, O woman, great is thy faith: be it unto thee even as thou wilt.
And her daughter was made whole from that hour. (Matthew 15: 21–28)

The kingdom of heaven, as conceived of by Christ, was not only populated by the foreigner—an inclusion sufficiently unacceptable—but by all who were deemed superfluous or sinful according to the terms of prevailing morality: by the prostitute, the tax collector, the diseased, the insane and, most radically, by the enemy. This of course did not mean the end of morality—did not mean establishment of an anarchic "community" where everything was equal, and therefore equally valueless (where the practising and unrepentant torturer and the authentic saint share would be worthy of equal distinction)—but the portrayal of a state where the life of the past or the conditions of birth, no matter how wretched, did not finally determine the value of the present, or the possibilities of the future.

The extremely radical nature of this viewpoint profoundly disturbed the traditionalists in Christ's community. His example served as reproach to their actions; his philosophy, as threat to the integrity of their most dearly held positions. In consequence, they constantly attempted to trap him into making an irrevocably criminal or heretic statement. This strategy tended to backfire badly:

Then went the Pharisees, and took counsel how they might entangle him in his talk.

And they sent out unto him their disciples with the Herodians, saying, Master, we know that thou are true, and teachest the way of God in truth, neither carest thou for any man: for thou regardest not the person of men.

Tell us therefore, What thinkest thou? Is it lawful to give tribute unto Caesar, or not?

But Jesus perceived their wickedness, and said, Why tempt me, ye hypocrites?

Shew me the tribute money. And they brought unto him a penny.

And he saith unto them, Whose is this image and superscription?

They say unto him, Caesar's. Then saith he unto them, Render therefore unto Caesar the things which are Caesar's; and unto God the things that are God's.

When they heard these words, they marvelled, and left him, and went their way. (Matthew 22:15–22)

also

And as he spake, a certain Pharisee besought him to dine with him: and he went in, and sat down to meat.

And when the Pharisee saw it, he marvelled that he had not first washed before dinner.

And the Lord said unto him, Now do ye Pharisees make clean the outside of the cup and the platter; but your inward part is full of ravening and wickedness.

Ye fools, did not he that made that which is without make that which is within also?

But rather give alms of such things as ye have: and, behold, all things are clean unto you.

But woe unto you, Pharisees! for ye tithe mint and rue and all manner of herbs, and pass over judgment and the love of God: these ought ye to have done, and not to leave the other undone.

Woe unto you, Pharisees! for ye love the uppermost seats in the synagogues, and greetings in the markets.

Woe unto you, scribes and Pharisees, hypocrites! for ye are as graves which appear not, and the men that walk over them are not aware of them.

Then answered one of the lawyers, and said unto him, Master, thus saying thou reproachest us also.

And he said, Woe unto you also, ye lawyers! for ye lade men with burdens grievous to be borne, and ye yourselves touch not the burdens with one of your fingers.

Woe unto you! for ye build the sepulchres of the prophets, and your fathers killed them.

Truly ye bear witness that ye allow the deeds of your fathers: for they indeed killed them, and ye build their sepulchres.

Therefore also said the wisdom of God, I will send them prophets and apostles, and some of them they shall slay and persecute:

That the blood of all the prophets, which was shed from the foundation of the world, may be required of this generation;

From the blood of Abel unto the blood of Zacharias, which perished between the alter and the temple: verily, I say unto you, it shall be required of this generation.

Woe unto you lawyers! for ye have taken away the key of knowledge: ye entered not in yourselves, and them that were entering in ye hindered.

And as he said these things unto them, the scribes and the Pharisees began to urge him vehemently, and to provoke him to speak of many things:

Laying wait for him, and seeking to catch something out of his mouth, that they might accuse him. (Luke 11:33–54)

Christ's ability to weave his way through verbal traps continually inflamed the Pharisees, who attempted, with ever more sophistry, to corner him:

Then one of them, which was a lawyer, asked him a question, tempting him, and saying,

Master, which is the great commandment in the law?

Jesus said unto him, Thou shalt love the Lord thy God with all thy heart, and with all thy soul, and with all thy mind.

This is the first and great commandment.

And the second is like unto it, Thou shalt love thy neighbour as thyself.

On these two commandments hang all the law and the prophets. (Matthew 22:35–40)

However, the power of Christ's unexpected answers—in conjunction with his evident mastery of traditional knowledge (Matthew 22: 42–45)—temporarily silenced his critics:

And no man was able to answer him a word, neither durst any man from that day forth ask him any more questions. (Matthew 22:46)

Christ's replies signified transition of morality from reliance on tradition to reliance on individual conscience—from rule of law to rule of spirit—from prohibition to exhortation. To love God means to listen to the voice of truth[571] and to act in accordance with its messages; to love thy neighbour, *as thy self*. This means, not merely to be pleasant, polite and friendly, but to attribute to the other a value equivalent to the value of the self—which, despite outward appearances, is a representative of God—and to act in consequence of this valuation. This transition means establishment of an active dynamic balance of competing subjectively based motivational demands, while maintaining and creatively modifying the social and natural environment. It means fulfillment of personal and interpersonal needs in accordance with meta-moral principles, rather than in accordance with the demands of power or dogmatic tradition. Thus, the process by which tradition is generated in brought into inevitable contrast with tradition itself:

Suppose ye that I am come to give peace on earth? I tell you, Nay; but rather division:
For from henceforth there shall be five in one house divided, three against two, and two against three.
The father shall be divided against the son, and the son against the father: the mother against the

> daughter, and the daughter against the mother; the mother in law against her daughter in law,
> and the daughter in law against her mother in law. (Luke 12: 51–53)

This is most truly death of unthinking adherence to authority—for as in the archaic society, the past rules:

In normal times, when culture is stable and the paternal canon remains in force for generations, the father-son relationship consists in handing down these values to the son and impressing them upon him, after he has passed the tests of initiation in puberty. Such times, and the psychology that goes with them, are distinguished by the fact that there is no father-son problem, or only the barest suggestion of one. We must not be deceived by the different experience of our own "extraordinary" age. The monotonous sameness of fathers and sons is the rule in a stable culture. This sameness only means that the paternal canon of rites and institutions which make the youth an adult and the father an elder holds indisputed sway, so that the young man undergoes his prescribed transition to adulthood just as naturally as the father undergoes his to old age.

   There is, however, one exception to this, and the exception is the creative individual—the hero. As Barlach says, the hero has to "awaken the sleeping images of the future which can come forth from the night, in order to give the world a new and better face." This necessarily makes him a breaker of the old law. He is the enemy of the old ruling system, of the old cultural values, and the existing court of conscience, and so he necessarily comes into conflict with the fathers. In this conflict the "inner voice," the command of the transpersonal father or father archetype who wants the world to change, collides with the personal father who speaks for the old law. We know this conflict best from the Bible story of Jehovah's command to Abraham: "Get thee out of thy country, and from thy kindred, and from thy father's house, unto a land that I will show thee (Genesis 12:1), which the midrash (Bin Gorion, Sagen der Juden, Vol. II, "Die Erzvater," XI) interprets as meaning that Abraham is to destory the gods of his father. The message of Jesus is only an extension of the same conflict, and it repeats itself in every revolution. Whether the new picture of God and the world conflicts with an old picture, or with the personal father, is unimportant, for the father always represents the old order and hence also the old picture current in his cultural canon.[572]

   What principle is rule of spirit, rather than law, predicated upon? Respect for the innately heroic nature of man. The "unconscious" archaic man mimics particular adaptive behaviors—integrated, however, into a procedural structure containing all other adaptive behaviors, capable of compelling imitation, and accompanied by episodic/semantic representation, in myth. Pre-experimental cultures regard the act of initial establishment of adaptive behavior as divine first because it follows an archetypal and therefore transpersonal pattern—that governing creative exploration—and second because it compels imitation, and therefore appears possessed of power. All behaviors that change history, and compel imitation, follow the same pattern—that of the divine hero, the embodiment of creative human potential. For the primitive individual, it is the consequences of such heroism and the particular acts themselves that constitute the essence of the past. The process of imitation and abstracted variants thereof, however, allow for the nature of this essence to be continually clarified—until, finally, representation of abstracted but specific heroic actions give way to representation of the process of heroism, *per se*. At this point, it becomes possible for the creative individual to mimic, consciously incarnate, the process of world-redemption itself.

Law is a necessary precondition to salvation, so to speak; necessary, but insufficient. Law provides the borders that limit chaos, and allows for the protected maturation of the individual. Law disciplines possibility, and allows the disciplined individual to bring his or her potentialities—those intrapsychic spirits—under voluntary control. The law allows for the application of such potentiality to the task of creative and courageous existence—allows spiritual water controlled flow into the valley of the shadow of death. Law held as absolute, however, puts man in the position of the eternal adolescent, dependent upon the father for every vital decision; removes the responsibility for action from the individual, and therefore prevents him or her from discovering the potential grandeur of the soul. Life without law remains chaotic, affectively intolerable. Life that is pure law becomes sterile, equally unbearable. The domination of chaos or sterility equally breeds murderous resentment and hatred.

Christ presented the kingdom of heaven (the archetypal goal) as a spiritual kingdom, which is to say, a psychological, then interpersonal, state. This state differed from the hypothetical promised land described in the Old Testament, in a number of vitally important manners. First, its construction was a matter of voluntarily chosen alteration in personal attitude and outlook, rather than a culmination of material labor and natural resource. Second, it was predicated upon revolutionary and paradoxical reconceptualization of the nature of the goal—of paradise itself. Christ's life and words—as archetypal exemplars of the heroic manner of being—place explicit stress on the process of life, rather than upon its products. The point of a symphony is not its final note, although it proceeds inexorably to that end. Likewise, the purpose of human existence is not the establishment of some static, perfect manner of being—man would find such perfection intolerable, as Dostoyevsky was at pains to illustrate. Rather, human purpose is generation of the ability to concentrate on the innately interesting and affectively significant events of the present, with sufficient consciousness and clarity, to render concern about the past and future unnecessary. "Consider the lilies of the valley" says Christ,

how they grow; they toil not, neither do they spin:
And yet I say unto you, That even Solomon in all his glory was not arrayed like one of these.
Wherefore, if God so clothe the grass of the field, which to day is, and to morrow is cast into the oven, shall he not much more clothe you, O ye of little faith?
Therefore take not thought, saying What shall we eat? or, What shall we drink? or, Wherewithal shall we be clothed?
for your Heavenly Father knoweth that ye have need of all these things.

But seek ye first the kingdom of God, and his righteousness; and all these things shall be added unto
    you.
Take therefore no thought for the morrow: for the morrow shall take thought for the things of itself.
    Sufficient unto the day is the evil thereof. (Matthew 6: 28–34)

"Sufficient unto the day is the evil thereof" does not mean "live the life of the grasshopper
instead of the ant, sing in the summer and starve in the winter," but concentrate on the task
at hand. Respond to error, when committed. Pay attention, and when your behavior produces a consequence you find intolerable, modify it—*no matter what it takes to produce such a
modification.* Allow consciousness of your present insufficiency to maintain a constant presence, so that you do not commit the error of pride, and become unbending, rigid, and dead
in spirit. Live in full recognition of your capacity for error—and your capacity to rectify such
error. Advance in confidence and faith; do not shrink back, avoiding inevitable contact with
the terrible unknown, to live in a hole that grows smaller and darker.

   The significance of the Christian passion is the transformation of the process by which
the goal is to be attained into the goal itself: the making of the "imitation of Christ"—the
duty of every Christian citizen—into the embodiment of courageous, truthful, individually
unique existence:

Then said Jesus unto his disciples, If any man will come after me, let him deny himself, and take up
    his cross, and follow me.
For whosoever will save his life shall lose it; and whosoever will lose his life for my sake shall find it.
For what is a man profited, if he shall gain the whole world, and lose his own soul? or what shall a
    man give in exchange for his soul? (Matthew 16:24–26)

Christ said, put truth and regard for the divine in humanity above all else, and everything
you need will follow—not everything you think you need, as such thought is fallible, and
cannot serve as an accurate guide, but everything actually necessary to render acutely
(self)conscious life bearable, without protection of delusion and necessary recourse to deceit,
avoidance or suppression, and violence. This idea is presented in imagistic form in *Figure 61:
World-Tree of Death and Redemption,*[573] which portrays the "host" as the second fruit of the
world-tree. Ingestion of the first fruit produced the fall; ingestion of the second redeems
those who have fallen. The negative feminine, in the form of Eve, hands out the apple, in the
form of a skull; the positive feminine, in the form of the church, distributes the wheaten
wafers that characterize the redeemer. The incorporation of "Christ's mystic body," during
the ritual of the mass, is dramatic representation of the idea that the hero must be incorporated into each individual—that everyone must partake of the essence of the savior.

   Existence characterized by such essence takes place, from the Oriental viewpoint, on the
path of meaning, in Tao, balanced on the razor's edge between mythic masculine and mythic
feminine—balanced between the potentially stultifying safety of order, and the inherently

Figure 61: World-Tree of Death and Redemption

destructive possibility of chaos. Such existence allows for introduction of sufficiently bearable meaning into blessed security; makes every individual a stalwart guardian of tradition and an intrepid explorer of the unknown; ensures simultaneous advancement and maintenance of stable, dynamic social existence; and places the individual firmly on the path to intrapsychic integrity and spiritual peace:

Therefore whosoever heareth these sayings of mine, and doeth them, I will liken him to a wise man,
    which built his house upon a rock;
And the rain descended, and the floods came, and the winds blew, and beat upon that house; and it
    fell not: for it was founded upon a rock. (Matthew 7:24–25)

The Alchemical Procedure and the Philosopher's Stone

Introductory Note

Part One

The western alchemists followed the scenario, known already in the Hellenistic period, of the four phases of the process of transmutation: that is, of the procurement of the Philosopher's Stone. The first phase (the *nigredo*)—the regression to the fluid state of matter—corresponds to the death of the alchemist. According to Paracelsus, "he who would enter the Kingdom of God must first enter with his body into his mother and there die." The "mother" is the *prima materia*, the *massa confusa*, the *abyssus*.[574] Certain texts emphasize the synchronism between the *opus alchymicum* and the intimate experience of the adept. "Things are rendered perfect by their similars and that is why the operator must take part in the operation."[575] "Transform yourself from dead stones into living philosopher's stones," writes Dorn. According to Gichtel, "we not only receive a new soul with this regeneration but also a new Body. The Body is extracted from the divine word or from the heavenly Sophia." That it is not solely a question of laboratory operations is proven by the insistence on the virtues and qualities of the alchemist: the latter must be healthy, humble, patient, chaste; he must be of free spirit and in harmony with his work; he must both work and meditate.

For our purposes, it will be unnecessary to summarize the other phases of the *opus*. Let us note, however, the paradoxical character of the *materia prima* and of the Philosopher's Stone. According to the alchemists, they both are to be found everywhere, and under all forms; and they are designated by hundreds of terms. To cite only a text of 1526, the Stone "is familiar to all men, both young and old; it is found in the country, in the village, and in the town, in all things created by God; yet it is despised by all. Rich and poor handle it every day. It is thrown into the street by servant maids. Children play with it. Yet no one prizes it, though, next to the human soul, it is the most beautiful and most precious thing upon earth [and has power to pull down kings and princes. Nevertheless, it is esteemed the vilest and meanest of earthly things....]"[576] It is truly a question of a "secret language" that is at once both the expression of experiences otherwise intransmissible by the medium of ordinary language, and the cryptic communication of the hidden meaning of symbols.

The Stone makes possible the identification of opposites.[577] It purifies and "perfects" the metals. It is the Arabic alchemists who imparted therapeutic virtues to the Stone, and it is through the intermediary of Arabic alchemy that the concept of the *Elixir vitae* arrived in the West.[578] Roger Bacon speaks of a "medicine which makes the impurities and all the corruptions of the most base metal disappear," and which can prolong human life for several centuries. According to Arnold of Villanova, the Stone cures all ills and makes the old young.

As regards the process for the transmutation of metals into gold, attested already in Chinese alchemy, it accelerates the temporal rhythm and thus contributes to the work of nature. As is written in the *Summa Perfectionis*, an alchemical work of the sixteenth century, "what Nature cannot perfect in a vast space of time we can achieve in a short space of time by our art." The same idea is expounded by Ben Jonson in his play *The Alchemist* (Act 2, Scene 2). The alchemist affirms that "lead and other metals . . . would be gold if they had time"; and another character adds: "And that our art doth further."[579] In other words, the alchemist substitutes himself for Time.[580]

The principles of traditional alchemy—that is, the growth of minerals, the transmutation of metals, the Elixir, and the obligation to secrecy—were not contested in the period of the Renaissance and the Reformation.[581] However, the horizon of medieval alchemy was modified under the impact of Neoplatonism and Hermeticism. The certitude that alchemy can second the work of Nature received a christological significance. The alchemists now affirmed that just as Christ had redeemed humanity by his death and resurrection, so the *opus alchymicum* could assure the redemption of Nature. Heinrich Khunrath, a celebrated Hermeticist of the sixteenth century, identified the Philosopher's Stone with Jesus Christ, the "Son of the Macrocosm"; he thought besides that the discovery of the Stone would unveil the true nature of the macrocosm, in the same way that Christ had brought spiritual plenitude to man—that is, to the microcosm. The conviction that the *opus alchymicum* could save both man and Nature prolonged the nostalgia for a radical *renovatio*, a nostalgia which had haunted western Christianity since Joachim of Floris.[582]

Carl Jung devoted a tremendous amount of attention to the writings of the alchemists in the latter part of his life. These efforts merely added fuel to the fire of those who had branded him eccentric, because of his interest in the psychology of religion (which is, after all, a fundamental aspect of human psychology and culture). Even the Pulitzer Prize–winning sociologist Ernest Becker, who was favorably (and critically) predisposed to the claims of psychoanalytic thought, stated, "I can't see that all [Jung's] tomes on alchemy added one bit to the weight of his psychoanalytic insight."[583]

Many people—some with an outstanding academic reputation—have cautioned me against discussing Jung, warned me about even mentioning his name in the academic context. This warning was presented, no doubt, with my best professional interests in mind. I once read a story about Paul Ricoeur, the French philosopher and literary critic, which may be apocryphal. Someone mentioned the specific relevance of Jung's work to Ricoeur's field of inquiry. Ricoeur replied, "I haven't read Jung. He's on the Index in France." This ironical response was, of course, made in reference to the Catholic Index of books—a listing of readings forbidden to devout followers of that creed.

I have never met someone, however, who actually understood what Jung was talking about and who was simultaneously able to provide valid criticism of his ideas. Often, Jung's notions are confused with Freud's—insofar as Freud's are understood. Freud himself certainly did not make this error. It was in fact Freud's apprehension of Jung's profound and irreconcilable differences in thought that led to their professional and private alienation.[584] Jung's ideas are *not* primarily Freudian. He places little emphasis on sexuality or on the role of past trauma in determining present mental state. He rejected the idea of the Oedipus complex (actually, he reinterpreted that complex in a much more compelling and complete manner). He viewed religion not as mere neurotic defense against anxiety, but as a profoundly important means of adaptation. It is much more accurate to view him as an intellectual descendant of Goethe and Nietzsche—influenced in his development, to be sure, by the idea of the unconscious—than as a Freudian "disciple."[585] Jung in fact spent much of his life answering, and attempting to answer, Nietzsche's questions about morality.

Furthermore, Jung was *not* a "mystic." He merely delved into areas that were forbidden, because of their religious association, to devout scientists, and was possessed of sufficient intellect and education to do so. It is incorrect, and evidence of one-sided thinking, to label him pejoratively. It is incorrect because Jung was, in fact, an experimental scientist of no small ability, particularly at the beginning of his career. Many of the word-association tests he helped pioneer are still used extensively, with some technical modifications (and little recognition of original source), in the fields of cognitive neuroscience and social psychology. The boxing and filing away of Jung is one-sided because experimental procedure constitutes, at best, one pole of the bipolar scientific process. A well-designed experiment allows for the testing of ideas, when it is undertaken properly. However, ideas to test must be generated—a truism often overlooked in the course of modern academic education. It was at this endeavor that Jung excelled. Some might object: his ideas cannot be tested. But they have been: the card-classification experiment by Jerome Bruner, described previously,[384] provides a classic and striking example (although the results of that experiment have not generally been interpreted from the perspective of Jung's thought). Furthermore, one axis of the personality dichotomy he proposed—that of introversion-extroversion—has stood up well, appears robust, in the face of repeated experimental inquiry.[586] In addition, the "unconscious" is clearly full of "complexes"—although now they go by different names.[587] Perhaps we will become sophisticated enough in the future, in our ability to experiment, and in our understanding of Jung's ideas, to test more of them.

Jung was primarily a physician, which meant that he was concerned with the promotion of mental health. He believed that such promotion was impossible—perhaps even counterproductive—in the absence of comprehension of value, and of the processes by which value is generated. His investigation into the nature of value led him to consideration of fantasy and myth. The world of value is a world in imagination, the internalized result of the historically determined social contract that provides fixed determination of affective and motivational significance. Apprehension of this fact led Jung to analysis of the fantasies generated by his seriously disturbed psychiatric patients, and comparison of these fantasies—which he was unwilling to define, *a priori*, as meaningless—with ideas generated by religious mystics from a variety of "primitive" and sophisticated cultures, with a vast body of literary productions in the Eastern and Western traditions, with imagery generated in dreams (more than 25,000 dreams, by his own estimate) and by diligent investigation into alchemical symbolism. This cross-cultural and multidisciplinary approach to the problem of value seems at least empirical, if not experimental—and remains eminently reasonable, in the absence of more appropriate methodology. (In fact, the prominent sociobiologist E.O. Wilson has recently recommended adoption of such a "cross-level" analytical procedure in the guise of "consilience"—to unite the natural sciences, the social sciences and the humanities.[588])

Jung's ideas—particularly his "alchemical" ideas—have been inappropriately, unfairly and dangerously ignored. They have been ignored because his students were outside the academic mainstream (and, perhaps, because they were frequently women). They have been ignored because they present a serious challenge—an absolutely fatal challenge, in my estimation— to Freudian psychoanalytic preconceptions. They have been overlooked because Jung took the frightening and mysterious statements of religion seriously. He presumed that such statements, which guided human adaptation successfully for thousands of years, had some significance, some meaning. Jung's ideas have remained unexamined because psychology, the youngest, most rational, and most deterministic of sciences, is most afraid of religion. They have been ignored, additionally, because they are difficult to understand, from the conceptual and affective points of view. What the ideas are is hard to specify, initially; what they signify, once understood, is emotionally challenging. Jung essentially described the nature of the "language" of imagination, that ancient process—of narrative, of the episodic memory system—which he thought of, fundamentally, as the *collective unconscious*. Comprehension of this language is perhaps more difficult than development of fluency in an foreign language, because such comprehension necessarily and inevitably alters modern moral presumption. It is this latter point that constitutes the core rationale for dismissal of Jung's ideas. Jung was no less revolutionary, from the moral perspective, than Martin Luther; he may reasonably be considered a figure in the tradition of Luther. Furthermore, moral revolution is the most dreadfully uncomfortable of all intrapsychic and social processes. It is the frightening *content* of Jung's thought that has led most fundamentally to its rejection.

Jung essentially discovered, in the course of his analysis of alchemy, the nature of the general human pattern of adaptation, and the characteristic expression of that pattern, in fantasy, and affect. Specific representation of this pattern, in the narrower domain of scientific endeavor, was outlined much later—to much wider comprehension and academic acclaim— by Thomas Kuhn. Jung's student Marie-Louise von Franz—who provided a cogent summary of Jung's complex alchemical ideas—states:

If you read the history of the development of chemistry and particularly of physics, you will see that even ... exact natural sciences [such as chemistry and physics] could not, and still cannot, avoid basing their thought systems on certain hypotheses. In classical physics, up to the end of the 18th century, one of the working hypotheses, arrived at either unconsciously, or half-consciously, was that space had three dimensions, an idea which was never questioned. The fact was always accepted, and perspective drawings of physical events, diagrams, or experiments, were always in accordance with that theory. Only when this theory is abandoned does one wonder how such a thing could have ever been believed.

How did one come by such an idea? Why were we so caught that nobody ever doubted or even discussed the matter? It was accepted as a self-evident fact, but what was at the root of it? Johannes Kepler, one of the fathers of modern or classical physics, said that naturally space must have three dimensions because of the Trinity! So our readiness to believe that space has three dimensions is a more recent offspring of the Christian trinitarian idea.

Further, until now the European scientific mind has been possessed by the idea of causality, an idea hitherto accepted without question: everything was causal, and the scientific attitude was that investigations should be made with that premise in mind, for there must be a rational cause for everything. If something appeared to be irrational, it was believed that its cause was not yet known. Why were we so dominated by that idea? One of the chief fathers of natural sciences—and a great protagonist of the absoluteness of the idea of causality—was the French philosopher Descartes, and he based his belief on the immutability of God. The doctrine of this immutability of God is one of the Christian tenets: the Divinity is unchanging, there must be no internal contradictions in God, or new ideas or conceptions. That is the basis of the idea of causality! From the time of Descartes onwards this seemed so self-evident to all physicists that there was no question about it. Science had merely to investigate the causes, and we still believe this. If something falls down then one must find out why—the wind must have blown it, or something like that, and if no reason is discovered I am sure that half of you will say that we do not yet know the cause, but that there must be one! Our archetypal prejudices are so strong that one cannot defend oneself against them, they just catch us.

The late physicist, Professor Wolfgang Pauli, frequently demonstrated the extent to which modern physical sciences are in a way rooted in archetypal ideas. For instance, the idea of causality as formulated by Descartes is responsible for enormous progress in the investigation of light, of biological phenomena, and so on, but that thing which promotes knowledge becomes its prison. Great discoveries in natural sciences are generally due to the appearance of a new archetypal model by which reality can be described; that usually precedes big developments, for there is now a model which enables a much fuller explanation than was hitherto possible.

So science has progressed, but still, any model becomes a cage, for if one comes across phenomena difficult to explain, then instead of being adaptable and saying that the phenomena do not conform to the model and that a new hypothesis must be found, one clings to one's hypotheses with a kind of emotional conviction, and cannot be objective. Why shouldn't there be more than three dimensions, why not investigate and see where we get? But that people could not do.

I remember a very good illustration given by one of Pauli's pupils. You know that the theory of ether played a great role in the 17th and 18th centuries—namely, that there was a kind of great air-like pneuma in the cosmos in which light existed, etc. One day when a physicist at a Congress proved that the theory of ether was quite unnecessary, an old man with a white beard got up and in a quavering voice said: "If ether does not exist, then everything is gone"! This old man had unconsciously projected his idea of God into ether. Ether was his god, and if he did not have that then there was nothing left. The man was naive enough to speak of his ideas, but all natural scientists have ultimate models of reality in which they believe, just like the Holy Ghost.

It is a question of belief, not of science, and therefore something which cannot be discussed, and people get excited and fanatical if you present them with a fact which does not fit the frame.[589]

Also:

So the archetype is the promoter of ideas and is also responsible for the emotional restrictions which prevent the renunciation of earlier theories. It is really only a detail or specific aspect of what happens everywhere in life, for we could not recognize anything without projection; but it is also the main obstacle to the truth. If one meets an unknown woman, it is not possible to make contact without projecting something; you must make a hypothesis, which of course is done quite unconsciously: the woman is elderly and is probably a kind of mother figure, and a normal human being, etc. You make assumptions and then you have a bridge. When you know the person better, then many earlier assumptions must be discarded and you must admit that your conclusions were incorrect. Unless this is done, then you are hampered in your contact.

At first, one has to project, or there is no contact; but then one should be able to correct the projection, and it is the same not only as regards human beings, but everything else also. The projection apparatus must of necessity work in us. Nothing can even be seen without the unconscious projection factor. That is why, according to Indian philosophy, the whole of reality is a projection—which it is, in a subjective matter of speaking.[590]

The idea of projection—that is, the idea that systems of thought have "unconscious" axioms—is clearly related to the notion of "paradigmatic thinking," as outlined by Kuhn, to wide general acclaim. Jung described the psychological consequences of paradigmatic thinking in great detail as well. He first posed the question—"What happens to the (paradigmatic) representational structure in someone's mind (in the human psyche, in human society) when anomalous information, of revolutionary import, is finally accepted as valid?"—and then answered it (my summary): "What happens has a pattern; the pattern has a biological, even genetic, basis, which finds its expression in fantasy; such fantasy provides subject material for myth and religion. The propositions of myth and religion, in turn, help guide and stabilize revolutionary human adaptation." These answers have been rejected prematurely and without sufficient consideration.

Part Two

> Where is what you most want to be found?
> Where you are least likely to look.
>
> "In sterquiliniis invenitur"[591]

King Arthur's knights sit at a round table, because they are all equal. They set off to look for the Holy Grail—which is a symbol of salvation, container of the "nourishing" blood of Christ, keeper of redemption. Each knight leaves on his quest, individually. Each knight enters the forest, to begin his search, at the point that looks darkest to him.

When I was about halfway through writing this manuscript, I went to visit my sister-in-law and her family. She had a son—my nephew—who was about five years old, very verbal and intelligent. He was deeply immersed in a pretend world and liked to dress up as a knight, with a plastic helmet and sword.

He was happy during the day, to all appearances, but did not sleep well, and had been having nightmares for some time. He would regularly scream for his mom in the middle of the night, and appeared quite agitated by whatever was going on in his imagination.

I asked him one morning after he had woken up what he had dreamed about. He told me, in the presence of his family, that dwarflike, beaked creatures who came up to his knees had been jumping up at him and biting him. Each creature was covered with hair and grease, and had a cross shaved in the hair on the top of its head. The dream also featured a dragon, who breathed fire. After the dragon exhaled, the fire turned into the dwarves, who multiplied endlessly, with each breath. He told the dream in a very serious voice to his parents and to my wife and me, and we were shocked by its graphic imagery and horror.

The dream occurred at a transition point in my nephew's life. He was leaving his mother to go to kindergarten, and was joining the social world. The dragon, of course, served as symbol for the source of fear itself—the unknown, the *uroboros*—while the dwarves were individual things to be afraid of, particular manifestations of the general unknown.

I asked him, "What could you do about this dragon?"

He said, without hesitation, and with considerable excitement, "I would take my dad, and we would go after the dragon. I would jump on its head, and poke out its eyes with my sword. Then I would go down its throat, to where the fire came out. I would cut out the box the fire came from, and make a shield from it."

I thought this was a remarkable answer. He had reproduced an archaic hero myth, in perfect form. The idea of making a shield from the firebox was nothing short of brilliant. This gave him the power of the dragon to use against the dragon.

His nightmares ended then and did not return, even though he had been suffering from them almost every night for a number of months. I asked his mother about his dreams, more than a year later, and she reported no further disturbance.

The little boy, guided by his imagination, adopted identification with the hero, and faced his worst nightmare. If we are to thrive, individually and socially, each of us must do the same. Our great technological power makes the consequences of our individual errors and weaknesses increasingly serious; if we wish to continually expand our power, we must also continually expand our wisdom. This is, unfortunately, a terrible thing to ask.

"*In sterquiliniis invenitur*"—in filth it will be found. This is perhaps the prime "alchemical" dictum. What you need most is always to be found where you least wish to look. This is really a matter of definition. The more profound the error, the more difficult the revolution—the more fear and uncertainly released as a consequence of restructuring. The things that are most informative are also frequently most painful. Under such circumstances, it is easy to run away. The act of running away, however, transforms the ambivalent unknown into that which is too terrifying to face. Acceptance of anomalous information brings terror and possibility, revolution and transformation. Rejection of unbearable fact stifles adaptation and strangles life. We choose one path or another at every decision point in our lives, and emerge as the sum total of our choices. In rejecting our errors, we gain short-term security—but throw away our identity with the process that allows us to transcend our weaknesses and tolerate our painfully limited lives:

There was a good man who owned a vineyard. He leased it to tenant farmers so that they might work it and he might collect the produce from them. He sent his servant so that the tenants might give him the produce of the vineyard. They seized his servant and beat him, all but killing him. The servant went back and told his master. The master said, "Perhaps they did not recognize him." He sent another servant. The tenants beat this one as well. Then the owner sent his son and said, "Perhaps they will show respect to my son." Because the tenants knew that it was he who was heir to the vineyard, they seized him and killed him. Let him who has ears hear.

Jesus said, "Show me the stone which the builders have rejected. That one is the cornerstone."[592]

Face what you reject, accept what you refuse to acknowledge, and you will find the treasure that the dragon guards.

The "Material World" as Archaic "Locus of the Unknown"

> All these myth pictures represent a drama of the human psyche on the
> further side of consciousness, showing man as both the one to be
> redeemed and the redeemer. The first formulation is Christian, the sec-
> ond alchemical. In the first case man attributes the need of redemption
> to himself and leaves the work of redemption ... to the autonomous
> divine figure; in the latter case man takes upon himself the redeeming
> opus, and attributes the state of suffering and consequent need of
> redemption to the anima mundi (world spirit) imprisoned in matter.[593]

Alchemy can be most simply understood as the attempt to produce the philosopher's stone—the lapis philosophorum. The lapis philosophorum had the ability to turn "base" metals into gold; furthermore, it conferred upon its bearer immortal life, spiritual peace and good health. The alchemical "procedure" stretched some twenty centuries, in the West, coming to an end with Newton; it had an equally lengthy and elaborate history in the Orient.

It is impossible to understand the essence of alchemical thought—or its relevance for modern psychology—without entering into the categorical system of the alchemist. The "stuff" with which the alchemist worked, although bearing the same name, was only vaguely akin to our modern matter. There are many ways to cut the world up, and they are not necessarily commensurate. Much of what the alchemist considered "thing" we would not think of as characteristic of the objective world; furthermore, what he considered unitary we would think of as evidently diverse. There are two major reasons for this difference in opinion.

First: the categorical system used to parse up the world derives its nature in large part from the nature of the end toward which activity is currently devoted. The ends pursued by the alchemist were by no means identical to those considered worthwhile today. In large part, they were much more comprehensive (the "perfection of nature"); in addition, they were "contaminated" with psychological formulations (the "redemption" of "corrupt" matter). Insofar as the alchemical procedure was psychological—that is, driven by apprehension of an "ideal state"—the categories it produced were *evaluative*. Phenomena that emerge in the course of goal-directed behavior are classified most fundamentally with regard to their relevance, or irrelevance, to that end. Those that are relevant are further discriminated into those that are useful and "good," and those that exist as impediments and are "bad." Since our behavior is motivated—since it serves to regulate our emotions—it is very difficult to construct a classification system whose elements are devoid of evaluative significance. It is only since the emergence of strict empirical methodology that such construction has been made possible. This means that pre-experimental systems of classification such as those employed in the alchemical procedure include evaluative appraisal, even when they consist of terms such as "matter" or "gold" that appear familiar to us.

Second: it seems that the more poorly something has been explored, the broader the category used to "encapsulate" or describe it. As exploration proceeds, finer discrimination becomes possible. Apparently unitary things fall apart, in this manner, into their previously implicit constituent elements (as nature is "carved at her joints"). We no longer consider the traditional four elements of the world, for example—fire, water, earth and air—either as irreducible elements or even as categories extant at the same level of analysis. Further investigation has reconfigured our systems of classification; we have transformed the comparative simple "material world" of our ancestors into something much more complex, useful and diverse. We believe, in consequence, that the primordial elements of the world were not really elements at all (failing to realize that an element is a tool, and that an incompletely fashioned tool is still much better than no tool at all).

The overwhelmingly evaluative dimension of pre-experimental classification, in combination with relatively poor capacity for discrimination, produced archaic categories of great generality—from the modern perspective. We can identify many "discriminable phenomena" within each of these categories, as a consequence of the centuries of increasingly efficient exploration that separate us from our medieval and pre-medieval forebears. Our viewpoint has in fact changed to such a degree that our use of the same word is in many cases only a historical accident. We might therefore make this discussion more concrete by first examining the "matter" of the alchemist, and by comparing it to what we think of as matter.

Alchemical matter was the "stuff" of which experience was made—and more: the stuff of which the experiencing creature was made. This "primal element" was something much more akin to "information" in the modern sense (or to Tao, from the Oriental perspective);

something like matter in the phrases "that *matters*" [594] (that *makes a difference*, that *we care about*, that *cannot be ignored*, that is *informative)* or "what is the *matter?*" We derive "information" as a consequence of our exploratory behavior, undertaken in the "unknown," attending to *things that matter;* from that information, we build ourselves (our behaviors and schemas of representation) and the "world," as experienced. As Piaget states:

Knowledge does not begin in the I, and it does not begin in the object; it begins in the interactions. . . . then there is a reciprocal and simultaneous construction of the subject on the one hand and the object on the other.[595]

The primal element of alchemy was something embedded or implicit in the world: something often hidden that could emerge unexpectedly. This unexpected emergence can be regarded as the "capacity" of the object to "transcend" its categorical representation (to "become" something new) as a consequence of its position in a new situation, or its "reaction" to a new exploratory procedure. This new thing "announces itself" first in terms of the affect it generates: failure of the previously understood (previously categorized) thing to behave as predicted elicits emotion from the observer. This is the "spirit of transformation" making itself manifest. The emotion so generated—fear/hope—may produce exploratory behavior, designed to specify the "new" properties of the transforming object. These new properties then become incorporated into the previous categorization system—become "attributes" now seen as "in the same class"; alternatively, the newly transformed substance may have to "shift categories" because it is now seen as so much different from "what it was." (The former case constitutes a normal shift, of course; the latter is revolutionary.)

When a novel thing has been explored, and placed within a certain sociohistorically determined context, it has been classified in accordance with its currently evident motivational status: promise, threat, satisfaction, punishment (or none of the above), as determined, situationally. This is evidently true with regards to the classification system of the individual animal, who cannot derive an empirical model of reality, because it cannot communicate—but equally true with regard to man, whose capacity for abstraction has blurred the essential nature and purpose of classification. What a thing *is* is most fundamentally its motivational significance—its relevance for the attainment of some affectively significant goal. Classification of the phenomenon (which means, determination of how to act in its presence) restricts its motivational significance to a particular domain (most frequently, to nothing, to irrelevance). Nonetheless, it is a fact that the phenomenon itself (which is of infinite complexity) is always capable of transcending its representation. This capacity for transcendence is a property of the "object" (a property of experience, from the phenomenological viewpoint), but can be exploited by the activity of man.

The alchemists regarded the "transcendent capacity of the object"—that is, the capacity of the familiar and explored in one context to become the unfamiliar and unexplored in another—as a *spirit*, embedded in matter. Jung cites Basilius Valentinus, an ancient alchemical authority:

The earth as material is not a dead body, but is inhabited by a spirit that is its life and soul. All created things, minerals included, draw their strength from the earth-spirit. This spirit is life, it is nourished by the stars, and it gives nourishment to all the living things it shelters in its womb. Through the spirit received from on high, the earth hatches the minerals in her womb as the mother her unborn child. This invisible spirit is like the reflection in a mirror, intangible, yet it is at the same time the root of all the substances necessary to the alchemical process or arising therefrom.[596]

The "spirit that inhabits the earth" was Mercurius, the *shape-shifter* (the reflected image of God in matter,[597] from the alchemical viewpoint) who both "guided" the alchemical process and was "released" by the activities of the alchemist. Mercurius was the spirit that made the "matter" investigated by the adept interesting, compelling—and interest is a "spirit" that moves from place to place, as knowledge changes and grows. Mercurius is the incarnation of transformation, the *uroboros*, who existed (and did not exist) as the most primal deity, before the creation of things (before the division of the world into subject and object, spirit and matter, known and unknown). The *uroboros* is, of course, the tail-eater, the dragon of chaos: an image of the embeddedness of the totality of things across time, in the particular manifest phenomenon. The image of the spirit Mercurius was an intimation of the infinite potential "trapped" in every particular aspect of experience.[598] Identification of this "potential"—that is, its classification—posed a constant problem to the medieval imagination:

All through the Middle Ages [Mercurius] was the object of much puzzled speculation on the part of the natural philosophers: sometimes he was a ministering and helpful spirit, an [assistant, comrade or familiar]; and sometimes the *servus* or *cervus fugitivus* (the fugitive slave or stag), an elusive, deceptive, teasing goblin who drove the alchemists to despair and had many of his attributes in common with the devil. For instance he is dragon, lion, eagle, raven to mention only the most important of them. In the alchemical hierarchy of gods Mercurius comes lowest as *prima materia* and highest as *lapis philosophorum*. The *spiritus mercurialis* is the alchemists' guide (Hermes Psychopompos), and their tempter: he is their good luck and their ruin.[599]

The alchemists conflated what we would think of as matter with what we might regard as the unknown. This is hardly surprising, since "matter" was the unknown to the pre-scientific mind (and is still something that retains much of its mystery today). As the unknown, matter possessed an attraction, which was the affective valence of what had not yet been explored. The ability of the unknown to "attract" provided impetus for its personification as "spirit"—as that which *motivates* or *directs*. Matter—even in its modern form—can easily revert to the unknown, even under modern conditions; can then exercise a similar force (that of a "stimulus") on the modern psyche. It does so, for example, when it manifests something anomalous—some unforeseen property, as a consequence of its placement in a new context, or its subjection to more creative exploration. The anomalous manifestation—the recurrence of the unknown—comes inevitably to attract increasing interest (or, conversely, attracts attempts to avoid, suppress or otherwise conjure it out of existence). All objects, even "explored objects," retain their connection with "that from which all things are made," even after they have been boxed and filed away (been categorized), in theory, "once and for all."

Take a rat, for example, who has habituated to a cage (who has explored the cage and become comfortable there). If a small object—say, an iron block—is dropped in front of it, it will first freeze, and then cautiously begin to investigate. The rat will use its capacity for motoric action to interact with the block—smelling it, looking at it, scratching it, perhaps gnawing it—to assess the motivational significance of the novel object. For the rat, limited by its lack of communicative ability to its own experience, limited by its restricted animal nature to fundamental processes of exploration, the block soon becomes irrelevant. It signifies no danger, in the course of interaction; it cannot be eaten; it is useless as nesting material. The block therefore "becomes" its lack of relevant properties for the no-longer-exploratory rat, and will henceforth be ignored. The process of exploration-predicated classification has eliminated the motivational significance of the novel—as is its function. From the mythic perspective, this is replacement of the "Great Mother" by the "Great Father"; replacement of ambivalent threat and promise by determinate valence (including irrelevance).

The sensory properties of the block—which are the relevant features of the object, as far as the spirit of scientific inquiry extends—have no intrinsic importance for the rat, except as they signify something of affective import. This more fundamental mode of thought, concerned with behavioral adaptation to circumstance, is how man thought, prior to the formalization of scientific methodology—and how man thinks still, insofar as he values and acts. The general case is, however, more complex. *Homo sapiens* is capable of observing a practically infinite series of novel properties emerge from the particular object, because he is capable of apprehending an object from a virtually unlimited number of points of perspective, spatial and temporal—or, it might be considered, equivalently, that the "object" is something so complex that it can manifest entirely different properties, merely in consequence of being viewed from alternative perspectives. The iron block was once, "of its own accord," something qualitatively different from what it is now, and will be something different, once again, in the future. In the earliest stages of its existence, considered as an independent object, the exemplary block was part of an undifferentiated totality, prior to the beginning of all things; then, the interplay of four fundamental forces; then simple hydrogen, coalescing into a star; then, matter transformed by gravity and nuclear processes; then a stone on Earth; finally, something transformed by man—with a still uncompleted and equally extensive developmental history before it. This transformation of the object is temporality itself—the manifestation of Tao, the flux of being. The capacity of human beings to apprehend variable spatial-temporal spans "turns" the object into something more complex than its mere present appearance; this increase in "complexity" is compounded by the extended active capacity for exploration also typical of our species.

What is an iron block for man? Shaped, a spear, and therefore food and death and security; suspended, a pendulum, key to detection of the earth's rotation; dropped, significant of gravity; reduced to its constituent particles, with sufficient patience and ingenuity, representative of molecular and atomic structure—a part like the whole. The question might be more accurately presented—what is an iron block not, for man? The pre-experimental mind of the alchemist, pondering the nature of the *prima materia*—the "fundamental constituent element of experience"—easily became possessed by intimations of the infinite possibility of "matter": of the boundless significance of the finite object; of the endless utility of the object and its inexhaustible capacity to reveal (become) the unknown.

When an object is explored, its motivational significance is constrained [generally, as a consequence of the specific goal-directed nature of the exploratory process, inevitably predicated upon a specific hypothesis—is this thing good for (a particular function?—but not any number of other potential functions)]. The question in mind, implicit or explicitly formulated, determines in part the answer "given" by the object. The object is always capable of superseding the constraint, in some unpredictable fashion. This infinite potential finds its symbolic expression in the self-devouring serpent, the mercurial spirit of transformation—the spirit that draws interest inexorably to itself.

While considering these ideas, I dreamed that a small object was traveling above the surface of the Atlantic Ocean. It moved along in the center of a procession of four immense hurricanes, configured as a square divided into quadrants, one hurricane per quadrant, tracked by satellites, monitored carefully and apprehensively by scientists manning the latest in meteorological equipment, in stations all over the world.

The dream scene shifted. The object, a sphere of about eight inches in diameter, was now contained and exhibited in a small glass display case, like that found in a museum. The case itself was in a small room, with no visible exit or entry points. The American president, symbol of social order, and the crippled physicist Stephen Hawking, representative of scientific knowledge (and of disembodied rationality), were in the room with the object. One of them described the features of the room. Its walls were seven feet thick and made of some impervious substance [titanium dioxide (?)]—which sounded impressive in the context of the dream. These walls were designed to permanently contain the object. I wasn't in the room, although I was there as an observer, like the audience in a movie. The object in the display case appeared alive. It was moving, and distorting its shape, like a chrysalis or a cocoon in its later stages of development. At one point, it transformed itself into something resembling a meerschaum pipe. Then it re-formed itself into a sphere, and shot out through one wall of the case, and the room, leaving two perfectly round, smooth holes—one in the case, and the other in the wall. It left with no effort whatsoever, as if the barriers designed to restrain its movement were of no consequence, once the "decision" had been made.

The object was an image of God, the uroboric serpent, embodied in matter (powerful enough to require the accompaniment of four hurricanes, as attendants).[600] The room was a classification system, something designed (by the most powerful representatives of the social and scientific worlds) to constrain the mysterious phenomenon. The object transformed itself into a pipe in reference to the famous painting (by Magritte) of a pipe, entitled (in translation) "This is not a pipe"—the map is not the territory, the representation not the phenomenon. The capacity of the object to escape, "at will," referred to the eternal transcendence of the phenomenal world, of its infinite capacity to unexpectedly supersede its representation, scientific and mythic.

I dreamed, much later (perhaps after a year) of a man suspended, equidistant from the floor, ceiling and walls, in a cubic room—about arm's length from each. The surfaces of the cube curved inward, toward the man (as if the room was constructed of the intersection of six spheres). All surfaces of the cube remained at the same distance from the man, regardless of his pattern of movement. If he walked forward, the cube moved forward with him. If he walked backward, the cube moved backward, at precisely the same rate, with no discontinuity whatsoever. The surfaces themselves were covered with circular patterns, about four inches in diameter, inscribed within squares of about the same size. Out of the center of each circle dangled the tip of a reptile's tail. The man could reach in any direction, grasp a tail, and pull it out of the surface, into the room.

This dream referred to the capacity of man to (voluntarily) pull the future into the present. The serpent— evident only in the form of his tail—was the uroboros, embedded implicitly in the phenomenal world.[601] *The potential for the emergence of something new was present in every direction the man could look, inside the cube. He could determine what aspect of being would reveal itself, as a consequence of his voluntary action.*

The act of exploration produces/elicits discriminable phenomena. These phenomena are mapped by the episodic and semantic "memory" systems. The exploratory process, however, is guided by the maps produced by the episodic system—particularly by its maps of the future. A desired end is posited, in fantasy. The motoric/abstract exploration system endeavors to bring about a match between emergent phenomena (produced in the course of activity) and that "map of the desired future." Mismatches between production and goal elicit (re)appearance of "the base matter of the world"—the unknown, manifest in negative affect, and curiosity.

The individual attempts to transform his wishes (rooted, in the final analysis, in emotion) into reality, suffering—and learning—when that process is disrupted. Exploration is deemed sufficient and may justly come to an end when the current affective state is deemed optimal: when knowledge, translated into action, has adjusted the world such that it is (once again) "paradisal." In the absence of such a paradisal state (in the absence of current security, happiness), exploration is or has been, by definition, incomplete. The "residual mysteries" that still accompany current being—which manifest themselves in the intrinsic attractiveness of the thing or situation—must therefore become the focus of active attention, so that the "information" embedded in them can be "pulled out," and transformed into subjective being and the world. The alchemical "base matter" of the world was, therefore, "the stuff of which determinate experience (subject and object) was made"; was something, in addition, capable of endless transformation; was something, finally, "corrupt"—as the material world was corrupt—incomplete, unrealized, fallen and suffering.

Analysis of the pre-experimental category of "gold" helps shed light on the relevance, importance and meaning of this archaic complex of undiscriminated ideas. Gold, as ultimate contrast to mere base matter, was the *ideal*, as it could be perceived in the concrete world. For the pre-experimental man, as well as for the modern, gold served as a medium of economic exchange. But the value of the metal did not, and still does not, consist solely in its economic utility. Gold has always been associated, in episodic representation, with divinity. Prior to development of the scientific worldview, this association made perfect sense. Gold, in contrast to "lesser" metals or substances, does not tarnish, dull or rust. It therefore appears imperishable, "immortal" and incorruptible. Gold is rare, rather than common. It shines like the sun, the evident source of life. The "category" of gold therefore tended to subsume everything Apollonian, sunlike, divine (in the patriarchal/heroic sense we have become familiar with). Jung describes the characteristic presumptions of the alchemist Michael Maier:

The sun, by its many millions of revolutions, spins the gold into the earth. Little by little the sun has imprinted its image on the earth, and that image is the gold. The sun is the image of God, the heart is the sun's image in man, just as gold is the sun's image in the earth, and God is known in the gold.[602]

The light of the sun is a "symbol" of power and the transcendence of clarity and consciousness, of heroism and permanence, and of victory over the forces of darkness, disintegration and decay. The earliest patriarchal gods and leaders of men combined the life-giving attributes of the sun with the heroic ideals of man, and the coins that bore their likeness were round and golden, in imitation of the solar disk.

Gold was, furthermore, the ideal *end* toward which all ores progressed—was the target of material progression. As it "ripened" in the womb of the earth, lead, for example, base and promiscuous [willing to "mate" (combine) with many other substances], aimed at the state characterized by gold, perfect and inviolable. This made the "gold state" the goal of the Mercurial "spirit of the unknown," embedded in matter. Eliade states:

If nothing impedes the process of gestation, all ores will, in time, become gold. "If there were no exterior obstacles to the execution of her designs," wrote a Western alchemist, "Nature would always complete what she wished to produce...." That is why we have to look upon the births of imperfect metals as we would on abortions and freaks which come about only because Nature has been, as it were, misdirected, or because she has encountered some fettering resistance....

Belief in the natural metamorphosis of metals is of very ancient origin in China and it is also found in Annam, in India and in the Indian archipelago. The peasants of Tonkin have a saying: "Black bronze is the mother of gold." Gold is engendered naturally by bronze. But this transmutation can materialize only if the bronze has lain a sufficiently long period in the bosom of the earth.[603]

The alchemist viewed himself as midwife to nature—bringing to fruition what nature endeavored slowly to produce—and therefore as aid to a transformation aimed at producing something ideal. "Gold" is that ideal. Eliade continues:

The "nobility" of gold is thus the fruit at its most mature; the other metals are "common" because they are crude; "not ripe." In other words, Nature's final goal is the completion of the mineral kingdom, its ultimate "maturation." The natural transmutation of metals into gold is inscribed in their destiny. The tendency of Nature is to perfection. But since gold is the bearer of a highly spiritual symbolism ("Gold is immortality," say the Indian texts repeatedly), it is obvious that a new idea is coming into being: the idea of the part assumed by the alchemist as the brotherly savior of Nature. He assists Nature to fulfil her final goal, to attain her "ideal," which is the perfection of its progeny—be it mineral, animal or human—to its supreme ripening, which is absolute immortality and liberty.[604]

The alchemists lived in a world that had theoretically been redeemed by the sacrifice of Christ—at least from the Christian perspective. But they did not feel at all redeemed—they remained unsatisfied with the present still-too-mortal condition. So they turned their attention to those aspects of the world that had been defined, in accordance with prevailing morality, as "unworthy of examination," as corrupt and contemptible. Presuming—or hoping—that things might yet be better, they *explored* (as we explore now, hoping to extract from the "unknown" new and useful tools). The alchemists assumed, implicitly, that further exploration might bring redemptive knowledge. This search was driven by their admission of the "unbearable present," by their identification with a "still-fallen world." The alchemists believed that the "desirable transmutation of matter" could be brought about by the "release" of Mercurius from matter. This meant that they implicitly recognized that (interest-guided) exploration was key to the (redemptive) expansion of being.

In participating in this process, the alchemists identified with the exploratory hero, and turned themselves unconsciously (that is, in procedure, if not always in representation) into "that which redeems." This identification was complicated by the fact that the alchemist also considered himself as partaking of the state of matter—as belonging in the "state necessitating redemption." This basically meant that the alchemist viewed himself, at least in part, as occupying the same *category* as "matter" (as well as being that which could become "gold," and which could aid in that transformation). For the pre-experimental mind, with its more general and conflated categories, there is no necessary distinction between the "thing being acted on" and the "thing doing the acting." Eliade describes, for example, the "sympathetic magic" necessary to complete a grafting operation between two different species of plants necessary (to induce "unlike to mate with unlike," from a broader perspective):

Ibn Washya—and he is not the only oriental writer to allow himself to be carried away by such images—speaks of fantastic graftings ("contrary to Nature") between differing vegetable species. He says, for instance, that the grafting of a branch of a lemon tree on to a laurel or olive tree would produce very small lemons, the size of olives. But he makes it clear that the graft could succeed only if it was performed in the ritual manner and at a certain conjunction of the sun and moon. He explains the rite thus: "the branch to be grafted must be held in the hands of a very beautiful maiden, while a man is having shameful and unnatural sexual intercourse with her; during coitus the girl grafts the branch onto the tree." The significance is clear: in order to ensure an "unnatural" union in the vegetable world an unnatural sexual union between human beings was necessary.[605]

Such ideas are far from rare. Virtually every process undertaken by pre-experimental individuals—from agriculture to metallurgy—was accompanied by rituals designed to "bring about the state of mind" or "illustrate the procedure" necessary to the successful outcome desired. This is because the action precedes the idea. So ritual sexual unions accompanied sowing of the earth, and sacrificial rituals and their like abounded among miners, smiths and potters. Nature had to be "shown what to do"; man led, not least, by example. The correct procedure could only be brought about by those who had placed themselves in the correct state of mind. This idea was taken to its logical conclusion during the alchemical procedure, which had as its fantastical end state or desired future the most profound and far-reaching notion of transformation ever conceptualized: the final perfection or "redemption" of matter.

To induce disparate elements to combine harmoniously in the production of the *lapis philosophorum*—that which will transmute base metals into gold—it was necessary to become unified oneself. To engender perfection from nature, therefore, man had to become perfect. The necessity for the perfection of the alchemist—and the relation of the alchemical procedure to his own being—was further strengthened by the alchemist's identity with the material world (that is, by his occupation of the same "categorical space" as "matter"). Man— a fallen, corrupt, material being, yet capable of endless transformation—partook of the essence of the fallen, corrupt, yet transformable material world. Those things relevant to the transformation of the being of "objects" were therefore also, by logical necessity, relevant to the transformation of his own being. The transformation of base matter into gold, *writ large*, was the redemption of the world—its transformation into the "state of gold." The *lapis philosophorum* was means to that end. This extension of the theories of sympathetic magic to the domain of "chemistry" meant that alchemy became increasingly rife with (primarily imagistic) speculation regarding the nature of perfection, as it developed over the centuries.

It is difficult for moderns to realize why any of this might be relevant. Our psychology and psychiatry—our "sciences of the mind"—are devoted, at least in theory, to "empirical" evaluation and treatment of mental "disorders." But this is mostly smoke and screen. We are aiming, always, at an ideal. We currently prefer to leave the nature of that ideal "implicit," because that helps us sidestep any number of issues that would immediately become of overwhelming difficulty, if they were clearly apprehended. So we "define" health as that state consisting of an absence of "diseases" or "disorders" and leave it at that—as if the notion of disease or disorder (or of the absence thereof) is not by necessity a medieval concatenation of moral philosophy and empirical description. It is our implicit theory that a state of "non-anxiety" is possible, however—and desirable—that leads us to define dominance by that state as "disordered." The same might be said for depression, for schizophrenia, for personality "disorders," and so on. Lurking in the background is an "implicit" (that is, *unconscious*) ideal, against which all "insufficient" present states are necessarily and detrimentally compared. We do not know how to make that ideal *explicit*, either methodologically or practically (that is, without causing immense dissent in the ranks); we know, however, that we must have a concept of "not ideal" in order to begin and to justify "necessary" treatment. Sooner or later, however, we will have to come to terms with the fact that we are in fact attempting to produce the ideal man—and will have to define explicitly what that means. It would be surprising, indeed, if the ideal we come to posit bore no relationship to those constructed painstakingly, over the course of centuries of effort, in the past. Something very similar happened in the case of alchemy, at least in the West: as the philosophy developed, through the Christian era, the *lapis* was increasingly identified with Christ. There is no reason to presume that this came as anything but a surprise to the alchemists themselves. We are in for a shock at least as great.

To perfect nature, it was necessary to harbor the correct attitude—to undertake the appropriate rituals and processes of spiritual purification; to become pure as the thing desired. The worker stood as example to nature, in small things and great. In the case of alchemy, which ambitiously desired to "redeem" the fallen material world, the alchemist himself had to become great. Thus the alchemical literature might be regarded, in part, as one long "meditation" on the nature of the ideal man.

Episodic Representation in Medieval Christendom

Science is predicated upon the axiomatic presupposition that it is worthwhile to analyze the material or collectively apprehensible sensory world and its transformations. This belief, which first manifested itself in (alchemical) fantasy, is so much a part of the modern world, so much its primary assumption, that it is difficult to realize what a remarkable achievement its formalization represented. It took thousands of years of cultural development to formulate the twin notions that empirical reality existed (independent of the motivational significance of things) and that it should be systematically studied (and these ideas only emerged, initially, in the complex societies of the Orient and Europe). The alchemists were the first to risk this attribution, or something similar to it; but they still studied "matter" in the absence of explicit empirical methodology. Jung states:

The concept of the "psychic," as we understand it today, did not exist in the Middle Ages, and even the educated modern man finds it difficult to understand what is meant by "reality of the psyche." So it is not surprising that it was incomparably more difficult for medieval man to imagine something between "esse in re" and "esse in intellectu solo." The way out lay in "metaphysics." The alchemist was therefore compelled to formulate his quasichemical facts metaphysically too.[606]

The lack of scientific methodology—the inability to conduct formalized comparison of behavior-predicated experience, to determine its generalizability—meant the inextricable admixture of the purely sensory and the subjective, affective, mythological aspects of experience. The purpose of scientific methodology is, in large part, to separate the empirical facts from the motivational presumption. In the absence of such methodology, the intermingling of the two domains is inevitable:

[The alchemists] ... believed that they were studying the unknown phenomenon of matter ... and they just observed what came up and interpreted that, somehow, but without any specific plan. There would be a lump of some strange matter, but as they did not know what it was they conjectured something or other, which of course would be unconscious projection, but there was no definite intention or tradition. Therefore one could say that in alchemy, projections were made [hypotheses were generated] most naively and unprogrammatically, and completely uncorrected.

Imagine an old alchemist's situation. A man in a certain village would build an isolated hut and cook things which caused explosions. Quite naturally, everyone calls him a witch doctor! One day someone comes and says he has found a queer piece of metal and would the alchemist be interested in buying it? The alchemist does not know the value of the metal, but gives the man some money at a guess. He then puts what has been brought him in his stove and mixes it with sulphur, or something similar, to see what happens, and if the metal was lead, he would be badly poisoned by the vapours. He concludes, therefore, that this particular matter makes one feel sick if approached, and nearly kills you, and therefore he says that there is a demon in lead! Afterwards, when he writes his recipes, he adds a footnote saying: "Beware of lead, for in it is a demon which will kill people and make them mad," which would be quite an obvious and reasonable explanation for someone of that time and level. Therefore lead was a wonderful subject for the projection of destructive factors, since in certain connections its effects are poisonous. Acid substances were also dangerous, but, on the other hand, being corrosive and a means of dissolving things, were highly important for chemical operations. Thus if you wished to melt something or have it in liquid form it could be melted or dissolved in acid solutions, and for this reason the projection was that acid was the dangerous substance which dissolves, but which also makes it possible to handle certain substances. Or else it is a medium of transformation— you open up, so to speak, a metal with which you can do nothing and make it accessible to transformation by the use of certain liquids. The alchemists therefore wrote about it in the naive form which I am now describing and did not notice that that was not natural science but contained a lot of projection, if looked at from a modern chemical standpoint.

Thus there exists in alchemy an astonishing amount of material from the unconscious, produced in a situation where the conscious mind did not follow a definite program, but only searched.[607]

Alchemy flourished for almost two thousand years, and only faded from view in the late eighteenth century. It developed (at least in the Middle Ages) as a movement compensating that embodied in absolutist Christianity, which emphasized the ultimate reality and value of the spirit, dogmatically concretized; which presumed that everything worth knowing had already been discovered, and which cast the material world into disrepute.

For the medieval mind, the body, the sensory, physical world—"matter," in general—was valued as immoral and as corrupt, as ruled by demonic, unknown forces. The story of Genesis—serpent and Eve conspiring to bring about the descent of mankind "into the profane and fallen (material) world"—in part provided the mythological basis for this union of category. The attractions of the material world also posed a threat to identification with the church, as the pull of sensuality, for example, or the desire for material instead of spiritual wealth. Furthermore, the fact of the fallen material state undermined faith in church dogma: apprehension of the unresolved suffering of man made it difficult to attribute to Christ's actions the final state of redemption they theoretically guaranteed. In consequence, contact with the "matriarchal underworld of matter" (that is, with the "unknown") seemed very threatening to the church authorities—and for very good reason (at least from the perspectives of conservation and tradition).

Alchemical fascination with "matter" developed antithetically to the early Christian valuation of the "spiritual" and the "established" (developed as an antithesis to the domain of the *known*). The suppression of the sensory material world by the church and the simultaneous establishment of an absolute body of knowledge meant rejection or denial of anomalous sensory/emotional experience, and therefore of the value contained in such experience. Alchemical preoccupation with matter arose as a consequence of this lost value asserting itself, in the attraction of the accumulated "rejected and unknown"—in the inevitable attraction of the "forbidden fruit."

Observing what he did not understand, the alchemist had recourse only to speculation, which he used to interpret that unknown. These speculations look like fantasies to the modern mind—like the fantasies of the medieval Christian (and pre-Christian, in some cases):

So, in a sense, they are, and for this reason they lend themselves to decipherment by the method of complex psychology. [The alchemical approach] ... is so patently a spiritual and moral attitude that one cannot doubt its psychological nature. To our way of thinking, this immediately sets up a dividing wall between the psychic and the chemical process. For us the two things are incommensurable, but they were not so for the medieval mind. It knew nothing of the nature of chemical substances and their combination. It saw only enigmatic substances which, united with one another, inexplicably brought forth equally mysterious new substances. In this profound darkness the alchemist's fantasy had free play and could playfully combine the most inconceivable things. It could act without restraint and, in so doing, portray itself without being aware of what was happening.[608]

The alchemist thought in a medieval or pre-medieval fashion, using archaic preconceptions and ideas. Analysis of that thought, "projected" upon matter (just as we interpret "matter" in the light of our own, current and therefore invisible theories), therefore means interpretation of fantasy, analysis of the spontaneous productions of the exploring mind. Such analysis means increased capacity to understand the workings of mind. The entire corpus of alchemy contains seventeen hundred years of fantasy regarding the nature of (moral) transformation, assumed to take place in matter (a category that included man) "striving" as it did "naturally" toward perfection. Central to this movement toward perfection was the dissolution, transformation and reconstitution of unredeemed primal matter, the *prima materia*:

As is indicated by the very name which he chose for it—the "spagyric" art—or by the oft-repeated saying "solve et coagula" [dissolve and reconstitute], the alchemist saw the essence of his art in separation and analysis on the one hand and synthesis and consolidation on the other. For him there was first of all an initial state in which opposite tendencies or forces were in conflict; secondly there was the great question of a procedure which would be capable of bringing the hostile elements and qualities, once they were separated, back to unity again. The initial state, named the *chaos*, was not given from the start, but had to be sought for as the *prima materia*. And just as the beginning of the work was not self-evident, so to an even greater degree was its end. There are countless speculations on the nature of the endstate, all of them reflected in its designations. The commonest are the ideas of its permanence (prolongation of life, immortality, uncorruptibility), its androgyny, its spirituality and corporeality, its human qualities and resemblance to man (homunculus), and its divinity.[609]

The alchemists began their work, their *opus*, by determining to face the unknown, locked away in the material world, in the pursuit of an ideal. Their ideal was symbolized by the *lapis philosophorum*, which was a unitary substance characterized by its ability to transform base metals into gold, and more—which could confer upon its bearer complete knowledge, immortal life, and impeccable mental and physical health. The medieval individual had no idea that the creation of such a "substance" was not possible, and was aware of many substances that had transformative properties.

Identification of what motivated such a pursuit appears straightforward. There were undoubtedly many who tried their hand at alchemy, purely for its potential economic benefit, just as there are many today who pursue their occupation solely for material gain. (Even this is a form of desire for redemption, however—through material means—and may be unexpectedly transformed into a more purely spiritual pursuit in the course of maturation, or through the unpredictable actions of fantasy and circumstance.) There were also alchemists who more clearly embodied the spirit of devout curiosity, and who worked with the same serious discipline as the later natural scientist. It is naive to underestimate the power and the mystery of the fantasy of the philosopher's stone. This idea provided the motive power underlying disciplined investigation into the secrets of matter—a difficult, painstaking, expensive procedure. The idea that matter contained locked within it the secret to wisdom, health and wealth underlies the entire *opus* of modern science. The fact that such an idea could arise, and be seriously entertained in spite of grandiosity and conflict with church dogma, is difficult enough to believe. It becomes truly incomprehensible when consideration is given to the additional fact that the procedure extended over seventeen centuries, despite the fact that no alchemist every reached his goal. Jung states: .

In view of the fact that ... a miracle never did occur in the retort, despite repeated assertions that someone had actually succeeded in making gold, and that neither a panacea nor an elixir has demonstrably prolonged a human life beyond its due, and that no homunculus has ever flown out of the furnace—in view of this totally negative result we must ask on what the enthusiasm and infatuation of the adepts could possibly have been based.

In order to answer this difficult question one must bear in mind that the alchemists, guided by their keenness for research, were in fact on a hopeful path since the fruit that alchemy bore after centuries of endeavor was chemistry and its staggering discoveries. The emotional dynamism of alchemy is largely explained by a premonition of these then unheard-of possibilities. However barren of useful or even enlightening results its labours were, these efforts, notwithstanding their chronic failure, seem to have had a psychic effect of a positive nature, something akin to satisfaction or even a perceptible increase in wisdom. Otherwise it would be impossible to explain why the alchemists did not turn away in disgust from their almost invariably futile projects.[610]

The alchemical fantasy provided (and still provides) *the motive power for the empirical endeavor*, just as the dream of Judeo-Christianity provided motive power for the civilization of the West. In this manner, myth, mysterious, absurd and incomprehensible, stands at the vanguard of the adaptive process. Eliade states (with specific regard to the origin of science):

Until recently, few were aware of Isaac Newton's role in this general [alchemical] movement, whose goal was the *renovatio* of European religion and culture by means of an audacious synthesis of the occult traditions and the natural sciences. It is true that Newton never published the results of his alchemical experiments, although he declared that some of them were crowned with success. His innumerable alchemical manuscripts, ignored until 1940, have recently been meticulously analyzed by Betty Jo Teeter Dobbs in her book *The Foundations of Newton's Alchemy* (1975). Dobbs affirms that Newton experimented in his laboratory with the operations described in the immense alchemical literature, probing the latter "as it has never been probed before or since" (p. 88). With the aid of alchemy, Newton hoped to discover the structure of the microuniverse in order to homologize it with his cosmological system. The discovery of gravity, the force which keeps the planets in their orbits, did not completely satisfy him. But although he pursued the experiments indefatigably from 1669 to 1696, he did not succeed in identifying the forces which govern the corpuscles. Nevertheless, when he began to study the dynamics of orbital movement in 1679–80, he applied his "chemical" conceptions of attraction to the universe.

As McGuire and Rattansi have shown, Newton was convinced that in the beginning, "God had imparted the secrets of natural philosophy and of true religion to a select few. The knowledge was subsequently lost but partially recovered later, at which time it was incorporated in fables and mythic formulations where it would remain hidden from the vulgar. In modern days it could be more fully recovered from experience."[611] For this reason, Newton examined the most esoteric sections of the alchemical literature, hoping that they would contain the true secrets. It is significant that the founder of modern mechanics did not reject the tradition of a primordial and secret revelation, just as he did not reject the principle of transmutation. As he wrote in his *Optics* (1704), "the change of Bodies into Light and of Light into Bodies is entirely in conformity with the Laws of Nature, for Nature seems ravished by Transmutation." According to Dobbs, "Newton's alchemical thoughts were so securely established that he never came to deny their general validity, and in a sense the whole of his career after 1675 may be seen as one long attempt to integrate alchemy and the mechanical philosophy" *(Foundations,* p. 230).

After the publication of the *Principia,* opponents declared that Newton's "forces" were in reality "occult qualities." As Dobbs recognizes, in a certain sense these critics were right: "Newton's forces were very much like the hidden sympathies and antipathies found in much of the occult literature of the Renaissance period. But Newton had given forces an ontological status equivalent to that of matter and motion. By so doing, and by quantifying the forces, he enabled the mechanical philosophies to rise above the level of imaginary impact mechanisms" (p. 211). In analyzing the Newtonian conception of force, Richard Westfall arrives at the conclusion that modern science is *the result of the wedding of the Hermetic tradition with the mechanical philosophy.*[612]

In its spectacular flight, "modern science" has ignored, or rejected, the heritage of Hermeticism. Or to put it differently, the triumph of Newtonian mechanics has ended up by annihilating its own scientific ideal. In effect, Newton and his contemporaries expected a different type of scientific revolution.

In prolonging and developing the hopes and objectives (the first among these being the redemption of Nature) of the neo-alchemist of the Renaissance, minds as different as those of Paracelsus, John Dee, Comenius, J. V. Andreae, Fludd, and Newton saw in alchemy the model for a no less ambitious enterprise: the perfection of man by a new method of knowledge. In their perspective, such a method had to integrate into a nonconfessional Christianity the Hermetic tradition and the natural sciences of medicine, astronomy, and mechanics. In fact, this synthesis constituted a new Christian creation, comparable to the brilliant results obtained by the earlier integrations of Platonism, Aristotelianism, and Neoplatonism. This type of "knowledge," dreamed of and partially elaborated in the eighteenth century, represents the last enterprise of Christian Europe that was undertaken with the aim of obtaining a "total knowledge."[613]

Not precisely the last.

Formulation of the idea that God might be known in material form meant positing the possibility that the highest conceivable value might be embodied concretely in "matter"—rather than in the established, patriarchal, "spiritual" world. This meant that the "nature of God" was something that could be made subject to material (and thoughtful) investigation. However, matter remained comparatively unknown to the medieval mind—and was therefore "contaminated" with everything else unknown, repressed and rejected. The assignment of value to matter was therefore attribution of value to unknown experience. This assignment was heretical, because it implied the fallibility or incomplete nature of church dogma (the formalized medieval European general model of expectation and desire), and was therefore dangerous, from the intrapsychic and the social viewpoints. Such heresy was compounded in severity, because the church explicitly regarded matter—representative of the inadmissible unknown—as degraded, corrupt, imperfect and demonic.

The alchemist was an unredeemed, suffering man, in search of an inexpressible ideal. He formulated that ideal, and its process of generation, using terms that referred to "the physical world," at least from the modern perspective. However, the alchemist made no clear distinction between psychological and objective. His "search for the ideal" was therefore as much psychological as chemical (more, actually, since he worked in the absence even of the basic measurement devices of modern science). The alchemist posited that the answer lay outside the church, in the unknown. Exploration of the unknown and forbidden meant generation of redemptive knowledge (then, as it does now). Incorporation of such knowledge meant movement toward perfection. Broadly speaking, the alchemist wanted to transform every subordinate element in the category "matter" (the unknown, fallen, corrupt world, including man as "material" being) into the category "gold" (the Apollinian, spiritual, sunlike, incorruptible state). He was searching for a transformative agent to bring about that change (the *lapis philosophorum*); but also viewed himself as that agent (since he was integrally involved in the transformative *opus* of alchemy). This relatively straightforward conceptualization of "movement toward the ideal" is schematically presented in *Figure 62: The Alchemical Opus as "Normal Story."*

The alchemist courageously posited that the work of redemption held up as absolute by

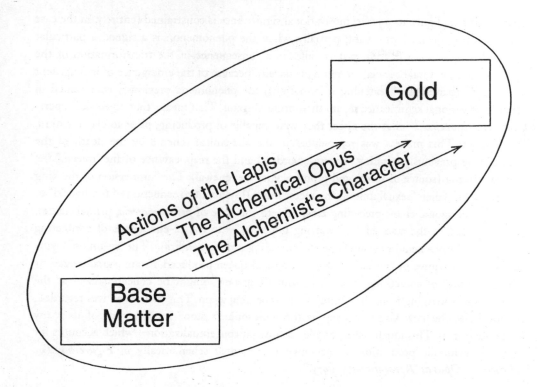

Figure 62: The Alchemical Opus as "Normal Story"

the church was not yet complete—or at least acted "as if" there was still work to be done. So he hoped to turn what was still unredeemed into gold. The problem is, of course, that turning base matter into gold is not possible, as a "normal" act of movement from point "a"—the unbearable present—to point "b"—the desirable future. The attempt to produce the "most ideal state possible," however—something akin to paradise on earth—is particularly unlikely, without a revolution. So the alchemical "story" rapidly turned into something more complex; something that essentially recapitulated the "union of the gods" (something like a process of initiation or spiritual transformation). The alchemists soon came to realize that movement toward the ideal did not mean an unbroken journey uphill; soon came to realize that a large leap forward was necessarily preceded by a radical descent.

Once the alchemist had decided to look into the unknown for salvation, rather than to the church (or at least in addition to the church), he placed himself outside the protective confines of his previous system of classification. Outside that dogmatic system, things took on new meaning (or at least new potential meaning). Once you have decided that you don't know absolutely everything about something, it is possible to learn something new. However, when an "object" has been placed in a system of classification (within the constraints of a particular paradigm), its *a priori* motivational significance is constrained (entirely, in the case of something deemed irrelevant; partially, when the phenomenon is assigned a particular use). When the classification system fails [as a consequence of the transformation of the environment, natural or social, or, more prosaically, because of the emergence of incongruent experience (paradigm-threatening information)] the phenomena previously constrained in their motivational significance regain their original status. That means that objects of experience are *renovelized*—that the affect they were capable of producing, prior to classification, re-emerges. This process was represented in the alchemical scheme by the death of the "king" (the previously reigning "system of order") and the reappearance of the "queen," the Great Mother (source of threat and promise, vital for renewal). The "immersion of the king in the queen" [their "sexual union" (the incest motif)] symbolized reduction of the "world" to the precosmogonic chaos preceding creation—to the state of *prima materia*, primal matter. This also led to the state of the "warring opposites"—to the re-emergence of conflicting "substances" once held in harmony by the preceding orderly condition. The "re-union" (symbolized as a creative or "sexual" union) of king and queen produced a state characterized by the possibility of something new. This "something new" might be conceptualized as the "divine son" emerging from that union, soon to be king again. This divine son was regarded, variously, as the new king—or even as the philosopher's stone itself, in one of its many potential forms. This much more complex process of conceptualization—which accounts for the vast symbolic production of alchemy—is presented schematically in *Figure 63: The Alchemical Opus as "Revolutionary Story."*

The Prima Materia

"It offers itself in lowly form. From it there springs our eternal water."[614]

The *prima materia* (alternatively: the "round chaos" or the alchemical *uroboros*) is the unknown as matter and, simultaneously, as effect upon imagination and behavior (inseparable pre-experimentally): is God as *substance* and *effect of substance*. The *prima materia* is the "precosmogonic egg," the dragon of chaos—the eternal source from which spirit and knowledge and matter and world arise. It is the unknown that simultaneously generates new phenomena, when explored; the unknown that serves as the source of the "information" that comes to constitute the determinate experiencing subject. The alchemists therefore granted the *prima materia* a "half-chemical, half-mythological" definition: For one alchemist, it was quicksilver, for others it was ore, iron, gold, lead, salt, sulphur, vinegar, water, air, fire, earth, blood, water of life, *lapis*, poison, spirit, cloud, dew, sky, shadow, sea, mother, moon, serpent.... Jung states:

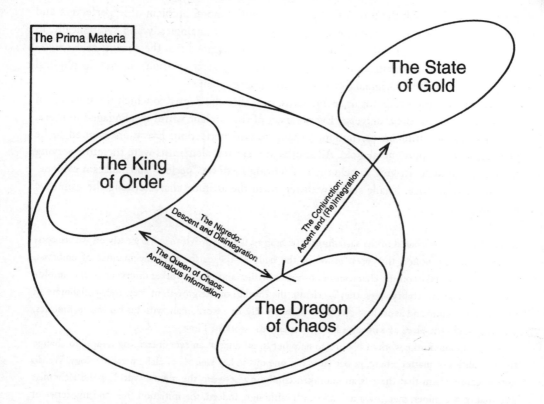

Figure: 63: The Alchemical Opus as "Revolutionary Story"

The autonomy and everlastingness of the *prima materia* in Paracelsus [for example] suggest a principle equal to the Deity, corresponding to a *dea mater*. . . . The following texts, for example, are applied to the *prima materia*: "and his going forth is from the beginning, from the days of eternity" (Micah 5:2) and "before Abraham was made, I am" (John 8:58). This is supposed to show that the stone is without beginning and has its [primary existence] from all eternity, and that it too is without end and will exist in all eternity. . . .

And in the same way, continues the author, that the stone together with its material has a thousand names and is therefore called "miraculous," all these names can in eminent degree be predicated of God, and the author thereupon proceeds to this application. A Christian can hardly believe his ears. . . . "That from which things arise is the invisible and immovable God."[615]

The alchemists understood the *prima materia* to be as of yet "unredeemed," however, and "base." The notion of corrupted matter was a moral notion, and the "imperfection" of matter therefore a moral imperfection. The alchemists' reflections on the nature of this imperfect matter inevitably took the form of reflections upon the moral problem of imperfection and material corruption as such. Since the alchemist thought analogically and symbolically, in the absence of the empirical method, he fantasized or imagined that the corrupt *prima materia* shared the characteristics of other corrupt and imperfect creations, including physical man, contaminated by Original Sin and his own transgressions.

It is virtually impossible for us, as moderns, to realize the degree to which the universe of our forebears was a *moral* universe. Every aspect of that archaic world was engaged in moral endeavor, participating in corruption, striving toward perfection. Every ore wanted to be pure metal, every pure metal, gold. All smiths, miners and alchemists were therefore serving the role of midwife, striving to help the Earth bring forth the "perfect" substances it evidently desired to produce. Eliade states, with regard to the attitude characterizing the primitive metalworker:

Mineral substances shared in the sacredness attaching to the Earth-Mother. Very early on we are confronted with the notion that ores "grow" in the belly of the earth, after the manner of embryos. Metallurgy thus takes on the character of obstetrics. Miner and metal-worker intervene in the unfolding of subterranean embryology: they accelerate the rhythm of the growth of ores, they collaborate in the work of Nature and assist it to give birth more rapidly. In a word, man, with his various techniques, gradually takes the place of Time: his labours replace the work of Time.

To collaborate in the work of Nature, to help her to produce at an ever-increasing tempo, to change the modalities of matter—here, in our view, lies one of the key sources of alchemical ideology. We do not, of couse, claim that there is an unbroken continuity between the mental world of the alchemist and that of the miner, metal-worker and smith (although, indeed, the initiation rites and mysteries of the Chinese smiths form an integral part of the traditions later inherited by Chinese Taoism and alchemy). But what the smelter, smith and alchemist have in common is that all three lay claim to a particular magico-religious experience in their relations with matter; this experience is their monopoly, and its secret is transmitted through the initiatory rites of their trades. All three work on a Matter which they hold to be at once alive and sacred, and in their labours they pursue the transformation of matter, its perfection and its transmutation.[616]

In the unredeemed *prima materia* the alchemist understood matter to be trapped in an imperfect state; just as man himself was trapped in a corrupt and perishable state by his sinful, demonic physical material nature. The transformation of this *prima materia* into gold or into the philosopher's stone therefore signified a moral transformation, which could be brought about through moral means. The alchemists were searching for a method to redeem corruption. They applied their fantastical reasoning to redemption of corrupt matter, which seems absurd from the modern viewpoint. However, experience of the physical world had been formally damned by the church—for reasons which had their own logic—and the lost value this experience represented stood therefore in dire need of redemption. The search for lost value led the alchemists deep into consideration of the nature of corruption, or limitation, and past that, into its transformation, and redemption. Their devoted concentration upon the nature of this problem set in motion fantasies associated with the archetype of the way, which always emerges of its own accord, when individuals face their limitations and come into contact with the unknown. And it must be understood: although the alchemists *conflated* "the psyche" and "objective reality," *their conflation was meaningful.* The alchemist did "redeem" himself by studying the "redemptive" transformations of matter—most simply, because exploration "releases" information that can be used to construct personality; more complexly, because the act of voluntary exploration, outside the domain allowed by tradition, *constitutes identification with the creative hero.*

The first alchemical transformation took the form of disintegration: chemical solution or putrefaction of the *prima materia,* in its "solid" form—in its *patriarchal* incarnation, in its manifestation as "ordered" or "stable" or "rigid" substance. The archetypal first stage in any moral transformation (which the alchemist was striving to produce) constitutes tragic disruption of the previous state of being. The disintegration of the *prima materia* was analogically equivalent to the degeneration of the alchemist's previous socially determined intrapsychic state, consequential to his decision to pursue the unknown:

The chemical putrefaction is compared to the study of the philosophers, because as the philosophers are disposed to knowledge by study, so natural things are disposed by putrefaction to solution. To this is compared philosophical knowledge.[617]

The nature of the previous state of the alchemist, its symbolic equivalent in the *prima materia,* and the consequences of its disintegration, can be placed in context by examination of the medieval worldview.

The investigation of matter was absolutely heretical, for medieval *Homo sapiens.* Mere suggestion that the unknown still existed, and therefore required investigation, threatened the absolute authority of the Christian dogma, as historically formulated by the fathers of the church. Questioning this authority meant that the alchemist placed himself outside the protection of his cultural canon, in the psychological sense, and at the mercy of the ecclesiastical authorities in the practical world. Investigation of matter and its transformations was therefore an exceptionally dangerous undertaking, both from the intrapsychic and social viewpoints. The standard punishment for heretical undertakings was exceedingly horrible—torture and excommunication—and the potential psychic consequences scarcely less dangerous.

The alchemist, in beginning his pursuit, placed himself outside the protective enclave of conformity, and risked the investigation of an aspect of experience which, according to the worldview of his time, was characterized by absolute demonism. The apprehension such an undertaking must have engendered in the minds of those who adopted it is scarcely imaginable (although such apprehension re-emerges in the mind of the modern, threatened by revolutionary ideas). The alchemical search of the unknown, for the ideal, had as its prerequisite or its immediate consequence abandonment or disorganization of the reigning individual and social worldview. To investigate matter, for the ideal, meant to investigate corruption, evil itself, in the pursuit of value. The alchemist who undertook this investigation already believed he was in need of redemption, that he was incomplete, or he would have never dared step outside the boundaries drawn by the church. His need of redemption, of completion, paralleled that of the corrupt *prima materia* and further strengthened the "unconscious" analogous identity between the adept and his material.

The King of Order

In the absence of formal empirical methodology, the alchemical adept could only investigate the transformations of matter with the preconceptions of his imagination. The products of alchemy were therefore necessarily structured according to the myth of the way, the primary archetypal manifestation of imaginative fantasy. The alchemist worked alone, concentrating on his procedure for months and years at a time, and in this solitary pursuit his fantasy had free reign. Once he had the courage to admit to his own ignorance, his own insufficiency, his investigations into "matter" took the form of contact with the unknown. Admission of personal ignorance presents a challenge to the cultural canon (to the degree that the ignorant one is identified with that canon) and sets the stage for moral transformation, which manifests itself in symbolic form. The alchemist was searching for comprehension of the nature of material corruption and for a method whereby it could be perfected. Christian dogma stated that the world had been finally redeemed by the Passion of Christ; but it appeared evident to the alchemist that material substances, including himself, remained "morally" corrupt and incomplete. Admission of imperfection was therefore tantamount to admission that Christian dogma, as presented by the authoritarian church—and, in consequence, as represented intrapsychically—was incomplete. This incompleteness, manifested as absolute authority, served a tyrannical function, which had to be eliminated, prior to the creation of new knowledge. The aspect of the *prima materia*, which was first in need of redemption, was therefore its patriarchal/tyrannical aspect, which appeared in imagination as the Great Father—as the King, or his symbolic equivalent. Jung states:

The conscious mind often knows little or nothing about its own transformation, and does not want to know anything. The more autocratic it is, and the more convinced of the eternal validity of its truths, the more it identifies with them. Thus the kingship of Sol, which is a natural phenomenon, passes to the human king who personifies the prevailing dominant idea and must therefore share its fate. In the phenomenal world the Heraclitean law of everlasting change, *panta ret*, prevails; and it seems that all the true things must change and that only that which changes remains true....

In this alchemical procedure we can easily recognize the projection of the transformation process: the aging of a psychic dominant is apparent from the fact that it expresses the psychic totality in an ever-diminishing degree. One can also say that the psyche no longer feels wholly contained in the dominant, whereupon the dominant loses its fascination and no longer grips the psyche so completely as before. On the other hand its content and meaning are no longer properly understood, or what is understood fails to touch the heart. A "sentiment d'incompletude" of this kind produces a compensatory reaction which attracts other regions of the psyche and their contents, so as to fill up the gap. As a rule this is an unconscious process that always sets in when the attitude and orientation of the conscious mind have proved inadequate. I stress this point because the conscious mind is a bad judge of its own situation and often persists in the illusion that its attitude is just the right one and is only prevented from working because of some external annoyance. If the dreams were observed it would soon become clear why the conscious assumptions have become unworkable. And if, finally, neurotic symptoms appear, then the attitude of consciousness, its ruling idea, is contradicted, and in the unconscious there is a stirring up of those archetypes that were the most suppressed by the conscious attitude. The therapist then has no other course than to confront the ego with its adversary and thus initiate the melting and recasting process. The confrontation is expressed, in the alchemical myth of the king, as the collision of the masculine, spiritual father ruled over by king Sol with the feminine, chthonic mother-world symbolized by the *aqua permanens* or by the chaos.[618]

The process represented symbolically as "disintegration of the king" found its analogical material equivalent in the chemical process of dissolution—in the immersion of a solid substance or compound (the *prima materia*) in a solvent, or in its decay, its return to earth (hence the still extant metaphor for the wastrel: the *dissolute* personality). The "solid substance," the king, represented the mythic core of the historically determined hierarchy of behavioral pattern and representation with which the adept had been previously identified and which had to be abandoned or challenged, before the investigation of matter/unknown could begin in earnest. Destruction of the culturally determined patriarchal system is typically represented in fantasy, symbolically, as the death of the old (sterile, sick) king, which becomes necessary when the land is no longer fruitful. Such a sacrifice—which was once a ritual—means rejection of reliance on a particular pattern of behavioral adaptation and representational presumption; means potential for reintroduction of new ideas (or even a new pattern of ideas), when adaptation is threatened by anomaly.[619] The idea of the king—the central representative of culture—was expressed in a multitude of symbolic images during the centuries alchemy flourished. The eagle, the sun, the lion, heaven, fire, height, and spirit all symbolized different aspects of the patriarchal system, which served to represent the initial condition of the *prima materia*, prior to its dissolution. Such symbolic representations emerge as a matter of course, if the conditions which originally elicited them re-emerge. Encounter with the unknown constitutes one such condition.

The Queen of Chaos

The solvent in which the *prima materia* as king dissolves, or the earth to which it returns, is imaginatively represented in the alchemical process by symbols characteristic of the matriarchal system. The *prima materia* dissolves in water, salt water, tears, or blood, as the old king dissolves in the matriarchal system—dissolves in (previously confined) sensuality, affect and imagination, which threaten and transcend knowledge in its concrete form, and serve simultaneously as the matrix from which it is borne. The matriarchal system is the intrapsychic representative of the queen, the Great and Terrible Mother, who is sea, toad, fish or dragon, lioness, earth, depth, the cross, death and matter:

It is the moon, the mother of all things, the vessel, it consists of opposites, has a thousand names, is an old woman and a whore, as Mater Alchimia it is wisdom and teaches wisdom, it contains the elixir of life in potentia and is the mother of the Savior and of the *filius Macrocosma*, it is the earth and the serpent hidden in the earth, the blackness and the dew and the miraculous water which brings together all that is divided.[620]

The *prima materia* as king dissolves in the *prima materia* as salt water, or as sea, which represent the matrix and emotion, as bitter salt water constitutes tears and tragic affect (the consequence of desire's failure). The heat which promotes chemical solution is the symbolic equivalent of passion, emotion or sensuality—aspects of the intrapsychic world, outside the domain of rational thought. The dissolution of the king in the matriarchal system thematically recreates the heroic/sacrificial motif of incest, creative (sexual) re-union with the mother. This creative reunion manifests itself first of all as psychological chaos, depression and anxiety, and only then as re-creation. The king is the son of God, in previously incarnated form, who loses his effectiveness in the course of time. Ritualistic primitive regicide is predicated upon the belief that the magic power of the king, his ability to renew his subjects and the land, decreases with age. Subjection to (intrapsychic and/or social) tyranny inevitably promotes stagnation and depression, dissolution. Nonetheless, challenge presented to the prevailing spirit of the times means the removal of knowledge from the context within which it has relevance, and the subsequent return of what is conditionally known to the domain of the terrible and promising unknown:

In order to enter into God's Kingdom the king must transform himself into the prima materia in the body of his mother, and return to the dark initial state which the alchemists called the "chaos." In this *massa confusa* the elements are in conflict and repel one another; all connections are dissolved. Dissolution is the prerequisite for redemption. The celebrant of the mysteries had to suffer a figurative death in order to attain transformation.[621]

The matriarchal "realm," which Jung personified in imagistic representation as the *anima*, is the source of new knowledge, as the unknown. This makes the matriarchal realm "mother/wisdom," matrix of the revelation that renews. Such revelation necessarily threatens the stability of previous knowledge, however, and "releases" previously "inhibited" affect (in consequence of the dissolution of predictability and certainty). Jung states:

The anima becomes creative when the king renews himself in her. Psychologically the king stands first of all for Sol, whom we have interpreted as consciousness. But over and above that he represents a dominant of consciousness, such as a generally accepted principle or a collective conviction or a tradition. These systems and ruling ideas "age" and thereby forcibly bring about a "metamorphosis of the gods." ... [This] seldom occurs as a definite collective phenomenon. Mostly it is a change in the individual which may, under certain conditions, affect society "when the time is fulfilled." In the individual it only means that the ruling idea is in need of renewal and alteration if it is to deal adequately with the changed outer or inner conditions.[622]

The king's original state of sickness is the certain eventual fate of concrete knowledge, posited as absolute. Since the unknown always transcends the limits of the known, no final statement about the nature of existence is possible. Attempts to limit knowledge to what is presently known must therefore necessarily result in eventual social and psychological stagnation. It is an unfortunate fact that attempts to overcome such stagnation must first result in the production of affective, motivational and ideational chaos. Dissolution of the patriarchal in the matriarchal system, even in the voluntary pursuit of an ideal, culminates in the creation of a psychically chaotic state, symbolized in alchemy as the sickbed of the king, as the pregnancy of the queen, or as some analogical equivalent thereof. The chaotic state engendered consequent to decision to pursue the unknown is accompanied by emergence of various constitutive psychological factors in fantasy, embodied, personified, as opposing forces, lacking mediating principle. This is precisely equivalent to an "internal" return to a state of polytheism, where the "gods who rule humanity" war without subjection to a higher-order "power." The alchemists described this stage of their *opus* as the *nigredo*, or blackness—a condition we would associate with depression, psychological chaos, uncertainty, impulsivity and anxiety.

Blackness descends when the motivational significance of events and processes, previously held in check through adherence to a central, paradigmatically structured set of beliefs, implicit and explicit, becomes once again indeterminate and novel. The dissolution of their previously held beliefs allowed the fundamental constituent structures of the alchemists' psyche to become actively personified in fantasy. Elements of individual, patriarchal, and matriarchal systems vie in competition, lacking uniting principle, abandoned in the pursuit of the unknown. Jung states:

This [initial] battle is the *separatio, divisio, putrefactio, mortificatio,* and *solutio,* which all represent the original chaotic state of conflict....[623] Dorn describes this vicious, warlike [state] allegorically as the four-horned serpent, which the devil, after his fall from heaven, sought to "infix" in the mind of man. Dorn puts the motif of war on a moral plane and thereby approximates it to the modern concept of psychic dissociation, which, as we know, lies at the root of the psychogenic psychoses and neuroses. In the "furnace of the cross" and in the fire, says the "Aquarium sapientum," "man, like the earthly gold, attains to the true black Raven's head; that is, he is utterly disfigured and is held in derision by the world, and this not only for forty days and nights, or years, but often for the whole duration of his life; so much so that he experiences more heartache in his life than comfort and joy, and more sadness than pleasure.... Through this spiritual death his soul is entirely freed." Evidently the *nigredo* brought about a deformation and a psychic suffering which the author compared to the plight of the unfortunate Job.

Job's unmerited misfortune, visited upon him by God, is the suffering of God's servant and a prefiguration of Christ's passion.[624]

Identification with the pre-existent cultural canon—or pretence of such identification—provides protection against the unknown, and context for knowledge, but promotes tyranny. The final cost of this identification is the lie—denial of deviance and the unknown. When such identification is abandoned voluntarily or rendered impossible by circumstantial change, the affects "held in check" by the integrity of the previous classification system are free once again to manifest themselves. The "dissolution of the king" means that much of what was previously understood reverts to the unknown. This might be regarded as the reversal of the historical process that made of all gods one supreme god or, speaking more psychologically, as the war of conflicting drives, desires and "subpersonalities" that ensues when an overarching hierarchy of values has collapsed. Such a "reversion" places the individual in a state characterized by great uncertainty, frustration, depression and turmoil.

The Peregrination

The alchemists believed that perfection was characterized by a state of unity, in which all "competing opposites" were united. The final stage of the alchemical procedure—the conjunction—was therefore preceded first by recognition and identification of all the diverse "aspects of the psyche" warring in opposition in the "belly of the uroboric dragon":

What, then, do the statements of the alchemists concerning their arcanum mean, looked at psychologically? In order to answer this question we must remember the working hypothesis we have used for the interpretation of dreams: the images in dreams and spontaneous fantasies are symbols, that is, the best possible formulation for still unknown or unconscious facts, which generally compensate the content of consciousness or the conscious attitude. If we apply this basic rule to the alchemical arcanum, we come to the conclusion that its most conspicuous quality, namely, *its unity and uniqueness*—one is the stone, one the medicine, one the vessel, one the procedure, and one the disposition—presupposes a *dissociated consciousness*. For no one who is one himself needs oneness as a medicine—nor, we might add, does anyone who is unconscious of his dissociation, for a *conscious* situation of distress is needed in order to activate the archetype of unity. From this we may conclude that the more philosophically minded alchemists were people who did not feel satisfied with the then prevailing view of the world, that is, with the Christian faith, although they were convinced of its truth. In this latter respect we find in the classical Latin and Greek literature of alchemy no evidences to the contrary, but rather, so far as Christian treatises are concerned, abundant testimony to the firmness of their Christian convictions. Since Christianity is expressly a system of "salvation," founded moreover on God's "plan of redemption," and God is unity *par excellence*, one must ask oneself why the alchemists still felt a disunity in themselves, or not at one with themselves, when their faith, so it would appear, gave them every opportunity for unity and unison. (This question has lost nothing of its topicality today, on the contrary!)[625]

This global recognition was conceptualized, variously, as a "journey to the four corners of the earth"—the *peregrination*—or as familiarization with every aspect of being, as a vast expansion of self-knowledge. The incorporation of all competing states of motivation into a single hierarchy of value presupposes recognition of all diverse (painful, uncomfortable, difficult to manage) desires, and the "forging" of an agreement between them. This can be most accurately viewed as a potentially "redemptive" expansion of self-consciousness. It might be said: emergence of the limited self-consciousness symbolically represented in myths of the Fall constituted grounds for the descent of man. The alchemical philosophers—meditating endlessly on the nature of perfection, or the transformative processes necessary for the production of perfection—came to "realize" that increased self-consciousness might constitute recompense for expulsion from paradise. But identification of all competing desires meant clear-headed recognition of the truly tragic situation of man, and of all the "sinful" and mortal weaknesses, sins and insufficiencies associated with individual being—and then the attempt to come to real terms with that situation, and those limitations.

If you are a miserable and disorganized fool, producing chaos wherever you go, it is tremendously painful to recognize yourself—and to see the enormity of the job ahead of you. It is very difficult to replace delusional identification with the *persona* with clear-headed apprehension of the real (and insufficient) individual personality. This clearer vision or conception is something attained at no small cost (and this says nothing about the cost of transforming that conception into action). The *"savage, wild animal"*—the "hungry robber, the wolf, lion, and other ravening beasts"[626] served as apt representative of the "unredeemed individual," from the alchemical perspective. Emergence and recognition of this animal is necessary precondition to his transformation. This idea is represented imagistically in *Figure 64: The Wolf as Prima Materia, Devouring the Dead King*.[627] This "devouring of the dead king" by the now-recognized "beast of the underworld" is very much akin to Solzhenitsyn's discovery of his personal responsibility for the Gulag that imprisoned him. For a typical modern, an equal shock might be produced by his discovery of identity with the Nazi. The barbarians of Hitler's state were *normal men—normal men, like you (and me)*. This cannot be emphasized strongly enough. But the "normal man" does not conceptualize himself as Nazi. This means, merely, that his self-concept provides him with illusory security (as a prison protects its inmates from the outside). But the Nazi actions—that is, the willful torture of innocents, and enjoyment of such—is well within the normal man's range of capacities (and does not likely exhaust them). The individual *is a terrible force for evil*. Recognition of that force—real recognition, the kind that comes as a staggering blow—is a precondition for any profound improvement in character. By such improvement, I mean the capacity to bear the tragedy of existence, to transcend that tragedy—and not to degenerate instead into something "unconsciously" desirous of disseminating pain and misery. Jung states:

It is worth noting that the animal is the symbolic carrier of the self [the psychic totality]. This hint in Maier is borne out by modern individuals who have no notion of alchemy. It expresses the fact that the structure of wholeness was always present but was buried in profound unconsciousness, where it can

Figure 64: The Wolf as Prima Materia, Devouring the Dead King

always be found again if one is willing to risk one's skin to attain the greatest possible range of consciousness through the greatest possible self-knowledge—a "harsh and bitter drink" usually reserved for hell. The throne of God seems to be no unworthy reward for such trials. For self-knowledge—in the total meaning of the word—is not a one-sided intellectual pastime but a journey through the four continents, where one is exposed to all the dangers of land, sea, air and fire. Any total act of recognition worthy of the name embraces the four—or 360!—aspects of existence. Nothing may be "disregarded." When Ignatius Loyalus recommended "imagination through the five senses" to the meditant, and told him to imitate Christ "by use of his senses," what he had in mind was the fullest possible "realization" of the object of contemplation. Quite apart from the moral or other effects of this kind of meditation, its chief effect is the training of consciousness, of the capacity for concentration, and of attention and clarity of thought. The corresponding forms of Yoga have similar effects. But in contrast to these traditional modes of realization, where the meditant projects himself into some prescribed form, the self-knowledge alluded to by Maier is a projection into the empirical self as it actually is. It is not the "self" we like to imagine ourselves to be after carefully removing all the blemishes, but the empirical ego just as it is, with everything that it does and everything that happens to it. Everybody would like to be quit of this odious adjunct, which is precisely why in the East the ego is explained as illusion and why in the West it is offered up in sacrifice to the Christ figure.

By contrast, the aim of the mystical peregrination is to understand all parts of the world, to achieve the greatest possible extension of consciousness, as though its guiding principle were the Carpocratic idea that one is delivered from no sin which one has not committed. Not a turning away from its empirical "so-ness," but the fullest possible experience of the ego as reflected in the "ten thousand things"—that is the goal of the peregrination.[628]

The mask each person wears in society is based upon the pretence that the individual is identical with his culture (usually, with the "best elements" of that culture). The fool, hiding behind the mask, is composed of individual deviance, which is deceitfully avoided, lied about, out of fear. This deviant, unlived life contains the worst and the best tendencies of the individual, suppressed by cultural opinion because they threaten the norm; forced underground by the individual himself, because they threaten personal short-term psychological stability (which means group identification and ongoing inhibition of fear). In the absence of an integrated hierarchical moral (patriarchal) system, competing values and viewpoints tend toward disintegration, as each pursues its own end—as greed might make the pursuit of lust difficult, as hunger might render love impossible. When a moral system undergoes dissolution and loses its absolute validity—its higher moral structure—the values which it held in union revert to incompatability, at least from the conscious viewpoint. This war of conflicting values—of which each is in itself a necessary "divine force"—engenders confusion, disorientation, and despair. Such despair—which can be truly unbearable—might be considered the first pitfall of moral transformation. Mere contemplation of the possibility of such a state usually engenders sufficient discomfort to bring further moral development to a halt. The alchemist, however, implicitly adopted a heroic role when he voluntarily determined to pursue the unknown, in search of the ideal. His unconscious identification with this eternal image, his active incarnation of the mythological role, enabled him to persevere in his quest, in the face of grave difficulty. Jung states:

Only the living presence of the eternal images can lend the human psyche a dignity which makes it morally possible for a man to stand by his own soul, and be convinced that it is worth his while to persevere with it. Only then will he realize that the conflict is *in him*, that the discord and tribulation are his riches, which should not be squandered by attacking others; and that, if fate should exact a debt from him in the form of guilt, it is a debt to himself. Then he will recognize the worth of his psyche, for nobody can owe a debt to a mere nothing. But when he loses his own values he becomes a hungry robber, the wolf, lion, and other ravening beasts which for the alchemists symbolized the appetites that break loose when the black waters of chaos—i.e., the unconsciousness of projection—have swallowed up the king.[629]

The unknown is contaminated with the psychoanalytic "unconscious," so to speak, because everything we do not know about ourselves, and everything we have experienced and assimilated but not accommodated to, has the same affective status as everything that exists merely as potential. All thoughts and impulses we avoid or suppress, because they threaten our self-conception or notion of the world—and all fantasies we experience, but do not admit to—exist in the same domain as chaos, the mother of all things, and serve to undermine our faith in our most vital presumptions. The encounter with the "unknown," therefore, is simultaneously encounter with those aspects of our selves heretofore defined as other (despite their indisputable "existence"). This integration means making behavioral potentialities previously disregarded available for conscious use; means (re)construction of a self-model that accurately represents such potential.

Experiences which are currently deemed taboo—forbidden, from the perspective of the currently extant moral schema—may therefore contain within them seeds of creative solution to problems that remain unsolved or may arise in the future. Taboo experience may yet constitute "unmined" and redemptive *possibility*. Tales of the "traveling sage," "wandering magician" or "courageous adventurer" constitute recognition of the utility of such potential. From the perspective of such narratives, a totality of experience and action is the necessary precondition for the attainment of wisdom. This "total immersion in life" is the mystical "peregrination" of the medieval alchemist, in search of the philosopher's stone, or the journey of Buddha through the complete sensory, erotic and philosophical realms, prior to his attainment of enlightenment. The ritual of pilgrimage—the "journey to the holy city"—constitutes half-ritual, half-dramatic enactment of this idea. The pilgrim voluntarily places him- or herself outside the "protective walls" of original culture and, through the difficult and demanding (actual) journey to "unknown but holy lands," catalyzes a psychological process of broadening, integration and maturation. It is in this manner that a true "quest" inevitably fulfills itself, even though its "final, impossible goal" (the Holy Grail, for example) may remain concretely unattained.

The necessity for experience as the precondition for wisdom may appear self-evident, once due consideration has been applied to the problem (since wisdom is obviously "derived" from experience)—but the crux of the matter is that those elements of experience that foster denial or avoidance (and therefore remain unencountered or unprocessed) always border on the maddening. This is particularly true from the psychological, rather than ritual, perspective. The holy pilgrimage in its abstract or spiritual version is the journey through "elements" of experience and personal character that constitute the subjective world of experience (rather than the shared social and natural world). The inner world is divided into familiar and unknown territory, much as the outer. The psychological purpose of the rite-of-passage adventure (and the reason for the popularity of such journeys, in actuality and in drama) is the development of character, in consequence of confrontation with the unknown. A "journey to the place that is most feared," however, can be undertaken spiritually much as concretely. What "spiritually" means, however, in such a context, is a "peregrination" through the rejected, hated and violently suppressed aspects of personal experience. This is most literally a voyage to the land of the enemy—to the heart of darkness.

When experience calls the absolute validity of a given belief system into question, the validity of the definitions of immorality—and of enmity—contained with that system also become questionable:

For one may doubt, first, whether there are any opposites at all, and secondly whether these popular valuations and opposite values on which the metaphysicians put their seal, are not perhaps merely foreground estimates, only provisional perspectives, perhaps even from some nook, perhaps from below, frog perspectives, as it were, to borrow an expression painters use. For all the value that the true, the truthful, the selfless may deserve, it would still be possible that a higher and more fundamental value for life might have to be ascribed to deception, selfishness and lust. It might even be possible that what constitutes the value of these good and revered things is precisely that they are insidiously related, tied to, and involved with these wicked, seemingly opposite things—maybe even one with them in essence.[630]

Recognition of potential in the transformation of the *prima materia* meant re-encounter with personal experience formerly suppressed by cultural pressure and personal decision. Such experience might have included hatred, cruelty, physical passion, greed, cowardice, confusion, doubt, flight of imagination, freedom of thought and personal talent. Things we avoid or deny are precisely those things that transcend our individual competence, as presently construed—the things or situations that define our limitations, and that represent inferiority, failure, decomposition, weakness and death. This means that everything despised and feared, every object of hatred and contempt, everything signifying cowardice, ruthlessness, ignorance—every experience that cries out for denial—may yet constitute information necessary for life. Jung states:

In general, the alchemists strove for a *total* union of opposites in symbolic form, and this they regarded as the indispensable condition for the healing of all ills. Hence they sought to find ways and means to produce that substance in which all opposites were united.[631]

Alchemy speaks of "the union of soul-sparks, to produce the gold." These sparks—*scintillae*—are "the light in the darkness," the consciousness associated with poorly integrated or even hostile elements of individual personality.[632] The germ or seed of unity may manifest itself symbolically at any time in the procedure, and comes to dominate later if procedure is successful. This center—Jung's "self"[633]—unites the disparate elements (the "compulsion of the stars") into *one*, in the course of a circular, cyclical journey (in the course of the revolutionary spiral path of the way). This emergent center was regarded by the alchemists as the spirit Mercurius (the trickster, who was "embedded" in matter), or as the mythical "pelican," who fed her offspring with her own body and blood, and was therefore an allegory both of Christ and the (self-nourishing) *uroboros*. The center was also regarded as the philosophical stone (the *solitaire*) "rejected by the builders," directly identified with Christ, and as the rock

Figure 65: Dragon of Chaos as "Birthplace" of Christ and the Lapis

upon which security itself might be founded. This stone, this immovable and indestructible center, "incorporated" the patriarchal and matriarchal principles (the king and the queen), and was also regarded as the "offspring" of chaos, fertilized by order. The emergence of the lapis/Christ/pelican from the domain of the dragon of chaos is represented in *Figure 65: Dragon of Chaos as "Birthplace" of Christ and the Lapis.*[634]

The mythological hero faces the unknown, voluntarily, cuts it up, and makes the world out of its pieces; identifies and overcomes evil, and rescues the ancestral father, languishing in the underworld; unites, consciously, with the virgin mother, and produces the divine child; and mediates between opposing and warlike kings. He is, therefore, explorer, creator, lover, judge and peacemaker. The hero is also he who has traveled everywhere—he who has "mastered strange territory" (even that inhabited by his enemy). This "traveling everywhere" and "mastering of strange territory" has a psychological significance and a social meaning: the divine hero knows and understands the "ways of the enemy" and can use them to advantage.

The Conjunction

The process of complete recognition, symbolized or dramatized as the peregrination, sets the stage for activation of the final alchemical sequence, which consisted of the (hypothetical) union of all now-manifest "things." Jung outlines the "Arisleus vision," in his text *Psychology and Alchemy.* This vision contains all the elements of the alchemical "theory," portrayed in episodic/narrative form. Its sequential analysis helps shed dramatic light on the nature of the "conjunction":

Arisleus (a Byzantine alchemist of the 8th or 9th century) tells of his adventures with the *Rex Marinus,* in whose kingdom nothing prospers and nothing is begotten. Moreover, there are no philosophers there. Only like mates with like, consequently there is no procreation. The king must seek the counsel of the philosophers and mate Thabritius with Beya, his two children whom he has hatched in his brain.[635]

Jung comments:

Thabritius is the masculine, spiritual principle of light and Logos which, like the Gnostic Nous, sinks into the embrace of physical nature.[636]

This is an elaboration of an idea presented earlier:

Nous seems to be identical with the God Anthropos: he appears alongside the demiurge and is the adversary of the planetary spheres. He rends the circle of the spheres and leans down to earth and water (i.e., is about to project himself into the elements). His shadow falls upon the earth, but his image is reflected in the water. This kindles the love of the elements, and he himself is so charmed with the reflected image of divine beauty that he would fain take up his abode within it. But scarcely has he set foot upon the earth when Physis locks him in a passionate embrace.[637]

It is important to understand this commentary, as well, to completely appreciate the nature of the *prima materia*. The *prima materia*—Physis—contains *spirit*, the masculine principle, as well as *matter*, the feminine *(Beya,* in this narrative). The prima materia—dragon of chaos—

serves simultaneously as the source of things, the subject to whom things appear, and the representations of the things characteristic of that subject. This is not a mere material "source"; it is the absolute unknown itself, in whose embrace spirit "sleeps," until it is released (in the course of the exploration that transforms the self, as well as producing something "real" and new). Jung continues, with an idea that we are now familiar with:

When we are told that the King is ... inanimate, or that his land is unfruitful, it is equivalent to saying that the hidden state is one of latency and potentiality. The darkness and depths of the sea [which stand for the unknown] symbolize the unconscious state of an invisible content that is projected. Inasmuch as such a content belongs to the total personality, and is only apparently severed from its context by projection, there is always an attraction between conscious mind and projected content. Generally it takes the form of a fascination. This, in the alchemical allegory, is expressed by the King's cry for help from the depths of his unconscious, dissociated state. The conscious mind should respond to this call: one should ... render service to the King, for this would be not only wisdom, but salvation as well.

Yet this brings with it the necessity of a descent into the dark world of the unconscious ["the unknown"] ... the perilous adventure of the night sea journey, whose end and aim is the restoration of life, resurrection, and the triumph over death.[638]

Despite the risk, Arisleus and his imaginary "companions" brave the quest into the kingdom of the submerged king. This quest ends terribly, with the death of Thabritius. His death echoes that of Osiris, and symbolizes the completion of the spirit's descent into "matter" or the unconscious or the unknown (where it then lies "implicit" or "unrevealed," and "calls for rescue," offering riches to its redeemer). Jung continues with the story:

The death of the King's son is naturally a delicate and dangerous matter. By descending into the unconscious, the conscious mind puts itself in a perilous position, for it is apparently extinguishing itself. It is in the situation of the primitive hero who is devoured by the dragon....

The deliberate and indeed wanton provocation of this state is a sacrilege or breach of taboo attended by the severest punishments. Accordingly, the King imprisons Arisleus and his companions in a triple glass house together with the corpse of the King's son. The heroes are held captive in the underworld at the bottom of the sea, where, exposed to every kind of terror, they languish for eighty days in an intense heat. At the request of Arisleus, Beya is imprisoned with them. [The *Rosarium* version of the "Visio" interprets the prison as Beya's womb.]

Clearly, they have been overpowered by the unconscious ["the unknown"] and are helplessly abandoned, which means that they have volunteered to die in order to beget a new and fruitful life in that region of the psyche which has hitherto lain fallow in darkest unconsciousness, and under the shadow of death.[639]

The "purpose" of the story, in describing this descent, is demonstration that "only in the region of danger (watery abyss, cavern, forest, island, castle, etc.) can one find the 'treasure hard to attain' (jewel, virgin, life-potion, victory over death)."[640] Jung ends his commentary:

The dread and resistance which every natural human being experiences when it comes to delving too deeply into himself is, at bottom, the fear of the journey to Hades. If it were only resistance that he felt, it would not be so bad. In actual fact, however, the psychic substratum, that dark realm of the unknown, exercises a fascinating attraction that threatens to become the more overpowering the further he penetrates into it. The psychological danger that arises here is the disintegration of personality into its functional components, i.e., the separate functions of consciousness, the complexes, hereditary units, etc. Disintegration—which may be functional or occasionally a real schizophrenia—is the fate which overtakes Gabricus (in the *Rosarium* version): he is dissolved into atoms in the body of Beya.... So long as consciousness refrains from acting, the opposites will remain dormant in the unconscious. Once they have been activated, the *regius filius*—spirit, Logos, Nous—is swallowed up by Physis.... In the hero myth this state is known as being swallowed up in the belly of the whale or dragon.

The heat there is usually so intense [a consequence of the war of affects; anxiety, anger] that the hero loses his hair, and is reborn as bald as a babe.... The philosopher makes the journey to hell as a "redeemer."[641]

The story continues:

Earlier on, we left Arisleus and his companions, together with Beya and the dead Thabritius, in the triple glass house where they had been imprisoned by the *Rex Marinus*. They suffer from the intense heat, like the three whom Nebuchadnezzar cast into the fiery furnace. King Nebuchadnezzar had a vision of a fourth, "like the son of God," as we are told in Daniel 3:25.

This vision is not without bearing on alchemy, since there are numerous passages in the literature stating that the stone is *trinus et unus*. It consists of the four elements, with fire representing the spirit concealed in matter. This is the fourth, absent and yet present, who always appears in the fiery agony of the furnace and symbolizes the divine presence—succour and the completion of the work.

And, in their hour of need, Arisleus and his companions see their master Pythagoras in a dream and beg him for help. He sends them his disciple Harforetus, the "author of nourishment." So the work is completed and Thabritius comes to life again. We may suppose that Harforetus brought them the miraculous food [akin to the host], though this only becomes clear through a discovery of Ruska's, who gave us access to the text of the Codex Berolinensis. There, in an introduction that is missing from the printed versions of the "Visio," we read: "Pythagoras says, 'Ye write and have written down for posterity how this most precious tree is planted, and how he that eats of its fruits shall hunger no more.'"[642]

The alchemical opus meant, at one level of analysis, the complete integration of "unknown" and "known," insofar as that could be attained—but more profoundly, participation in the process that made "one thing" of unknown and known. This construct and act of construction typically had twin "final" aims, insofar as it constituted the pursuit of perfection: The first of these aims was union of the feminine, maternal background of the "unknown 'material' world," seething with danger, passion and sensuality, into harmony with the ordering principle of the spirit. [This was represented symbolically as dissolution of the dead king and his subsequent regeneration, after eating the miraculous food (which is the beneficial aspect of the unknown, and the hero, simultaneously)]. The second final aim was *re-introduction of the integrated psychic structure to the physical body*—the conscious "incarnation" of the now-more-complete spirit. So this meant that the union attained by the (re)incorporation of the "material unknown" was not complete, if it was still a matter of philosophy or abstract conceptualization: the well-integrated spirit also had to be realized in behavior. And this was not necessarily yet even the final stage. The alchemist Dorn states:

> We conclude that meditative philosophy consists in the overcoming of the body by mental union (*unio mentalis*). This first union does not as yet make the wise man, but only the mental disciple of wisdom. The second union of the mind with the body shows forth the wise man, hoping for and expecting that blessed third union with the first unity [the *unus mundus*, the latent unity of the world]. May Almighty God grant that all men be made such, and may He be one in All.[643]

Dorn's ideas refer to a conjunction conceptualized as a *three-stage* process. The first stage was "union of the mind" (the "overcoming the body by mental union"). This stage refers to the integration of "states of motivation" (drives, emotions) into a single hierarchy, dominated by the figure of the exploratory hero. The second stage was (re)union of the united mind with the body. This is analogous to the "second stage" of the hero's journey. After the treasure is released, consequential to the battle with the dragon, the purely personal aspect of the hero's journey is completed. After all, he has found the "treasure hard to attain." But the hero must return to the community. This is equivalent to Buddha's determination to retire from the state of Nirvana, until all who were living could make their home there; is tantamount to the Buddha's belief that the redemption of the one was impossible, in the presence of the unredeemed many. The reunion of the united mind with the body is inculcation of proper attitude in action (and is, therefore, the effect of the hero on the world).

The third stage is particularly difficult to comprehend. Reconsideration of the theme of the "tailor who mends"[644]—and who can therefore sew up the hole in the sky, made by the dying king—might help with initial comprehension. Things that are wrong must be set right. This is a psychological process, even if it undertaken purely as a consequence of actions conducted "in the outside world." The union of the united spirit/body with the world means recognition of the essential equivalence of all experience, or consideration of all aspects of experience as literally equivalent to the self. We presume the existence of a final barrier between "subject" and "object," but a standpoint exists that gives to all aspects of individual experience—whether "subjective" or "objective"—equal status, *as aspects of experience*. Redeeming any aspect of that experience, then—whether "material" or "psychological"; whether "self" or "other" *is then regarded as the same act*—as the act whose purpose is establishment of the "kingdom of god" (which is simultaneously psychological and social state). "Spiritual work" may therefore be regarded as indistinguishable from "work on the external circumstances of existence": redeem yourself, redeem the world. Or, alternatively: the attempt to bring about the perfection of the external world may be regarded as equivalent to the attempt to perfect oneself. After all, dedication to an ideal necessitates development of self-discipline. This is voluntary apprenticeship. The world and the self are not different places; from this perspective, "all is experience." The attempt to redeem either necessarily brings about redemption in the other.

All three of these conjunctions may be represented symbolically by the *syzygy*, the "divine union of opposites," most generally considered as male and female:

1. first, "known" (previous knowledge, subsumed under patriarchal/spiritual category) + "unknown" (anomaly, subsumed under matriarchal/affective/material/physical category) = "united spirit";
2. then "united spirit" (in this context, subsumed into the patriarchal/spiritual category) + "body" (subsumed under matriarchal/material category) = "united spirit/body";
3. then "united spirit/body" (in this context, subsumed under the patriarchal/spiritual category) + "world" (matriarchal/material/category) = "united spirit/body/world."

All three of these unions can be considered variants of the "incest motif," (brother/sister, son/mother, king/queen pairings). Stage one, the "mental union," was construed as necessary, valuable, but incomplete: the attainment of an ordered subjective state (stage two) was another important step along the way:

Learn therefore, O Mind, to practise sympathetic love in regard to thine own body, by restraining its vain appetites, that it may be apt with thee in all things. To this end I shall labour, that it may drink with thee from the fountain of strength and, when the two are made one, that ye find peace in their union. Draw nigh, O Body, to this fountain, that with thy Mind thou mayest drink to satiety and hereafter thirst no more after vanities. O wondrous efficacy of this fount, which maketh one of two, and peace between enemies! The fount of love can make *mind* out of spirit and soul, but this maketh *one man* out of mind and body.[645]

The third step, however, was critical: philosophical knowledge and ordered intrapsychic structure—even when embodied—was regarded as insufficient. That embodied union must be extended to all the world—regarded as "an aspect of experience" and, therefore, as equivalent (even identical) to the self.

The alchemical procedure was based on the attempt to redeem "matter," to transform it into an ideal. This procedure operated on the assumption that matter was originally corrupted—like man, in the story of Genesis. The study of the transformations of corruption and limitation activated a mythological sequence in the mind of the alchemist. This sequence followed the pattern of the way, upon which all religions have developed. Formal Christianity adopted the position that the sacrifice of Christ brought history to a close, and that "belief" in that sacrifice guaranteed redemption. Alchemy rejected that position, in its pursuit of what remained unknown. In that (heroic) pursuit the alchemist found himself transformed:

Whereas the Christian belief is that man is freed from sin by the redemptory act of Christ, the alchemist was evidently of the opinion that the "restitution to the likeness of original and incorrupt nature" had still to be accomplished by the art, and this can only mean that Christ's work of redemption was regarded as incomplete. In view of the wickedness which the "Prince of this world," undeterred, goes on perpetrating as liberally as before, one cannot withhold all sympathy from such an opinion. For an alchemist who professed allegiance to the Ecclesia spiritualis it was naturally of supreme importance to make himself an "unspotted vessel" of the Paraclete and thus to realize the idea "Christ" on a plane far transcending a mere imitation of him.[646]

This "realization of Christ on a plane transcending imitation" is an overwhelming idea. It makes of religious "belief" something far more than belief—something far more terrifying, and far more promising. The sequence of the alchemical transformation paralleled Christ's Passion, paralleled the myth of the hero and his redemption. The essential message of alchemy is that individual rejection of tyranny, voluntary pursuit of the unknown and terrifying—predicated upon faith in the ideal—may engender an individual transformation so overwhelming that its equivalent can only be found in the most profound of religious myths:

The Son of the great World who is Theocosmos, i.e., a divine power and world (but whom even today, unfortunately, many who teach nature in a pagan spirit and many builders of medical science reject in the high university schools), is the exemplar of the stone which is Theanthropos, i.e., God and man (whom, as Scripture tells us, the builders of the Church have also rejected); and from the same, in and from the Great World Book of Nature, [there issues] a continuous and everlasting doctrine for the wise and their children: indeed, it is a splendid living likeness of our Savior Jesus Christ, in and from the Great World which by nature is very similar to him (as to miraculous conception, birth, inexpressible powers, virtues, and effects); so God our Lord, besides his Son's Biblical histories, has also created a specific image and natural representation for us in the Book of Nature.[647]

It was in pursuit of the unknown that the alchemist experienced this psychological transformation, just as it was originally in contact with the unknown that the (monotheistic) patriarchal system developed, in the furthest reaches of history. It is the symbolic expression of *the action of instinct*, which manifests itself in some variant of the hero myth, whenever the unknown is pursued, without avoidance, in the attempt to improve life. The alchemist experienced what the individual always experiences when he determines to face every aspect of his existence (individual and collective) without denial or recourse to sterile preconceptions.

The passion that vibrates in [the alchemical texts] is genuine, but would be totally incomprehensible if the lapis were nothing but a chemical substance. Nor does it originate in contemplation of Christ's Passion; it is the real experience of a man who has got involved in the compensatory contents of the unconscious by investigating the unknown, seriously and to the point of self-sacrifice. He could not but see the likeness of his projected contents to the dogmatic images [which were in fact likely utilized by the instinctual procedure], and he might have been tempted to assume that his ideas were nothing else than the familiar religious conceptions, which he was using in order to explain the chemical procedure. But the texts show clearly that, on the contrary, a real experience of the opus had an increasing tendency to assimilate the dogma or to amplify itself with it.[648]

In Christianity, spirit descends to matter, and the result of the union is the birth of Christ (and, unfortunately, the formal realization of his opponent). In alchemy, which compensated for the one-sided view of Christianity, matter rises to spirit, with analogous result: creation of the *lapis* or philosophical stone, which bears an unmistakable resemblance to Christ, embodied in abstractly material form. This form, the philosopher's stone, the *lapis*, was composed of the most paradoxical elements: it was base, cheap, immature and volatile; perfect, precious, ancient and solid; visible to all yet mysterious; costly, dark, hidden and evident, having one name and many names. The *lapis* was also the renewed king, the wise old man, and the child. The wise old man posesses the charisma of wisdom, which is the knowledge that transcends the limits of history. The child represents the creative spirit, the possibility in man, the Holy Ghost. He is not the child of ignorance, but the innocence of maturity. He precedes and antedates history in the subjective and collective sense:

The "child" is all that is abandoned and exposed and at the same time divinely powerful; the insignificant, dubious beginning, and the triumphal end. The "eternal child" in man is an indescribable experience, an incongruity, a handicap, and a divine prerogative; an imponderable that determines the ultimate worth or worthlessness of a personality.[649]

This final value, the goal of the pursuit of the alchemists, is discovery and embodiment of the meaning of life itself: integrated subjective being actively expressing its nature through manipulation of the possibilities inherent in the material/unknown world. This final goal is the production of an integrated intrapsychic condition—identical to that of the mythological hero—"acted out" in a world regarded as equivalent to the self. Production of this condition—the *lapis philosophorum*—constitutes the "antidote" for the "corruption of the world," attendant upon the Fall [attendant upon the emergence of (partial) self-consciousness]. The *lapis* is "agent of transformation," equivalent to the mythological redemptive hero—able to to turn "base metals into gold." It is, as such, *something more valuable than gold*—just as the hero is more valuable than any of his concrete productions. The "complete" alchemical opus—with production of the lapis as goal—is presented schematically in *Figure 66: The Alchemical Opus as Myth of Redemption.*

Alchemy was a living myth: the myth of the individual man as redeemer. Organized Christianity had "sterilized itself," so to speak, by insisting on the worship of some external truth as the means to salvation. The alchemists (re)discovered the error of this presumption, and came to realize that *identification* with the redeemer was in fact necessary, not his worship; that myths of redemption had true power when they were incorporated, and acted out, rather than believed, in some abstract sense. This meant: to *say* that Christ was "the greatest man in history"—a combination of the divine and mortal—was *not sufficient expression of*

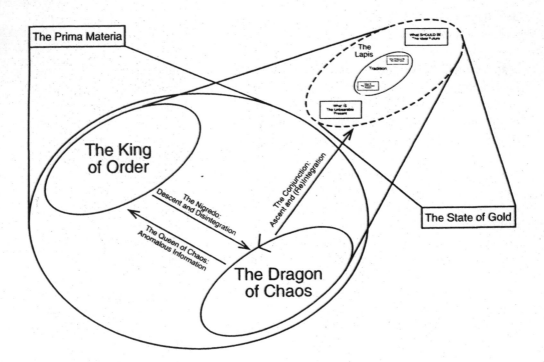

Figure 66: The Alchemical Opus as Myth of Redemption

faith. Sufficient expression meant the attempt to live out the myth of the hero, within the confines of individual personality—to voluntarily shoulder the cross of existence, to "unite the opposites" within a single breast, and to serve as active conscious mediator between the eternal generative forces of known and unknown.

CONCLUSION: THE DIVINITY OF INTEREST

Anomalies manifest themselves on the border between chaos and order, so to speak, and have a threatening and promising aspect. The promising aspect dominates, when the contact is voluntary, when the exploring agent is up-to-date—when the individual has explored all previous anomalies, released the "information" they contained, and built a strong personality and steady "world" from that information. The threatening aspect dominates, when the contact is involuntary, when the exploring agent is not up-to-date—when the individual has run away from evidence of his previous errors, failed to extract the information lurking behind his mistakes, weakened his personality, and destabilized his world.

The phenomenon of interest—that precursor to exploratory behavior—signals the presence of a potentially beneficial anomaly. Interest manifests itself where an assimilable but novel phenomenon exists: where something new hides in a partially comprehensible form. Devout adherence to the dictates of interest—assuming a suitably disciplined character—therefore ensures stabilization and renewal of personality and world.

Interest is a spirit beckoning from the unknown, a spirit calling from outside the "walls" of society. Pursuit of individual interest means hearkening to this spirit's call, journeying outside the protective walls of childhood dependence and adolescent group identification, and returning to rejuvenate society. This means that pursuit of individual interest—development of true individuality—is equivalent to identification with the hero. Such identification renders the world bearable, despite its tragedies, and reduces neurotic suffering, which destroys faith, to an absolute minimum.

This is the message that everyone wants to hear. Risk your security. Face the unknown. Quit lying to yourself, and do what your heart truly tells you to do. You will be better for it, and so will the world.

Introduction

Where does one not encounter that veiled glance which burdens one with a profound sadness, that inward-turned glance of the born failure which betrays how such a man speaks to himself—that glance which is a sigh! "If only I were someone else," sighs this glance: "but there is no hope of that. I am who I am: how could I ever get free of myself? And yet—I *am sick of myself!*"

It is on such soil, on swampy ground, that every weed, every poisonous plant grows, always so small, so hidden, so false, so saccharine. Here the worms of vengefulness and rancor swarm; here the air stinks of secrets and concealment; here the web of the most malicious of all conspiracies is being spun constantly—the conspiracy of the suffering against the well-constituted and victorious, here the aspect of the victorious is *hated.* And what mendaciousness is employed to disguise that this hatred is hatred! What a display of grand words and postures, what an art of "honest" calumny! These failures: what noble eloquence flows from their lips! How much sugary, slimy, humble submissiveness swims in their eyes! What do they really want? At least to *represent* justice, love, wisdom, superiority—that is the ambition of the "lowest," the sick. And how skillful such an ambition makes them! Admire above all the forger's skill with which the stamp of virtue, even the ring, the golden-sounding ring of virtue, is here counterfeited. They monopolize virtue, these weak, hopelessly sick people, there is no doubt of it: "we alone are the good and just," they say, "we alone are *homines bonae voluntatis.*" They walk among us as embodied reproaches, as warnings to us—as if health, well-constitutedness, strength, pride and sense of power were in themselves necessarily vicious things for which one must pay some day, and pay bitterly: how ready they themselves are at bottom to *make* one pay; how they crave to be *hangmen.*[650]

I was reading Jeffrey Burton Russell's *Mephistopheles: The Devil in the Modern World*,[650] when I came across his discussion of Dostoyevsky's *The Brothers Karamazov*. Russell discusses Ivan's argument for atheism, which is perhaps the most powerful ever mounted:

Ivan's examples of evil, all taken from the daily newspapers of 1876, are unforgettable: the nobleman who orders his hounds to tear the peasant boy to pieces in front of his mother; the man who whips his struggling horse "on its gentle eyes"; the parents who lock their tiny daughter all night in the freezing privy while she knocks on the walls pleading for mercy; the Turk who entertains a baby with a shiny pistol before blowing its brains out. Ivan knows that such horrors occur daily and can be multiplied without end. "I took the case of children," Ivan explains, "to make my case clearer. Of the other tears with which the earth is soaked, I will say nothing."[652]

Russell states:

The relation of evil to God has in the century of Auschwitz and Hiroshima once again become a center of philosophical and theological discussion. The problem of evil can be stated simply: God is omnipotent; God is perfectly good; such a God would not permit evil to exist; but we observe that evil exists; therefore God does not exist. Variations on this theme are nearly infinite. The problem is not only abstract and philosophical, of course; it is also personal and immediate. Believers tend to forget that their God takes away everything that one cares about: possessions, comforts, success, profession or craft, knowledge, friends, family and life. What kind of God is this? Any decent religion must face this question squarely, and no answer is credible that cannot be given in the face of dying children.[653]

It seems to me that we use the horrors of the world to justify our own inadequacies. We make the presumption that human vulnerability is a sufficient cause of human cruelty. We blame God, and God's creation, for twisting and perverting our souls, and claim, all the time, to be innocent victims of circumstance. What do you say to a dying child? You say, "You can do it; there is something in you that is strong enough to do it." And you don't use the terrible vulnerability of children as an excuse for the rejection of existence, and the perpetration of conscious evil.

I do not have much experience as a clinical psychologist. Two of my patients, however, stay in my mind. The first was a woman, about thirty-five years old. She looked fifty. She reminded me of a medieval peasant—of my conception of a medieval peasant. She was dirty—clothes, hair, teeth; dirty with the kind of filth that takes months to develop. She was unbearably shy; she approached anyone who she thought was superior in status to her— which was virtually everyone—hunched over, with her eyes shaded by her hands, both hands, as if she could not tolerate the light emanating from her target.

She had been in behavioral treatment in Montreal, as an outpatient, before, and was in fact a sight known to the permanent staff at the clinic. Others had tried to help her overcome her unfortunate manner of self-presentation, which made people on the street shy away from her; made them regard her as crazy and unpredictable. She could learn to stand or sit up, temporarily, with eyes unguarded, but she reverted to her old habits as soon as she left the clinic.

She may have been intellectually impaired, in consequence of some biological fault; it was difficult to tell, because her environment was so appalling it may have caused her ignorance. She was illiterate, as well. She lived with her mother, whose character I knew nothing about, and with an elderly, desperately ill, bed-ridden aunt. Her boyfriend was a violent alcoholic schizophrenic who mistreated her psychologically and physically, and was always muddling her simple mind with tirades about the devil. She had nothing going for her—no beauty, no intelligence, no loving family, no skills, no creative employment. Nothing.

She didn't come to therapy to resolve her problems, however, nor to unburden her soul, nor to describe her mistreatment and victimization at the hands of others. She came because she wanted to do something for someone who was worse off than her. The clinic where I was interning was associated with a large psychiatric hospital. All of the patients that still remained after the shift to community care in the aftermath of the sixties were so incapacitated that they could not survive, however poorly, on the streets. My client had done some volunteer work, of some limited type, in that hospital, and decided that she might befriend a patient—take him or her outside for a walk. I think she got this idea because she had a dog, which she walked regularly, and which she liked to take care of. All she wanted from me was help arranging this—help finding someone whom she could take outside; help finding someone, in the hospital bureaucracy, who would allow this to happen. I was not very successful in aiding her, but she didn't seem to hold that against me.

It is said that one piece of evidence that runs contrary to a theory is sufficient to disprove that theory. Of course, people do not think this way, and perhaps should not. In general, a theory is too useful to give up, easily—too difficult to regenerate—and the evidence against should be consistent and believable before it is accepted. But the existence of this woman made me think. She was destined for a psychopathological end, from the viewpoint of biological and environmental determinism—fated as surely as anyone I had even met. And maybe she kicked her dog sometimes and was rude to her sick aunt. Maybe. I never saw her vindictive, or unpleasant—even when her simple wishes were thwarted. I don't want to say that she was a saint, because I didn't know her well enough to tell. But the fact was that in her misery and simplicity she remained without self-pity and could still see outside of herself. Why wasn't she corrupted—cruel, unbalanced and miserable? She had every reason to be. And yet she wasn't.

In her simple way, she had made the proper choices. She remained bloody but unbowed. And she seemed to me, rightly or wrongly, to be a symbol of suffering humanity, sorely afflicted, yet capable of courage and love:

> Such I created all the Ethereal Powers
> And Spirits, both them who stood and them who failed;
> Not free, what proof could they have given sincere
> Of true allegiance, constant faith, or love,

Where only what they needs must do appeared,
Not what they would? What praise could they receive
What pleasure I, from such obedience paid,
When Will and Reason (Reason also is Choice),
Useless and vain, of freedom both despoiled,
Made passive both, had served Necessity,
Not me? They, therefore, as to right belonged,
So were created, nor can justly accuse
Their Maker, or their making, or their fate,
As if Predestination overruled
Their will, disposed by absolute decree
Or high foreknowledge; they themselves decreed
Their own revolt, not I: if I foreknew,
Foreknowledge had no influence on their fault,
Which had no less proved certain unforeknown.
So without least impulse or shadow of fate,
Or aught by me immutably foreseen,
They trespass, authors to themselves in all,
Both what they judge and what they choose; for so
I formed them free, and free they must remain
Till they enthrall themselves: I else must change
Their nature, and revoke the high decree
Unchangeable, eternal, which ordained
Their freedom; they themselves ordained their fall.[654]

The other patient I wish to describe was a schizophrenic in a small inpatient ward at a different hospital. He was about twenty-nine when I met him—a few years older than I was at the time—and had been in and out of confinement for seven years. He was, of course, on anti-psychotic medication, and participated in occupational-therapy activities on the ward—making coasters and pencil holders and so on—but he could not maintain attention for any amount of time, and was not even much good at craft. My supervisor asked me to administer an intelligence test to him—the standard WAIS-R[655] (more for the sake of my experience than for any possible diagnostic good). I gave my patient some of the red-and-white blocks that made up the Block Design Subtest. He was supposed to arrange the blocks so they matched a pattern printed on some cards. He picked them up, and started to rearrange them on the desk in front of him, while I timed him, stupidly, with a stopwatch. The task was impossible for him, even at the simplest of stages. He looked constantly distracted and frustrated. I asked, "What's wrong?" He said, "The battle between good and evil in heaven is going on in my head."

I stopped the testing at that point. I didn't know exactly what to make of his comment. He was obviously suffering, and the testing seemed to make it worse. What was he experiencing? He wasn't lying, that was for sure. In the face of such a statement it seemed ridiculous to continue.

I spent some time with him that summer. I had never met someone who was so blatantly mentally ill. We talked on the ward, and occasionally I would take him for a walk through the hospital grounds. He was the third son of first-generation immigrants. His firstborn brother was a lawyer; the other, a physician. His parents were obviously ambitious for their children, hardworking and disciplined. He had been a graduate student, working toward a degree (in immunology, perhaps—I don't precisely remember). His brothers had set him a daunting example, and he felt pressure to succeed. His experimental work had not turned out as he had expected, however, and he apparently came to believe that he might not graduate—not, at least, when he had hoped to. So he faked his experimental results, and wrote up his thesis anyway.

He told me that the night he finished writing, he woke up and saw the devil standing over him, at the foot of his bed. This event triggered the onset of his mental illness, from which he had never recovered. It might be said that the Satanic apparition merely accompanied the expression of some pathological stress-induced neural development, whose appearance was biologically predetermined, or that the devil was merely personification of his culture's conception of moral evil, manifesting itself in imagination, as a consequence of his guilt. Both explanations have their merits. But the fact remains that he saw the devil, and that the vision accompanied or even was the event that destroyed him.

He was afraid to tell me much of his fantasy, and it was only after I had paid careful attention to him that he opened up. He was not bragging, or trying to impress me. He was terrified about what he believed; terrified as a consequence of the fantasies that impressed themselves upon him. He told me that he could not leave the hospital, because someone was waiting to shoot him—a typical paranoid delusion. Why did someone want to kill him?

Well, he was hospitalized during the Cold War—not at its height, perhaps, but still during a time when the threat of purposeful nuclear annihilation seemed more plausible than it does now. Many of the people I knew used the existence of this threat to justify, to themselves, their failure to participate fully in life—a life which they thought of, romantically, as doomed, and therefore as pointless. But there was some real terror in the pose, and the thought of the countless missiles pointed here and there around the world sapped the energy and faith of everyone, hypocritical or not.

My schizophrenic patient believed that he was, in fact, the incarnation of the world-annihilating force; that he was destined, upon his release from the hospital, to make his way south to a nuclear missile silo, on American territory; that he was fated to make the decision that would launch the final war. The "people" outside the hospital knew this, and that is why they were waiting to shoot him. He did not want to tell me this story, in consequence—although he did—because he thought I might then want to kill him too.

My friends in graduate school thought it ironic that I had contact with a patient of this type. My peculiar interest in Jung was well known to them, and it seemed absurdly fitting that I would end up talking to someone with delusions of this type. But I didn't know what to do with his ideas. Of course, they were crazy, and they had done in my patient. But it still seemed to me that they were *true*, from the metaphorical viewpoint.

His story, in totality, linked his individual choice, between good and evil, with the cumulative horror then facing the world. His story implied that because he had given in to temptation, at a critical juncture, he was in fact responsible for the horror of the potential of nuclear war. But how could this be? It seemed insane to me to even consider that the act of one powerless individual could be linked in some manner to the outcome of history as a whole.

But I am no longer so sure. I have read much about evil, and its manner of perpetration and growth, and I am no longer convinced that each of us is so innocent, so harmless. It is of course illogical to presume that one person—one speck of dust, among six billion motes— is in any sense responsible for the horrible course of human events. But that course in itself is not logical, far from it, and it seems likely that it depends on processes that we do not understand.

The most powerful arguments for the non-existence of God (at least a good God) are predicated on the idea that such a Being would not allow for the existence of evil in its classical natural (diseases, disasters) or moral (wars, pogroms) forms. Such arguments can be taken further even, than atheism—can be used to dispute the justice of the existent world itself. Dostoyevsky states, "Perhaps the entire cosmos is not worth a single child's suffering." How can the universe be constructed such that pain is permitted? How can a good God allow for the existence of a suffering world?

These difficult questions can be addressed, in part, as a consequence of careful analysis of evil. First, it seems reasonable to insist upon the value of the natural/moral distinction. The "tragic circumstances of life" should not be placed in the same category as "willfully undertaken harm." Tragedy—subjugation to the mortal conditions of existence—has an ennobling aspect, at least in potential, and has been constantly exploited to that end in great literature and mythology. True evil, by contrast, is anything but noble.

Participation in acts whose sole purpose is expansion of innocent pain and suffering destroys character; forthright encounter with tragedy, by contrast, may increase it. This is the meaning of the Christian myth of the crucifixion. It is Christ's full participation in and freely chosen acceptance of his fate (which he shares with all mankind) that enables him to manifest his full identity with God—and it is that identity which enables him to bear his fate, and which strips it of its evil. Conversely—it is the voluntary demeaning of our own characters that makes the necessary tragic conditions of existence appear evil.

But why is life tragic? Why are we subject to unbearable limitation—to pain, disease, and death; to cruelty at the hands of nature and society? Why do terrible things happen to everyone? These are, of course, unanswerable questions. But they must be answered, somehow, if we are to be able to face our own lives.

The best I can make of it is this (and this has helped me): Nothing can exist without preconditions. Even a game cannot be played without rules—and the rules say what cannot be done, as much as what can. Perhaps the world is not possible, as a world, without its borders, without its rules. Maybe existence wouldn't be possible in the absence of our painful limitations.

Think of it this way: If we could have everything we wanted, merely for wishing it; if every tool performed every job, if all men were omniscient and immortal—then everything would be the same, the same all-powerful thing, God, and creation would not exist. It is the difference between things, which is a function of their specific limitations, that allows them to exist at all.

But the fact that things *do* exist does not mean that they *should* exist, even if we are willing to grant them their necessary limitations.

Should the world exist? Are the preconditions of experience so terrible that the whole game should be called off? (There is never any shortage of people working diligently toward this end.)

It seems to me that we answer this question, implicitly but profoundly, when we lose someone we loved, and grieve. We cry not because they existed, but because they are lost. This presupposes a judgment rendered, at a very fundamental level. Grief presupposes having loved, presupposes the judgment that this person's specific, bounded existence was valuable, was something that should have been (even in its inevitably imperfect and vulnerable form). But still the question lingers—why should things, even loved things, exist at all, if their necessary limitations cause such suffering?

Perhaps we could reserve answer to the question of God's nature, his resp ... ability for the presence of the evil in creation, until we have solved the problem of our own. Perhaps we could tolerate the horrors of the world if we left our own characters intact, and developed them to the fullest; if we took full advantage of every gift we have been granted. Perhaps the world would not look horrible then.

I dreamed I was walking up out of a deep valley, along a paved two-lane highway. The highway was located in northern Alberta, where I grew up, and came out of the only valley for miles around, in the endlessly flat prairie. I passed a man, hitchhiking, and could see another in the distance. As I approached him I could see that he was in the first stages of old age; but he still looked terribly strong. Someone passed him in a car, driving the opposite direction, and a female voice yelled, "Look out—he has a knife!"

He was carrying what looked like a wooden-handled kitchen knife, well-worn and discolored, but it had a blade at least two and a half feet long. Across his shoulder he had strapped a large leather sheaf. He was walking along the edge of the highway, muttering to himself, and swinging the blade in a jerky and chaotic fashion.

He looked like the landlord who lived next door to me when I was a graduate student, living in a poor district in Montreal. My landlord was a powerful, aging ex-biker—former president of the local Hell's Angels chapter, by his own account—who had spent some time in prison as a younger man. He had settled down somewhat, typically, as he became more mature, and had brought his drinking under control for a long time. His wife committed suicide when I lived there, however, and he went back to his wilder ways. He often went on binge drinking sprees, and spent all the money he earned in the electronics shop he ran out of his small apartment. He would drink forty or even fifty beers in a single day, and would return home in the evening, blind drunk, howling at his little dog, laughing, hissing between his teeth, incoherent, good-natured still, but able to become violent at the slightest provocation. He took me once to a favourite haunt on his 1200 cc Honda, which had the acceleration of a jet plane for short distances—me perched precariously on the back of his bike, clinging to him, wearing his wife's helmet, which sat on my head ridiculously, uselessly, at least five sizes too small. Drunk, he was almost innocently destructive, and ended up in fights constantly, unavoidably—as he would take slights from people whose paths he crossed, who were insufficiently cautious in their conversation with him.

I hurried by this figure. He seemed upset that no one would stop and pick him up, as if he were unaware of the danger he posed. As I went by, his gaze fell on me, and he started after me—not from anger, but from desire for companionship. He was not fast, however, and I easily stayed ahead of him on the road.

The scene shifted. The knife-wielding figure and I were now on opposite sides of an immense tree—perhaps a hundred yards in diameter—on a spiral staircase emerging from the dark below and ascending equally far above. The staircase was made of old, worn dark wood. It reminded me of the pews in the church I attended with my mother as a child, and where I was eventually married. The figure was looking for me, but he was a long way back, and I had hidden myself from his view as I ascended the staircase. I remember wanting to continue on my original journey, out of the valley, onto the flat surrounding plain, where the walking would be easy. The only way to stay away from the knife, however, was to continue up the staircase— up the axis mundi.

It is thus that awareness of death, the grim reaper—the terrible face of God—compels us inexorably upwards, toward a consciousness sufficiently heightened to bear the thought of death.

The point of our limitations is not suffering; it is existence itself. We have been granted the capacity to voluntarily bear the terrible weight of our mortality. We turn from that capacity and degrade ourselves because we are afraid of responsibility. In this manner, the necessarily tragic preconditions of existence are made intolerable.

It seems to me that it is not the earthquake, the flood or the cancer that makes life unbearable, horrible as those events appear. We seem capable of withstanding natural disaster, even of responding to that disaster in an honorable and decent manner. It is rather the pointless suffering that we inflict upon each other—our evil—that makes life appear corrupt beyond acceptability; that undermines our ability to manifest faith in our central natures. So why should the capacity for evil exist?

I have been teaching my six-year old daughter to play the piano. I am trying to teach her hard lessons—that is, trying to show her that there is really a right way to play the piano, and a wrong way. The right way involves paying attention to each phrase, each written note, every sound she makes, every finger motion. I taught her what rhythm meant, a few weeks ago, in one difficult lesson. And what does difficult mean? Well, she will sit at the piano and work so hard that she cries—but she won't stop. And she is really interested in learning to play. She spends time in the car listening to music, working out the rhythms. She uses the metronome by herself, playing songs she likes faster and slower. Yesterday I taught her the difference between playing loudly and playing softly. She found this challenging, experimenting carefully with each key on our old piano (which has many idiosyncracies), trying to determine exactly how hard it had to be pressed to emit a whispery tone.

I woke up the morning after one of her lessons, and caught a fragment of the end of a dream revery. This is the idea revealed by that revery: it is the fact that differences exist between paths of action that makes actions worthwhile. I know that what is believed determines the value of things. But I had never taken this argument to its logical conclusion. If belief determines value, then the distance between good and evil gives life its meaning. The more worthwhile a path of action (which is to say, the "better" it is—the more it is good, rather than evil) the more positive emotional valence that path "contains." This means that things have no meaning, because no differential value, for those who do not believe in good and evil.

I have certainly known people in that position (although I did not know explicitly that they were in that position because they did not know the difference between good and evil). They were unable to do anything, because they could not tell the difference between one path and another. And it is the case that in the absence of differences between one thing and another, life begins to appear a "cruel and meaningless joke," to use Tolstoy's phrase. I guess that is because the burdens of life do not appear worth sustaining in the absence of evidence that striving and work have any real value.

This all means: value is a continuum, a line stretching from necessary point "a" to necessary point "b." "A" and "b" are defined in relation to one another, as two points define a line. The polarity between the two determines the valence of the goal. The more polarity (that is, the more tension) between the two points, the more worthwhile the enterprise. Good cannot be defined—cannot exist—in the absence of evil. Value cannot exist in the absence of polarity. So, for the world to be worthwhile (that is, for the choice between two things to constitute a real choice) both good and evil have to exist.

But then it would be possible to only choose good, at least in the ideal—and then evil would not exist, except in potential. So it appears that the world could be valuable (could justify the burden it requires to maintain) if evil were only to exist in potential—if everyone chose to act properly, that is. This seems to me to be the most optimistic thought I have ever encountered.

But how can we put an end to our errors? What path can we follow to eliminate our blindness and stupidity, to bring us closer to the light? Christ said, Be ye therefore perfect, as your Father in heaven is perfect.[656] But how? We seem stymied, as always, by Pontius Pilate's ironic query: What is truth? (John 18:38)

Well, even if we don't know precisely what the truth is, we can certainly tell, each of us, what it isn't. It isn't greed, and the desire, above all else, for constant material gain; it isn't denial of experience we know full well to be real, and the infliction of suffering for the purpose of suffering. Perhaps it is possible to stop doing those things which we know, beyond doubt, to be wrong—to become self-disciplined and honest—and to therefore become ever more able to perceive the nature of the positive good.

The truth seems painfully simple—so simple that it is a miracle, of sorts, that it can every be forgotten. Love God, with all thy mind, and all thy acts, and all thy heart. This means, serve truth above all else, and treat your fellow man as if he were yourself—not with the pity that undermines his self-respect, and not with the justice that elevates you above him, but as a divinity, heavily burdened, who could yet see the light.

It is said that it is more difficult to rule oneself than a city, and this is no metaphor. This is truth, as literal as it can be made. It is precisely for this reason that we keep trying to rule the city. It is a perversion of pride to cease praying in public and to clean up the dust under our feet, instead; it seems too mundane to treat those we actually face with respect and dignity, when we could be active, against, in the street. Maybe it is more important to strengthen our characters than to repair the world. So much of that reparation seems selfish anyway; is selfishness and intellectual pride masquerading as love, creating a world polluted with good works that don't work.

Who can believe that it is the little choices we make, every day, between good and evil, that turn the world to waste and hope to despair? But it is the case. We see our immense capacity for evil, constantly realized before us, in great things and in small, but can never seem to realize our infinite capacity for good. Who can argue with a Solzhenitsyn when he states: "One man who stops lying can bring down a tyranny"?

Christ said, the kingdom of heaven is spread out upon the earth, but men do not see it.[657] What if it was nothing but our self-deceit, our cowardice, hatred and fear, that pollutes our experience and turns the world into hell? This is a hypothesis, at least—as good as any other, admirable and capable of generating hope. Why can't we make the experiment, and find out if it is true?

The Divinity of Interest

The central ideas of Christianity are rooted in Gnostic philosophy, which, in accordance with psychological laws, simply had to grow up at a time when the classical religions had become obsolete. It was founded on the perception of symbols thrown up by the unconscious individuation process which always sets in when the collective dominants of human life fall into decay. At such a time there is bound to be a considerable number of individuals who are possesed by archetypes of a numinous nature that force their way to the surface in order to form new dominants.

This state of possession shows itself almost without exception in the fact that the possessed identify themselves with the archetypal contents of their unconscious, and, because they do not realize that the role which is being thrust upon them is the effect of new contents still to be understood, they exemplify these concretely in their own lives, thus becoming prophets and reformers.

In so far as the archetypal content of the Christian drama was able to give satisfying expression to the uneasy and clamorous unconscious of the many, the *consensus omnium* raised this drama to a universally binding truth—not of course by an act of judgment, but by the irrational fact of possession, which is far more effective.

Thus Jesus became the tutelary image or amulet against the archetypal powers that threatened to possess everyone. The glad tidings announced: "It has happened, but it will not happen to you inasmuch as you believe in Jesus Christ, the Son of God!"

Yet it could and it can and it will happen to everyone in whom the Christian dominant has decayed. For this reason there have always been people who, not satisfied with the dominants of conscious life, set forth—under cover and by devious paths, to their destruction or salvation—to seek direct experience of the eternal roots and, following the lure of the restless unconscious psyche, find themselves in the wilderness where, like Jesus, they come up against the son of darkness....

Thus an old alchemist—and he a cleric!—prays ... "Purge the horrible darknesses of our minds, light a light for our senses!" The author of this sentence must have been undergoing the experience of the *nigredo*, the first stage of the work, which was felt as melancholia in alchemy and corresponds to the encounter with the shadow in psychology.

When, therefore, modern psychotherapy once more meets with the activated archetypes of the collective unconscious, it is merely the repetition of a phenomenon that has often been observed in moments of great religious crisis, although it can also occur in individuals for whom the ruling ideas have lost their meaning. An example of this is the *descensus ad inferos* in Faust which, consciously or unconsciously, is an *opus alchymicum*.

The problem of opposites called up by the shadow plays a great—indeed, the decisive—role in alchemy, since it leads in the ultimate phase of the work to the union of opposites in the archetypal form of the hierosgamos or "chymical wedding." Here the supreme opposites, male and female (as in the Chinese ying and yang), are melted into a unity purified of all opposition and therefore incorruptible.[658]

November, 1986

Dear Dad

I promised you that one day I would tell you what the book I am trying to write is supposed to be about. I haven't been working on it much in the last month, although in some regards it is always on my mind and everything I learn, in my other work, has some bearing upon it. Because I have abandoned it, temporarily, I thought perhaps I could tell you about it, and that would help me organize my thoughts.

I don't completely understand the driving force behind what I have been working on, although I understand it better now than I used to, three or four years ago, when it was literally driving me crazy. I had been obsessed with the idea of war for three or four years prior to that, often dreaming extremely violent dreams, centered around the theme of destruction. I believe now that my concern with death on a mass scale was intimately tied into my personal life, and that concerns with the meaning of life on a personal level (which arise with the contemplation of death) took a general form for me, which had to do with the value of humanity, and the purpose of life in general.

Carl Jung has suggested that all personal problems are relevant to society, because we are all so much alike, and that any sufficiently profound solution to a personal problem may, if communicated, reduce the likelihood of that problem existing in anyone's experience in the future.[659] This is in fact how society and the individual support one another. It was in this way that my concern with war, which is the application of death on the general level, led me into concepts and ideas concerning the meaning of life on the personal level, which I could never have imagined as relevant, or believable, prior to learning about them—and which I still believe border on what might normally be considered insanity.

The reasons for war, many believe, are rooted in politics. Since it is groups of men that fight, and since groups indulge in politics, this belief seems well-founded and in fact contains some truth. It is just as true, however, that it is a good thing to look for something you don't want to find in a place where you know it won't be—and the modern concern with global politics, and the necessity to be involved in a "good cause," rather than to live responsibly, seems to me to be evidence that the desire not to find often overpowers the real search for truth. You see, it is true that people don't want the truth, because the truth destroys what lack of faith erects, and the false comfort it contains. It is not possible to live in the world that you wish could be, and in the real world at the same time, and it often seems a bad bargain to destroy fantasy for reality. It is desire for lack of responsibility that underlies this evasion, in part—but it is also fear of possibility. At least this is how it seems to me.

Because everyone is a product of their times, and because that applies to me as well, I looked for what I wanted to find where it was obvious to everyone it would be—in politics, in political science, in the study of group behavior. This took up the years I spent involved with the NDP, and in studying political science, until I learned that the application of a system of thought, like socialism (or any other ism, for that matter) to a problem, and solving that problem, were not the same thing. In the former case, you have someone (who is not you) to blame—the rich, the Americans, the white people, the government, the system—whatever, as long as it is someone else.

I came to realize, slowly, that a problem of global proportions existed as a problem because everyone on the globe thought and acted to maintain that problem. Now what that means is that if the problem has a solution, then what everyone thinks is wrong—and that meant, too, that what I thought had to be fundamentally wrong. Now the problem with this line of reasoning is simple. It leads inexorably to the following conclusion: the more fundamental the problem, the more fundamental the error—in my own viewpoint.

I came to believe that survival itself, and more, depended upon a solution to the problem of war. This made me consider that perhaps everything I believed was wrong. This consideration was not particularly pleasant, and was severely complicated by the fact that I had also come to realize that, although I definitely believed a variety of things, I did not always know what I believed—and when I knew what, I did not know why.

You see, history itself conditioned everything I believed, even when I did not know it, and it was sheer unconscious arrogance that made me posit to begin with that I had half a notion of who or what I was, or what the process of history had created, and how I was affected by that creation.

It is one thing to be unconscious of the answers, and quite another to be unable to even consider the question.

I had a notion that confronting what terrified me—what turned my dreams against me—could help me withstand that terrible thing. This idea—granted me by the grace of God—allowed me to believe that I could find what I most wanted (if I could tolerate the truth; if I was willing to follow wherever it led me; if I was willing to devote my life to acting upon what I had discovered, whatever that might be, without reservation—knowing somehow that once started, an aborted attempt would destroy at least my self-respect, at most my sanity and desire to live).

I believe now that everyone has this choice in front of them, even when they do not know or refuse to admit it; that everyone makes this choice, with every decision and action they take.

I mentioned earlier that history conditioned what I think and acted. Pursuit of this realization—which is rather self-evident, once realized—has led me to the study of history, as a psychological phenomenon. You see, if what I think and am is a product of history, that means that history must take form inside me, so to speak, and from inside me determine who I am. This is easier to understand if you consider that I carry around inside me an image of you—composed of memories of how you act, and what you expected, and depictions of your behavior. This image has had profound impact on how I behaved, as a child—when, even in your absence, I was compelled to follow the rules which you followed (and which I learned through imitation, and which you instilled into me, through praise and punishment). Sometimes that image of you, in me, even takes the form of a personality, when I dream about you.

So it is a straightforward matter to believe, from the psychological point of view, that each individual carries around an image of his parents, and that this image governs his behavior, at least in part.

But you see it is the case that the rules that you followed—and which I learned from you—were not rules that you yourself created, but rather those that you handed to me just as you had been handed them while still a child.

And it is more than likely true that the majority of what I learned from you was never verbalized—that the rules which governed the way you acted (and that I learned while watching you) were implicit in your behavior, and are now implicit in mine. It was exactly in this manner that I learned language—mostly from watching and listening, partly from explicit instruction. And just as it is certainly possible (and most commonly so) to speak correctly and yet to be unable to describe the rules of grammar that "underlie" the production of language, it is possible to act upon the world and make assumptions about its nature without knowing much about the values and beliefs that necessarily underlie those actions and assumptions.

The structure of our language has been created in a historical process, and is in a sense an embodiment of that process. The structure of that which governs our actions and perceptions has also been created during the course of history, and is the embodiment of history.

The implications of this idea overwhelmed me. I have been attempting to consider history itself as a unitary phenomenon—as a single thing, in a sense—in order to understand what it is, and how it affects what I think and do. If you realize that history is in some sense in your head, and you also realize that you know nothing of the significance of history, of its meaning—which is almost certainly true—then you must realize that you know nothing of the significance of yourself, and of your own meaning.

I am writing my book in an attempt to explain the psychological significance of history—to explain the meaning of history. In doing so, I have "discovered" a number of interesting things:

1. *All cultures, excepting the Western, do not possess a history based on "objective events." The history of alternative cultures—even those as highly developed as the Indian, Chinese, and ancient Greco-Roman—is mythological, which means that it describes what an event meant, in psychological terms, instead of how it happened, in empirical terms.*

2. *All cultures, even those most disparate in nature, develop among broadly predictable lines, and have, within their mythological history, certain constant features (just as all languages share grammatical structure, given a sufficiently abstract analysis). The lines among which culture develops are determined biologically, and the rules which govern that development are the consequence of the pyschological expression of neurophysiological structures. (This thesis will be the most difficult for me to prove, but I have some solid evidence in its favor, and as I study more neuroanatomy and neuropsychology, the evidence becomes clearer).*

3. *Mythological renditions of history, like those in the Bible, are just as "true" as the standard Western empirical renditions, just as literally true, but how they are true is different. Western historians describe (or think they describe) "what" happened. The traditions of mythology and religion describe the significance of what happened (and it must be noted that if what happens is without significance, it is irrelevant).*

Anyway—I can't explain in one letter the full scope of what I am planning to do. In this book, I hope to describe a number of historical tendencies, and how they affect individual behavior—in the manner I have attempted in this letter. More importantly, perhaps, I hope to describe not only what the problem is (in historical terms), but where a possible solution might lie, and what that solution conceivably could be—and I hope to describe it in a manner that makes its application possible.

If you're interested in me telling you more (I can't always tell if someone is interested) then I will, later. I don't know, Dad, but I think I have discovered something that no one else has any idea about, and I'm not sure I can do it justice. Its scope is so broad that I can see only parts of it clearly at one time, and it is exceedingly difficult to set down comprehensibly in writing. You see, most of the kind of knowledge that I am trying to transmit verbally and logically has always been passed down from one person to another by means of art and music and religion and tradition, and not by rational explanation, and it is like translating from one language to another. It's not just a different language, though—it is an entirely different mode of experience.

Anyways

I'm glad that you and Mom are doing well. Thank you for doing my income tax returns.

Jordan

It has been almost twelve years since I first grasped the essence of the paradox that lies at the bottom of human motivation for evil: People need their group identification, because that identification protects them, literally, from the terrible forces of the unknown. It is for this reason that every individual who is not decadent will strive to protect his territory, actual and psychological. But the tendency to protect means hatred of the other, and the inevitability of war—and we are now too technologically powerful to engage in war. To allow victory to the *other*, however—or even continued existence, on his terms—means subjugation, dissolution of protective structure, and exposure to that which is most feared. For me, this meant "damned if you do, damned if you don't": belief systems regulate affect, but conflict between belief systems is inevitable.

Formulation and understanding of this terrible paradox devastated me. I had always been convinced that sufficient understanding of a problem—any problem—would lead to its resolution. Here I was, however, possessed of understanding that seemed not only sufficient but complete, caught nonetheless between the devil and the deep blue sea. I could not see how there could be any alternative to either *having* a belief system or to *not having* a belief system—and could see little but the disadvantage of both positions. This truly shook my faith.

I turned, in consequence, to my dreams—acting on a tip from Jung, who had proposed that the dream might contain information, when no other source could suffice. But my dreams dried up, just when I needed them most, and no information was forthcoming. I was in a kind of stasis, at a standstill. This was very painful for me. I had spent several years intensely working and thinking—trying to understand individual human motivation for the worst possible human actions. I was doing what I truly believed was best, to the best of my ability, and doing it despite its substantial interference with my personal and professional lives. I had decided to devote my life to the problem of evil—to the development of a true understanding of evil, in the hopes of finding some means of combating it—yet my search had come to a stop, a dead end. This did not seem reasonable. I truly believed I deserved better.

Then, one night, my dreams came back, with a vengeance. I had the following nightmare, as terrible and potent as the dreams of destruction that had started my quest several years before:

I dreamed I was living in a two-story house. After a bout of heavy drinking, I went to the attic and fell asleep. After I fell asleep, I had the following dream—nested within a dream:

I was trapped in a huge chandelier, which was hanging directly beneath the dome of an immense darkened cathedral. The chandelier hung hundreds of feet below its point of connection on the dome, and was still so high off the ground that the people below, on the floor, looked like ants. These people were in charge of the cathedral, and I could tell that they were angry at me for being where I was. I did not feel guilty, because I was not there by choice—I just happened to have arrived there, and I wanted to leave.

I realized I was dreaming, and shook myself "awake," as a means of removing myself from my uncomfortable position. But when I "awoke," I still found myself suspended in the same place. I tried to deny this, by falling back asleep—reasoning that it was better to be dreaming about this circumstance than to actually be trapped in it. However, I could not return to my previous unconscious state, and stayed painfully awake.

Then I found myself on the ground, as a consequence of some process whose nature I cannot recall. The people of the cathedral protested my presence—but that didn't really bother me. All I wanted to do was get home, where it was familiar, and go back to sleep.

When I returned home, I went into a small, windowless room—I think it was the furnace room—in the middle of the ground floor of my house. This room was surrounded by other rooms; it had no contact point with the outside. There was a small single bed there, that was actually much like my real bed. I crawled into it, and tried to fall asleep, but a strange wind descended on me. Under its influence, I started to dissolve. I knew beyond a shadow of a doubt that it was going to transport me back to the chandelier, in the center of the cathedral. I attempted to fight the wind, but found that I was virtually paralyzed, and in some sort of convulsion. I tried desperately to yell for help, and actually awakened, at least partially.

The windows behind my bed in my real room were wide open, and a wind was blowing through them. I frantically closed them, and then turned around. I was awake, but in front of me appeared a huge double door, like that on a gothic cathedral, between my bedroom and the adjacent room, which were only partially separated. I shook myself, and the apparition disappeared. The terror I was experiencing vanished much more slowly.

I had read much of the Gospels earlier that day—which might account for the initial reference to the drinking bout (the imbibing of spirit, so to speak). I formulated an interpretation of the dream—an essentially unacceptable formulation—immediately upon awakening. I knew that the word "spirit" had been derived from the Greek *pneuma*—which meant wind: the wind, for example, that moved upon the water, in Genesis; the wind or breath that God blew into the *adamah*, the matter, to make man.

I found myself at the central point of a cathedral, in my dream—and I could not escape. A cathedral is "sacred space," designed to keep the forces of chaos at bay; it has the same layout as the cross. The central point of a cathedral is, symbolically, the place where Christ was crucified, and the center of the universe, simultaneously. All the forces embodied in my dream were conspiring to put me there, awake, despite my best efforts to the contrary. I could not, at that time, accept the implications of that dream (could not believe the implications), and it has taken me a long time to assimilate its meaning:

"He that believeth on Me, the works that I do he shall do also, and greater works than these shall he do" (John 14:12).

The quotation from John is taken from the fourteenth chapter, where Christ teaches that whoever sees him sees the Father. He is in the Father and the Father is in him. The disciples are in him and he in them; moreover they will be sent the Holy Ghost as Paraclete and will do works that are greater

Figure 67: The Restitution of [Christ] the Mystic Apple to the Tree of Knowledge

than his own. This fourteenth chapter broaches a question that was to have great repercussions for the future: the problem of the Holy Ghost who will remain when Christ has gone, and who intensifies the interpenetration of the divine and the human to such a degree that we can properly speak of a "Christification" of the disciples....

It is easy to see what happens when the logical conclusion is drawn from the fourteenth chapter of John: the *opus Christi* is transferred to the individual. He then becomes the bearer of the mystery, and this development was unconsciously prefigured and anticipated in alchemy, which showed clear signs of becoming a religion of the Holy Ghost and of the Sapientia Dei.[660]

Myths of the origin commonly portray the condition of paradise as the source of all things. The paradisal initial condition, disrupted by the events of the Fall, also serves as the goal toward which history proceeds. Stories of the Fall describe the introduction of uncontrollable anxiety into human experience, as the consequence of traumatically heightened consciousness (as the result of irrevocably attained knowledge of human vulnerability and mortality). Re-establishment of paradise, in the aftermath of such attainment, becomes dependent upon manifestation of an exemplary way of behaving, directed toward a meaningful end—becomes dependent upon establishment of a particular mode of redemption:

An old English legend reports what Seth saw in the Garden of Eden. In the midst of paradise there rose a shining fountain, from which four streams flowed, watering the whole world. Over the fountain stood a great tree with many branches and twigs, but it looked like an old tree, for it had no bark and no leaves. Seth knew that this was the tree of whose fruit his parents had eaten, for which reason it now stood bare. Looking more closely, Seth saw that a naked snake without a skin had coiled itself round the tree. It was the serpent by whom Eve had been persuaded to eat of the forbidden fruit. When Seth took a second look at paradise he saw that the tree had undergone a great change. It was now covered with bark and leaves, and in its crown lay a little new-born babe wrapped in swaddling clothes, that wailed because of Adam's sin. This was Christ, the second Adam. He is found in the top of the tree that grows out of Adam's body in representations of Christ's genealogy.[661]

The tree, the *axis mundi*, is without bark and leaves because has been mortally affected, as a consequence of the fall. Production of its first fruit—self-reference—placed it in shock, exhausted its resources. Its second fruit, associated with re-emergence into lush life and health, is the savior, the hero who redeems mankind from the consequences of the Fall,[662] the divine individual whose path of being leads back to paradise.[663] This notion is represented imagistically (it has never really proceeded much past the image) in *Figure 67: The Restitution of [Christ] the Mystic Apple to the Tree of Knowledge*.[664] A similar pattern of redemptive ideation informs the East. For Gautama, suffering and disenchantment are the (necessary) preconditions of adult enlightenment (the name of the Buddha—Siddhartha—literally means "goal attained"[665]). The "enlightened one," whose mode of being in the world transcends the suffering in that world, is an eternal spirit (that is, an eternally recurrent spirit), despite "historical" incarnation in the figure of the Buddha. *Figure 68: The Eternal*

Figure 68: The Eternal Return of the Boddhisatva

Return of the Boddhisatva[666] portrays this spirit, forever dominant over the mass of mankind and the Great and Terrible Mother. The *boddhisatva*, the central "character" in this figure, is an Oriental Christ-equivalent (or, perhaps, an image of the paraclete or Holy Ghost). The creator of this work has superimposed the bodhisattva on a "tunnel" in the sky, ringed by transformative fire. This tunnel adds a temporal dimension to the spatial dimensions represented in the image, and allows for the portrayal of the constant recurrence, throughout time, of the heroic spirit. This is a reflection of the same idea that drove Christian thinkers to attribute prehistorical (and eternal) reality to Christ, despite his "historical" nature; as well, that drove speculation about the "spirit of truth" Christ left behind, after his death.

Myths of the fall and redemption portray the emergence of human dissatisfaction with present conditions—no matter how comfortable—and the tendency or desire for movement toward "a better future." Such myths describe, in narrative format, how human beings think and will always think—regardless of time or place. The most profound of such cyclical myths portray heightening of consciousness as cause for emergent unrest. Simultaneously, such myths portray *qualitatively transformed consciousness as cure for that unrest (more profoundly, portray participation in the act of qualitative transformation of consciousness as cure for that unrest).*

Proclivity to posit an ideal, implicitly or explicitly, to work toward its attainment, to become dissatisfied with its establishment, as new "matter" makes itself manifest, and to thereby re-enter the cycle—this constitutes the centrally defining pattern of human abstraction and behavior. The simplest and most basic day-to-day human activities, invariably goal-directed, are necessarily predicated upon conscious or tradition-bound acceptance of a value hierarchy, defining the desired future in positive contrast to the insufficient present. To live, from the human perspective, is to act in light of what is valued, what is desired, what should be—and to maintain sufficient ignorance, in a sense, to allow belief in such value to flourish. Collapse of faith in the value hierarchy—or, more dangerously, collapse of faith in the idea of such hierarchies—brings about severe depression, intrapsychic chaos and re-emergence of existential anxiety.

The myth of the Fall describes the development of human self-consciousness as a great tragedy, the greatest conceivable anomaly, an event that permanently altered the structure of the universe and doomed humanity to suffering and death. But it was this same Fall that enabled the individual to adopt the redemptive role of the hero, the creator of culture; the same Fall that lifted the curtain on the drama of human history. Whether or not it would have been better for humanity to have remained unconscious is no longer a point that can be usefully considered—although that path does not seem particularly productive for those who take it now. Original Sin has tainted everyone; there is no way back.

For much of human history—after the Fall, so to speak—the individual remained firmly ensconced within the confines of a religious dream: a dream that gave meaning to the tragedy of existence. Many modern thinkers, including Freud, viewed that dream in retrospect as defensive, as a barrier of fantasy erected against the existential anxiety generated by knowledge of mortality. However, the dividing line between fantasy and reality is not so easily drawn. It is certainly possible to disappear voluntarily into the mists of delusion; to withdraw into the comforts of denial from a world terrible beyond what can be borne. Imagination is not always insanity, however; its use does not always imply regression. Imagination and fantasy allow each of us to deal with the unknown, which must be met before it is comprehended. Fantasy applied to consideration of the unknown is therefore not delusory. It is, instead, the first stage in the process of understanding—which eventually results in the evolution of detailed, empirical, communicable knowledge. Fantasy can be used to create the real world, as well as the world of illusion. It all depends on who is doing the imagining, and to what end.

When pre-experimental man conceived of the unknown as an ambivalent mother, he was not indulging in childish fantasy. He was applying what he knew to what was unfamiliar but could not be ignored. Man's first attempts to describe the unknown cannot be faulted because they lacked empirical validity. Man was not originally an empirical thinker. This does not mean he was self-deluded, a liar. Likewise, when the individual worships the hero, he is not necessarily hiding from reality. It may also be that he is ready and willing to face the unknown, as an individual; that he is prepared to adopt the pattern of heroic endeavour in his own life, and to further creation in that manner.

The great myths of Christianity—the great myths of the past, in general—no longer speak to the majority of Westerners, who regard themselves as educated. The mythic view of history cannot be credited with reality, from the material, empirical point of view. It is nonetheless the case that all of Western ethics, including those explicitly formalized in Western law, are predicated upon a mythological worldview, which specifically attributes divine status to the individual. The modern individual is therefore in a unique position: he no longer believes that the principles upon which all his behaviors are predicated are valid. This might be considered a second fall, in that the destruction of the Western mythological barrier has re-exposed the essential tragedy of individual existence to view.

It is not the pursuit of empirical truth, however, that has wreaked havoc upon the Christian worldview. It is confusion of empirical fact with moral truth that has proved of great detriment to the latter. This has produced what might be described as a secondary gain, which has played an important role in maintaining the confusion. That gain is abdication of the absolute personal responsibility imposed in consequence of recognition of the divine in man. This responsibility means acceptance of the trials and tribulations associated with expression of unique individuality, as well as respect for such expression in others. Such acceptance, expression and respect requires courage in the absence of certainty, and discipline in the smallest matters.

Rejection of moral truth allows for rationalization of cowardly, destructive, degenerate self-indulgence. This is one of the most potent attractions of such rejection, and constitutes primary motivation for the lie. The lie, above all else, threatens the individual—and the interpersonal. The lie is predicated upon the presupposition that the tragedy of individuality is unbearable—that human experience itself is evil. The individual lies because he is afraid— and it is not the lies he tells another that present the clearest danger, but the lies he tells himself. The root of social and individual psychopathology, the "denial," the "repression" is the lie. The most dangerous lie of all is devoted to denial of individual responsibility—denial of individual divinity.

The idea of the divine individual took thousands of years to fully develop, and is still constantly threatened by direct attack and insidious counter-movement. It is based upon realization that the individual is the locus of experience. All that we can know about reality we know through experience. It is therefore simplest to assume that all there is of reality is experience—its being and progressive unfolding. Furthermore, it is the *subjective* aspect of individuality—of experience—that is divine, not the objective. Man is an animal, from the objective viewpoint, worthy of no more consideration than the opinion and opportunities of the moment dictate. From the mythic viewpoint, however, every individual is unique—is a new set of experiences, a new universe; has been granted the ability to bring something new into being; is capable of participating in the act of creation itself. It is the expression of this capacity for creative action that makes the tragic conditions of life tolerable, bearable—remarkable, miraculous.

The paradise of childhood is absolute meaningful immersion. That immersion is a genuine manifestation of subjective interest. Interest accompanies the honest pursuit of the unknown, in a direction and at a rate subjectively determined. The unknown, in its beneficial guise, is the ground of interest, the source of what matters. Culture, in its supportive role, extends the power with which the unknown can be met, by disciplining the individual and expanding his range of ability. In childhood, the parent serves as cultural surrogate, and the child explores under the umbrella of protection provided by his parents. The parental mechanism has its limits, however, and must be superseded by the internalization of culture—by the intrapsychic incorporation of belief, security, and goal. Adoption of this secondary protective structure dramatically extends and shapes individual capability.

The great dragon of chaos limits the pursuit of individual interest. The struggle with the dragon—against the forces that devour will and hope—constitutes the heroic battle in the mythological world. Faithful adherence to the reality of personal experience ensures contact with the dragon, and it is during such contact that the great force of the individual spirit makes itself manifest, if it is allowed to. The hero voluntarily places himself in opposition to the dragon. The liar pretends that the great danger does not exist, to his peril and to that of others, or abdicates his relationship with his essential interest, and abandons all chance at further development.

Interest is meaning. Meaning is manifestation of the divine individual adaptive path. The lie is abandonment of individual interest—hence meaning, hence divinity—for safety and security; is sacrifice of the individual to appease the Great Mother and Great Father.

The lie is fear's statement, in the face of genuine experience: "That could not really be the case; that did not really happen." The lie weakens the individual—who no longer extends the range of his competence by testing his subjectivity against the world—and drains his life of meaning. Life without meaning is mortal limitation, subjection to pain and suffering without recourse. Life without meaning is tragedy, without hope of redemption.

The abandonment of meaning ensures the adoption of a demonic mode of adaptation, because the individual hates pointless pain and frustration and will work toward its destruction. This work constitutes revenge against existence, rendered unbearable by pride.

Rebirth is re-establishment of interest, after adoption of culturally determined competence. The rebirth of interest moves the individual to the border between the known and the unknown and thereby expands the social world. In this manner, God acts through the individual, in the modern world, and extends the domain of history.

Self-consciousness means knowledge of individual vulnerability. The process by which this knowledge comes to be can destroy faith in individual worth. This means, in concrete terms, that an individual may come to sacrifice his own experience, in the course of development, because its pursuit creates social conflict or exposes individual inadequacy. However, it is only through such conflict that change takes place, and weakness must be recognized, before it can be transformed into strength. This means that the sacrifice of individuality eliminates any possibility that individual strength can be discovered or developed, and that the world itself might progress.

Individuals whose life is without meaning hate themselves for their weakness and hate life for making them weak. This hatred manifests itself in absolute identification with destructive power, in its mythological, historical and biological manifestations; manifests itself in the desire for the absolute extinction of existence. Such identification leads man to poison whatever he touches, to generate unnecessary misery in the face of inevitable suffering, to turn his fellows against themselves, to intermingle earth with hell—merely to attain vengeance upon God and his creation.

The human purpose, if such a thing can be considered, is to pursue meaning—to extend the domain of light, of consciousness—despite limitation. A meaningful event exists on the boundary between order and chaos. The pursuit of meaning exposes the individual to the unknown in gradual fashion, allowing him to develop strength and adaptive ability in proportion to the seriousness of his pursuit. It is during contact with the unknown that human power grows, individually and then historically. Meaning is the subjective experience associated with that contact, in sufficient proportion. The great religious myths state that continued pursuit of meaning, adopted voluntarily and without self-deception, will lead the individual to discover his identity with God. This "revealed identity" will make him capable of withstanding the tragedy of life. Abandonment of meaning, by contrast, reduces man to his mortal weaknesses.

Meaning is the most profound manifestation of instinct. Man is a creature attracted by the unknown; a creature adapted for its conquest. The subjective sense of meaning is the instinct governing rate of contact with the unknown. Too much exposure turns change to chaos; too little promotes stagnation and degeneration. The appropriate balance produces a powerful individual, confident in the ability to withstand life, ever more able to deal with nature and society, ever closer to the heroic ideal. Each individual, constitutionally unique, finds meaning in different pursuits, if he has the courage to maintain his difference. Manifestation of individual diversity, transformed into knowledge that can be transferred socially, changes the face of history itself, and moves each generation of man farther into the unknown.

Social and biological conditions define the boundaries of individual existence. The unfailing pursuit of interest provides the subjective means by which these conditions can be met, and their boundaries transcended. Meaning is the instinct that makes life possible. When it is abandoned, individuality loses its redeeming power. The great lie is that meaning does not exist, or that it is not important. When meaning is denied, hatred for life and the wish for its destruction inevitably rules:

If you bring forth what is within you, what you bring forth will save you.
If you do not bring forth what is within you, what you do not bring forth will destroy you.[667]

The wisdom of the group can serve as the force that mediates between the dependency of childhood and the responsibility of the adult. Under such circumstances, the past serves the present. A society predicated upon belief in the paramount divinity of the individual allows personal interest to flourish and to serve as the power that opposes the tyranny of culture and the terror of nature. The denial of meaning, by contrast, ensures absolute identification with the group—or intrapsychic degeneration and decadence. The denial of meaning makes the absolutist or degenerate individual desperate and weak, when the great maternal sea of chaos threatens. This desperation and weakness makes him hate life, and to work for its devastation—in him, as well as in those around him. The lie is the central act in this drama of corruption:

These are the secret sayings which the living Jesus spoke and which Didymos Judas Thomas wrote down..
And he said, "Whoever finds the interpretation of these sayings will not experience death."
Jesus said, "Let him who seeks continue seeking until he finds. When he becomes troubled, he will be astonished, and he will rule over the all."
Jesus said, "If those who lead you say to you, 'See, the kingdom is in the sky,' then the birds of the sky will precede you. If they say to you, 'It is in the sea,' then the fish will precede you. Rather, the kingdom is inside of you, and it is outside of you. When you come to know yourselves, then you will become known, and you will realize that it is you who are the sons of the living father. But if you will not know yourselves, you dwell in poverty and it is you who are that poverty."
Jesus said, "The man old in days will not hesitate to ask a small child seven days old about the place of life, and he will live. For many who are first will become last, and they will become one and the same."
Jesus said, "Recognize what is in your sight, and that which is hidden from you will become plain to you. For there is nothing hidden which will not become manifest."
His disciples questioned him and said to him, "Do you want us to fast? How shall we pray? Shall we give alms? What diet shall we observe?"
Jesus said, "Do not tell lies, and do not do what you hate, for all things are plain in the sight of heaven. For nothing hidden will not become manifest, and nothing covered will remain without being uncovered."[668]

NOTES

Preface

1. Jung, C.G. (1978a), p. 78; also:

> Although seemingly paradoxical,
> the person who takes upon himself,
> the people's humiliation,
> is fit to rule;
> and he is fit to lead,
> who takes the country's disasters upon himself. (Lao-ızu, 1984c)

2. "Sometimes I look at a Socialist—the intellectual, tract-writing type of Socialist, with his pullover, his fuzzy hair, and his Marxian quotation—and wonder what the devil his motive really *is*. It is often difficult to believe that it is a love of anybody, especially of the working class, from whom he is of all people the furthest removed." (Orwell, G. [1981], pp. 156–157)

3. Jung, C.G. (1970a), p. 157.

4. Ibid., p. 158.

5. Ten years later, when I was finishing this manuscript, a student of mine, Heidi Treml, wrote:

> During the journey from Egypt to Canaan the impatient Israelites accused God and Moses of leading them into the desert to die. As a consequence of this complaining, Yahweh sent venomous serpents among the Israelites. Those Israelites who were not bitten by the serpents repented and asked Moses to intervene with God. Yahweh instructed Moses to make a bronze [or fiery] serpent and to place it on top of a pole so that those who were bitten could behold it and live. Moses did as he was commanded and, whenever a snake bit someone, that person would look at the bronze statue and live [Numbers 21:5–10].... John the Evangelist has Jesus explaining to Nicodemus, "And as Moses lifted up the serpent in the wilderness, even so must the Son of man be lifted up; That whosoever believeth in him should not perish, but have eternal life." [John 3:13–15]

Treml pointed out that the serpent has been widely regarded both as an agent of death (because of its venom) and as an agent of transformation and rebirth (because it could shed its skin). This intense ambivalence of feature makes it an apt representative of the "numinous" (following Rudolf Otto, whose ideas are described later in this manuscript). The *numinous* is able to invoke trembling and fear (*mysterium tremendum*) and powerful attraction and fascination (*mysterium fascinans*). Treml commented, further: "if a person could sustain the gaze of the serpent—which symbolized his greatest fear—he would be healed."

Why was Christ assimilated to the serpent, in my painting and in the New Testament? (Understand that I knew nothing whatsoever of this relationship when I originally constructed the sketch.) It has something to do with his representation as judge in Revelations:

I know thy works, that thou art neither cold nor hot: I would thou wert cold or hot.

So then because thou art lukewarm, and neither cold nor hot, I will spue thee out of my mouth.

Because thou sayest, I am rich, and increased with goods, and have need of nothing; and knowest not that thou art wretched, and miserable, and poor, and blind, and naked:

I counsel thee to buy of me gold tried in the fire, that thou mayest be rich; and white raiment, that thou mayest be clothed, and *that* the shame of thy nakedness do not appear; and anoint thine eyes with eye-salve, that thou mayest see. (Revelations 3:15–19)

The idea of the Savior necessarily implies the Judge—and a judge of the most implacable sort—because the Savior is a mythological representation of that which is ideal, and the ideal always stands in judgment over the actual. The archetypal image of the Savior, who represents perfection or completion, is therefore terrifying in precise proportion to personal distance from the ideal.

6. Joyce, J. (1986), p. 28.

7. Jung, C.G. (1968b), p. 32.

8. Ibid., pp. 32–33.

Chapter 1

9. Jacqueline Kennedy Onassis' tape measure sold for $45,000 in 1996 (Gould, L., Andrews, D., & Yevin, J. [1996 December], p. 46).

10. Jung, C.G. (1976b), pp. 92–93.

11. Ibid., pp. 10–11.

12. Eliade, M. (1978b).

13. Jung, C.G. (1933), p. 204.

14. Nietzsche, F. (1981), pp. 69–70. Nietzsche referred to "the English" in the original; the viewpoint he was criticizing is so widely held now that my substitution of "modern Westerners" seems perfectly apropos.

15. Fukuyama, F. (1993).

16. Nietzsche, F. In Kaufmann, W. (1975), p. 126.

17. There are at least four independent Sumerian narratives (including the *Enuma elish*, detailed later) that describe the origin of the cosmos. Eliade assumes a "plurality of traditions," most likely deriving from the peoples whose union produced Sumerian civilization. Eliade, M. (1978b), p. 59.

18. Eliade, M. (1978b), pp. 57–58.

19. Nietzsche, F. (1966), pp. 97–98.

20. Nietzsche, F. (1968a), pp. 77–78.

21. Dostoevsky, F. In Kaufmann, W. (1975), pp. 75–76.

22. Frye, N. (1990), pp. 90–92.

23. Richard Wilhelm translated the Chinese Tao, the ground of being, the *way*, as "sinn," the German equivalent of "meaning" (Wilhelm, R. [1971], p. iv). The *way* is a path of life, guided by processes manifested outside the area circumscribed by defined, logic, internally consistent cognitive structures. From such a perspective, meaningful experiences might be considered "guideposts" marking the path to a new mode of being. Any form of art that produces an aesthetic seizure, or intimation of meaning, might therefore serve as such a guidepost—at least in principle (see Solzhenitsyn, A.I. [1990], pp. 623–630).

24. See, for example, Eliade, M. (1975).

Chapter 2

25. Gray, J.A. (1982; 1987); Gray, J.A. & McNaughton, N. (1996); Gray, J.A., Feldon, J., Rawlins, J.N.P., Hemsley, D.R., & Smith, A.D. (1996).

26. Sokolov, E.N. (1969), p. 672.

27. Ibid. (1969), p. 673.

28. These "maps" are so important to us, so vital, that their mere abstract description (acted, orally transmitted or written) is intrinsically interesting, capable of engaging us in a simulated world (see Oatley, K. [1994]).

29. Vinogradova, O. (1961; 1975); Luria, A.R. (1980).

30. Lao Tzu (1984b).

31. Ohman, A. (1979); Vinogradova, O. (1961).

32. Ibid.

33. Obrist, P.A., Light, K.C., Langer, A.W., Grignolo, A., & McCubbin, J.A. (1978).

34. Gray, J.A. (1982).

35. Nietzsche, F. (1968a), p. 88.

36. Gray, J.A. (1982).

37. I use the term "promise" here partly because it makes a good (that is, symmetric) counterpart to "threat." The term "promise" used here means "incentive reward" or "cue for satisfaction" or "cue for consummatory reward." Furthermore, neither the former term, nor the latter, appear particularly appropriate for the positive state induced by contact with novelty. Novelty does not seem reasonably categorized as a "reward"; also, positive affect can be generated through contact with novelty, in the absence of any conditioning whatsover (Gray, J.A. [1982]), so the term "cue" seems inappropriate.

38. Eliade, M. (1978b); Jung, C.G. (1969).

39. Gray, J.A. (1982; 1987); Gray, J.A., & McNaughton, N. (1996).

40. Kuhn, T.S. (1970).

41. Ibid., (1970).

42. Obrist, P.A., Light, K.C., Langer, A.W., Grignolo, A., & McCubbin, J.A. (1978).

43. Kuhn, T.S. (1970).

44. Jung, C.G. (1976b), pp. 540–541.

45. Jung, C.G. (1967a; 1968; 1967b); Ellenberger, H. (1970); Campbell, J. (1968); Eliade, M. (1964; 1978b; 1982; 1985); Piaget, J. (1977).

46. Bruner, J. (1986).

47. Eliade, M. (1965).

48. Jung, C.J. (1967a; 1968b; 1969; 1967b); Eliade, M. (1978b; 1982; 1985).

49. Nietzsche, F. (1968a), pp. 203–204.

50. Eliade, M. (1978b), p. 59.

51. Frankl, V. (1971), pp. 70–72.

52. Skinner, B.F. (1966; 1969).

53. Solzhenitsyn, A.I. (1975), pp. 605–606.

54. Gray, J.A. (1982); Gray, J.A. & McNaughton, N. (1996); Pihl, R.O. & Peterson, J.B. (1993; 1995); Tomarken, A.J., Davidson, R.J., Wheeler, R.E., & Doss, R.C. (1992); Wheeler, R.E., Davidson, R.J., & Tomarken, A.J. (1993); Tomarken, A.J., Davidson, R.J., & Henriques, J.B. (1990); Davidson, R.J. & Fox, N.A. (1982).

55. Gray, J.A. (1982); Ikemoto, S. & Panksepp, J. (1996).

56. Wise, R.A. (1988); Wise, R.A. & Bozarth, M.A. (1987).

57. Gray, J.A. (1982).

58. Mowrer, O.H. (1960).

59. Wise, R.A. (1988); Wise, R.A. & Bozarth, M.A. (1987).

60. Gray, J.A. (1982).

61. Reviewed in Gray, J.A. (1982).
62. Skinner, B.F. (1966; 1969).
63. Panksepp, J., Siviy, S. & Normansell, L.A. (1985).
64. Gray, J.A. (1982).
65. Ibid.; Dollard, J. & Miller, N. (1950).
66. Gray, J.A. (1982).
67. Ibid.
68. Ibid.
69. Ibid.
70. Ibid.
71. Ibid.
72. Reviewed in Gray, J.A. (1982).
73. Reviewed in Gray, J.A. (1982); Wise, R.A. & Bozarth, M.A. (1987).
73. Dollard, J. & Miller, N. (1950).
74. Wise, R.A. (1988); Wise, R.A. & Bozarth, M.A. (1987).
75. Dollard, J. & Miller, N. (1950).
76. Ibid., (1950).
77. Luria, A.R. (1980).
78. Goldman-Rakic, P.S. (1987); Shallice, T. (1982); Milner, B., Petrides, M., & Smith, M.L. (1985).
79. Oatley, K. (1994).
80. Patton, M.F. (1988), p. 29.
81. Gray, J.A. (1982).
82. Dollard, J. & Miller, N. (1950).
83. Gray, J.A. (1982); Gray, J.A., Feldon, J., Rawlins, J.N.P., Hemsley, D.R., & Smith, A.D. (1991).
84. Dollard, J. & Miller, N. (1950).
85. Gray, J.A. (1982); Fowles, D.C. (1980; 1983; 1988; 1994).
86. Wise, R.A. (1988); Wise, R.A. & Bozarth, M.A. (1987); Gray, J.A. (1982).
87. Gray, J.A. (1982).
88. Wise, R.A. (1988); Wise, R.A. & Bozarth, M.A. (1987).
89. Gray, J.A. (1982).
90. Damasio, A.R. (1994; 1996); Bechara, A., Tranel, D., Damasio, H. & Damasio, A.R. (1996); Bechara, A., Damasio, H., Tranel, D., & Damasio, A.R. (1997).
91. Bechara, A., Damasio, H., Tranel, D., & Damasio, A.R. (1997); Damasio, A.R. (1994); Bechara, A., Tranel, D., Damasio, H. & Damasio, A.R. (1996).
92. Luria, A.R. (1980); Nauta, W.J.H. (1971).
93. Luria, A.R. (1980); Granit, R. (1977).
94. Luria, A.R. (1980).
95. Ibid.
96. Sokolov, E.N. (1963); Vinogradova, O. (1975); Gray, J.A. (1982; 1987); Gray, J.A. & McNaughton, N. (1996).
97. Gray, J.A. (1982; 1987); Gray, J.A. & McNaughton, N. (1996); Sokolov, E.N. (1969); Vinogradova, O. (1975); Halgren, E., Squires, N.K., Wilson, C.L., Rohrbaugh, J.W., Babb, T.L., & Crandell, P.H. (1980); Watanabe, T. & Niki, H. (1985).
98. See Aggleton, J.P. (1993).
99. Halgren, E. (1992), p. 205.
100. Ibid.
101. Ibid., p. 206.
102. Halgren, E. (1992).
103. Ohman, A. (1979; 1987).
104. Halgren, E. (1992), p. 206.

105. For reviews of supporting evidence, see Tucker, D.M. & Williamson, P.A. (1984); Davidson, R.J. (1984a; 1984b; 1992); Goldberg, E., Podell, K., & Lovell, H. (1994); Goldberg, E. (1995); Goldberg, E. & Costa, L.D. (1981); for some indication of why two different systems may in fact be necessary, see Grossberg, S. (1987).

106. Dollard, J. & Miller, N. (1950).

107. Ohman, A. (1979; 1987).

108. Brown, R. (1986); Rosch, E., Mervis, C.B., Gray, W., Johnson, D., & Boyes-Braem. (1976); Lakoff, G. (1987); Wittgenstein, L. (1968).

109. Eliade, M. (1978b).

110. Sokolov, E.N. (1969); Vinogradova, O. (1975); Gray, J.A. (1982); Gray, J.A. (1987); Gray, J.A. & McNaughton, N. (1996).

111. Aggleton, J.P. (1993).

112. For a discussion of the simplicity and general utility of "default-on" systems (as opposed to "default-off") see Brooks, A. & Stein, L.A. (1993); Brooks, A. (1991).

113. LeDoux, J.E. (1992).

114. Ibid.

115. Luria, A.R. (1980).

116. This point appears to me to be related to the stability-plasticity dilemma outlined by Grossberg, S. (1987).

117. Blanchard, D.C. & Blanchard, R.J. (1972); Bouton, M.E. & Bolles, R.C. (1980); LeDoux, J.E., Sakaguchi, A., & Reis, D.J. (1984).

118. Blanchard, D.C. & Blanchard, R.J. (1972).

119. Kapp, B.S., Pascoe, J.P., & Bixler, M.A. (1984); Iwata, J., Chida, K., & LeDoux, J.E. (1987).

120. LeDoux, J.E. (1992).

121. Recent work conducted on three related phenomena—latent inhibition, prepulse inhibition of startle, and negative priming—illustrates the essential validity of this viewpoint. "Latent inhibition" (LI) is the difficulty in learning that "a" signifies "b," when "a" previously signified "c" (where "c" is, most frequently, nothing) (see Lubow, R.E. [1989], for a review of the literature; Gray, J.A. & McNaughton, N. [1996] and Gray, J.A., Feldon, J., Rawlins, J.N.P., Hemsley, D.R., & Smith, A.D. [1991] for a discussion of putative neuropsychology). If you expose a caged rat repeatedly to a intermittent light, paired with a shock, he becomes afraid of the light. In classical terms, the light has become a *conditioned stimulus* for shock, and therefore elicits fear. However, if you have pre-exposed the rat to the same light, repeatedly, in the absence of any consequence, then he takes substantially longer to learn the new light/shock connection. Latent inhibition (LI) provides an example of the difficulty in learning (new valence), consequential to previous learning (an alternative is provided by the related Kamin blocking effect [Jones, S.H., Gray, J.A., & Hemsley, D.R. (1992)]). The capacity for LI characterizes a variety of animal species, as well as man; the phenomena itself can be elicited using a number of different experimental paradigms (using differently valenced "unconditioned" stimuli). Acute first-onset schizophrenics and their close "cousins" (schizotypes)—overwhelmed by their everyday experience—manifest decreased LI, as do individuals taking amphetamines or other dopaminergic agonists (which produce heightened exploratory behavior [Wise, R.A. & Bozarth, M.A. (1987)]). Antipsychotic medications, which damp down the *a priori* significance of things, normalize this decrease.

Prepulse inhibition (PPI) of startle occurs when the magnitude of a startle response to an intense, unexpected "stimuli" (such as a loud noise) is attenuated as a consequence of a "hint" (such as a similar, but less intense noise) given 30–500 msec earlier. The fact of the hint apparently decreases the relative novelty (the unpredictability) of the more intense subsequent stimuli, at least among normal individuals; alternatively, it might be regarded as an analog of graduated exposure (the procedure by which behavioral therapists "desensitize" the fear responses of their clients). Schizophrenics, once again, or persons with related cognitive abnormalities (Swerdlow, N.R, Filion, D., Geyer, M.A., & Braff, D.L. [1995]) manifest decreased PPI, indicating, perhaps, that they cannot effectively use the capacity to predict, on the basis of past experience, to modulate their affective/psychophysiological responses to stimuli that "intrinsically demand" response (to "uncondi-tioned stimuli," in the old terminology).

Individuals participating in the "negative priming" paradigm must learn to respond to a stimulus that appears in the same place recently (< 350 msec) occupied by an irrelevant or distractor stimulus. Normals are better at defining a place as irrelevant than are schizophrenic or schizotypic individuals—hence, their reaction times, when "negatively primed," are longer (Swerdlow, N.R., Filion, D., Geyer, M.A., & Braff, D.L. [1995]). The negative priming paradigm, like the others, demonstrates that irrelevancy (the cardinal charac-teristic of the unassociated "conditioned stimulus") must be learned and may be unlearned (sometimes with devastating consequences). At issue, therefore, is the *a priori* status of the conditioned stimulus with regard to valence and how that status might be altered or "explored away."

The experiments utilizing LI (and related procedures) are fascinating—and critically important—because they demonstrate that the irrelevance of most context-dependent irrelevant things is not given. Irrelevance must be learned; furthermore, such learning is sufficiently powerful to interfere with subsequent learning, when motivational contingencies shift. The original curiosity- or hope-inducing aspect of now-familiar things appears to be driven by disinhibited amygdalic-driven dopaminergic activation in the nucleus accumbens (Gray, J.A., Feldon, J., Rawlins, J.N.P., Hemsley, D.R., & Smith, A.D. [1991]) which is the same center acti-vated by most, if not all, "positively reinforcing" drugs of abuse (Wise, R.A. & Bozarth, M.A. [1987]). The "fear-inducing aspect"—which must logically co-exist—has received less attention (although the role of the amygdala in producing novelty-driven fear is well-established, as described previously). It is these twin aspects—threat and promise, inducing *a priori* fear and hope (relevance, in its most fundamental guise)—that normally lie beyond (Huxley, A. [1956]) William Blake's "doors of perception" and that lend to existence itself its "intrinsic" (and sometimes overwhelming) meaning:

If the doors of perception were cleansed everything would appear to man as it is, infinite.
For man has closed himself up, till he sees all things thro' narrow chinks of his cavern.

(Blake, W. [1946], p. 258)

Physiological or environmental events that open these doors allow us insight into the original nature of things; such insight, when involuntary (as it appears to be in the case of schizophrenia, for example) is of suf-ficient power to terrify and, perhaps, to destroy. The *a priori* valence of the object is potent, and potentially terrifying (as terrifying, literally, as anything imaginable). Our normal circumstances, our prior learning, pro-tect us from this valence; they shield us, restrict our access to meaning *as such*. Events that interfere with the stability of that learning or with its conditional validity have the capacity to allow meaning to re-emerge, with its awful force unshielded.

122. Luria, A.R. (1980), pp. 30–32.
123. Hebb, D.O. & Thompson, W.R. (1985), p. 766.
124. Blanchard, R.J. & Blanchard, D.C. (1989).
125. Blanchard, D.C., Blanchard, R.J., & Rodgers, R.J. (1991).
126. Pinel, J.P.J. & Mana, M.J. (1989).
127. Blanchard, R.J., Blanchard, D.C., & Hori, K. (1989).
128. Blanchard, R.J. & Blanchard, D.C. (1989).
129. Blanchard, D.C., Veniegas, R., Elloran, I., & Blanchard, R.J. (1993).
130. Lorenz, K. (1974).
131. Goodall, J. (1990).
132. Exploration is *not* merely specification of the "inherent" properties of the unexpected thing or situa-tion. The actual nature of things or situations (from the perspective of valence and objective classification) is dependent upon the behavioral strategies employed in its presence, and on the ends that are currently being pursued. This means that determinate experience must be considered an emergent property of behavior to a degree that is presently unspecifiable. This appears as true for the purely objective aspects of experience

(which constitute the subject matter for science) (see Kuhn, T. [1970]; Feyeraband, P.K. [1981]) as for the subjective.

The *word* itself, as case in point, can no longer reasonably be regarded as a "label" for a "thing" (Wittgenstein, L. [1968], pp. 46e–47e). The notion that a *concept* is a label for an *object* is nothing but a slightly higher-order version of the same error. Wittgenstein pointed out, essentially, that our sense of unified "thing" is not simply given (Wittgenstein, L. [1968]). We tend to think of the objects we perceive as "being there" in some essential sense; but we see the tree before the branches. Despite this conceptual phenomenon, the tree has no objective precedence over the branches (or the leaves, or the cells that make up the leaves, or the forest, for that matter). Roger Brown, following Wittgenstein's lead, demonstrated that "objects" have their "basic levels"—their levels of resolution, essentially, that appear most easily and rapidly learned by children, and constant across cultures (Brown, R. [1986]).

Wittgenstein solved the "words are not labels for objects" problem by positing that a word was a tool. A word plays a role in a game and is akin to a chessman in chess (Wittgenstein, L. [1968], pp. 46e–47e). "The meaning of a piece is its role in the game" (Wittgenstein, L. [1968], p. 150e). He noted, furthermore, that the "game" has "not only rules, but a point" (Wittgenstein, L. [1968], p. 150e).

Wittgenstein was driving at a general principle: an object is *defined*, even perceived (categorized as a unity, rather than a multiplicity), with regard to its utility as a means to a given end. In a basic sense, an object is a tool or an obstacle. What we perceive as objects are phenomena that may be easily utilized (to grant our desires)—at least in principle (or things that may well interfere with our attainment of desired ends). Facilitators are valenced, positively (as incentive rewards); obstacles, negatively (as punishments or threats). Normal facilitators and obstacles have minor valence, relatively speaking; their revolutionary counterparts may produce overwhelming emotion (think of Archimedes' "Eureka!"). What can reasonably be parsed out of the environmental flux as an object is therefore determined in large part by the goal we have in mind while interacting with that flux. This complex situation is further complicated by the fact that the valence of objects, once given as objects, may still change with alteration in the ends we are pursuing (because tools in one situation may easily become obstacles—or something irrelevant—in another). Finally, many things that could manifest themselves as objects, at a given time or place, will not (because they are apparently irrelevant to the task at hand, and remain invisible).

133. Luria, A.R. (1980).
134. Granit, R. (1977).
135. Agnew, N.M. & Brown, J.L. (1990).
136. Holloway, R.L. & Post, D.G. (1982).
137. Jerison, H.J. (1979).
138. Ridgeway, S.H. (1986).
139. Lilly, J.C. (1967).
140. Penfield, W. & Rasmussen, T. (1950).
141. Brown, R. (1986).
142. Garey, L.J. & Revishchin, A.V. (1990).
143. Granit, R. (1977).
144. Ibid.
145. Wise, R.A. & Bozarth, M.A. (1987).
146. Granit, R. (1977).
147. Oatley, K. (1994).
148. For reviews of supporting evidence, see Tucker, D.M. & Williamson, P.A. (1984); Davidson, R.J. (1984a, 1984b, 1992); Goldberg, E., Podell, K., & Lovell, H. (1994); Goldberg, E. (1995); Goldberg, E. & Costa, L.D. (1981); for some indication of why two different systems may in fact be necessary, see Grossberg, S. (1987).
149. Fox, N.A. & Davidson, R.J. (1986, 1988).
150. Maier, N.R.F. & Schnierla, T.C. (1935).

151. Schnierla, T.C. (1959).

152. See review by Springer, S.P. & Deutsch, G. (1989).

153. Goldberg, E. (1995); Goldberg, E. & Costa, L.D. (1981); Goldberg, E., Podell, K., & Lovell, H. (1994).

154. Springer, S.P. & Deutsch, G. (1989).

155. Fox, N.A. & Davidson, R.J. (1986, 1988).

156. Goldberg, E. & Costa, L.D. (1981).

157. Goldberg, E. (1995).

158. Donald, M. (1993).

159. "We believe that the internal and external states which constitute the response to the stimulus are identical with the 'evaluation' of the stimulus" (Kling, A.S. & Brothers, L.A. [1992], p. 372); "affect is no more and no less than the confluence and integration of sensory information in several modalities, combined with immediate coactivation of somatic effector systems (motor, autonomic and endocrine)" (p. 371); "reciprocal connections between amygaloid nuclei and the hippocampal formation may serve to link affective response patterns with the encoding of perceptions in memory, thus providing rapid access to appropriate motivational states when complex social situations or particular individuals are re-encountered" (p. 356).

160. Vitz, P.C. (1990).

161. Ibid.

162. Ibid.

163. Ryle, G. (1949).

164. Milner, B. (1972); Zola-Morgan, S., Squire, L.R., & Amaral, D.G. (1986); Teylor, T.J. & Discenna, P. (1985, 1986).

165. Squire, L.R. & Zola-Morgan, S. (1990).

166. Ibid.

167. Squire and Zola-Morgan state:

> The term declarative, which we have used, captures the notion that one kind of memory can be "declared"; it can be brought to mind explicitly, as a proposition or image. The capacity for declarative memory may be a relatively recent feat of evolution, appearing early in the vertebrates with the development of the hippocampus, and the capacity for declarative memory may be ontogenetically delayed. Procedural knowledge, by contrast, can be expressed only through performance, and the contents of this knowledge are not accessible to awareness. Procedural knowledge is considered to be phylogenetically primitive and ontogenetically early.... We agree with Tulving and his colleagues that the episodic-semantic distinction, which has something interesting to say about the structure of normal memory, is a subset of declarative (propositional) memory. (Squire, L.R. & Zola-Morgan, S. [1990], p. 138.).

My presupposition is that a story is a semantic representation of an episodic representation of the outputs of the procedural system: a verbal description of an image of behavior (and the consequences of that behavior).

168. Schachter, D.L. (1994).

169. Kagan, J. (1984).

170. Piaget, J. (1962), p. 3.

171. Ibid., p. 5.

172. Ibid.

173. Ibid., p. 6.

174. Adler, A. (1959); Vaihinger, H. (1924).

175. Oatley, K. (1994).

176. Donald, M. (1993).

177. An idea is (in part) an abstracted action, whose consequences can be analyzed in abstracted fantasy. The distance between the idea and the action has widened within the course of recent evolutionary history. Medieval people, unused to rhetorical speech, were easily seized emotionally or inspired to action by pas-

sionate words (see Huizinga, J. [1967]). In the modern world, flooded by meaningless speech, words have lost much of their immediate procedural power, under normal conditions. However, music still unconsciously compels movement, dance or, at least, the compulsion to keep the beat. Even chimpanzees seem capable of becoming possessed by simple rhythms (see Campbell, J. [1987], pp. 358–359). In addition, modern individuals are still easily seized and motivated by drama, like that portrayed in motion pictures—much like the "primitive" seized by ritual—and can easily lose themselves in the act of acting "as if" the drama is actually happening. In the absence of this seizure, which is meaningful, drama loses its interest. Rhetoric—the call to action—also still dominates advertising, with evident effect.

178. "Meaningful" drama, or meaningful information, per se, has that characteristic because it produces affect, indicative of occurrence outside of predictability, and because it implies something for alteration of behavior. The phenomenon of meaning occurs when information can be translated from one "level" of memory to another, or to all others.

179. Piaget, J. (1932).
180. Piaget, J. (1962).
181. Nietzsche, F. (1966), p. 98.
182. Nietzsche, F. (1968a), p. 217.
183. Ibid., p. 203.
184. Wittgenstein, L. (1968).
185. Eliade, M. (1978b).
186. Wittgenstein, L. (1968), p. 16e.
187. An analogous notion of "goal-hierarchy" was presented by Carver, C.S. & Scheier, M.F. (1982).
188. Eysenck, H.J. (1995).
189. Shallice, T. (1982).
190. Milner, B., Petrides, M., & Smith, M.L. (1985).
191. Petrides, M. & Milner, B. (1982).
192. Milner, B. (1963).
193. "And he dreamed, and behold a ladder set up on the earth, and the top of it reached to heaven: and behold the angels of God ascending and descending on it. And, behold, the LORD stood above it, and said, I *am* the LORD God of Abraham thy father, and the God of Isaac." (Genesis 28:12–13)
194. Frye, N. (1982), p. 220.
195. Eliade, M. (1957), pp. 107–108.
196. Brown, R. (1965), p. 476.
197. Ibid., p. 478.
198. Goethe, J.W. (1976).
199. Frazier, J.G. (1994).
200. Brown, R. (1986), p. 470.
201. Lakoff, G. (1987), pp. 12–13.
202. Brown, R. (1965), p. 321.
203. Wittgenstein, L. (1968), pp. 66–71.
204. See Armstrong, S.L., Gleitman, L.R., & Gleitman, H. (1983).
205. Eliade, M. (1978b), pp. 57–58.
206. Heidel, A. (1965).
207. Eliade, M. (1978b).
208. "**Logos** λόγος. Theol. and Philos. [Gr. *logos* word, speech, discourse, reason, f. *log-*, ablaut-variant of *leg-* in *leg-ein* to say.] A term used by Greek (esp. Hellenistic and Neo-Platonist) philosophers in certain metaphysical and theological applications developed from one or both of its ordinary senses 'reason' and 'word'; also adopted in three passages of the Johannine writings of the N.T. (where the English versions render it by 'Word') as a designation of Jesus Christ; hence employed by Christian theologians, esp. those who were versed in Greek philosophy, as a title of the Second Person of the Trinity. By mod. writers the Gr. word is used untranslated in historical expositions of ancient philosophical speculation, and in discussions of the doctrine of the Trinity in its philosophical aspects." (*Oxford English Dictionary: CD-ROM for Windows*

[1994]).

209. Eliade, M. (1978b); Jung, C.G. (1967b).

210. Shakespeare, W. (1952).

211. Neumann, E. (1955, 1954); Jung, C.G. (1976b, 1967b, 1968b, 1967a); Eliade, M. (1978b).

212. Ibid.

213. Ibid.

214. Brown, R. (1986).

215. Brown, R. (1986); Rosch, E., Mervis, C.B., Gray, W., Johnson, D., & Boyes-Braem, P. (1976); Lakoff, G. (1987).

216. Lao Tzu (1984a).

217. *Vierge Ouvrante*, reproduced as plate 177 in Neumann, E. (1955).

218. Eliade, M. (1978b), pp. 88–89.

219. See Frye, N. (1990).

220. Heidel, A. (1965).

221. Frye, N. (1982), p. 146.

222. Frye, N. (1990).

223. Tablet 1:4; Heidel, A. (1965), p. 18.

224. Tablet 1:5; Heidel, A. (1965), p. 18.

225. Tablet 1:6–8; Heidel, A. (1965), p. 18.

226. Tablet 1:9; Heidel, A. (1965), p. 18.

227. Ea is also known as Nudimmud, in the original text. I have used the single appellation here, for the sake of simplicity.

228. Tablet 1:17; Heidel, A. (1965), p. 18.

229. Tablet 1:18-19; Heidel, A. (1965), p. 18.

230. Tablet 1: 20; Heidel, A. (1965), p. 18.

231. Tablet 1:23; Heidel, A. (1965), p. 19.

232. Tablet 1:80; Heidel, A. (1965), p. 21.

233. Tablet 1:86; Heidel, A. (1965), p. 21.

234. Tablet 1:90–102; Heidel, A. (1965), pp. 21–22.

235. Tablet 1:133–138; Heidel, A. (1965), p. 23.

236. Tablet 1:156; Heidel, A. (1965), p. 24.

237. Tablet 2:1–10; Heidel, A. (1965), p. 25.

238. Tablet 2:96–117; Heidel, A. (1965), pp. 28–29.

239. Tablet 2:118–129; Heidel, A. (1965), pp. 29–30.

240. Jacobsen, T. (1943).

241. Heidel, A. (1965), pp. 30–31.

242. Tablet 3:1–66; Heidel, A. (1965), pp. 30–33.

243. Tablet 3:131–138, 4:1–10; Heidel, A. (1965), pp. 35–36.

244. According to Campbell, J. (1964), p. 82.

245. Ibid.

246. Tablet 4:27–34; Heidel, A. (1965), pp. 37–38.

247. Tablet 4:87–94; Heidel, A. (1965), p. 40.

248. Tablet 4:129–144; Heidel, A. (1965), pp. 42–43.

249. Yahweh's role in creation is considered similarly, in relationship to Rahab, or Leviathan—the serpent from whom the world is constructed. Isaiah 51:9 states, for example "Awake, awake, put on strength, O arm of the Lord; awake, as in the ancient days, in the generations of old. *Art* thou not it that hath cut Rahab, *and* wounded the dragon?" Psalm 74 contains several comparable passages (14–17):

Thou brakest the heads of leviathan in pieces, *and* gavest him *to be* meat to the people inhabiting the wilderness. Thou didst cleave the fountain and the flood: thou driedst up mighty rivers. The day *is* thine, the night also *is* thine: thou hast prepared the light and the sun. Thou hast set all the borders of the earth: thou hast made summer and winter.

250. Tablet 6:8; Heidel, A. (1965), p. 46.

251. Tablet 6:49–51; Heidel, A. (1965), p. 48.

252. Eliade, M. (1978b), pp. 73–74.

253. Ibid., pp. 74–76.

254. Tablet 6:151; Heidel, A. (1965), p. 52.

255. Tablet 6:152–153; Heidel, A. (1965), p. 53.

256. Tablet 6:155–156; Heidel, A. (1965), p. 53.

257. Tablet 7:1–2; Heidel, A. (1965), p. 53.

258. Tablet 7:21; Heidel, A. (1965), p. 54.

259. Tablet 7:30; Heidel, A. (1965), p. 55.

260. Tablet 7:39; Heidel, A. (1965), p. 55.

261. Tablet 7:81; Heidel, A. (1965), p. 57.

262. Tablet 7:112, 7:115; Heidel, A. (1965), p. 58.

263. Eliade, M. (1978b), p. 89.

264. In Pritchard, J.B. (1955), p. 4.

265. Eliade, M. (1978b), pp. 89–90.

266. Eliade, M. (1978b), p. 91.

267. Eliade, M. (1978b), pp. 91–92. It is of additional interest to note that the Egyptians prohibited foreigners from entering their sanctuaries, which were "microcosmic images of the country"; native Egyptians were the only "rightful inhabitants" of Egypt, the "first country formed," and the "center of the world." Foreigners brought disorder (Eliade, M. [1978b], p. 90).

268. Eliade comments: "When Horus descended into the otherworld and resuscitated Osiris, he bestowed on him the power of "knowing." Osiris was an easy victim because he "did not know," he had no knowledge of Seth's true nature." (Eliade, M. [1978b], p. 100, footnote 41). The story of Osiris is in part a parable about the dangers of the incapacity to recognize evil.

269. Eliade, M. (1978b), p. 100.

270. Anaximander of Miletus (611 B.C. to 546 B.C.).

271. William James, in the throes of nitrous oxide intoxication. Quoted by Tymoczko, D. (May 1996), p. 100.

272. These myths express the fact that the unknown tends to manifest itself first in terrifying form.

273. Eliade, M. (1978b), pp. 205–207.

274. Derived from *The Self-Consuming Dragon*, an allegorical figure in the works of Lamspringk, reproduced as plate LIXa in Jung (1967a).

275. Neumann, E. (1954), pp. 10–11.

276. Eliade, M. (1978b), p. 145.

277. Evans, P.I. (1973). See also footnote 593.

278. Cornford, F.M. (1956).

279. Wilhelm, R. (1971), pp. liv–lvii.

280. Ibid.

281. Eliade, M. (1957), p. 29.

282. "Indra's combat served as model for the battles that the Aryans had to sustain against the Dasyus (also termed vrtani): 'he who triumphs in a battle, he truly kills Vītra' (Maitrayana-Samhita 2.1.3.)." (Eliade, M. [1978b], p. 207).

283. Eliade, M. (1978b), p. 104, footnote 48.

284. Eliade, M. (1978b), p. 320. Eliade also points out that the name Faridun is derived from Thraetona

(Thraetona - Freton - Faridun), and states: "in Iran as elsewhere, the process of historicization of mythical themes and personages is counterbalanced by a contrary process: the real adversaries of the nation or the empire are imagined as monsters, and especially as dragons."

285. Eliade, M. (1957), pp. 29–32.

286. Stevenson, M.S. (1920), p. 354.

287. Eliade, M. (1991b), p. 19.

288. Neumann, E. (1955), faceplate, Part II.

289. Whitehead, A.N. (1958), p. xx.

290. Otto, R. (1958).

291. Ibid., pp. 12–13.

292. Jung, C.G. (1971), p. 477.

293. There is good evidence for actual independence of subpersonalities in the human imagination (in the episodic and procedural memory systems [?]), and incontrovertible evidence for the use of metaphoric personality in ritual, art and literature. There are many forms of normal experience that involve the explicit participation of "foreign" personalities, or partial personalities. These include dreams, in which characters appear within experience in known and unknown guise, and follow what are apparently their own intrinsic and often incomprehensible laws of behavior (see Jung, C.G. [1968b] for an analysis of an extensive series of dreams [the physicist Wolfgang Pauli's, as it happens]).

Moods, arriving upon the stage of consciousness, influence perception, memory, cognition and behavior, producing perplexing outbursts of sadness and rage on the part of the person who is so influenced (Jung identified the "anima," the archetype of the feminine, with mood [see Jung, C.G. (1968a), p. 70]). "Active imagination" (see Jung, C.G. [1968a], p. 190), a process that might be compared to purposive daydreaming—with mood as the focus—can generate images and fantasies associated with that mood. Participation in this process helps illuminate the structure of the "personalities" associated with given states of emotional seizure.

In states of abnormal tension and in psychopathological or neurological breakdown the effects of foreign personalities are easily observable. Individuals afflicted with Tourette's syndrome appear "possessed" by a complex spirit, for want of a better description, whose personality uncannily matches that of the Trickster of the North American Indian (see Sacks, O. [1987]; Jung, C.G. [1968a], pp. 255–274).

Schizophrenic breakdown involves the apparent participation of many fragmented personalities who make their appearance in voices and urges "foreign" to the assaulted mind of the psychotic (see Jung, C.G. [1967a]; Romme, M.A. & Escher, A.D. [1989]). The physiologist and schizophrenia researcher Doty states:

Among the more widely recognized diagnostic criteria are the "first rank" symptoms identified by Kurt Schneider. As summarized by Crow and Johnstone, these are: "(1) hearing one's thoughts spoken aloud within one's head, (2) hearing voices arguing, (3) hearing voices that comment upon what one is doing, (4) experiences of bodily influence (that bodily functions are affected by an outside agency), (5) experiences that one's thoughts are being withdrawn from or inserted into one's head, (6) thought diffusion or the experience that one's thoughts are broadcast to others, (7) delusional perception (the attribution of special significance to a particular perception), and (8) feelings or volitions experienced as imposed on the patients by others." This list of first rank schizophrenic symptoms is uniquely fascinating in the present context, for, as Nasrallah astutely phrases it, they can all be summarized by the basic idea "that in the schizophrenic brain the unintegrated right-hemisphere consciousness may become an 'alien intruder' on the verbally expressive left hemisphere." In other words, they are prototypical of what one might expect were interhemispheric communication so distorted that the left hemisphere could no longer identify the origin of activities in the right hemisphere as belonging to the unified consciousness of self. (Doty, R.W. [1989], p. 3).

Cleghorn has reported that schizophrenics experiencing auditory hallucinations were characterized by increased glucose uptake (assessed with positron-emission tomography [PET] scan) in regions of the right hemisphere, corresponding to the language areas of the left hemisphere (Cleghorn, J.M. [1988]). Doty suggests that these right-hemisphere structures may have been released from tonic inhibition by the dominant left-hemisphere language center, in the course of the schizophrenic breakdown (Doty, R.W. [1989]).

Multiple personality disorder, a historically cycling condition (see Ellenberger, H.F. [1970]), emerges when "personalities" and representations thereof, external to the central ego, appear without union of memory, often in those with dissociative tendencies who were punished severely and arbitrarily early in life.

Obsessive-compulsive disorder reduces its victims to total domination by an object of experience or a thought, producing behavioral and cognitive patterns foreign to those afflicted (and to those concerned with the afflicted) (Rapoport, J. [1989]).

Shamanic and religious rituals, primitive initiation rites and psychoactive chemicals produce complex physiological changes within the individual brain, activating affectively based complexes that could not otherwise reach consciousness, producing insights and affects not otherwise attainable, with oft-dramatic consequences. (It is of interest to note, in this regard, that LSD and other psychotomimetic or hallucinogenic drugs are characterized by their effect on the phylogenetically ancient serotonergic brain-stem projections [see Doty, R.W. (1989)].)

Epileptic seizures, often accompanied by strange perceptual, emotional and cognitive changes, run the gamut from awe-inspiring and holy to demonic and terrifying (see Ervin, F. & Smith, M. [1986]). The discussion presented in this chapter is particularly interesting, insofar as it describes pathological alterations, not of systematic cognition, but of *meaning*. Ervin describes cases where epileptic patients refuse pharmacological treatment, risking their physiological and psychological well-being, because they are unwilling to forgo the pre-epileptic "aura"—a condition of altered experience, preceding the epileptic seizure *per se*. This "aura" may partake of the quality of revelation—producing apparently profound subjective insight into the deepest meaning of the universe, for example (although it is more commonly associated with extreme terror). Before such states are packaged as pathological, necessarily delusional, it should be remembered that Dostoyevski was epileptic, altered, and perhaps deepened in psychological insight by the processes of his illness. Such seizures can also induce violent outbursts, completely dissociated from the individual's normal behavioral state (see Mark, V.H. & Ervin, F.R. [1970]).

294. See Jung, C.G. (1967b).

295. With regard to the potentially four-dimensional structure of the human memory system, see Teylor, T.J. & Discenna, P. (1986).

296. See Russell, J.B. (1986).

297. Jung states: "All numinous contents ... have a tendency to self-amplification, that is to say they form the nuclei for an aggregation of synonyms" (Jung, C.G. [1976b], p. 458). Contents in memory with the same affective valence tend to group. This phenomena has long been recognized in the case of depression. Depressed people are characterized by a bias toward the perception, remembrance and conception of punishments: disappointment, frustration (absences of expected rewards), loneliness and pain (see Beck, A. [1979]).

298. See Gall, J. (1988).

299. This figure is derived from the tray painting *The Triumph of Venus*, reproduced as plate 62 in Neumann, E. (1955).

300. The *vesica pisces* is a very complex symbol, associated with the fish that is (water-dwelling) serpent, phallus and womb simultaneously. See Johnson, B. (1988), particularly *Part Nine: The Fish*.

301. Eliade, M. (1982), pp. 20–21.

302. Eliade, M. (1982), p. 21.

303. Neumann, E. (1955), pp. 31–32.

304. This classic Freudian state of affairs is intelligently and accurately portrayed in the movie *Crumb* (Zwigoff, T. [1995]).

305. Shelton, G. (1980), p. 45.

306 Neumann, E. (1955), pp. 12–13.

307. It is my understanding that this progression has not been demonstrated, and that the "patriarchal"

deities stand in a secondary "psychological," rather than historical, relationship to the matriarchal deities (as "things derived from the matrix"). Furthermore, as we have discussed, the "unknown" can also be regarded as "derived" from the "known" (as "things defined in opposition to the known"). For the purposes of this book, however, the precise temporal/historical relationship of the various deities to one another is of secondary importance, compared to the fact and meaning of their existence as eternal "categories" of imagination.

308. Neumann, E. (1955), pp. 153–157.

309. See Bowlby, J. (1969). Bowlby investigated the curious fact that a substantial proportion of orphaned or otherwise isolated babies, provided with adequate food, basic physical care and shelter, still "failed to thrive" and died. More recent research has been devoted toward investigation of the processes underlying social attachment in general, and maternal attachment in particular, and has focused in part on the role of the opiate system, which is also involved in governing reaction to pain, frustration and disappointment—broadly, to punishment (reviewed in Pihl, R. O. & Peterson, J.B. [1992]).

310. Neumann, E. (1955), pp. 149–150.

311. See Neumann, E. (1955).

312. *Kali, the Devourer*, reproduced as plate 66 in Neumann, E. (1955).

313. From MacRae, G.W. (Trans.) (1988), p. 297.

314. A similar—and illuminating—conflation of source with attitude also characterizes Christ's terminology with regard to himself. He is to be regarded both as model for subjective stance (I am the way, the truth, and the life [John 14:6]), but also as source of the "water of life" (If any man thirst, let him come unto me, and drink. He that believeth on me, as the scripture hath said, out of his belly shall flow rivers of living water [John 7:37–38]).

315. I am indebted to Mike McGarry for bringing these passages to my attention.

316. "Diana of Ephesus" (plate 35 in Neumann, E. [1955]).

317. See Neumann, E. (1955).

318. See, for example, the "Venuses" of Willendorf, Menton and Lespugne (portrayed in plate 1 in Neumann, E. [1955]).

319. Neumann, E. (1955), p. 39.

320. See McGlynn, F.D. & Cornell, C.C. (1985); Chambless, D.L. (1985).

321. Foa, E.B, Molnar, C., & Cashman, L. (1995). See also Pennebaker, J.W. (1997); Pennebaker, J., Mayne, T.J., & Francis, M.E. (1997).

322. Koestler, A. (1976).

323. Durga is Kali's benevolent counterpart.

324. Zimmer, H. (1982), pp. 74–75.

325. Rychlak, J. F. (1981), p. 767.

326. See Neumann, E. (1955, 1954).

327. Derived from unknown source.

328. Castle derived from a *temenos*, in Maier's Viatorium (1651) (plate 31 in Jung, C.G. [1968b]); St. George derived from Ripa, C. (1630) *Virtue* (Didi-Huberman, G., Garbetta, R. & Morgaine, M. [1994], p. 50).

329. This brief description is a summary of (isomorphic) information contained in the writings of Carl Jung (particularly in Jung, C.G. [1967a]; Joseph Campbell (particularly in Campbell, J. [1987 and 1968]); Northrop Frye (particularly in Frye, N. [1982 and 1990]); and Erich Neumann (particularly in Neumann, E. [1954 and 1955].

330. Bellini, J. (15th century), *St. George battling the dragon*. In Didi-Huberman, G., Garbetta, R., & Morgaine, M. (1994), p. 102. Dozens of representative examples are provided in this volume.

331. Didi-Huberman, G., Garbetta, R., & Morgaine, M. (1994), pp. 53, 59, 64, 65, 67, 69, 74, 77, 81.

332. Neumann, E. (1954), pp. 160–161.

333. Neumann, E. (1954, 1955); Jung, C.G. (1976b, 1967b, 1968b, 1967a); Eliade, M. (1978b).

334. See Jung, C.G. (1970a).

335. See Eliade, M. (1978b), p. 147.

336. Eliade, M. (1978b), pp. 145–146.

337. See Chapter 2.

338. Eliade, M. (1978b), pp. 147–149.

339. Binswanger, L. (1963), pp. 152–153.

340. See Eliade, M. (1978b), pp. 114–125.

341. Ibid., p. 123.

342. Ibid., p. 124.

343. Eliade, M. (1965), p. xi.

344. Borski, L.M. & Miller, K.B. (1956).

345. Eliade, M. (1991a).

346. L'Engle, M. (1997), p. 136.

347. Ibid., p. 142.

348. Lucas, B.V., Crane, L. & Edwards, M. (1945), pp. 171–178.

349. This is imagery of paradise.

350. Frye states, in keeping with this theme:

There is one theme that recurs frequently in the early books of the Bible: the passing over of the first-born son, who normally has the legal right of primogeniture, in favor of a young one. The firstborn son of Adam, Cain, is sent into exile, and the line of descent goes through Seth. Ham, the rejected son of Noah, is not said to be his eldest son, but the same pattern recurs. Abraham is told to reject his son Ishmael because a younger son (Isaac) is to be born to him. Isaac's eldest son Esau loses his birthright to Jacob through some rather dubious maneuvers on Jacob's part, some of them backed by his mother. Jacob's eldest son Reuben loses his inheritance for the reason given in Genesis 49:4. Joseph's younger son Ephraim takes precedence over the elder Manasseh. The same theme is extended, though not essentially changed, in the story of the founding of the monarchy, where the first chosen king, Saul, is rejected and his line passed over in favor of David, who is practically his adopted son (I Samuel 18:2). In later literature the theme is carried much farther back: if we look at the fifth book of *Paradise Lost*, for instance, we see an archetype of the jealousy of an older son, Lucifer or Satan, at the preference shown to the younger Christ. (Frye, N. [1982], pp. 180–181)

351. Frye, N. (1982).

352. Figure of God derived from anonymous Italian (15th century) *Saint George and the Dragon*. In Didi-Huberman, G., Garbetta, R., & Morgaine, M. (1994), p. 65.

353. Derived from *Figurea et emblemata* in Lambspringk's *Musaeum hermeticum* (1678) (plate 179 in Jung, C.G. [1968b]).

354. Smith, H. (1991), pp. 289–290.

355. Smith, H. (1991), p. 292.

356. There is some evidence, in our times, that would-be tyrants themselves are even beginning to realize this. Many of the "transitions to democracy" characteristic of the last 30 years have been voluntary transferrals of power on the part of military strongmen unable to believe in the justice of their own "strength." See Fukuyama, F. (1993).

357. Derived from *Figurea et emblemata* in Lambspringk's *Musaeum hermeticum* (1678) (plate 168 in Jung, C.G. [1968b]).

358. "Pregnant father" derived from derived from *Tabula smaragdina* in Maier, *Scrutinium chymicum* (1687) (plate 210 in Jung, C.G. [1968b]).

Chapter 3

359. Jung, C.G. (1968b), p. 86.

360. Morley, J. (1923), p. 127.

361. Nietzsche, F. (1966), pp. 100–102.

362. Field, T.M., Schanberg, S.M., Scafidi, F., Bauer, C.R., Vega-Lahr, N., Garcia, R., Nystrom, J., & Kuhn, C.M. (1986).

363. Polan, H.J. & Ward, M.J. (1994); Berkowitz, C.D. & Senter, S.A. (1987); also footnote 309.

364. Hyde, J.S. (1984); Saner, H. & Ellickson, P. (1996).

365. See Eliade, M. (1965).

366. This is something akin to Jung's *animus*. See Jung, C.G. (1968a).

367. Eliade, M. (1965).

368. See Neumann, E. (1955), p. 61.

369. Ibid., particularly chapter 15.

370. Eliade, M. (1965), pp. xii–xiv.

371. There is evidence, for example, that the dynastic cultures of ancient Egypt existed in virtually unchanged form over periods of time as long as fifteen hundred years (after the Fifth Dynasty, 2500–2300 B.C.). Eliade, M. (1978b), p. 86.

372. This is the mythic theme of Dostoyevsky's *Crime and Punishment* (1993). Raskolnikov, Dostoyevsky's "revolutionary" socialist protagonist, places himself above God (somewhat in the manner of Nietzsche's superman), and resolves to commit a crime (murder) justified elaborately and carefully by recourse to demythologized rationality. The crime succeeds, but Raskolnikov is unable to bear its burden, and confesses, as a consequence of intrapsychic compulsion (in the absence of objective necessity). In consequence, he is able to regain his (protective) identity with the common community.

This theme has been revisited, in recent years, by Woody Allen, a great admirer of Russian literature, in his motion picture *Crimes and Misdemeanors* (1989). Allen's protagonist, a respected physician, murders his mistress to prevent her from disrupting his family's (false) security. Unlike Raskolnikov, however, the good physician suffers no long-term psychic trauma, and everything "returns to normal" within the year. The movie, placid on the surface, is more horrifying than Dostoyevsky's tortured book. In the latter, moral order (predicated on respect for the intrinsic value of the individual) rules, in contrast to presumptuous rationality. In the former, rational meaninglessness prevails absolutely—although it remains thinly veiled by urban pleasantry and pretence.

373. Nietzsche, F. (1968a), p. 217.

374. Tablet 6:152–153; Heidel, A. (1965), p. 53. See footnote 255.

Chapter 4

375. Wittgenstein, L. (1958), p. 50.

376. Kuhn, T.S. (1970), p. viii.

377. Taken from Hofstadter, D.R. (1979), p. 89.

378. Polyani, M. (1958).

379. Kuhn, T.S. (1970), p. 44.

380. Nietzsche, F. (1968a), p. 213, section 16.

381. Frye, N. (1990), pp. 42–44.

382. Ibid., pp. 103–104.

383. See Peake, M. (1995) for a dramatic portrayal of this state of affairs.

384. Bruner, J.S. & Postman, L. (1949).

385. Kuhn, T.S.. (1970), pp. 62–64.

386. For elaborated description, see Jung, C.G. (1967a); Neumann, E. (1954). Jung states: "The purpose of the descent as universally exemplified in the myth of the hero is to show that only in the region of danger (watery abyss, cavern, forest, island, castle, etc.) can one find the "treasure hard to attain" (jewel, virgin, life-potion, victory over death.") (Jung, C.G. [1968b], p. 335).

387. Nietzsche states: "The unhistorical is like the surrounding atmosphere that can alone create life and in whose annihilation life itself disappears. It is true that man can only become man by first suppressing this unhistorical element in his thoughts, comparisons, distinctions, and conclusions, letting a clear sudden light break through these misty clouds by his power of turning the past to the uses of the present. But an excess of history makes him flag again." (Nietzsche, F. [1957]).

388. Frye, N. (1990), p. 256.

389. See Neumann, E. (1954 and 1955).

390. Tablet 7:112, 7:115; Heidel, A. (Trans.) (1965), p. 58 (see footnote 262).

391. Nietzsche, F. (1968a), p. 301.

392. Nietzsche, F. (1995).

393. This story was recently cited in Hawking, S. (1988).

394. Hofstadter, D.R. (1979), pp. 397–398.

395. Discussion, Bruner, J. (1986), pp. 27–28.

396. Jung, C.G. (1968b), p. 86.

397. Eliade, M. (1975), p. 155.

398. Kuhn, T.S. (1970), pp. 84–85.

399. Ibid., p. 113.

400. Quoted in Kuhn, T.S. (1957), p. 138.

401. Einstein, A. (1959), p. 45.

402. Kronig, R. (1960), pp. 22, 25–26.

403. Kuhn, T.S. (1970), pp. 82–84.

404. This, as Karl Popper pointed out, "permits our hypotheses to die in our stead."

405. Tolstoy, L. (1887 [1983]), p. 13.

406. Ibid., p. 54.

407. Ibid., pp. 26–29.

408. Another relevant Nietzschean comment:

The structure of the scenes and the visual images reveal a deeper wisdom than the [ancient Greek poets themselves could] put into words and concepts: the same is also observable in Shakespeare, whose Hamlet, for instance, similarly, talks more superficially than he acts, so that the previously mentioned lesson of Hamlet is to be deduced, not from his words, but from a profound contemplation and survey of the whole. (Nietzsche, F. [1967a], p. 105).

409. Nietzsche, F. (1967a), p. 60.

410. Dostoyevski, F. (1961), p. 21.

411. Nietzsche's epigrams: "A criminal is frequently not equal to his deed: he makes it smaller and slanders it." (Nietzsche, F. [1968a], p. 275); "The lawyers defending a criminal are rarely artists enough to turn the beautiful terribleness of his deed to his advantage." (Nietzsche, F. [1968a], p. 275).

412. Cited in Kaufmann, W. (1975), pp. 130–131.

413. See Eliade, M. (1965, 1975).

414. See Ambady, N., & Rosenthal, R. (1992).

415. As when Oedipus unwittingly sleeps with his mother, then blinds himself.

416. Nietzsche, F. (1968a), p. 320.

417. Nietzsche, F. (1967a), p. 75.

418. Binswanger, L. (1963), p. 157.

419. Tolstoy, L. (1983), pp. 57–58.

420. Frye, N. (1990), p. xvi.

421. Nietzsche, F. (1968a), pp. 260–261.

422. See footnote 26.

423. Eliade, M. (1972), p. 4.

424. Ellenberger, H.F. (1970), pp. 447–448.

425. Jung, C.G. (1971), p. 477.

426. Eliade, M. (1964).

427. Eliade, M. (1965), p. 89.

428. The symbol of the tree and the meanings of that symbol are discussed in detail in chapter 4.

429. Eliade, M. (1965), pp. 88–89.

430. "The Brazen Serpent of Moses on the Cross." From *serpens mercurialis* in Eleazar, *Uraltes chymisches Werk* (1760) (plate 238 in Jung, C.G. [1968b]).

431. When I first began the process that led me to understand these ideas, I painted a frightening picture of the crucified Christ, "glaring, judgmental, demonic, with a cobra wrapped around his naked waist, like a belt" (as described in the Preface). I was struggling with problems of identity, in an world that had apparently gone insane. The image of the exploratory hero manifested itself to me in imagistic representation, contaminated with the figure of the Dragon of Chaos—"and as Moses lifted up the serpent in the wilderness, even so must the Son of man be lifted up" (John 3:14). This contamination might be regarded as indicative of the danger the development of full understanding of that hero and the "world" he inhabited posed to my then-extant personality structure (which in fact dissolved and regenerated, over a lengthy period, thereafter). The "identity" of the revolutionary hero with the serpent of chaos, however, accounts for the hatred and fear his necessary actions produce among the population he is striving to help.

432. Origen, in Hodson, G. (1963), p. xii.

433. Eliade, M. (1975), p. 60.

434. Campbell, J. (1973), p. 25.

435. Neumann, E. (1968), p. 395.

436. Tao Te Ching 25 in Waley, A. (1934), p. 34.

437. See Part I. A. III: The separation of the world parents. In Neumann, E. (1954).

438. Tiuitchev, F.I. *Sviataia noch na nebosklon vzoshla*, translated by Vladimir Nabokov, cited in Joravsky, D. (1989), p. 173. I am indebted to Carolyn Butler for bringing this poem to my attention.

439. Frye states:

A descent into a world below consciousness involves some break in the continuity of conscious memory, or some annihilation of the previous conditions of existence, corresponding to falling asleep. The lower world is often a world of greatly enlarged time, where a few moments may correspond to many years in the upper world. (Frye, N. [1990], p. 266).

This is reminiscent of Jung's notion that time is relativized in the collective unconscious.

440. Frye, N. (1982), p. 108.

441. Cited in Neumann, E. (1968), p. 395.

442. Wheeler, J. (1980), p. 341.

443. Nietzsche generated a hypothesis that seems relevant:

Suppose nothing else were "given" as real except our world of desires and passions, and we could not get down, or up, to any other "reality" besides the reality of our drives—for thinking is merely a relation of these drives to each other: is it not permitted to make the experiment and to ask the question whether this "given" would not be *sufficient* for also understanding on the basis of this kind of thing the so-called mechanistic or ("material") world? I mean, not as a deception, as "mere appearance," and "idea" (in the sense of Berkeley and Schopenhauer) but as holding the same rank of reality as our affect—as a more primitive form of the world of affects in which everything still lies contained in a powerful unity before it undergoes ramifications and developments in the organic process (and, as is only fair, also becomes tenderer and weaker)—as a kind of instinctive life in which all organic functions are still synthetically intertwined along with self-regulation, assimilation, nourishment, excretion, and metabolism—as a *preform* of life. (Nietzsche, F. [1966], pp. 47–48).

444. Eliade, M. (1982), p. 75.

445. For a modern illustration, see Tchelitchew, P. (1992), p. 49.

446. See tree and snake discussion in Jung, C.G. (1988), pp. 1431–1450; Jung, C.G. (1967b), pp. 251–350.

447. Radha, Swami S. (1978), pp. 16–20.

448. Frye, N. (1990), pp. 284–285.

449. Eliade, M. (1975), p. 64.

450. "Yggdrasil, the world tree of the Edda." From the *Elder Edda* by Magnusson, F. (18th century) (figure 55 in Neumann, E. [1955]).

451. In Bellows, H.A. (1969), p. 60.

452. Neumann, E. (1954), pp. 30–31.

453. Jung, C.G. (1976b), p. 117.

454. See Jung, C.G. (1967b), pp. 240, 315; Jung, C.G. (1968b), p. 317.

455. The role of the reticular activating system in regulating consciousness was established by Morruzzi, G. & Magoun, H.W. (1949). The precise mechanisms by which such regulation take place are still under debate.

456. Goethe, J.W. (1979a), p. 99.

457. Voltaire (1933), p. 450.

458. Frye elaborates on the myth of Narcissus:

The beautiful youth paralyzed by the mirror-reflection of himself and hence unable to love. Mythologists very early made Narcissus a type of the fall of Adam, as Adam, like Narcissus, identified himself with his own parody-reflection in a lower world. Paul's conception of Christ as the second Adam makes Christ the double of the Narcissus-Adam who delivers the original one from what Lacan calls the *stade de miroir* and Eliot a wilderness of mirrors. (Frye, N. [1990], p. 271).

459. Consider Nietzsche's statement:

On his way to becoming an "angel" (to employ no uglier word) man has evolved that queasy stomach and coated tongue through which not only the joy and innocence of the animal but life itself has become repugnant to him—so that he sometimes holds his nose in his own presence, and, with Pope Innocent the Third, disapprovingly catalogues his own repellent aspects ("impure begetting, disgusting means of nutrition in his mother's womb, baseness of the matter out of which man evolves, hideous stink, secretion of saliva, urine, and filth"). (Nietzsche, F. [1967b], p. 67).

460. The human neocortex developed at an unprecedented rate, from the evolutionary standpoint. This expansion and the extension of consciousness to the self were synchronous phenomena. One factor limiting this expansion, which increases head size dramatically, is the diameter of the female pelvic girdle, which must allow the infant passage during birth. The frequently traumatic nature of human birth is a consequence, at least in part, of the conflict between neonate cranial circumference and maternal pelvic structure.

461. Eve taken from "the Tree of Knowledge: Church and Synagogue" from a Swiss manuscript (15th century) (Figure 56, Neumann, E. [1955]).

Chapter 5

462. Eliade, M. (1978b), pp. 62–63.

463. See chapter 3: Apprenticeship and Enculturation: Adoption of a Shared Map.

464. Nietzsche, F. (1966), p. 228.

465. Plate 36 in Jung, C.G. (1968b).

466. Milton, J., (1961), 1:40–43, p. 38.
467. Frye, N. (1990), pp. 272–273.
468. See footnote 350.
469. Pagels, E. (1995).
470. Eliade, M. (1978b), p. 302.
471. These can be reasonably considered akin to the "elder gods" in the *Enuma elish* (see chapter 2).
472. Eliade, M. (1978b), p. 310.
473. Milton, J. (1961), 3:96–99, p. 95.
474. Goethe, J.W. (1979a), p. 75.
475. Goethe, J.W. (1979b), p. 270.
476. See chapter 4: The revolutionary hero.
477. Tolstoy, L. (1983), pp. 49–52.
478. Shakespeare, W. (1952b), 3:5: 78–83, p. 104.
479. Milton, J. (1961), 4:40, p. 116.
480. Detailed in Russell, J.B. (1986), p. 103.
481. Milton, J. (1961), 1:159–165, p. 41.
482. Nietzsche, F. (1967b), p. 333.
483. Nietzsche, F. (1981), p. 125.
484. Cited in Kaufmann, W. (1975), pp. 122–123.
485. Edwardes, A. & Masters, R.E.L. (1963), p. 124.
486. Durnin, R. (1994).
487. Wilhelm, R. (1971), p. lv.
488. Solzhenitsyn, A.I. (1975), p. 390.
489. Joyce, J. (1992).
490. Solzhenitsyn, A.I. (1975), pp. 4–7.
491. Solzhenitsyn, A.I. (1974), pp. 5–7.
492. Frankl, V. (1971), pp. 20–21.
493. Solzhenitsyn, A.I. (1975), pp. 602–603.
494. Solzhenitsyn estimates that 250,000 destructive labor camp inmates lost their lives to build the Volga-Moscow canal, which, when finished, was much too shallow to serve any of its intended uses. Much of the canal was dug by hand, with the most primitive tools, in the midst of the winter (see Solzhenitsyn, A.I. [1975], pp. 80–102).
495. Frankl, V. (1971), p. 50.
496. Solzhenitsyn, A.I. (1975), p. 201.
497. Blake, W. in Keynes, G. (1966), p. 213.
498. "Zek" is Russian slang for prison-camp inmate.
499. Solzhenitsyn, A.I. (1975), pp. 195–197.
500. See, for example, Browning, C.R. (1993).
501. Solzhenitsyn, A.I. (1975), pp. 147–149.
502. Milton, J. (1961), 2: 380–385, p. 71.
503. Solzhenitsyn, A.I. (1975), p. 603.
504. Ibid., pp. 619–620.
505. Lao Tzu (1984d).
506. Solzhenitsyn, A. I. (1975), pp. 338, 341–342.
507. Ibid., p. 626.
508. Frankl, V. (1971), pp. 117–120.
509. Ibid., p. 7.
510. Ibid., p. 4.
511. Solzhenitsyn, A.I. (1975), p. 622.
512. Milton, J. (1961), 1:249–253, p. 44.

513. Milton, J. (1961), 4:109–123, p. 118.

514. Shakespeare, W. (1952c). *Titus Andronicus.* 5:3:184–190, p. 198.

515. Milton, J. (1961), 9: 119–130, p. 37.

516. Shakespeare (1952b), *Richard III,* 5:3:200–203, p. 145.

517. Solzhenitsyn, A.I. (1975), pp. 326–328.

518. Ibid., p. 347.

519. Milton, J. (1961), 1:54–74, p. 38.

520. Ibid., 1:44–48, p. 38.

521. Nietzsche, F. (1966), p. 86.

522. From the *Gospel of Thomas,* in Robinson, J.R. (1988), pp. 133–134.

523. Milton, J. (1961), 4:75–78, p. 117.

524. Ibid., 4:79–105, p. 117.

525. Frye, N. (1982), p. 130.

526. Solzhenitsyn, A.I. (1975), pp. 610–612.

527. Frankl, V. (1971), p. 104.

528. Solzhenitsyn, A.I. (1975), pp. 624–626.

529. Ibid., p. 615.

530. Arendt, H. (1994).

531. "All nature is through fire renewed." Occult/gnostic interpretation of the meaning of the initials traditionally depicted at the top of the cross of Christ: I.N.R.I. (*Iesus Nasaremus Rex Iudaeorum* [Jesus of Nazareth, King of the Jews]); see Dee, J. (1993) for amplification.

532. Jaeger, W. (1968), p. 35.

533. Niebuhr, R. (1964), pp. 6–7.

534. Ibid., pp. 13–14.

535. Piaget, J. (1932), pp. 16–18.

536. See footnote 463.

537. Frye states, with regard to the role of criticism in illuminating narrative meanings:

The poetic imagination constructs a cosmos of its own, a cosmos to be studied not simply as a map but as a world of powerful conflicting forces. This imaginative cosmos is neither the objective environment studied by natural science nor a subjective inner space to be studied by psychology. It is an intermediate world in which the images of higher and lower, the categories of beauty and ugliness, the feelings of love and hatred, the associations of sense experience, can be expressed only by metaphor and yet cannot be dismissed or reduced to projections of something else. Ordinary consciousness is so possessed by the either-or contrast of subject and object that it finds difficulty in taking in the notion of an order of words that is neither subjective nor objective, though it interpenetrates with both. But its presence gives a very different appearance to many elements of human life, including religion, which depend on metaphor but do not become less "real" or "true" by doing so.

Of course "metaphorical" is as treacherous a conception as "truth" or "reality" could ever be. Some metaphors are illuminating; some are merely indispensable; some are misleading or lead only to illusion; some are socially dangerous. Wallace Stevens speaks of "the metaphor that murders metaphor." But for better or worse it occupies a central area—perhaps *the* central area—of both social and individual experience. It is a primitive form of awareness, established long before the distinction of subject and object became normal, but when we try to outgrow it we find that all we can really do is rehabilitate it.

At this point another recent critical observation comes to hand, from Italo Calvino's posthumous Norton lectures, also a paradox, but an exhilarating one: "Literature remains alive only if we set ourselves immeasurable goals, far beyond any hope of achievement." Strictly speaking the writer does not set the

(Frye, N. [1990], pp. xxii–xxiii).

538. The attempts by the pharaoh to control the Jewish "threat" through means of infanticide provides a (noncoincidental) narrative parallel to the actions of Herod, who killed all the Jewish children under two in Bethlehem and environs, for similar reasons, many centuries later (see Matthew 2:1–16).

539. Frye continues:

And yet Canaan seems a rather shrunken and anticlimactic form of the paradisal land of promise flowing with milk and honey that was originally promised to Israel. Perhaps Moses was really the only person to see the Promised Land: perhaps the mountain outside it he climbed in his last hours was the only place from which it could be seen. (Frye, N. [1990], p. 299).

540. Frye, N. (1990), p. 299.

541. The Hebrews feed on *manna* during their desert sojourn. Such "spiritual bread"—made, in its profane condition, from wheat, the metaphorical body of eternally dying and resurrecting corn god—is later offered by and equated with Christ, to aid, ritually (procedurally), in the incorporation of heroic faith and courage. Frye states:

Christ is constantly associated with the miraculous provision of food. Miracles of feeding large multitudes with very small amounts of food (fish, as contents drawn up from the [unconscious, maternal] deep, as well as bread) are recorded in all four Gospels, sometimes more than once, and such miracles are explicitly antitypes of the provision of manna in the wilderness (John 6:49–51). The imagery of eating Christ's flesh and drinking his blood meets us in the Gospels even before the institution of the Eucharist. That Christ's body is an unfailing source of food and drink is asserted on both physical and spiritual levels (the "daily {epiousion} bread" of the Lord's Prayer might also be regarded as "supersubstantial" bread). The body of Christ is not only "to be eaten, to be divided, to be drunk," in the words of Eliot's "Gerontion," but is the source of the continuity of the life of his people, hidden within their bodies. It was so in Old Testament times too, according to Paul, who says that the Israelites in the wilderness all ate the same spiritual food and drank the same spiritual drink, the latter from a rock which was Christ (I Corinthians 10:3). (Frye, N. [1990], p. 257).

542. Reference to mythic narrative as source for explicit rule is explicitly presented here.

543. Frye states:

I have noted (Great Code 18) the passage in *Faust* where Faust deliberately alters "In the beginning was the Word" to "In the beginning was the Act." I should have added that Faust was simply following the established Christian practice up to his time. In the beginning God did something, and words are descriptive servomechanisms telling us what he did. This imports into Western religion what poststructural critics call the "transcendental signified," the view that what is true or real is something outside the words that the words are pointing to. (Frye, N. [1990], p. 34).

544. Frye states:

If it is true that creative verbal power is associated with something in the mind supplementary to ordinary consciousness, we have inched a little closer to the writer's social context. Such a mind would often be baffled by the arbitrary conventions of behavior that consciousness more easily masters: one often finds a naivete in the writer that may sometimes incapacitate him from almost anything except writing. But he might have in compensation an insight into social phenomena that would give him, not merely an intense vision of the present, but an unusual ability to see a conditional future, the consequence of tendencies in the present. This in turn may give the sense of a distinctive kind of knowledge hidden from most of society. The element of the *prophetic* in literature is often spoken of very vaguely, but is tangible enough to be worth looking into. In any case, the word comes closer to anything we have stumbled over so far to indicate the quality of the poet's authority, and to indicate also the link between secular and sacred literature that is one of our main themes.

If we look at the prophetic writers of the Old Testament, beginning with Amos, the affiliation of primitive and prophetic emerges at once. Amos has the refusal to compromise with polite conventions, a social reputation in northern Israel for being a fool and a madman, and an ability to derive the substance of what he says from unusual mental states, often allied to trance. Such prophets also foretell a future which is an inevitable result of certain foolish policies, like the policy of the king of Judah toward Babylon that led, as Jeremiah told him it would, to the destruction of Jerusalem. The principle involved here is that honest social criticism, like honest science, extends the range of predictability in society.

In modern times the writers that we instinctively call prophetic—Blake, Dostoevsky, Rimbaud—show similar features. Such writers are as deeply pondered by readers as the Greek and Hebrew oracles were: like them, they shock and disturb; like them, they may be full of contradictions and ambiguities, yet they retain a curiously haunting authority. As early as Elizabethan times there were critics who suggested that the distinction between sacred and secular inspiration might be less rigid than generally assumed. George Puttenham, writing in the 1580s, pointed to the etymology of poet as "maker," which implied for him an analogy between the poet's creative power and the creative power of God in making the world. He quotes Ovid's phrase in the *Fasti*, 'est deus in nobis,' which would mean either God or *a* god. In the sixteenth century it would certainly have been safer to settle for a Muse or a God of Love or something sanctioned by convention and not taken seriously as doctrine, but the analogy is still there, though latent until the time of Coleridge. It has been frequently observed that the arts are prophetic also in the sense of indicating symbolically the social trends that become obvious several generations later.

The term prophetic in itself might apply to some writers (Luther, Condorcet, Marx) whom we normally would place outside literature. This troublesome apparatus of inside and outside will not go away even when so many aspects of it vanish under examination. It appears to be the connection with the psychologically primitive that characterizes the prophetic writer who is generally thought of as inside literature or at least (as with Rousseau, Kierkegaard, or Nietzsche) impossible to ignore as a literary figure. (Frye, N. [1990], pp. 52–54).

545. See Jung, C.G. (1978a) for a complete work on the topic of prophecy; also Jung's prescient comments (1918) on the "Blond Beast" (Jung, C.G. [1978a], pp. 3–28).

546. The pleroma.

547. Nietzsche, F. (1981), p. 97.

548. Dostoevsky, F. (1981), pp. 299–301.

549. Ibid., p. 309.

550. Ibid., p. 313.

551. Ibid., p. 316.

552. James, W. (1880, October), p. 100.

553. Frye, N. (1982), pp. 132–133.

554. Frye, N. (1990), p. 104.

555. Frye, N. (1982), p. 56.

556. Frye also states:

We referred earlier to the structure of the Book of Judges, in which a series of stories of traditional tribal heroes is set within a repeating *mythos* of the apostasy and restoration of Israel. This gives us a narrative structure that is roughly U-shaped, the apostasy being followed by a descent into disaster and bondage, which in turn is followed by repentance, then by a rise through deliverance to a point more or less on the level from which the descent began. This U-shaped pattern, approximate as it is, recurs in literature as the standard shape of comedy, where a series of misfortunes and misunderstandings brings the action to a threateningly low point, after which some fortunate twist in the plot sends the conclusion up to a happy ending. The entire Bible, viewed as a "divine comedy," is contained within a U-shaped story of this sort, one in which man, as explained, loses the tree and water of life at the beginning of Genesis and gets them back at the end of Revelation. In between, the story of Israel is told as a series of declines into the power of heathen kingdoms, Egypt, Philistia, Syria, Rome, each followed by a rise into a brief moment of independence. The same U-narrative is found outside the historical sections also, in the account of the disasters and restoration of Job and in Jesus' parable of the prodigal son. This last, incidentally, is the only version in which the redemption takes place as the result of a voluntary decision on the part of the protagonist (Luke 15:18).

It would be confusing to summarize all the falls and rises of the Biblical history at once. In honor of the days of creation, let us select six, with a seventh forming the end of time. The first fall, naturally, is that of Adam from Eden, where Adam goes into a wilderness that modulates to the heathen cities founded by the family of Cain. Passing over the story of Noah, which adds the sea to the images of disaster, the first rise is that of Abraham, called out of the city of Ur in Mesopotamia to a Promised Land in the west. This introduces the pastoral era of the patriarchs, and ends at the end of Genesis, with Israel in Egypt. This situation again changes to an oppressive and threatening servitude; Israel again passes through a sea and a wilderness, and under Moses and Joshua reaches its promised land again, a smaller territory where the main images are agricultural. There succeed the invaders in the Book of Judges, of whom the most formidable were the Philistines, probably a Greek-speaking people from Crete (if that is the "Caphtor" of Amos 9:7) who gave their name to Palestine. They held the mastery of Israel after the defeat and death of Saul and his son Jonathan. The third rise begins with David and continues with Solomon, where the imagery is urban, concerned with cities and buildings. After Solomon, however, another disaster begins with the splitting of the kingdom. The northern kingdom was destroyed by Assyria in 722 B.C.; the southern kingdom of Judah had a reprieve until after Assyria was destroyed in its turn (Nahum 2:3ff.); but with the capture of Jerusalem by Nebuchadnezzar in 586 the Babylonian captivity began.

The fourth rise in the fortunes of the Israelites, now the Jews, begins with the permission—perhaps the encouragement—given the Jewish captives in Babylon by Cyrus of Persia to return and rebuild their temple. Two returns are prominently featured in the Old Testament, and there were probably more, but symbolically we need only one. Some flickering hopes of a restored Israel clustered around the chief figure of the first return, Zerubbabel of the line of David. After several changes of masters, the next dramatic descent was caused by the savage persecution of the non-Hellenized Jews by Antiochus Epiphanes of the Seleucian empire, which provoked the rebellion of the Maccabees, five brothers of a priestly family who finally gained independence for Judea and established a royal dynasty. This lasted until the Roman legions under Pompey rolled over the country in 63 B.C., and began the Roman domination that lasts throughout the New Testament period. At this point Jewish and Christian views of the sixth deliverance of Israel diverge. For Christianity, Jesus achieved a definitive deliverance for all mankind with his revelation that the ideal kingdom of Israel was a spiritual kingdom. For Judaism, the expulsion from their homeland by the edict of Hadrian in 135 A.D. began a renewed exile which in many respects still endures.

This is a sequence of mythoi, only indirectly of historical events, and our first step is to realize that all the high points and all the low points are metaphorically related to one another. That is, the garden of Eden, the Promised Land, Jerusalem, and Mount Zion are interchangeable synonyms for the home of the soul, and in Christian imagery they are identical, in their "spiritual" form (which we remember means metaphorically, whatever else it may mean) with the kingdom of God spoken of by Jesus. Similarly, Egypt, Babylon and Rome are all spiritually the same place, and the Pharaoh of the Exodus, Nebuchadnezzar, Antiochus Epiphanes, and Nero are spiritually the same person. And the deliverers of Israel—Abraham, Moses, and Joshua, the judges, David, and Solomon—are all prototypes of the Messiah or final deliverer.... As the various declines of Israel through apostasy and the like are not acts so much as failures to act, it is only the rises and restorations that are real events, and as the Exodus is the definitive deliverance and the type of all the rest, we may say that mythically the Exodus is the only thing that really happens in the Old Testament. On the same principle the resurrection of Christ, around which the New Testament revolves, must be, from the New Testament's point of view, the antitype of the Exodus. The life of Christ as presented in the Gospels becomes less puzzling when we realize that it is being presented in this form.

Like that of many gods and heroes, the birth of Jesus is a threatened birth: Herod orders a massacre of infants in Bethlehem from which Jesus alone escapes. Moses similarly escapes from an attempt to destroy Hebrew children, as they in turn escape later from a slaughter of Egyptian firstborn. The infant Jesus is taken down into Egypt by Joseph and Mary, and his return from there, Matthew (2:15) says, fulfills the prophecy of Hosea (11:1) "I called my son out of Egypt," where the reference is quite explicitly to Israel. The names Mary and Joseph recall the Miriam who was the sister of Moses and the Joseph who led the family of Israel into Egypt. The third Sura of the Koran appears to be identifying Miriam and Mary; Christian commentators on the Koran naturally say this is ridiculous, but from the purely typological point of view from which the Koran is speaking, the identification makes good sense.

Moses organizes the twelve tribes of Israel; Jesus gathers twelve disciples. Israel crosses the Red Sea and achieves its identity as a nation on the other side; Jesus is baptized in the Jordan and is recognized as the Son of God. The baptism is the point at which Mark and John begin, the infancy stories of Matthew and Luke being probably later material. Israel wanders forty years in the wilderness; Jesus, forty days. Miraculous food is provided for Israel and by Jesus for those gathered around him (see John 6:49–50). The law is given from Mount Sinai and the gospel preached in the Sermon on the Mount. A brazen serpent is placed on a pole by Moses as preservation against the fatal bites of "fiery serpents" (Numbers 21:9); this brazen serpent was accepted by Jesus as a type of his crucifixion (John 3:14) with an underlying association between the lethal serpents and the serpent of Eden. Moses dies just outside the Promised Land, which in Christian typology signifies the inability of the law alone to redeem man, and the Promised Land is conquered by Joshua. The hidden link here is that Jesus and Joshua are the same word, hence when the Virgin Mary is told to call her child Jesus or Joshua, the typological meaning is that the reign of the law is over, and the assault on the Promised Land has begun (Matthew 1:21). (Frye, N. [1982], pp. 169–172).

557. Frye, N. (1982), p. 131.
558. See Chapter 2: The Great Father: Images of the Known, or Explored Territory.
559. The *Dialogue of the Savior*, in Robinson, J.R. (1988), p. 525.
560. It is not merely the Gnostic Gospels that lay stress on the psychological nature of the Kingdom of God:

> And when he was demanded of the Pharisees, when the kingdom of God should come, he answered them and said, The kingdom of God cometh not with observation:
> Neither shall they say Lo here! or lo there! for, behold, the kingdom of God is within you. (Luke 17:20–21).

561. Referring to Exodus 31:12–15:

> And the Lord spake unto Moses, saying,
> Speak thou also unto the children of Israel, saying, Verily my sabbaths ye shall keep: for it *is* a sign between me and you throughout your generations: that *ye* may know that I *am* the Lord that doth sanctify you.
> Ye shall keep the sabbath therefore: for it *is* holy unto you: every one that defileth it shall surely be put to death: for whosoever doeth *any* work therein, that soul shall be cut off from among his people.

Six days may work be done; but in the seventh *is* the sabbath of rest, holy to the Lord: whosoever doeth *any* work in the sabbath day, he shall surely be put to death.

562. There is an apocryphal insertion at Luke 6:4. The insertion reads "Man, if indeed thou knowest what thou doest, thou art blessed; but if thou knowest not, thou art cursed, and a transgressor of the law" (Codex Bezae ad Lucam [to Luke] 6:4). More information is available in James M.R. (1924). Jung notes that the moral of this story is analogous to that in the parable of the unjust steward:

It is the task of the Paraclete, the "spirit of truth," to dwell and work in individual human beings, so as to remind them of Christ's teachings and lead them into the light. A good example of this activity is Paul, who knew not the Lord and received his gospel not from the apostles but through revelation. He is one of those people whose unconscious was disturbed and produced revelatory ecstasies. The life of the Holy Ghost reveals itself through its own activity, and through effects which not only confirm the things we all know, but go beyond them. In Christ's sayings there are already indications which go beyond the traditionally "Christian" morality—for instance the parable of the unjust steward (Luke 16:1–8), the moral of which agrees with the Logion of the Codex Bezae, and betrays an ethical standard very different from what is expected. Here the moral criterion is *consciousness*, and not law or convention. One might also mention the strange fact that it is precisely Peter, who lacks self-control and is fickle in character, whom Christ wishes to make the rock and foundation of his church. (Jung, C.G. [1969], pp. 433–444).

Jung also makes reference to the Oxyrhynchus papyrus, which is "older than the first conception of the gospels" (Jung, C.G. [1969], p. 444): Christ says, "Wherever there are two they are not without God, and wherever there is one alone I say I am with him." Jung notes that this is in contrast to the standard version: "For where two or three are gathered together in my name, there am I in the midst of them" (Matthew 18:20). The latter part of the former statement is strikingly reminiscent of Kierkegaard's notion:

For a "crowd" is the untruth. In a godly sense it is true, eternally, Christianly, as St. Paul says, that "only one attains the goal"—which is not meant in a comparative sense, for comparison takes others into account. It means that every man can be that one, God helping him therein—but only one attains the goal. And again this means that every man should be chary about having to do with "the others," and essentially should talk only with God and with himself—for only one attains the goal. And again this means that man, or to be a man, is akin to deity. In a worldly and temporal sense, it will be said by the man of bustle, sociability, and amicableness, "How unreasonable that only one attains the goal; for it is far more likely that many, by the strength of united effort, should attain the goal; and when we are many success is more certain and it is easier for each man severally." True enough, it is far more *likely*; and it is true also with respect to all earthly and material goods. If it is allowed to have its way, this becomes the only true point of view, for it does away with God and eternity and with man's kinship with deity. It does away with it or transforms it into a fable, and puts in its place the modern (or, we might rather say, the old pagan) notion that to be a man is to belong to a race endowed with reason, to belong to it as a specimen, so that the race or species is higher than the individual, which is to say that there are no more individuals but only specimens. But eternity which arches over and high above the temporal, tranquil as the starry vault at night, and God in heaven who in the bliss of that sublime tranquillity holds in survey, without the least sense of dizziness at such a height, these countless multitudes of men and knows each single individual by name—He, the Great Examiner, says that only one attains the goal. (cited in Kaufmann, W. [1975], pp. 94–95).

563. Piaget, J. (1965), p. 197.
564. Ibid., p. 13.
565. Ibid., p. 398.

566. Ibid., p. 111.
567. Ibid., p. 102.
568. Ibid., p. 362.
569. Rychlak, J. (1981), p. 699.
570. Lao Tzu (1984c).
571. Socrates' comments on the internal oracle are of interest here. He states, in *The Apology*, after (voluntarily) accepting his sentence of death:

And now, O men who have condemned me, I would fain prophesy to you; for I am about to die, and in the hour of death men are gifted with prophetic power. And I prophesy to you who are my murderers, that immediately after my departure punishment far heavier than you have inflicted on me will surely await you. Me you have killed because you wanted to escape the accuser, and not to give an account of your lives. But that will not be as you suppose: far otherwise. For I say that there will be more accusers of you than there are now; accusers whom hitherto I have restrained: and as they are younger they will be more inconsiderate with you, and you will be more offended at them. If you think that by killing men you can prevent some one from censuring your evil lives, you are mistaken; that is not a way of escape which is either possible or honourable; the easiest and the noblest way is not to be disabling others, but to be improving yourselves. This is the prophecy which I utter before my departure to the judges who have condemned me. Friends, who would have acquitted me, I would like also to talk with you about the thing which has come to pass, while the magistrates are busy, and before I go to the place at which I must die. Stay then a little, for we may as well talk with one another while there is time. You are my friends, and I should like to show you the meaning of this even which has happened to me. O my judges—for you I may truly call judges—I should like to tell you of a wonderful circumstance. Hitherto the divine faculty of which the internal oracle is the source has constantly been in the habit of opposing me even about trifles, if I was going to make a slip or error in any matter; and now you may see there has come upon me that which may be thought, and is generally believed to be, the last and worst evil. But the oracle made no sign of opposition, either when I was leaving my house in the morning, or when I was my way to the court, or while I was speaking, at anything which I was going to say; and yet I have often been stopped in the middle of a speech, but now in nothing I either said or did touching the matter in hand has the oracle opposed me. What do I take to be the explanation of this silence? I will tell you. It is an intimation that what has happened to me is a good, and that those of us who think that death is an evil are in error. For the customary sign would surely have opposed me had I been going to evil and not to good. (Plato, in Hutchins, R.M. [1952], pp. 210–211).

572. Neumann, E. (1954), pp. 173–174.
573. "The Tree of Knowledge: Church and Synagogue" from a Swiss manuscript (15th century). (Figure 56, Neumann, E. [1955]).
574. Eliade comments: "See (Eliade, M. [1978a], pp. 154–155) for other citations on the 'philosophical incest.' The acrostic constructed by Basil Valentine with the term *vitriol* underscores the implacable necessity of the decensus ad inferos: *Visita Interiora Terrae Recflficando invenies Occultum Lapidem* ('Visit the interior of the Earth, and by purification you will find the secret Stone')." (Eliade, M. [1985], p. 256, footnote 89).
575. Eliade comments: "*Liber Platonis quartorum* (of which the Arabic original cannot be later than the tenth century), cited in (Eliade, M. [1978a], p. 158). One will find the same doctrine among the Chinese alchemists (see Eliade, M. [1982], pp. 37–43)." (Eliade, M. [1985], p. 256, footnote 90).
576. Additional quotation in parenthesis from Eliade, M. (1978a), pp. 163–164.
577. Eliade comments: "According to Basil Valentine, 'evil must become the same as good.' Starkey describes the stone as 'the reconciliation of Contraries, a making of friendship between enemies' (Eliade, M. [1978a], p. 166)." (Eliade, M. [1985], p. 256, footnote 91).
578. Eliade comments: (see Multhauf, R.F. [1967], p. 135 and following). (Eliade, M. [1985], p. 257, footnote 92).

579. Eliade comments: (see Eliade, M. [1978a], p. 51). (Eliade, M. [1985], p. 257, footnote 93).

580. Eliade comments: "We have discussed the consequences of this Promethean gesture, in (Eliade, M. [1978a], pp. 169–178)." (Eliade, M. [1985], p. 257, footnote 94).

581. Eliade comments: "Even in the eighteenth century, the learned did not question the growth of minerals. They asked themselves, however, whether alchemy could assist nature in this process, and above all whether 'those alchemists who claimed to have done so already were honest men, fools, or impostors' (see Dobbs, B.J.T. [1975], p. 44). Herman Boerhaave (1664–1739), considered the greatest 'rationalist' chemist of his time and famous for his strictly empirical experiments, still believed in the transmutation of metals. And we will see the importance of alchemy in the scientific revolution accomplished by Newton." (Eliade, M. [1985], p. 257, footnote 95).

582. Eliade, M. (1985), pp. 255–258.

583. Becker, E. (1973), p. xiv.

584. After the publication of Jung, C.G. (1912).

585. See Ellenberger, H.F. (1970).

586. Costa, P.T., Jr. & McCrae, R.R. (1992a); Goldberg, L.R. (1993b).

587. Representative samples of modern exemplars of "complex" and "unconscious": Banaji, M.R., Hardin, C., & Rothman, A.J. (1993); Nader, A., McNally, R.J., & Wiegartz, P.S. (1996); Watkins, P.C., Vache, K., Verney, S.P., & Mathews, A. (1996); Gabrieli, J.D.E., Fleischman, D.A., Keane, M., Reminger, M., Sheryl, L., et al. (1995).

588. Wilson, E.O. (1998).

589. von Franz, M.L. (1980), pp. 32–34.

590. Ibid., p. 34.

591. Translation: "It is found in cesspools," cited in Jung, C.G. (1976b), p. 35.

592. The Gospel of Thomas. In Robinson, J.R. (Ed.). (1988), p. 134.

593. Jung, C.G. (1968b), p. 306.

594. I am indebted to Erin Driver-Linn for bringing this phrase to my attention, in this context.

595. Cited in Evans, P.I. (1973), p. 126.

596. Jung, C.G. (1968b), pp. 342–343.

597. See Jung, C.G. (1968b), p. 253, for illustration.

598. See Jung, E. & von Franz, M.L. (1980), pp. 369–370. The authors describe the apprentice Taliesen's description of Merlin, spirit of transformation:

> I am the wind that blows upon the sea;
> I am the ocean wave;
> I am the murmur of the surges;
> I am seven battalions;
> I am a strong bull;
> I am an eagle on a rock;
> I am a ray of the sun;
> I am the most beautiful of herbs;
> I am a courageous wild boar;
> I am a salmon in the water;
> I am a lake upon the plain;
> I am a cunning artist;
> I am a gigantic, sword-wielding champion;
> I can shift my shape like a god.

599. Jung, C.G. (1968b), pp. 66–67.

600. At least two years after experiencing this dream (and a year or so after writing it down) I was reading Dante's *Inferno* (Dante, A. [1982]). In the ninth Canto, a messenger from God appears in hell to open

the Gate of Dis, which is barring the divinely ordained way of Virgil and Dante. The approach of this messenger is preceded by a great storm, described in the following manner (p. 90):

> Suddenly there broke on the dirty swell
> of the dark marsh a squall of terrible sound
> that sent a tremor through both shores of Hell;
> a sound as if two continents of air,
> one frigid and one scorching, clashed head on
> in a war of winds that stripped the forests bare,
> ripped off whole boughs and blew them helter skelter
> along the range of dust it raised before it
> making the beasts and shepherds run for shelter.

The similarity of imagery and meaning in my dream and this poem struck me as very interesting.

601. This was actually an image I had used in therapeutic discussions, previously. I told my clients that an unresolved anomaly was like the tip of a monster's tail: it looked harmless enough, viewed only as a tail—but that meant pretending that the part did not imply the whole.

602. Jung, C.G. (1968b), p. 343.

603. Eliade, M. (1978a), p. 50.

604. Ibid., pp. 51–52.

605. Ibid., p. 35.

606. Jung, C.G. (1976b), p. 439.

607. von Franz, M.L. (1980), pp. 21–22.

608. Jung, C.G. (1976b), pp. 482–483.

609. Ibid., p. xiv.

610. Jung, C.G. (1976b), pp. 319–320.

611. Eliade comments: "(Dobbs, B.J.T. [1975], p. 90), citing the article of E. McGuire & P. M. Rattansi, 'Newton and the "Pipes of Pan,"' pp. 103–143." (In Eliade, M. [1985], p. 260, footnote 104).

612. Eliade comments: (Westfall, R.S. [1971], pp. 377–391; Dobbs, R.J.T. [1975], p. 211). (In Eliade, M. [1985], p. 260, footnote 104).

613. Eliade, M. (1985), pp. 259–261.

614. Jung, C.G. (1968b), p. 324.

615. Ibid., pp. 322–323.

616. Eliade, M. (1978a), pp. 8–9.

617. Dorn, in Jung, C.G. (1976b), p. 271.

618. Jung, C.G. (1976b), pp. 358–359.

619. Frye states:

Prophet, high priest, and king are all figures of authority, but prophets are often martyred and even kings ... have scapegoat and victim imagery attached to them. Joshua was a type of Christ as the conqueror of the Promised Land: his enemies included five kings who were hung on trees and then buried in a cave with great stones rolled against it (Joshua 10:16ff.). Solomon, the king who succeeded David, is a type of Christ as a temple builder and wise teacher: Absalom, equally a son of David, rebelled against his father and was caught in a tree, traditionally by his golden hair, hanging there "between heaven and earth" until David's general Joab came up and thrust darts into his side (II Samuel 18:14). Absalom's curious helplessness in what seems a relatively easy situation to get out of suggests a ritual element in the story of his death. The writers of the Gospels found that in telling the story of Jesus they needed the imagery of the executed kings and Absalom quite as much as that of the figures of glory and triumph. (Frye, N. [1982], p. 180).

620. Jung, C.G. (1976b), p. 21.

621. Ibid., p. 283.

622. Ibid., p. 308.

623. Jung describes this state as a quaternity of opposites; I have eliminated this particular reference in the attempt to simplify an already sufficiently complex discussion.

624. Jung, C.G. (1976b), pp. 353–354.

625. Ibid., pp. 540–541.

626. Ibid., pp. 363–364.

627. From Maier, *Scrutinium chymicum* (1687) (plate 175 in Jung, C.G. [1968b]).

628. Ibid., pp. 214–215.

629. Ibid., pp. 363–364.

630. Nietzsche, F. (1966), p. 10.

631. Jung, C.G. (1976b), p. 475.

632. Jung, C.G. (1976a).

633. It seems possible that Piaget's hypothetical "organ of equilibration" is equivalent to Jung's "self"—the highest regulator of intrapsychic activity:

The organism has special organs of equilibrium. The same is true of mental life, whose organs of equilibrium are special regulatory mechanisms. This is so at all levels of development, from the elementary regulators of motivation (needs and interests) up to will for affectivity and from perceptual and sensorimotor regulations up to operations for cognition. (Piaget, J. [1967], p. 102).

Piaget also points out (a) that *consciousness* arises in personality "when the environmental situation in which some person finds himself or herself blocks some ongoing (goal-directed) activity. Children act in accordance with their needs and everything takes place without conscious awareness or the equilibrations going on until there is a frustration [Piaget's terminology, likely equivalent to emergence of the unexpected (and punishing ?)].... Each of these frustrating circumstances serves to focus the child's attention on the reasons for the disequilibration rather than simply on the desired goal" (Rychlak, J. [1981], p. 688—see Piaget, J. [1967]; Piaget, J. [1962]), and (b) that *will* arises when there is a conflict in behavioral tendencies (Jung would say, when there is a conflict in duty). Piaget believed that the will (the will to power; the heroic principle) could be considered the consequence of integration of affect and motivation:

To the extent that the emotions become organized, they emerge as regulations whose final form of equilibrium is none other than the will. Thus, will is the true affective equivalent of the operation in reason. Will is a late-appearing function. The real exercise of will is linked to the function of the autonomous moral feelings, which is why we have waited until this [late-childhood] stage to discuss it. (Piaget, J. [1967], p. 58).

This idea, in passing, is very much like Jung's notion of the integration of feeling-toned complexes into the ego. Piaget elaborates, elsewhere:

The act of will does not consist of following the inferior and stronger tendency; on the contrary, one would then speak of a failure of will or "lack of will power." Will power involves reinforcing the superior but weaker tendency so as to make it triumph. (Piaget, J. [1965], p. 59).

634. "Fabulous monster containing the *mass confusa*, from which rises the pelican (symbol of Christ and the *lapis*)." In *Hermaphroditisches Sonn- und Mondskind* (1752) (plate 256 in Jung, C.G. [1968b]).

635. Jung, C.G. (1968b), pp. 327–329.

636. Ibid., p. 331.

637. Ibid., p. 302.

638. Ibid., p. 329.

639. Ibid., pp. 332–334.

640. Ibid., p. 335.

641. Ibid., p. 336–339.

642. Ibid., p. 346–348.

643. Dorn, in Jung, C.G. (1976b), p. 465.

644. See chapter 2: The Great Father: Images of the Known, or Explored Territory.

645. Dorn, in Jung, C.G. (1976b), p. 41.

646. Jung, C.G. (1976b), pp. 34–35.

647. Khunrath, in Jung, C.G. (1976b), p. 329.

648. Jung, C.G. (1976b), p. 349.

649. Jung, C.G. (1968a), p. 179.

650. Nietzsche, F. (1967a), p. 122–123.

651. Russell, J.B. (1986).

652. Ibid., p. 246.

653. Ibid., p. 300.

654. Milton, J. (1961), 3:100–128, p. 95.

655. Wechsler, D. (1981).

656. See, for example, The Gospel of Mary, in Robinson, J.R. (Ed.) (1988), p. 527.

657. See, for example, The Gospel of Thomas, in Robinson, J.R. (Ed.) (1988), pp. 132, 138.

658. Jung, C.G. (1968b), pp. 35–37.

659. See footnote 1. ➔

660. Jung, C.G. (1976b), pp. 374–375.

661. Jung, C.G. (1967b), p. 304.

662. There are (at least) two (major) alternative dogmatic formulations of Original Sin, in Christian tradition: (1) the source of eternal guilt; (2) the fortunate error, which leads to the incarnation of Christ. Toni Wolff notes:

There are early medieval representations of the genealogical tree of Christ. On the branches, as the fruits of the tree, are the prophets and all Christ's ancestors. The roots of the tree grow out of the skull of Adam, and Christ is its central and more precious fruit.

Jung amplifies this comment:

Well, the tree sometimes grows out of Adam's navel, and on the branches, as you say, sit the prophets and kings of the Old Testament, Christ's ancestors, and then on top of the tree is the triumphant Christ. That life begins with Adam and ends with Christ is the same idea. (Jung, C.G. [1988], p. 1440).

663. Neumann states:

Originally, Messianism was bound up with a historical process ending in the emergence of a savior who, after the transformation crisis of the apocalypse, ushers in the eschatological age of redemption. This conception can easily be shown to be a projection of an individuation process, the subject of which, however, is the people, the chosen collectivity, and not the individual. In the collective projection, history appears as the collective representation of destiny; the crisis is manifested in the projection of the ways which characterize the Last Days; and the transformation, as the Last Judgment, death and resurrection. Similarly, the transfiguration and conquest of the self corresponds to transfiguration in the celestial paradise which in the shape of a mandala gathers in mankind, or else it is projected as life in a re-created and renewed world governed by the king-Adam-*anthropos*-self at its center. (Neumann, E.

[1968], p. 408).

664. "The restitution of the mystic apple to the tree of knowledge." Giovanni da Modena (15th century) (plate 116 In Neumann, E. [1955]).

665. Eliade, M. (1982), p. 73.

666. "The Boddhisatva" (plate XII in Campbell, J. [1973]).

667. From the Gospel of Thomas. Cited in Pagels, E. (1979), p. xv.

668. The Gospel of Thomas. In Robinson, J.R. (Ed.) (1988), pp. 126–127.

REFERENCES

Adler, A. (1958). *What life should mean to you.* New York: Capricorn Books.

Aggleton, J.P. (Ed.). (1993). *The amygdala: Neurobiological aspects of emotion, memory, and mental dysfunction.* New York: Wiley-Liss.

Agnew, N.M. & Brown, J.L. (1990). Foundations for a model of knowing: Constructing reality. *Canadian Journal of Psychology, 30,* 152–183.

Ambady, N. & Rosenthal, R. (1992). Thin slices of expressive behavior as predictors of interpersonal consequences: A meta-analysis. *Psychological Bulletin, 111,* 256–274.

Arendt, H. (1994). *Eichmann in Jerusalem : A report on the banality of evil.* New York: Penguin.

Armstrong, S.L., Gleitman, L.R., & Gleitman, H. (1983). What some concepts might not be. *Cognition, 13,* 263–308.

Banaji, M.R., Hardin, C., & Rothman, A.J. (1993). Implicit stereotyping in person judgment. *Journal of Personality and Social Psychology, 65,* 272–281.

Bechara, A., Damasio, H., Tranel, D., & Damasio, A.R. (1997). Deciding advantageously before knowing the advantageous strategy. *Science, 275,* 1293–1295.

Bechara, A., Tranel, D., Damasio, H., & Damasio, A.R. (1996). Failure to respond autonomically to anticipated future outcomes following damage to prefrontal cortex. *Cerebral Cortex, 6,* 215–225.

Beck, A. (1979). *Cognitive therapy of depression.* New York: Guilford Press.

Becker, E. (1973). *The denial of death.* New York: The Free Press.

Bellows, H.A. (1969). *The poetic Edda.* New York: Biblo and Tannen.

Berkowitz, C.D. & Senter, S.A. (1987). Characteristics of mother-infant interactions in nonorganic failure to thrive. *Journal of Family Practice, 25,* 377–381.

Binswanger, L. (1963). *Being in the world.* New York: Basic Books.

Blake, W. (1793/1946). The marriage of heaven and hell. In A. Kazin (Ed.), *The portable Blake* (pp. 249–266). New York: Viking.

Blanchard, D.C. & Blanchard, R.J. (1972). Innate and conditioned reactions to threat in rats with amygdaloid lesions. *Journal of Comparative Physiology and Psychology, 81,* 281–290.

Blanchard, D.J. & Blanchard, D.C. (1989). Antipredator defensive behaviors in a visible burrow system. *Journal of Comparative Psychology, 103,* 70–82.

Blanchard, D.C., Blanchard, R.J., & Rodgers, R.J. (1991). Risk assessment and animal models of anxiety. In B. Olivier, J. Mos & J.L. Slangen (Eds.), *Animal models in psychopharmacology* (pp. 117–134). Boston: Birkhauser Verlag.

Blanchard, D.C., Veniegas, R., Elloran, I., & Blanchard, R.J. (1993). Alcohol and anxiety: Effects on offensive and defensive aggression. *Journal of Studies on Alcohol, Supplement Number 11,* 9–19.

Blanchard, R.J., Blanchard, D.C., & Hori, K. (1989). Ethoexperimental approach to the study of defensive behavior. In R.J. Blanchard, P.F. Brain, D.C. Blanchard, & S. Parmigiani, (Eds.), *Ethoexperimental approaches to the study of behavior* (pp. 114–136). Boston: Kluwer-Nijhoff Publishing.

Borski, L.M. & Miller, K.B. (1956). The jolly tailor who became king. In P.R. Evans (Ed.), *The family treasury of children's stories: Book two* (pp. 60–68). New York: Doubleday and Company.

Bouton, M.E. and Bolles, R.C. (1980). Conditioned fear assessed by freezing and by the suppression of three different baselines. *Animal Learning and Behavior, 8*, 429–434.

Bowlby, J. (1969). *Attachment and loss: Vol. 1. Attachment.* New York: Basic Books.

Brooks, A. (1991). *Intelligence without reason.* MIT Artificial Intelligence Laboratory: Artificial Intelligence Memo 1293.

Brooks, A., and Stein, L.A. (1993). *Building brains for bodies.* MIT Artificial Intelligence Laboratory: Artificial Intelligence Memo 1439.

Brown, R. (1965). *Social psychology.* New York: The Free Press.

Brown, R. (1986). *Social psychology: The second edition.* New York: Macmillan.

Browning, C.R. (1993). *Ordinary men: Reserve police battalion 101 and the final solution in Poland.* New York: Harper Perennial.

Bruner, J. (1986). *Actual minds, possible worlds.* Cambridge: Harvard University Press.

Bruner, J.S. & Postman, L. (1949). On the perception of incongruity: A paradigm. *Journal of Personality, 18*, 206–223.

Campbell, J. (1964). *Occidental mythology: The masks of God.* London: Penguin Books.

Campbell, J. (1968). *The hero with a thousand faces.* Princeton: Princeton University Press.

Campbell, J. (1973). *Myths to live by.* New York: Bantam Books.

Campbell, J. (1987). *The masks of God: Vol. 1. Primitive mythology.* New York: Penguin.

Carver, C.S. & Scheier, M.F. (1982). Control theory: A useful conceptual framework for personality, social, clinical, and health psychology. *Psychological Bulletin, 92*, 111–135.

Cornford, F.M. (1956). *Plato's cosmology: The timaeus of Plato.* London: Routledge.

Costa, P.T., Jr. & McCrae, R.R. (1992a). Four ways five factors are basic. *Personality and Individual Differences, 13*, 653–665.

Damasio, A.R. (1994). *Descartes' error.* New York: Putnam.

Damasio, A.R. (1996). The somatic marker hypothesis and the possible functions of the prefrontal cortex. *Philosophical Transactions of the Royal Society of London (Biological Science), 351*, 1413–1420.

Dante, A. (1982). *The inferno: Dante's immortal drama of a journey through hell* (J. Ciardi, Trans.). New York: Mentor Books.

Davidson, R.J. (1984a). Affect, cognition, and hemispheric specialization. In C.E. Izard, J. Kagan, & R. Zajonc (Eds.), *Emotion, cognition, and behavior* (pp. 320–365). New York: Cambridge University Press.

Davidson, R.J. (1984b). Hemispheric asymmetry and emotion. In K. Scherer & P. Ekman (Eds.), *Approaches to emotion* (pp. 39–57). Hillsdale, NJ: Erlbaum.

Davidson, R.J. (1992). Anterior cerebral asymmetry and the nature of emotion. *Brain and Cognition, 20*, 125–151.

Davidson, R.J. and Fox, N.A. (1982). Asymmetrical brain activity discriminates between positive and negative affective stimuli in human infants. *Science, 218*, 1235–1237.

Dee, J. (1993). *Diary of Doctor John Dee: Together with a catalogue of his library of manuscripts.* New York: Holmes.

Didi-Huberman, G., Garbetta, R., & Morgaine, M. (1994). *Saint-Georges et le dragon: Versions d'une legende.* Paris: Societe Nouvelle Adam Biro.

Dobbs, B.J.T. (1975). *The foundations of Newton's alchemy.* New York: Cambridge University Press.

Dollard, J. & Miller, N. (1950). *Personality and psychotherapy: An analysis in terms of learning, thinking, and culture.* New York: McGraw-Hill.

Donald, M. (1993). *The origins of the modern mind.* Cambridge: Harvard University Press.

Dostoyevsky, F. (1961). *Notes from underground.* New York: Penguin Group.

Dostoyevsky, F. (1981). *The brothers Karamazov* (A.H. MacAndrew, Trans.). New York: Bantam Books.

Dostoyevsky, F. (1993). *Crime and punishment.* New York: Vintage Classics.

Doty, R.W. (1989). Schizophrenia: A disease of interhemispheric processes at forebrain and brainstem levels? *Behavioural Brain Research, 34,* 1–33.

Durnin, R. (1994). *Letter to Thomas.* Unpublished manuscript.

Edwardes, A. & Masters, R.E.L. (1963). *The cradle of erotica.* New York: Julian Press.

Einstein, A. (1959). Autobiographical note. In P.A. Schilpp (Ed.), *Albert Einstein: Philosopher scientist.* New York: Harper.

Eliade, M. (1957). *The sacred and the profane: The nature of religion.* New York: Harcourt Brace.

Eliade, M. (1964). *Shamanism: Archaic techniques of ecstasy* (W.R. Trask, Trans.). Princeton: Princeton University Press.

Eliade, M. (1965). *Rites and symbols of initiation: The mysteries of birth and rebirth* (W.R. Trask, Trans.). New York: Harper and Row.

Eliade, M. (1975). *Myths, dreams, and mysteries: The encounter between contemporary faiths and archaic realities* (P. Mairet, Trans.). New York: Harper Colophon, Harper and Row.

Eliade, M. (1978a). *The forge and the crucible* (S. Corrin, Trans.) (2nd ed.). Chicago: University of Chicago Press.

Eliade, M. *A history of religious ideas* (W.R. Trask, Trans.). Chicago: Chicago University Press.

(1978b). *Vol. 1. From the stone age to the Eleusinian mysteries.*

(1982). *Vol. 2. From Gautama Buddha to the triumph of Christianity.*

(1985). *Vol. 3. From Muhammad to the age of reforms.*

Eliade, M. (1991a). *Images and symbols: Studies in religious symbolism* (P. Mairet, Trans.). Princeton: Mythos.

Eliade, M. (1991b). *The myth of the eternal return, or, cosmos and history* (W.R. Trask, Trans.). Princeton: Princeton University Press.

Ellenberger, H. (1970). *The discovery of the unconscious: The history and evolution of dynamic psychiatry.* New York: Basic Books.

Ervin, F. & Smith, M. (1986). Neurophysiological bases of the primary emotions. In R. Plutchik & H. Kellerman (Eds.), *Emotion: Theory, research, and experience: Vol. 3. Biological foundations of emotion* (pp. 145–170). New York: Academic Press.

Evans, P.I. (1973). *Jean Piaget: The man and his ideas.* New York: E.P. Dutton and Company.

Eysenck, H.J. (1995). Creativity as a product of personality and intelligence. In D.H. Saklofske & M. Zeidner (Eds.), *International handbook of personality and intelligence* (pp. 231–247). New York: Plenum Press.

Feyeraband, P.K. (1981). *Realism, rationalism, and scientific method: Philosophical papers (Vol. 1).* New York: Cambridge University Press.

Field, T.M., Schanberg, S.M., Scafidi, F., Bauer, C.R., Vega-Lahr, N., Garcia, R., Nystrom, J., & Kuhn, C.M. (1986). Tactile-kinesthetic stimulation effects on preterm neonates. *Pediatrics, 77,* 654–658.

Fierz, M. & Weisskopf, V.F. (Eds.). (1960). *Theoretical physics in the twentieth century: A memorial volume to Wolfgang Pauli.* New York: Interscience Publishers.

Foa, E.B., Molnar, C., & Cashman, L. (1995). Change in rape narratives during exposure therapy for post-traumatic stress disorder. *Journal of Traumatic Stress, 8,* 675–690.

Fowles, D.C. (1980). The three arousal model: Implications of Gray's two factor learning theory for heart-rate, electrodermal activity, and psychopathy. *Psychophysiology 17,* 87–104.

Fowles, D.C. (1983). Motivational effects of heart rate and electrodermal activity: Implications for research on personality and psychopathology. *Journal of Research on Personality, 17,* 48–71.

Fowles, D.C. (1988). Psychophysiology and psychopathology: A motivational approach. *Psychophysiology, 25,* 373–391.

Fox, N.A. & Davidson, R.J. (1986). Taste-elicited changes in facial signs of emotion and the asymmetry of brain electrical activity in human newborns. *Neuropsychologia, 24,* 417–422.

Fox, N.A. & Davidson, R.J. (1988). Patterns of brain electrical activity during facial signs of emotion in

10–month old infants. *Developmental Psychology, 24,* 230–236.

Frankl, V. (1971). *Man's search for meaning: An introduction to logotherapy.* New York: Pocket Books.

Frazier, J.G. (1994). *The golden bough: A study in magic and religion (the world's classics).* Oxford: Oxford University Press.

Frye, N. (1982). *The great code: The Bible and literature.* London: Harcourt Brace Jovanovitch.

Frye, N. (1990). *Words with power: Being a second study of the Bible and literature.* London: Harcourt Brace Jovanovitch.

Fukuyama, F. (1993). *The end of history and the last man.* New York: Avon Books.

Gabrieli, J.D.E., Fleischman, D.A., Keane, M., Reminger, M., Sheryl, L., et al. (1995). Double dissociation between memory systems underlying explicit and implicit memory systems in the human brain. *Psychological Science, 6,* 76–82.

Gall, J. (1988). *Systemantics: The underground text of systems lore.* Ann Arbor: The General Systematics Press.

Garey, L.J. & Revishchin, A. V. (1990). Structure and thalamocortical relations of the cetacean sensory cortex: Histological, tracer, and immunocytochemical studies. In J.A. Thomas & R.A. Kastelein (Eds.), *Sensory abilities of Cetaceans: Laboratory and field evidence* (pp.19–30). New York: Plenum Press.

Goethe, J.W. (1979a). *Faust, part one* (P. Wayne, Trans.). London: Penguin Books.

Goethe, J.W. (1979b). *Faust, part two* (P. Wayne, Trans.). London: Penguin Books.

Goldberg, E. (1995). Rise and fall of modular orthodoxy. *Journal of Clinical and Experimental Neuropsychology, 17,* 193–208.

Goldberg, E. and Costa, L.D. (1981). Hemisphere differences in the acquisition and use of descriptive systems. *Brain and Language, 14,* 144–173.

Goldberg, E., Podell, K., and Lovell, M. (1994). Lateralization of frontal lobe functions and cognitive novelty. *Journal of Neuropsychiatry and Clinical Neuroscience, 6,* 371–378.

Goldberg, L.R. (1993). The structure of phenotypic personality traits. *American Psychologist, 48,* 26–34.

Goldman-Rakic, P.S. (1987). Circuitry of primate prefrontal cortex and regulation of behavior by representational memory. In F. Plum (Ed.), Handbook of physiology: Vol. 5: The nervous system (pp. 373–417). Baltimore: American Physiological Society. Society.

Goodall, J. (1990). *Through a window.* Boston: Houghton Mifflin Company.

Gould, L., Andrews, D. & Yevin, J. (1996, December). The spy 100 line-up. *Spy Magazine.*

Granit, R. (1977). *The purposive brain.* Cambridge: Cambridge University Press.

Gray, J.A. (1982). *The neuropsychology of anxiety: An enquiry into the functions of the septal-hippocampal system.* Oxford: Oxford University Press.

Gray, J.A. (1987). *The psychology of fear and stress: Vol. 5. Problems in the behavioral sciences.* Cambridge: Cambridge University Press.

Gray, J.A., & McNaughton, N. (1996). The neuropsychology of anxiety: Reprise. *Nebraska Symposium on Motivation, 43,* 61–134.

Gray, J.A., Feldon, J., Rawlins, J.N.P., Hemsley, D.R., and Smith, A.D. (1991). The neuropsychology of schizophrenia. *Behavioral and Brain Sciences, 14,* 1–84.

Grossberg, S. (1987). Competitive learning: From interactive activation to adaptive resonance. *Cognitive Science, 11,* 23–63.

Halgren, E. (1992). Emotional neurophysiology of the amygdala within the conyext of human cognition. In J.P. Aggleton (Ed.), *The amygdala: Neurobiological aspects of emotion, memory and mental dysfunction* (pp. 191–228). New York: Wiley-Liss.

Halgren, E., Squires, N.K., Wilson, C.L., Rohrbaugh, J.W., Babb, T.L., and Crandell, P.H. (1980). Endogenous potentials generated in the human hippocampal formation and amygdala by infrequent events. *Science, 210,* 803–805.

Hawking, S. (1988). *A brief history of time.* New York: Bantam.

Hebb, D.O. & Thompson, W.R. (1985). The social significance of animal studies. In G. Lindzey & E. Aronson (Eds.), *The handbook of social psychology* (pp. 729–774). New York: Random House.

Heidel, A. (1965). *The Babylonian genesis.* Chicago: Chicago University Press (Phoenix Books).

Hodson, G. (1963). *The hidden wisdom in the Holy Bible: Vol. 1.* Adyar, India: Theosophical Publishing House.

Hofstadter, D.R. (1979). *Godel, Escher, Bach: An eternal golden braid*. New York: Vintage.

Holloway, R.L. & Post, D.G. (1982). The relativity of relative brain measures and hominid mosaic evolution. In E. Armstrong & D. Falk (Eds.), *Primate brain evolution: Method and concepts* (pp. 57–76). New York: Plenum Press.

Huizinga, J. (1967). *The waning of the Middle Ages*. New York: St. Martin's Press.

Huxley, A. (1956). *The doors of perception, and heaven and hell*. New York: Harper and Row.

Hyde, J.S. (1984). How large are gender differences in aggression? A developmental meta-analysis. *Developmental Psychology, 20*, 722–736.

Ikemoto, S. & Panksepp, J. (1996). Dissociations between appetitive and consummatory responses by pharmacological manipulations of reward-relevant brain regions. *Behavioral Neuroscience, 110*, 331–345.

Iwata, J., Chida, K., & LeDoux, J.E. (1987). Cardiovascular responses elicited by stimulation of neurons in the central amygdaloid complex in awake but not anesthetized rats resemble conditioned emotional responses. *Brain Research, 36*, 192–306.

Jacobsen, Thorkild. (1943). Primitive democracy in ancient Mesopotamia. *Journal of Near Eastern Studies, 2*, 159–170.

Jaeger, W. (1968). *The theology of the early Greek philosophers: The Gifford lectures 1936*. London: Oxford University Press.

Jaffe, A. (1961). *Memories, dreams, and reflections*. New York: Random House.

James, M.R. (1924). *The apocryphal New Testament*. Oxford: Clarendon Press.

James, W. (1880). Great men and their environment. *Atlantic Monthly*, October.

Jerison, H.J. (1979) The evolution of diversity in brain size. In M.E. Hahn, C. Jensen, & B.C. Dudek (Eds.), *Development and evolution of brain size: Behavioral implications* (pp. 29–57). New York: Academic Press.

Johnson, B. (1988). *Lady of the beasts*. New York: Harper and Row.

Jones, S.H., Gray, J.A., & Hemsley, D.R. (1992). The Kamin blocking effect, incidental learning, and schizotypy: A reanalysis. *Personality and Individual Differences, 13*, 57–60.

Joravsky, D. (1989). *Russian psychology: A critical history*. Cambridge: Basil Blackwell.

Joyce, J. (1986). *Ulysses*. New York: Random House.

Joyce, J. (1992). *The portrait of the artist as a young man*. New York: Bantam Classics.

Jung, C.G. *The collected works of C.G. Jung* (R.F.C. Hull, Trans.). Bollingen Series XX. Princeton University Press.

(1967a). *Vol. 5. Symbols of transformation: an analysis of the prelude to a case of schizophrenia*.

(1971). *Vol. 6. Psychological types*.

(1970a). *Vol. 7. Two essays on analytical psychology*.

(1976a). *Vol. 8. The structure and dynamics of the psyche*.

(1968a). *Vol. 9. Part 1. The archetypes and the collective unconscious*.

(1978a). *Vol. 9. Part 2. Aion: researches into the phenomenology of the self*.

(1978b). *Vol. 10. Civilization in transition*.

(1969). *Vol. 11. Psychology and religion: west and east*.

(1968b). *Vol. 12. Psychology and alchemy*.

(1967b). *Vol. 13. Alchemical studies*.

(1976b). *Vol. 14. Mysterium Coniunctionis: an inquiry into the separation and synthesis of psychic opposites in alchemy*.

(1970b). *Vol. 17. The development of personality*.

Jung, C.G. (1912). *Wandlungen und symbole der libido*. Leipzig: F. Deuticke.

Jung, C.G. (1933). *Modern man in search of a soul*. New York: Harcourt Brace.

Jung, C.G. (1988). *Nietzsche's Zarathrustra: Notes of the seminar given in 1934* (J.L. Jarrett, Ed.). Princeton: Princeton University Press.

Jung, E. & von Franz, M.L. (1980). *The grail legend*. Boston: Sigo Press.

Kagan, J. (1984). Behavioral inhibition in young children. *Child Development, 55*, 1005–1019.

Kapp, B.S., Pascoe, J.P. & Bixler, M.A. (1984). The amygdala: A neuroanatomical systems approach to its contributions to aversive conditioning. In N. Butters & L.R. Squire (Eds.), *Neuropsychology of Memory*. (pp.

473–488). New York: Guilford.

Kaufmann, W. (Ed. and Trans.). (1968). *The basic writings of Nietzsche*. New York: Random House.

Kaufmann, W. (Ed. and Trans.). (1975). *Existentialism from Dostoevsky to Sartre*. New York: Meridian.

Keynes, G. (Ed.). (1966). *The complete works of William Blake, with variant readings*. London: Oxford University Press.

Kling, A.S. & Brothers, L.A. (1992). The amygdala and social behavior. In J.P. Aggleton (Ed.), *The amygdala: Neurobiological aspects of emotion, memory, and mental dysfunction* (pp. 353–377). New York: Wiley-Liss.

Koestler, A. (1976). *The ghost in the machine*. London: Hutchison.

Kronig, R. (1960). The turning point. In M. Fierz & V.F. Weisskopf (Eds.), *Theoretical physics in the twentieth century: A memorial volume to Wolfgang Pauli*. (pp. 5–39). New York: Interscience Publishers.

Kuhn, T.S. (1957). *The Copernican revolution: Planetary astronomy in the development of Western thought*. Cambridge: Harvard University Press.

Kuhn, T.S. (1970). *The structure of scientific revolutions*. Chicago: Chicago University Press.

L'Engle, M. (1997). *A wrinkle in time*. New York: Bantam Doubleday Yearling Newbery.

Lakoff, G. (1987). *Women, fire, and dangerous things: What categories reveal about the mind*. Chicago: University of Chicago Press.

Lao Tzu. (1984a). 64: Staying with the misery. In *Tao Te Ching* (S. Rosenthal, Trans.) [On-line]. Available: *http://www.warrior-scholar.com/text/tao.htm*.

Lao Tzu. (1984b). 78: Sincerity. In *Tao Te Ching* (S. Rosenthal, Trans.) [On-line]. Available: *http://www.warrior-scholar.com/text/tao.htm*.

Lao Tzu. (1984c). 38: The concerns of the great. In *Tao Te Ching* (S. Rosenthal, Trans.) [On-line]. Available: *http://www.warrior-scholar.com/text/tao.htm*.

Lao Tzu. (1984d). 50: The value set on life. In *Tao Te Ching* (S. Rosenthal, Trans.) [On-line]. Available: *http://www.warrior-scholar.com/text/tao.htm*.

LeDoux, J.E. (1992). Emotion and the amygdala. In J.P. Aggleton (Ed.), *The amygdala: Neurobiological aspects of emotion, memory, and mental dysfunction* (pp. 339–351). New York: Wiley-Liss.

LeDoux, J.E. (1993). Emotional networks in the brain. In M. Lewis and J.M. Haviland (Eds.), *Handbook of Emotions* (pp. 109–118). New York: Guilford.

LeDoux, J.E., Sakaguchi, A., & Reis, D.J. (1984). Subcortical efferent projections of the medial geniculate nucleus mediate emotional responses conditioned to acoustic stimuli. *Journal of Neuroscience, 4*, 683–698.

Lewis, M. & Haviland, J.M. (Eds.). (1993). *Handbook of emotions*. New York: Guilford.

Lilly, J.C. (1967). *The mind of the dolphin*. New York: Doubleday.

Lindzey, G. & Aronson, E. (1985). *The handbook of social psychology*. New York: Random House.

Lorenz, K. (1974). *On aggression*. New York: Harcourt Brace Jovanovitch.

Lubow, R.E. (1989). *Latent inhibition and conditioned attention theory*. Cambridge: Cambridge University Press.

Lucas, B.V., Crane, L. & Edwards, M. (Trans.) (1945). *Grimm's fairy tales* (pp. 171–178). New York: Grosset and Dunlap, Companion Library.

Luria, A.R. (1980). *Higher cortical functions in man*. New York: Basic Books.

MacRae, G.W. (Trans.). (1988). The thunder: Perfect mind. In J.M. Robinson (Ed.), *The Nag Hammadi library in English* (pp. 297–319). New York: Harper Collins.

Maier, N.R.F. & Schnierla, T.C. (1935). *Principles of animal psychology*. New York: McGraw-Hill.

Mark, V.H. & Ervin, F.R. (1970). *Violence and the brain*. New York: Harper and Row, Medical Division.

Melzack, R. (1973). *The puzzle of pain*. New York: Basic Books.

Melzack, R. & Wall, P.D. (1983). *The challenge of pain*. New York: Basic Books.

Milner, B. (1963). Effects of different brain lesions on card sorting. *Archives of Neurology, 9*, 100–110.

Milner, B. (1972). Disorders of learning and memory after temporal lobe lesions in man. *Clinical Neurosurgery, 19*, 421–446.

Milner, B., Petrides, M., & Smith, M.L. (1985). Frontal lobes and the temporal organization of memory.

PERMISSIONS

Brothers Grimm; excerpts from Grimm's Fairy Tales, translated by B.V. Lucas, L. Crane and M. Edwards. Translation copyright © 1945, Grosset and Dunlap. Reprinted with permission of Penguin Putnam Inc.

Dostoevsky, Fyodor; excerpts from The Brothers Karamazov by Fyodor Dostoevsky, translated by Andrew R. MacAndrew. Translation copyright © 1970 by Bantam, a division of Bantam Doubleday Dell Publishing Group, Inc. Used by permission of Bantam Books, a division of Bantam, Doubleday Dell Publishing Group, Inc.

Eliade, Mircea; excerpts from The Forge and the Crucible, second edition. Copyright © 1978 by The University of Chicago Press. Reprinted with permission.

Eliade, Mircea; excerpts from A History of Religious Ideas, Vol. 1. Copyright © 1978 by The University of Chicago Press. Reprinted with permission.

Eliade, Mircea; excerpts from A History of Religious Ideas, Vol. 2 Copyright © 1982 by The University of Chicago Press. Reprinted with permission.

Eliade, Mircea; excerpts from A History of Religious Ideas, Vol. 3 Copyright © 1985 by The University of Chicago Press. Reprinted with permission.

Eliade, Mircea; specified excerpts from Myths, Dreams and Mysteries. Copyright © 1957 by Librairie Gallimard. English translation copyright 1960 by Harville Press. Copyright Renewed. Reprinted by permission of HarperCollins Publishers, Inc.

Heidel, Alexander; excerpts from The Babylonian Genesis, second edition. Copyright © 1967 by The University of Chicago Press. Reprinted with permission.

Jung, C.G.; Mysterium Conjunctionis. Copyright © 1970 by Princeton University Press. Reprinted by permission of Princeton University Press.

Jung, C.G.; Psychology and Alchemy. Copyright © 1953 by Bollingen Foundation. Reprinted by permission of Princeton University Press.

Kuhn, Thomas; excerpts from The Structure of Scientific Revolutions. Copyright © 1970 by The University of Chicago Press. Reprinted with permission.

Neumann, Erich; The Great Mother. Copyright © 1955 by Princeton University Press. Reprinted by permission of Princeton University Press Copyright renewed 1983 by Bollingen Foundation.

Neumann, Erich; The Origins and History of Consciousness. Copyright © 1954 by Princeton University Press. Reprinted by Permission of Princeton University Press. Copyright renewed 1982 by Bollingen Foundation.

Nietzsche, Friedrich; excerpts from The Basic Writings of Nietzsche by Friedrich Nietzsche, translated by

INDEX